The Handbook of
Electrical Engineering

Staff of
Research & Education Association

Morteza Shafii-Mousavi, Ph.D.
Assistant Professor of Mathematics
Indiana University at South Bend
South Bend, Indiana

Alan D. Solomon, Ph.D.
Adjunct Professor of Mathematics
The University of Tennessee – Knoxville
Knoxville, Tennessee

Research & Education Association
Visit our website at
www.rea.com

Research & Education Association
61 Ethel Road West
Piscataway, New Jersey 08854
E-mail: info@rea.com

THE HANDBOOK OF
ELECTRICAL ENGINEERING

Copyright © 2006, 1996 by Research & Education Association, Inc. All rights reserved. No part of this book may be reproduced in any form without permission of the publisher.

Printed in the United States of America

Library of Congress Control Number 2005934539

International Standard Book Number 0-7386-0171-3

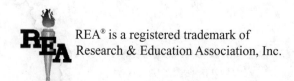

REA® is a registered trademark of
Research & Education Association, Inc.

What This Book Will Do For You

REA's *Handbook of Electrical Engineering* is a comprehensive yet concise review of the critical facts and concepts in electrical engineering.

This handy reference for the professional and student provides quick access to the important facts, principles, theorems, and equations in this field.

It condenses the vast amount of detail characteristic of the subject matter and summarizes the essentials of the field.

This handbook was carefully prepared by educators and professionals to ensure accuracy and maximum usefulness.

Larry B. Kling
Chief Editor

CONTENTS

ELECTRIC CIRCUITS

SECTION A

Chapter No.		Page No.
1	**UNITS AND NOTATION**	A-1
1.1	Systems of Units	1
1.1.1	The SI System	1
1.1.2	Standard Abbreviations for SI	1
1.1.3	The MKS and CGS System (Metric System)	2
1.1.4	The English System	2
1.1.5	Units Conversion Between English MKS and CGS Systems	2
1.2	Laws of Units	3
1.3	Unit of Charge and Coulomb's Law	3
1.4	Scientific Notation	4
2	**RESISTIVE CIRCUITS AND EXPERIMENTAL LAWS**	5
2.1	Current, Voltage, Power and Energy	5
2.1.1	Current	5
2.1.2	Voltage	6
2.1.3	Power and Energy	7
2.2	Types of Circuit Elements	8
2.2.1	Independent Voltage Source	8
2.2.2	Independent Current Source	9
2.2.3	Dependent Voltage and Current Sources	9
2.3	Resistance and Conductance	10
2.3.1	Resistance	10
2.3.2	Conductance	10

2.3.3	Resistivity	10
2.3.4	Resistor and Conductor Combinations	11
2.4	Voltage and Current Division	11
2.4.1	Voltage Division	11
2.4.2	Current Division	12
2.5	Ohm's Law	12
2.6	Kirchhoff's Law	13
2.6.1	Kirchhoff's Current Law	13
2.6.2	Kirchhoff's Voltage Law	13

3	**TRANSIENT CIRCUITS**	14
3.1	Capacitor and Corresponding Voltage, Current and Power Relationships	14
3.1.1	Parallel Plate Capacitor	14
3.2	Inductor and Transformer	15
3.2.1	Iron-core Inductor	15
3.2.2	Inductor	16
3.2.3	Mutual Inductance	17
3.2.4	Dot Notation	18
3.2.5	Iron-core Transformer	20
3.3	Simple RL and RC Circuits	21
3.3.1	Source Free RL Circuit	21
3.3.2	Source Free RC Circuit	21
3.4	Natural and Forced Response of RL and RC Circuits	22
3.4.1	Natural Response	22
3.4.2	Forced Response	22
3.4.3	A General Differential Equation	22
3.4.4	Procedure to Find the Complete Response $f(t)$ of RL and RC Circuits with DC Sources	23
3.5	The RLC Circuits	24
3.5.1	Parallel RLC Circuit (Source Free)	24
3.5.2	Special Response of Parallel RCL Circuit	25
3.5.3	Series RLC Circuit (Source Free)	26
3.5.4	Complete Response of RLC Circuit	27

4	**NETWORK THEOREMS**	29
4.1	Linearity and Superposition	29
4.1.1	Linearity	29
4.1.2	Superposition Theorem	29
4.2	Thevenin's and Norton's Theorems	30

4.2.1	Thevenin's Theorem	30
4.2.2	Norton's Theorem	30
4.3	Maximum Power Transfer Theorem	31
4.4	Millman's Theorem	31
4.5	Substitution (compensation) Theorem	32
4.6	Reciprocity Theorem	33

5 USEFUL TECHNIQUES OF CIRCUIT ANALYSIS 34

5.1	Matrices	34
5.1.1	A general form of a Rectangular Matrix R	34
5.1.2	Addition, Subtraction and Multiplication of Matrices	34
5.1.3	Properties of Matrices	35
5.2	Determinants	36
5.2.1	The Determinant	36
5.2.2	Second and Third Order Determinant Calculations	37
5.3	Single - Loop and Single - Node - Pair Circuit Analysis	38
5.3.1	Single - Loop Analysis	38
5.3.2	Node - Pair Analysis	39
5.4	Source Transformations	39
5.5	Nodal Analysis	40
5.5.1	Nodal Analysis - Format Approach	40
5.5.2	Nodal Analysis - General Approach	41
5.5.3	Mesh Analysis	41
5.6	Free and General Nodal Analysis	41

6 SINUSOIDAL ANALYSIS 43

6.1	Sinusoidal Current, Voltage and Phase Angle	43
6.1.1	Sinusoidal Forcing Function - General Form	43
6.1.2	Lead and Lag Concept of Phasor Angle θ	43
6.1.3	Sinusoidal Currents and Voltages	44
6.1.4	Characteristics of Phase Angle in Pure Element	45
6.2	Concept of Phasor	45
6.2.1	Phasor Notation	45
6.2.2	Time Domain to Frequency Domain Transformation, and Vice Versa	46

6.2.3	Time - Domain and Frequency - Domain Relationships of Voltage and Current for element R,L and C	47
6.3	Complex Numbers	47
6.3.1	Imaginary Numbers	47
6.3.2	Complex Numbers	47
6.3.3	Complex Numbers - Multiplication and Division	48
6.3.4	Powers of a Complex Number	48
6.3.5	Roots of a Complex Number	49
6.3.6	Commonly Used Functions of a Complex Number of the Form: $Z = x + iy$	49
6.3.7	Euler's Theorem	49
6.4	Impedance and Admittance	50
6.4.1	Impedance	50
6.4.2	Two General Forms of an Impedance	50
6.4.3	Admittance	51
6.4.4	Representation of Impedance and Admittance in terms of Phasor Voltage and Current	52
6.4.5	Conversion of Z to y and Vice Versa in Polar Form	53
6.5	AC Analysis	53
6.6	Average Power and rms Values	53
6.6.1	Instantaneous Power	53
6.6.2	Average Power	54
6.6.3	Special Cases of pf	55
6.6.4	Power Triangle for Inductive and Capacitive Load	55
6.6.5	Complex Power	56
6.6.6	Rms or Effective Value	56

7 POLYPHASE SYSTEMS 58

7.1	Single-Phase, 2-Phase and 3-Phase Systems	58
7.1.1	Single-Phase, Three-wire System	58
7.1.2	Two-Phase System	58
7.1.3	Three-Phase System	59
7.1.4	Three-Phase System Voltages	60
7.2	The WYE (Y) and Delta (Δ) Connections	61
7.2.1	WYE (Y) and Delta (Δ) Alternators	61
7.2.2	Three-Phase Y-Y Connection	61
7.2.3	Characteristics of Balanced Three-Phase Sources	62
7.2.4	Delta Connections	63

7.3	Power in Y- and Δ- Connected Loads	64
7.3.1	Y- Connected Load	64

8 FREQUENCY DOMAIN ANALYSIS — 66

8.1	Complex Frequencies	66
8.1.1	Complex Frequency	66
8.2	Complex Frequency Impedances and Admittances	67
8.3	The S-Plane (Complex-Frequency Plane)	68
8.3.1	The S-Plane	68
8.4	Poles and Zeros	68
8.5	Resonance (Series and Parallel)	70
8.5.1	Resonance	70
8.5.2	Parallel Resonance	70
8.5.3	Series Resonance	72
8.6	Quality Factor-Q	73
8.6.1	Quality Factor	73
8.6.2	Quality Factor Q for Some General Circuits	73
8.7	Scaling	74

9 STATE - VARIABLE ANALYSIS — 75

9.1	State-Variable Method	75
9.1.1.	Conditions for the State-Variable Method	75
9.2	State Equations for n-th Order Circuits	76
9.2.1	State Equations for the First and Second Order Circuits	76
9.2.2	General Steps to Obtain the State Equations for a Linear Time-Invariant Network	77
9.3	Normal-Form Equations	78
9.3.1	General Procedures to Obtain a Set of Normal-Form Equations	78
9.4	State Transition Matrix (e^{At})	79
9.4.1	State Transition (e^{At})	79
9.4.2	Cayley-Hamilton Theorem	80
9.4.3	General Procedures to Obtain the State Transition Matrix (e^{At}), Given Matrix A	80

10 FOURIER ANALYSIS — 81

10.1	Trigonometric Fourier Series	81

10.1.1	Dirichlet Conditions for the Existence of a Fourier Series	81
10.1.2	Trigonometric Form of a Fourier Series	81
10.1.3	Special Integration Properties for Sine and Cosine	82
10.2	Exponential Fourier Series	83
10.3	Complex Form of a Fourier Series	84
10.3.1	Special Case	84
10.4	Waveform Symmetry Properties	85
10.5	The Fourier Transform	86
10.5.1	Some Useful Fourier Transform Pairs	87
10.6	Parseval's Identity	88
10.7	Convolution Theorem for the Fourier Transforms	89

11 LAPLACE TRANSFORMATION 90

11.1	Definition of Laplace Transform	90
11.2	Definition of the Inverse Laplace Transform	91
11.3	Complex Inversion Formula	91
11.4	General Laplace Transform Pairs	92
11.5	Operators for the Laplace Transform	93
11.6	Heaviside Expansion Theorem	95
11.7	Final and Initial Value Theorem	96

12 TWO PORT NETWORK PARAMETERS 97

12.1	Z - Parameters	97
12.1.1	Impedance or Z - Parameters	97
12.2	Hybrid — Parameters	98
12.2.1	Hybrid or H - Parameters	98
12.3	Admittance Parameters	100
12.3.1	Admittance or Y - Parameters	100
12.4	Z, Y, and H Parameters and Relationships	101

13 DISCRETE SYSTEMS AND Z- TRANSFORMS 102

13.1	Discrete-Time Systems	102
13.1.1	A Discrete-Time System	102
13.2	First-Order Linear Discrete System	103
13.3	Closed-Form Identities	104
13.4	The Z - Transform	105
13.5	Properties of Z-Transform	107
13.6	Methods of Evaluating Inverse Z-Transforms	108

		Inverse Laplace Transform	108
13.7		Z-Transform Pairs	110

14 TOPOLOGICAL ANALYSIS — 111
- 14.1 Incident Matrix — 111
- 14.2 The Circuit (Loop) Matrix — 112
- 14.3 Fundamental (Loop) Matrix — 113

15 NUMERICAL METHODS — 114
- 15.1 Newton's Method — 114
- 15.2 Simpson's Rule — 115

ELECTRONICS

SECTION B

Chapter No.			Page No.
1	**FUNDAMENTALS OF SEMICONDUCTOR DEVICES**		B-1
	1.1	Charged Particles and the Energy Gap Concept	1
	1.2	Field Intensity, Potential and Energy	2
	1.3	Mobility and Conductivity	3
	1.4	Electrons and Holes in an Intrinsic Semiconductor	4
	1.5	Donor and Acceptor Impurities	5

1.6	Charge Densities in a Semiconductor	5
1.7	Diffusion and the Potential Variation within a Graded Semiconductor	6

2 JUNCTION DIODE: THEORY AND SIMPLE CIRCUIT ANALYSIS — 11

2.1	The Open-Circuited P-N Junction	11
2.2	The P-N Junction as a Rectifer	12
2.3	V-I Characteristics	12
2.3.1	Temperature Dependence of the V-I Characteristics	13
2.4	Diode Resistance, Transition and Diffusion Capacitance	13
2.5	Analysis of Simple Diode Circuits: Concept of Dynamic Resistance and A.C. Load Line	15
2.5.1	The D.C. Load Line	15
2.5.2	The A.C. Load Line	18
2.6	Schottky and Zener Diodes	18
2.7	Diode Logic Circuits	20
2.7.1	A Diode "AND" Gate	20
2.7.2	A Diode "OR" Gate	20

3 THE BIPOLAR JUNCTION TRANSISTOR — 21

3.1	The Junction Transistor Theory	21
3.1.1	Open-Circuited Transistor	21
3.1.2	Transistor Biased in the Active Region	22
3.1.3	Transistor Current Components	23
3.1.4	Transistor Configuration	24
3.2	The Junction Transistor: Small-Signal Models	25
3.2.1	The Hybrid-Pi Model	25
3.2.2	h-Parameters	27
3.2.3	The Concept of Bias Stability	28

4 POWER SUPPLIES — 31

4.1	Diode Rectifiers	31
4.1.1	Half-Wave Rectifier	31
4.1.2	Full-Wave Rectifier	32

4.1.3		PIV Rating	32
4.1.4		Ripple Factor	33
4.1.5		Rectifier Efficiency	33
4.1.6		Comparision of Rectifiers	34
4.2	Filters		34
4.2.1		Shunt-Capacity Filter	34
4.2.2		Pi-Filter	36
4.2.3		RC Filter	36
4.2.4		L - Section Filter	37
4.2.5		Voltage Multiplier	37
4.2.6		Comparison of Filters	38

5   MULTITRANSISTER CIRCUITS   39

5.1	The Difference Amplifier	39
5.1.1	Basic Difference Amplifier	39
5.1.2	Q-Point Analysis	40
5.1.3	Common-Mode Load Line	40
5.1.4	Difference Amplifier with Constant Current Source	43
5.1.5	Difference Amplifier with Emitter Resistors for Balance	45
5.2	The Darlington Amplifier	46
5.3	The Cascade Amplifier	47
5.4	The OP Amplifier	48

6   SMALL SIGNAL, LOW FREQUENCY ANALYSIS AND DESIGN   49

6.1	Hybrid Parameters	49
6.1.1	General Two-Port Network	49
6.1.2	Hybrid Equations	49
6.1.3	Terminal Definitions for H-Parameters	50
6.2	The C-E Configuration	50
6.3	The C-B Configuration	53
6.4	The C-C (Emitter-Follower) Configuration	54
6.5	Significant Parameters	56
6.6	Small-Signal Equivalent Circuit of the Fet	56
6.6.1	Equivalent Circuit	56
6.6.2	Transconductance	56
6.6.3	Drain Source Resistance	57
6.6.4	Amplication Factor	57
6.7	The Common-Source Voltage Amplifier	57
6.8	The Common-Drain Voltage Amplifier (The Source Follower)	58

6.9	The Phase-Splitting Circuit	60

7 AUDIO-FREQUENCY LINEAR POWER AMPLIFIERS 61

7.1	The Class A Common-Emitter Power Amplifier	61
7.2	The Transformer-Coupled Amplifier	65
7.3	Class B Push-Pull Power Amplifiers	67
7.4	Amplifiers Using Complementary Symmetry	69

8 FEEDBACK AMPLIFIERS 71

8.1	Classification of Amplifiers	71
8.1.1	Voltage Amplifier	71
8.1.2	Current Amplifier	71
8.1.3	Transconductance Amplifier	72
8.1.4	Transresistance Amplifier	72
8.1.5	Ideal Amplifier Characteristics	73
8.2	The Concept Of "FEEDBACK"	73
8.2.1	Sampling Network	73
8.2.2	Transfer Ratio or Gain	74
8.3	Transfer Gain with FEEDBACK	74
8.3.1	Single-Loop FEEDBACK Amplifier	74
8.3.2	FEEDBACK Amplifier Topologies	74
8.4	Features of the Negative FEEDBACK Amplifier	75
8.5	Input Resistance	76
8.6	Output Resistance	77
8.7	FEEDBACK Amplifier Analysis	78

9 FREQUENCY RESPONSE OF AMPLIFIERS 81

9.1	Frequency Distortion	81
9.2	The Effect of Coupling and Emitter Bypass Capacitors on Low-Frequency Response	82
9.3	The Hybrid-π Transistor Model at High Frequencies	84
9.4	The C-E Short-Circuit Current Gain	85
9.5	The Common-Drain Amplifier at High Frequencies	87
9.6	The Common-Source Amplifier at High Frequencies	88

9.7	The Emitter-Follower Amplifier at High Frequencies	88
9.8	Single-Stage CE Transistor Amplifier Response	90

10 OPERATIONAL AMPLIFIERS 92

10.1	The Basic Operational Amplifier	92
10.2	Offset Error Voltages and Currents	94
10.3	Measurement of Operational Amplifier Parameters	96

11 OPERATIONAL AMPLIFIER SYSTEMS 100

11.1	Basic Operational Amplifier Applications	100
11.1.1	Sign Changer, or Inverter	100
11.1.2	Scale Changer	100
11.1.3	Phase Shifter	101
11.1.4	Adder	101
11.1.5	Non-Inverting Adder	101
11.1.6	Voltage-To-Current Converter (Transconductive Amplifier)	102
11.1.7	Current-To-Voltage Converter	102
11.1.8	D.C. Voltage Follower	103
11.2	A.C.-Coupled Amplifier	103
11.3	Active Filters	105
11.3.1	Ideal Filters	105
11.3.2	Butterworth Filter	105

12 FEEDBACK AND FREQUENCY COMPENSATION OF OP AMPS 108

12.1	Basic Concepts of FEEDBACK	108
12.2	Frequency Response of a FEEDBACK Amplifier	111
12.3	Stabilizing Networks	114

13 MULTIVIBRATORS 117
13.1 Collector-Coupled Monostable Multivibrators 117
13.2 Emitter-Coupled Monostable Multivibrators 120
13.3 Collector-Coupled Astable Multivibrators 124
13.4 Emitter-Coupled Astable Multivibrators 125

14 LOGIC GATES AND FAMILIES 128
14.1 Logic Level Concepts 128
14.2 Basic Passive Logic 130
14.2.1 Resistor Logic (RL) 130
14.2.2 Diode Logic Circuits 131
14.2.3 Logic Circuit Current, Voltage, and Parameter Definitions and Notation 131
14.3 Basic Active Logic 132
14.3.1 Resistor-Transistor Logic (RTL) 132
14.3.2 Diode-Transistor Logic (DTL) 133
14.4 Advanced Active Logic Gates 134
14.4.1 Transistor-Transistor Logic (TTL) 134
14.4.2 Emitter-Couple Logic (ECL) 137

15 BOOLEAN ALGEBRA 138
15.1 Logic Functions 138
15.1.1 NOT Function 138
15.1.2 AND Function 138
15.1.3 OR Function 139
15.2 Boolean Algebra 139
15.3 The NAND and NOR Functions 142
15.4 Standard Forms for Logic Functions 143
15.5 The Karnaugh Map 144

16 REGISTERS, COUNTERS AND ARITHMETIC UNITS 148
16.1 Shift Registers 148
16.1.1 Serial-In Shift Registers 148

16.1.2	Parellel-In Shift Registers	149
16.1.3	Universal Shift Registers	150
16.2	Counters	151
16.2.1	The Ripple Counter	151
16.2.2	The Synchronous Counter	152
16.3	Arithmetic Circuits	153
16.3.1	Addition of Two Binary Digits, The Half Adder	153
16.3.2	The Full Adder	154
16.3.3	Parallel Addition	155
16.3.4	Look-Ahead-Carry Adders	155

17 OSCILLATORS 156

17.1	Harmonic Oscillators	156
17.1.1	The RC Phase Shift Oscillator	156
17.1.2	The Colpitts Oscillator	157
17.1.3	The Hartley Oscillator	158
17.1.4	The Clapp Oscillator	159
17.1.5	The Crystal Oscillator	159
17.1.6	Tunnel Diode Oscillators	160
17.2	Relaxation Oscillators	162

18 RADIO-FREQUENCY CIRCUITS 164

18.1	Non-Linear Circuits	164
18.2	Small-Signal RF Amplifiers	167
18.3	Tuned Circuits	168
18.4	Circuits Employing Bipolar Transistors	169
18.5	Analysis Using Admittance Parameters	171

19 FLIP-FLOPS 173

19.1	Types of Flip-Flops	173
19.1.1	The Basic Flip-Flop	173
19.1.2	R-S Flip-Flop	173
19.1.3	Synchronous R-S Flip-Flop (Clocked R-S Flip-Flop)	174
19.1.4	Preset and Clear	175
19.1.5	D-Type Flip Flop	175
19.1.6	J-K Flip-Flop	176

19.1.7	T-Type Flip-Flop	176
19.1.8	Master-Slave Flip-Flops	177
19.2	Flip-Flop Timing	177
19.3	Collector-Coupled Flip-Flops	178
19.4	Emitter-Coupled Flip-Flops	179
19.5	Switching Speed of a Flip-Flop	180
19.6	Regenerative Circuits	182

20	**WAVESHAPING AND WAVEFORM GENERATORS**	**184**
20.1	Common Waveforms	184
20.2	Linear Waveshaping Circuits	185
20.3	Sweep Generators	191

ELECTROMAGNETICS

SECTION C

Chapter No.		Page No.
1	**VECTOR ANALYSIS**	C-1
1.1	Vector Algebra	1
1.1.1	Notations and Unit Vector	1
1.1.2	Vector Operations	2
1.1.3	Vector Field	3

1.1.4	Vector Dot and Cross Product	3
1.2	Different Types of Coordinate Systems and Differential Volume	6

2 COULOMB'S LAW AND ELECTRIC FIELD 7

2.1	Coulomb's Law	7
2.2	Electric Field Intensity	8
2.3	Electric Field Intensity Due to: Point Charges, Volume Charge Distribution, Line of Charge and Sheet of Charge	9
2.4	Streamlines and Sketches of Fields	11

3 ELECTRIC FLUX DENSITY, GAUSS' LAW AND DIVERGENCE 13

3.1	Electric Flux and Flux Density	13
3.2	Gauss' Law and its Applications	14
3.3	Divergence	17
3.4	The Del Operator	19
3.5	The Divergence Theorem and Maxwell's First Equation	20

4 ENERGY AND POTENTIAL 22

4.1	Work	22
4.2	Potential	23
4.3	Gradient	25
4.4	Energy	26
4.5	The Dipole	27

5 CURRENT DENSITY, CAPACITANCE, DIELECTRICS, AND CONDUCTORS 29

5.1	Capacitance	29

5.1.1	Example of Capacitance	29
5.2	Dielectric and Current Density	33
5.3	Boundary Conditions	34
5.4	Conductors in Electrostatic Fields	36

6 POISSON'S AND LAPLACE'S EQUATIONS 37

6.1	Poisson's and Laplace's Equations	37
6.2	Laplace's Equation in Different Coordinate Systems	38
6.3	Uniqueness Theorem	38

7 THE STEADY MAGNETIC FIELD 41

7.1	Biot-Savart Law	41
7.2	Ampere's Circuital Law	43
7.3	Curl of a Vector Field	44
7.4	Stoke's Theorem	46
7.5	Magnetic Flux and Magnetic Flux Density	46
7.6	Scalar and Vector Magnetic Potentials	47

8 FORCES, TORQUES AND INDUCTANCE IN MAGNETIC FIELDS 49

8.1	Magnetic Force on Moving Particles	49
8.2	Magnetic Force on a Differential Current Element	50
8.3	Magnetic Force Between Differential Current Element	50
8.4	Force and Torque on a Closed Circuit	51
8.5	Magnetic Materials	53
8.6	Magnetic Circuit	55
8.7	Potential Energy and Forces on Magnetic Materials	56
8.8	Inductance and Mutual Inductance	57
8.9	Boundary Conditions	58

9 EMF IN A TIME-VARYING FIELD AND MAXWELL'S EQUATIONS — 59
9.1 Faraday's Law — 59
9.2 Displacement Current — 60
9.3 Maxwell's Equations — 61

10 UNIFORM PLANE WAVE AND POYNTING THEOREM — 63
10.1 Maxwell's and Wave Equations — 63
10.2 Wave Motion in Perfect (Lossless) Dielectric — 66
10.3 Wave Motion In Lossy Dielectrics — 68
10.4 The Poynting Theorem and Electromagnetic Power — 70

11 PROPAGATION OF ELECTROMAGNETIC WAVES — 73
11.1 Propagation in Conductive Medium — 73
11.2 Skin Depth — 74
11.3 Power Consideration — 74
11.4 Reflection of Uniform Plane Waves — 75
11.5 Standing Waves and SWR — 76

12 TRANSMISSION LINES — 78
12.1 Transmission-Line Equations — 78
12.2 Transmission-Line Parameters — 81
12.3 Transmission-Line Example — 83
12.4 Graphical Method (Smith Chart) — 85

13 GUIDED WAVES 87

13.1 Waves Between Parallel Planes 87
13.2 Transverse Electric Waves 88
13.3 Transverse Magnetic Waves 89
13.4 Transverse Electromagnetic Waves 90
13.5 Miscellaneous Conditions in TE, TM and TEM Modes 91

14 WAVE GUIDES 99

14.1 Properties of Wave Guides 99
14.2 Miscellaneous Examples 102

15 RADIATION 107

15.1 Near and Far Zone 107
15.2 Total Radiated Power 108
15.3 Radiation Resistance 109
15.4 Miscellaneous Examples 109

16 ANTENNAS 115

16.1 Properties of Antennas 115
16.2 Electric Field Intensity at a Point P Far From a Practical Half-Wave Antenna 117
16.3 Radiation Field of an Isolated, Full Wave Antenna and Radiation Pattern 118
16.4 Effective Length of an Antenna 120
16.5 Miscellaneous Examples 120

ELECTRONIC COMMUNICATIONS

SECTION D

Chapter No.		Page No.
1	**BASIC CIRCUIT PRINCIPLES**	D - 1
1.1	The Capacitor	1
1.2	RC Circuits	2
1.3	The Inductor	3
1.4	RL Circuits	4
2	**FOURIER SERIES AND FOURIER TRANSFORMS**	**8**
2.1	Signal Terminology	8
2.1	Fourier Series	10
2.3	Convergence of Fourier Series	12
2.4	Symmetry Conditions	13
2.5	Fourier Transform	16
2.6	Properties of Fourier Transforms	16
2.7	List of Fourier Transform Theorems	20
2.8	Fourier Transform Pairs	21

3	**LAPLACE TRANSFORM**	22
3.1	Laplace Transforms of Functions	22
3.2	Inverse Laplace Transforms	24
3.3	Initial and Final Value Theorem and Convolution	25

4	**SPECIAL ANALYSIS**	27
4.1	The Sampling Function	27
4.2	Response of a Linear System	28
4.3	Normalized Power	29
4.4	Power Spectral Density (PSD)	30
4.5	Relationship Between Input and Output in the Time Domain	31
4.6	Parseval's Theorem	31
4.7	Band Limiting of Waveforms	32
4.8	Correlation Between Waveforms	33
4.9	Autocorrelation	34

5	**TRANSFER FUNCTION AND FILTERING**	37
5.1	Concept of Transfer Function	37
5.2	Ideal Filter	38
5.3	Response of an Ideal Filter	39
5.4	Filter Approximations	44
5.5	Transmission of Energy and Power Spectrum	46

6	**RANDOM VARIABLES AND PROCESSES**	47
6.1	Discrete Random Variables	47
6.2	Set Theory	47
6.3	Algebra of Sets	48
6.4	Probability Theory	50

6.5	Probability Density Function	51
6.6	Continuous Random Variables	53
6.7	One Dimensional Probability Density Function	53
6.8	Moments of One-Dimensional Random Variables	56
6.9	Statistical Averages For One Dimensional Random Variables	57
6.10	Correlation Functions	58
6.11	Power Spectra Determination	59
6.12	The Gaussian Probability Density	61
6.13	Sum of Random Variables	61

7	**AMPLITUDE MODULATION**	**63**
7.1	Basic Principles	63
7.2	Square Law Demodulator	65
7.3	Balanced Modulator and Double-Sideband Modulator	66
7.4	Single Sideband Modulation	69

8	**FREQUENCY MODULATION**	**70**
8.1	Angle Modulation	70
8.2	Phase and Frequency Modulation	70
8.3	Bandwidth and Spectrum of an FM Signal	73
8.4	Bessel's Functions	76
8.5	Spectrum of an FM Signal With Sinusoidal Modulation	80
8.6	Frequency Multiplication and FM Generation	81
8.7	FM Demodulation	84

9	**PULSE MODULATION SYSTEMS**	**86**
9.1	Sampling Theorem	86
9.2	Pulse Amplitude Modulation	89

9.3	Natural Sampling and Flat Top Sampling	90
9.4	Signal Recovery Through Holding, Crosstalk	94
9.5	Pulse Time Modulation	97

10 PULSE CODE MODULATION 100

10.1	Quantization of Signals	100
10.2	Pulse Code Modulation	102
10.3	Companding	103
10.4	Delta Modulation	104
10.5	Binary Phase-Shift Keying (BPSK)	105
10.6	Differential Phase Shift Keying	106
10.7	Frequency Shift Keying	107

11 MATHEMATICAL REPRESENTATION OF NOISE 108

11.1	Frequency Domain Representation and Spectral Components of Noise	108
11.2	Filtering	110
11.3	Noise Bandwidth	113
11.4	Shot Noise	114

12 NOISE IN COMMUNICATION SYSTEMS 116

12.1	Noise in AM Systems	116
12.2	Noise in FM Systems	121
12.3	Noise in PCM Systems	125
12.4	Noise in Delta Modulation	130

13 NOISE CALCULATING 133

13.1	Resistor Noise	133
13.2	Noise in Network With Reactive Elements	134
13.3	Available Power	135

13.4	Noise Temperature	136
13.5	Noise Figure	137
13.6	Noise Bandwidth	139

14 DATA TRANSMISSION — 141

14.1	An Integrate and Dump PCM Receiver	141
14.2	The Matched Filter	142
14.3	Coherent Reception	143
14.4	Phase Shift Keying	144
14.5	Frequency Shift Keying	145
14.6	Four Phase PSK (QPSK)	146

15 INFORMATION THEORY AND CODING — 148

15.1	Concept of Information	148
15.2	Entrophy and Information Rate	148
15.3	Shannon's Theorem	149
15.4	Bandwidth and SNR Tradeoff	150
15.5	Coding and Error Detection	150
15.6	Block Codes	152

16 ANTENNAS — 155

16.1	Introduction	155
16.2	Concept of Gain and Beamwidth	157
16.3	Antenna Characteristics	158
16.4	Power Relations	159

17 TRANSMISSION LINES — 162

17.1	Equations for Line Voltage and Current	162
17.2	Distortionless Line	165
17.3	Lossless Lines	165
17.4	Input Impedance and Standing Waves (For Lossless Lines)	166
17.5	Skin Effect, High and Low Loss Approximation	167

LAPLACE TRANSFORMS

SECTION E

Chapter No.		Page No.
1	**THE LAPLACE TRANSFORM**	**E - 1**
1.1	Integral Transforms	1
1.2	Definition of Laplace Transform	2
1.3	Notation	2
1.4	Laplace Transformation of Elementary Functions	3
1.5	Sectionally or Piecewise Continuous Functions	4
1.6	Functions of Exponential Order	4
1.6.1	Sufficient Condition for Exponential Order Functions	5
1.7	Existence of Laplace Transform	6
1.7.1	Behavior of Laplace Transforms at Infinity	6
1.8	Some Important Properties of Laplace Transforms	7
1.8.1	Linearity Property	7
1.8.2	Changes of Scale Property	8
1.8.3	First Shift Property	8
1.8.4	Second Shift Property	9
1.8.5	Laplace Transform of Derivative	10
1.8.6	Derivative of Laplace Transforms	12
1.8.7	Periodic Functions	13
2	**INVERSE LAPLACE TRANSFORM**	**15**
2.1	Definition of Inverse Laplace Transform	15
2.2	Uniqueness of Inverse Laplace Transform	16
2.2.1	Null Functions	16
2.3	Some Inverse Laplace Transforms	17
2.4	Some Properties of Inverse Laplace Transforms	18

2.4.1	Linearity Property	18
2.4.2	First Translation or Shifting Property	18
2.4.3	Second Translation or Shifting Property	19
2.4.4	Change of Scale Property	20
2.4.5	Inverse Laplace Transform of Derivatives	20
2.4.6	Inverse Laplace Transform of Integrals	21
2.5	The Convolution Property	22
2.5.1	Definition	22
2.5.2	Properties of Convolution	23
2.5.3	Inverse Laplace Transform of the Convolution	23
3	**SOME SPECIAL FUNCTIONS**	**25**
3.1	The Gamma Function	25
3.1.1	Properties of the Gamma Function	26
3.2	Bessel Functions	27
3.2.1	Modified Bessel Function	28
3.2.2	Some Properties of Bessel Functions	28
3.3	The Error Function	29
3.3.1	Some Properties of erf(t)	30
3.4	The Complementary Error Function	30
3.4.1	A Property of erfc(t)	31
3.5	The Sine and Cosine Integrals	31
3.6	The Exponential Integral	31
3.7	The Unit Step Function or the Heaviside Function	31
3.7.1	Some Properties of the Unit Step Function	32
3.8	The Unit Impulse Function	35
3.8.1	Some Properties of the Unit Impulse Function	35
3.9	The Beta Function	36
3.10	More Properties of Inverse Laplace Transforms	37
3.10.1	Multiplication by s^n	37
3.10.2	Division by S	38
4	**APPLICATION OF ORDINARY LINEAR DIFFERENTIAL EQUATIONS**	**42**
4.1	Ordinary Differential Equation with Constant Coefficients	42

4.2	Ordinary Differential Equations with Variable Coefficients	46
4.3	Systems of Linear Differential Equations	48
4.4	The Vibration of Spring	51
4.4.1	Damped Vibrations	53
4.4.2	Undamped Vibration	54
4.4.3	Free Vibration	54
4.5	Resonance	56
4.6	The Simple Pendulum	58
4.7	Electric Circuits	60
4.8	Beams	63
5	**METHODS OF FINDING LAPLACE TRANSFORMS AND INVERSE TRANSFORMS**	**65**
5.1	Initial Value Theorem	65
5.2	Final Value Theorem	65
5.3	Methods of Finding Laplace Transforms	65
5.3.1	Direct Method	66
5.3.2	Power Series Method	66
5.3.3	Method of Differential Equations	68
5.3.4	Method of Differentiation with Respect to a Parameter	70
5.3.5	Miscellaneous Methods — Multiplication by T^H Property and Division by T Property	71
5.4	Methods of Finding Inverse Transforms	72
5.4.1	Partial Fraction Method	72
5.4.2	The Heaviside Expansion Formula	76
6	**FOURIER TRANSFORMS**	**77**
6.1	Fourier Series	77
6.2	Fourier Sine and Cosine Series	79
6.2.1	Odd Extension	80
6.2.2	Even Extension	80
6.3	Piecewise Smooth Functions	81
6.4	Periodic Extensions	81

6.5	Theorem 2 (Convergence Theorem)	81
6.5.1	First Criterion for Uniform Convergence	82
6.5.2	Second Criterion for Uniform Convergence	83
6.6	General Criterion for Differentiation	83
6.7	Parseval's Theorem and Mean Square Error	84
6.8	Complex Form of Fourier Series	85
6.9	Parseval's Identity	85
6.10	Fourier Integral Transforms	86
6.10.1	Parseval's Theorem	86
6.10.2	Inversion Theorem for Fourier Transforms	87
6.11	Fourier Cosine Formulas	88
6.12	Fourier Sine Formulas	88
6.13	The Convolution Theorem	89
6.14	Relationship of Fourier and Laplace Transforms	89
7	**APPLICATIONS OF LAPLACE TRANSFORMS TO INTEGRAL AND DIFFERENCE EQUATIONS**	**91**
7.1	Integral Equations	91
7.1.1	Fredholm Integral Equation	91
7.1.2	Volterra Integral Equation	91
7.1.3	Integral Equation of Convolution Type	92
7.2	Integro-Differential Equations	93
7.3	Difference Equations	95
7.4	Differential-Difference Equations	96
8	**APPLICATIONS TO BOUNDARY-VALUE PROBLEMS**	**98**
8.1	Functions of Two Variables	98
8.2	Partial Differential Equation	98
8.3	Some Important Partial Differential Equations	99
8.4	Classification of Second-Order Partial Differential Equations	99
8.5	Boundary Conditions	100
8.5.1	Dirichlet Problem	100

8.5.2	Cauchy Problem	101
8.6	Solution of Boundary-Value Problems by Laplace Transforms	101
8.6.1	One-Dimensional Boundary Value Problems	101
8.6.2	Two-Dimensional Boundary Value Problem	103
9	**TABLES**	**104**
9.1	Table of General Properties	104
9.2	Table of More Common Laplace Transforms	107
9.3	Table of Special Functions	119

AUTOMATIC CONTROL SYSTEMS / ROBOTICS

SECTION F

Chapter No.		Page No.
1	**SYSTEM MODELING: MATHEMATICAL APPROACH**	**F-1**
1.1	Electric Circuits and Components	1
1.2	Mechanical Translation Systems	3
1.3	Mechanical and Electrical Analogs	5
1.4	Mechanical Rotational Systems	6
1.5	Thermal Systems	7
1.6	Positive-Displacement Rotational Hydraulic Transmission	9

1.7	D-C and A-C Servomotor	10
1.8	Lagrange's Equation	14

2 SOLUTIONS OF DIFFERENTIAL EQUATIONS: SYSTEM'S RESPONSE 16

2.1	Standardized Inputs	16
2.2	Steady State Response	16
2.3	Transient Response	20
2.4	First and Second-Order System	24
2.5	Time-Response Specifications	28

3 APPLICATIONS OF LAPLACE TRANSFORM 29

3.1	Definition of Laplace Transform	29
3.2	Application of Laplace Transform to Differential Equations	32
3.3	Inverse Transform	33
3.4	Frequency Response from the Pole-Zero Diagram	38
3.5	Routh's Stability Criterion	39
3.6	Impulse Function: Laplace-Transform and its Response	41

4 MATRIX ALGEBRA AND Z-TRANSFORM 44

4.1	Fundamentals of Matrix Algebra	44
4.2	Z-Transforms	47

5 SYSTEM'S REPRESENTATION: BLOCK DIAGRAM, TRANSFER FUNCTIONS, AND SIGNAL FLOW GRAPHS 51

5.1 Block Diagram and Transfer Function 51
5.2 Transfer Functions of the Compensating Networks 53
5.3 Signal Flow Graphs 55

6 SERVO CHARACTERISTICS: TIME-DOMAIN ANALYSIS 58

6.1 Time-Domain Analysis Using Typical Test Signals 58
6.2 Types of Feedback Systems 59
6.3 General Approach to Evaluation of Error 60
6.4 Analysis of Systems: Unity Feedback 63

7 ROOT LOCUS 67

7.1 Roots of Characteristic Equations 67
7.2 Important Properties of the Root Loci 69
7.3 Frequency Response 78

8 SPECIAL POLE-ZERO TOPICS: DOMINANT POLES AND THE PARTITION METHOD 80

8.1 Transient Response: Dominant Complex Poles 80
8.2 Pole-Zero Diagram and Frequency and Time Response 85
8.3 Factoring of Polynomials Using Root-Locus 87

9 SYSTEM ANALYSIS IN THE FREQUENCY-DOMAIN: BODE AND POLAR PLOTS — 89

9.1 The Frequency Response and the Time Response: Relationship — 89
9.2 Frequency Response Plots — 90
9.3 Polar Plot — 92
9.4 Logarithmic Plots — 92
9.5 Log-Magnitude and Phase Diagram: Basic Approach — 94
9.6 Relation Between System Type, Gain and Log-Magnitude Curves — 97
9.7 Direct Polar Plots — 99
9.8 Inverse Polar Plots — 101
9.9 Dead Time — 105

10 NYQUIST STABILITY CRITERION — 107

10.1 Determining and Enhancing the System's Stability — 107
10.2 Inverse Polar Plots: Application of Nyquist's Criterion — 111
10.3 Phase Margin and Gain Margin: Definitions — 113
10.4 System Stability — 114
10.5 Stability — 115
10.6 Effect of Adding a Pole or a Zero: Effect on the Polar Plots — 117

11 PERFORMANCE EVALUATION OF A FEEDBACK CONTROL SYSTEM IN THE FREQUENCY-DOMAIN — 118

11.1 Performance Evaluation Using Direct Polar Plot — 118

11.2	Resonant Frequency and the Maximum Magnitude of C/R of a Second Order System	120
11.3	Plotting Maximum Magnitude and Resonant Frequency on the Complex Plane	122
11.4	Magnitude and Angle Curves in the Inverse Polar Plane	128
11.5	Gain Adjustment for a Desired Maximum Magnitude Using a Direct Polar Plot	131
11.6	Nichol's Chart	133

12 SYSTEM STABILIZATION: USE OF COMPENSATING NETWORKS AND THE ROOT LOCUS — 136

12.1	Function of a Compensating Network	136
12.2	Types of Compensations	137
12.2.1	PI (Integral and Proportional) Control	137
12.2.2	Lag Compensator	139
.2.3	Porportional Plus Derivative (PD) Compensator	140
12.2.4	Lead Compensation	141
12.2.5	Lead-Lag Compensation	142
12.2.6	Comparison of Compensators	145

13 FREQUENCY-RESPONSE PLOTS OF CASCADE COMPENSATED SYSTEMS — 146

13.1	Selecting a Proper Compensator	146
13.2	Analysis of a Lag Network	149
13.3	Analysis of a Lead Network	151
13.4	Analysis of a Lag-Lead Compensator	153

14	**FEEDBACK COMPENSATION: PARALLEL COMPENSATION**	155
14.1	Parallel Compensation: Pros and Cons of Selecting a Feedback Compensator	155
14.2	Effects of the Different Types of Feedback on the System's Time Response	157
14.3	Application of Log-Magnitude Curve: for Feedback Compensation	160

15	**SYSTEM SIMULATIONS: USE OF ANALOG COMPUTERS**	163
15.1	Analog Computer: Basic Components	163
15.2	Simulations Using Analog Computer	166
15.3	Application of Analog Computers for System Tuning	172
15.4	Setting of a Control System: Controller Setting	174

MATHEMATICS FOR ENGINEERS

SECTION G

Chapter No.		Page No.
1	VECTORS, MATRICES, AND EQUATION SYSTEMS	G-1
1.1	Vectors in Three Dimensions	1

1.2	Vectors in N-Dimensions	7
1.3	Matrices and Systems of Equations	9
1.4	Complex Vectors and Matrices	16

2	**ESSENTIALS OF CALCULUS**	**18**
2.1	Calculus of One Variable	18
2.2	Vector Functions of One Variable	23
2.3	Functions of Two or More Independent Variables	25
2.4	Vector Fields and Divergence	27
2.5	The Double Integral	28
2.6	Line Integrals	32
2.7	Green's Theorem	34
2.8	Surfaces in 3 Dimensions	35
2.9	Volume Integrals	38

3	**COMPLEX FUNCTIONS**	**40**
3.1	Basic Concepts	40
3.2	Sets in the Complex Plane	42
3.3	Functions of a Complex Variable	43
3.4	Limits, Continuity, and Derivatives	45
3.5	Harmonic Functions	46
3.6	The Elementary Complex Functions	47
3.7	Integrals of Complex Functions	51
3.8	Number Sequences and Series	55
3.9	Function Sequences and Series	56
3.10	Poles and Residues	57
3.11	Elementary Mappings and the Mobius Transformation	58
3.12	Conformal Mappings and Harmonic Functions	60
3.13	Zeros and Singular Points of Analytic Functions	62
3.14	Riemann Mapping Theorem	63

4	**ORDINARY DIFFERENTIAL EQUATIONS**	**64**
4.1	Ordinary Differential Equations of First Order	64
4.2	Linear Ordinary Differential Equation	66
4.3	Systems of First Order ODE's	69
4.4	Methods for Solving ODE's	71
4.5	Boundary Value Problems	72
5	**FOURIER ANALYSIS AND INTEGRAL TRANSFORMS**	**74**
5.1	Basic Ideas	74
5.2	Fourier Series	76
5.3	Fourier Series and Vector Space Concepts	79
5.4	Fourier Transforms	82
5.5	Special Functions	85
6	**PARTIAL DIFFERENTIAL EQUATIONS (PDE'S)**	**90**
6.1	Fundamental Ideas	90
6.2	The Laplace Equation	96
6.3	The Heat Equation	100
6.4	The First Order Wave Equation	103
6.5	The Wave Equation	106
7	**CALCULUS OF VARIATIONS**	**110**
7.1	Basic Theory of Maxima and Minima	111
7.2	The Simplest Problem of Variational Calculus	115
7.3	Some Classical Problems	119
7.4	Control	120
7.5	Dynamic Programming	122
7.6	Linear Programming	123

8	**NUMERICAL METHODS**	**125**
8.1	Solutions of Equations	125
8.2	Function Approximation	127
8.3	Numerical Integration	130
8.4	Numerical Linear Algebra	132
8.5	Solving Ordinary Differential Equations	138
8.6	Solving Partial Differential Equations	139
9	**STATISTICS AND PROBABILITY**	**145**
9.1	On Statistics and Probability	145

Handbook of Electrical Engineering

SECTION A
Electric Circuits

CHAPTER 1

UNITS AND NOTATION

1.1 SYSTEMS OF UNITS

1.1.1 THE SI SYSTEM

The International System of Units is based on six basic units: 1) meter, 2) kilogram, 3) second, 4) ampere, 5) degree kelvin, and 6) candela.

1.1.2 STANDARD ABBREVIATIONS FOR SI

A	-	ampere	Np	-	neper
ac	-	alternating current	PF	-	power factor
C	-	coulomb	rad	-	radian
cps	-	cycle per second	RLC	-	resistance-inductance-capacitance
dc	-	direct current	rms	-	root-mean-square
dB	-	decibel	rps	-	revolutions per second
eV	-	electron volt	s	-	second
F	-	farad	V	-	volt
ft	-	foot	VA	-	voltampere
g	-	gram	W	-	watt
H	-	henry	Wh	-	watthour
h	-	hour	^{0}F	-	degree Farenheit
Hz	-	hertz	^{0}C	-	degree celsius
J	-	joule	^{0}K	-	degree kelvin
kg	-	kilogram	Ω	-	ohm
m	-	meter	$\mho$	-	mho
min	-	minute			
mks	-	meter-kilogram-second			
N	-	newton			
N·m	-	newton-meter			

A-1

1.1.3 THE MKS AND CGS SYSTEM (METRIC SYSTEM)

	MKS (m,kg,sec)	CGS (cm,g,sec)
1. Length (ℓ)	meter	centimeter (cm)
2. Mass (m)	kilogram	gram (g)
3. Time (t)	second	second
4. Force (F or f) *Note: $F[N] = m[kg] \times$ Acceleration (a) $[m/s^2]$	newton	Dyne
5. Work and Energy (W or w)	newton-meter (joule)	Dyne-centimeter (or Erg)
6. Power (P or p)	joule/sec (watt)	

1.1.4 THE ENGLISH SYSTEM

1. Length → yard (yd)
2. Mass → slug
3. Time → second
4. Force → Pound (lb)
5. Energy → foot-pound (ft-lb)

1.1.5 UNITS CONVERSION BETWEEN ENGLISH MKS AND CGS SYSTEMS

	English	MKS	CGS
Length:	1 yd = 0.914m 1 in = 0.0254m	1m = 39.37 in = 100 cm	2.54 cm = 1 in
Mass:	1 slug = 14.6 kg	1 kg = 1000 g 0.45359237 kg = 1 lbm	
Force:	1 lb = 4.45 N	1N = 0.22481 lb_f 1N = 100,000 dynes	

	English	MKS	CGS
Energy:	1 ft-lb = 1.356 J	1J = 0.7376 ft-lb = 1N·m	1J = 10^7 ergs
Power:		1J/s = 0.7376 ft-lb$_f$/s = 11745.7 hp	
Temperature:	$°F = \frac{9}{5} C° + 32$	$°C = \frac{5}{9}(F° - 32)$	

1.2 LAWS OF UNITS

1. In each term of an equation, the units of measurement must be the same.
2. Only one system of units is used with any one equation; both sides of the equality must be of the same system.

1.3 UNIT OF CHARGE AND COULOMB'S LAW

Charge: Symbol → Q - constant charge
q - time-varying or instantaneous value of charge, i.e., g(t).

Unit → Coulomb (C). (Note: the charge of an electron = -1.60219×10^{-19} C, where 1 C (negative) = combined charge of about 6.24×10^{18} electrons. The charge of one proton = $+1.602 \times 10^{-19}$ C.)

Coulomb's Law:

$$F = \frac{k Q_1 Q_2}{d^2} = \frac{1}{4\pi\epsilon_0} \frac{Q_1 Q_2}{d^2} \quad [N]$$

where $k = 9 \times 10^9 \, N\text{-}m^2/C^2$ = proportionality constant

$= (4\pi\varepsilon_0)^{-1}$

$Q_1 + Q_2$ = charge of 2 bodies in coulomb

d = separated distance between 2 charged bodies

and ε_0 = permittivity of free space = $8.85 \times 10^{-12} \, C^2/N \cdot m^2$

Note: 1) For any material: ε = permittivity of the material = $K \varepsilon_0$, where K = dielectric constant
2) In a vaccum: $K = 1$ and $\varepsilon = \varepsilon_0$.

I.4 SCIENTIFIC NOTATION

	Power of Ten	Prefix	Abbreviation
	10^{-18}	atto	a
	10^{-15}	femto	f
*	10^{-12}	pico	p
*	10^{-9}	nano	n
*	10^{-6}	micro	μ
*	10^{-3}	milli	m
*	10^{-2}	centi	c
*	10^{-1}	deci	d
	10^{1}	deka	da
	10^{2}	hecto	h
*	10^{3}	kilo	k
*	10^{6}	mega	M
*	10^{9}	giga	G
	10^{12}	tera	T

(* - most frequently used)

CHAPTER 2

RESISTIVE CIRCUITS AND EXPERIMENTAL LAWS

2.1 CURRENT, VOLTAGE, POWER AND ENERGY

2.1.1 CURRENT

Definition:

The measurement of the rate of the number of charges moving through a given reference point in a circuit in 1 second. For steady current,

$$i = \frac{q}{t},$$

where q is the net charge passing through the point in t seconds.

Unit of current = ampere (A) = 1 coulomb of charge moving past a point in 1 second.

Instantaneous current (i) = time rate of change of charge = $\frac{dq}{dt}$.

Current Flow: The current flow in a wire is opposite to the motion of the electrons by convention. (See Fig.)

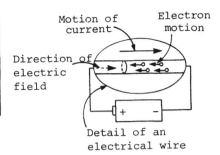

Detail of an electrical wire

Direct Current (DC): A current which is constant due to a steady, unchanging, unidirectional flow of charge.

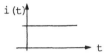

Alternating Current (AC): Sinusoidal time varying current, e.g., household current.

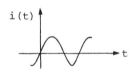

2.1.2 VOLTAGE

Definition:

The voltage (V or v), or the potential difference between two points, is the measure of the work required to move a unit charge from one point to another.

Unit of voltage = volt = 1 joule/coulomb

Voltage Sign Convention:

Assume a positive current supplied by an external source is entering terminal 1. Then,

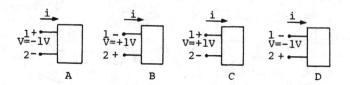

Terminal 1 is 1 volt positive with respect to terminal 2 - Figs. C & D and Terminal 2 is 1 volt positive with respect to terminal 1 - Figs. A & B.

2.1.3 POWER AND ENERGY

Definition:

(p)power [watts] = v[volts] × i[amperes]

Efficiency: $\eta = \dfrac{\text{power output}}{\text{power input}}$ $0 < \eta < 1$

Energy:

Since power (p) is the time rate of energy transfer ($p = \dfrac{dW}{dt}$)

$$W = \int_{t_1}^{t_2} p\, dt$$

(the energy transferred during a given time interval x) or, W(Energy in watt-seconds or joules) = p(power in watts) × t(time in seconds).

Energy and Voltage Relationships:

1. voltage drop across element → positive released energy

2. voltage rise ⟶ positive generated energy

2.2 TYPES OF CIRCUIT ELEMENTS

2.2.1 INDEPENDENT VOLTAGE SOURCE

Characteristics:

1. The voltage between the two terminals is independent of the current through it.

2. The same amount of voltage output is supplied continuously regardless of the amount of current drawn from it.

Types:

A) time-varying

B) time-invarying (independent DC voltage source)(i.e., constant terminal voltage).

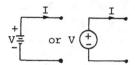

2.2.2 INDEPENDENT CURRENT SOURCE

Characteristic:

The current supplied by the source is fixed to a load and is completely independent of the voltage across it.

Circuit symbol:

Note: Both independent current and voltage sources are approximations for a physical element.

2.2.3 DEPENDENT VOLTAGE AND CURRENT SOURCES

Circuit symbol:

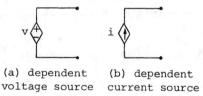

(a) dependent voltage source   (b) dependent current source

Characteristic:

The source quantity of a dependent source is determined by a voltage or current existing at some other location in the electrical system under consideration.

2.3 RESISTANCE AND CONDUCTANCE

2.3.1 RESISTANCE

The measure of the tendency of a material to impede the flow of electric charges through it.

Circuit symbol:

R = resistance of the resistor having units [volts/ampere] or ohm (Ω).

2.3.2 CONDUCTANCE

The reciprocal of resistance, or the ratio of current to voltage, i.e.,

$$G = \frac{1}{R} = \frac{i}{v} \; [\text{mho}(\mho)].$$

2.3.3 RESISTIVITY

ρ, the characteristic of a material which indicates how much a material impedes current flow.

$$\boxed{R = \frac{\rho \ell}{A}} \quad \text{(at constant temperature)}$$

R = resistance in ohms
ℓ = length [m]
A = cross-sectional area [m^2]
ρ = resistivity [Ω-m]

Note: Resistivity is low in a good conductor but high in a poor conductor (insulator).

2.3.4 RESISTOR AND CONDUCTOR COMBINATIONS

For Series Combination of N Resistors

$$R_{eq} = R_1 + R_2 + \ldots + R_N$$

or

$$\frac{1}{G_{eq}} = \frac{1}{G_1} + \frac{1}{G_2} + \ldots + \frac{1}{G_N}$$

For Parallel Combination of N Resistors

$$\frac{1}{R_{eq}} = \frac{1}{R_1} + \frac{1}{R_2} + \ldots + \frac{1}{R_N}$$

or

$$G_{eq} = G_1 + G_2 + \ldots + G_N$$

2.4 VOLTAGE AND CURRENT DIVISION

2.4.1 VOLTAGE DIVISION

Circuit diagram:

Formula:

$$V_b = \frac{R_b}{R_a + R_b} V$$

2.4.2 CURRENT DIVISION

Circuit diagram:

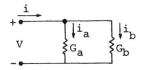

Formula:

$$i_b = \frac{G_b}{G_a + G_b} i = \frac{R_a}{R_a + R_b} i$$

2.5 OHM'S LAW

The voltage across a conducting material is directly proportional to the current through the material, i.e., $v = Ri$, where R(resistance) is the proportionality constant.

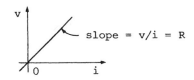

Hence, absorbed power in a resistor is given by

$$p = Vi = i^2R = V^2/R.$$

Note: This power is in the form of heat because a resistor is a passive element; it neither delivers power nor stores energy.

Short Circuit: Circuit as a zero ohms resistance, i.e., voltage across a short circuit = 0.

Open Circuit: Circuit as an infinite resistance, i.e., current across an open circuit = 0.

2.6 KIRCHHOFF'S LAW

2.6.1 KIRCHHOFF'S CURRENT LAW

The algebraic sum of all currents entering a node equals the algebraic sum of all currents leaving it, i.e., for a given node, Σ currents entering = Σ currents leaving or

$$\sum_{n=1}^{N} i_n = 0.$$

2.6.2 KIRCHHOFF'S VOLTAGE LAW

The algebraic sum of all voltages around a closed loop (or path) is zero, i.e., for a closed loop, Σ potential rises = Σ potential drops.

CHAPTER 3

TRANSIENT CIRCUITS

3.1 CAPACITOR AND CORRESPONDING VOLTAGE, CURRENT AND POWER RELATIONSHIPS

3.1.1 PARALLEL PLATE CAPACITOR

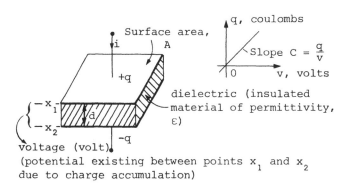

Capacitance (C) = $K\varepsilon_0 \frac{A}{d}$ (where K = $\varepsilon/\varepsilon_0$ = relative dielectric constant) = $\frac{q}{v}$ [Coulombs/volt or farad (F)]

or C = q(t)/v(t) (time variant)

(Note: ε_0 (for air or vaccum) = 8.854 pF/m = $\left(\frac{1}{36\pi}\right)$ nF/m)

A-14

Circuit Symbol:

$$\xrightarrow{i} \; \overset{C}{+\!\!\Vert\!\!-} \quad \text{or} \quad \xrightarrow{i} \; \overset{C}{+\!\!\!+\!\!\Vert\!\!-}$$

Capacitor Voltage, Current, Power and Energy:

$$\boxed{v(t) = \frac{1}{C} \int_{-\infty}^{t} i(\tau)\,d\tau}$$

$$\boxed{i(t) = \frac{dq(t)}{dt} = C\frac{dv(t)}{dt}}$$

$$p = Cv\left(\frac{dv}{dt}\right)$$

and

$$\boxed{W = \tfrac{1}{2}C v^2} \quad = \text{stored energy [joules]}$$

or

$$W = \frac{Q^2}{2C}$$

Characteristic:

A capacitor acts as an open circuit to dc.

3.2 INDUCTOR AND TRANSFORMER

3.2.1 IRON-CORE INDUCTOR

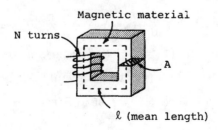

A-15

Inductance (L) = $\frac{\mu N^2 A}{\ell}$ where N = no. of turns of coil
µ = permeability of core
A = cross-sectional area of core
ℓ = mean length of core.

(Note: permeability of vacuum $\mu_0 = 4\pi \times 10^{-7}$.)

Magnetic, flux $\phi(t) = \left(\frac{\mu N^2 A}{\ell}\right) i(t) = L\, i(t)$.

Magnetic field intensity, $H = \frac{Ni}{\ell}$ [A·turn/m].

ℓ = length of material through which ϕ travels.

i = current flowing in the coil.

Flux density, $B = \mu H = \phi/A$ [telsa (T) or wb/m^2].

B-H Curve:

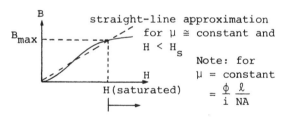

straight-line approximation for $\mu \cong$ constant and $H < H_s$

Note: for μ = constant
$= \frac{\phi}{i} \frac{\ell}{NA}$

3.2.2 INDUCTOR

Concept of Self-Inductance:

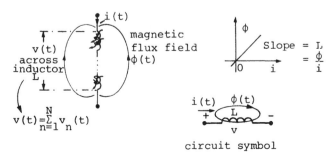

circuit symbol

A-16

Inductance $(L) = \frac{Nd\phi(i)}{di}$ [henry(H) or volt-second/ampere].

Voltage $v(t) = N$(no. of turns of coil) $\times \frac{d\phi(t)}{dt}$
(rate of change of ϕ with respect to time), or

$$v(t) = \frac{Nd\phi(i)}{di} \frac{di}{dt} = L\frac{di(t)}{dt}$$

$$i(t) = \frac{1}{L} \int_{-\infty}^{t} v\, dt$$

$$W = \frac{1}{2}Li^2$$

$p = vi = Li\frac{di}{dt}$ [W].

An inductor acts as a short circuit to dc.

3.2.3 MUTUAL INDUCTANCE

Definition:
   The coupling of two coils such that the change of flux produced by one will link the other, resulting in an induced voltage across each coil. (See Fig.)

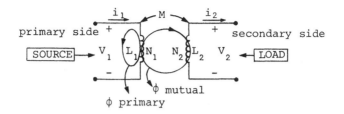

From the figure above,

$$v_2(t) = \frac{M di_1(t)}{dt} \quad \text{and} \quad v_1(t) = \frac{M di_2(t)}{dt}$$

or, by Faraday's law,

$$v_2(t) = \frac{N_2 d\phi_m}{dt} \quad \text{and} \quad v_1(t) = \frac{N_1 d\phi_p}{dt},$$

where

    M = proportionality constant between 2 coils

       = mutual inductance

       = $K\sqrt{L_1 L_2}$ [Henry (H) or volt · second/ampere];

    K = coupling coefficient = $\dfrac{\phi_m}{\phi_p}$

3.2.4 DOT NOTATION

Assigning the dots on a pair of coupled coils:

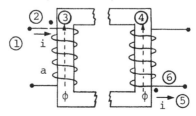

Procedure:

1. Select current direction in one of the coils.

2. Assign a dot to where the current enters the winding. (Note: This is the positive terminal with respect to point a.)

3. Use the right-hand rule to assign flux direction.

4. By Lenz's law, assign opposite flux direction for the second coil.

5. Use right-hand rule to assign current direction.

6. Assign a dot to where the current leaves the winding.

7. Obtain simplified diagram as shown:

Sign of mutual inductance M:

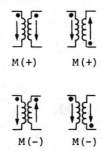

$M = +$ (Both currents pass through coils and are either leaving or entering dots.)

$M = -$ (If arrow indicating current direction through coil is entering the dot for one coil and leaving the dot for another.)

3.2.5 IRON-CORE TRANSFORMER

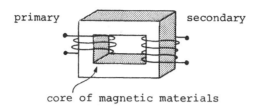

core of magnetic materials

Turns ratio: The determination of how much a transformer steps up or steps down a voltage.

$$\text{Turns ratio} = \frac{\text{no. of turns on the primary }(N_p)}{\text{no. of turns on the secondary}(N_s)}$$

or $\dfrac{V_p}{V_s} = \dfrac{N_p}{N_s}$ and $\dfrac{I_p}{I_s} = \dfrac{N_s}{N_p}$

Voltage step-up transformer $N_p < N_s$:

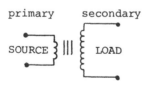

Voltage step-down transformer $N_p > N_s$:

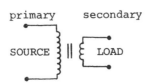

A-20

3.3 SIMPLE RL AND RC CIRCUITS

3.3.1 SOURCE FREE RL CIRCUIT

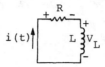

Properties: Assume initially $i(0) = I_0$.

a) $v_R + v_L = Ri + L\frac{di}{dt} = 0.$

b) $i(t) = I_0 e^{-Rt/L} = I_0 e^{-t/\tau}$, τ = time constant $= \frac{L}{R}$

c) Power dissipated in the resistor =
$P_R = i^2 R = I_0^2 R e^{-2Rt/L}.$

d) Total energy in terms of heat in the resistor =
$W_R = \frac{1}{2} L I_0^2.$

3.3.2 SOURCE FREE RC CIRCUIT

Properties: Assume initially $v(0) = V_0$

a) $C\frac{dv}{dt} + \frac{v}{R} = 0.$

b) $v(t) = v(0)e^{-t/RC} = V_0 e^{-t/RC}.$

A-21

(Note: RC = time constant = τ.)

c) $\dfrac{1}{C} \displaystyle\int_{-\infty}^{t} i(\tau)d\tau + i(t)R = 0$

$i(t) = i(0)e^{\frac{-t}{RC}}$

3.4 NATURAL AND FORCED RESPONSE OF RL AND RC CIRCUITS

3.4.1 NATURAL RESPONSE

The complementary solution of a linear differential equation.

3.4.2 FORCED RESPONSE

The particular solution of a linear differential equation.

3.4.3 A GENERAL DIFFERENTIAL EQUATION

For: $\dfrac{di}{dt} + Pi = Q$, where Q = forcing function,

P = general function of time.

Solution: $i = e^{-pt} \displaystyle\int Qe^{pt}dt + Ae^{-pt}$.

Note: For a source free circuit, $Q = 0$. $\underbrace{i_n = Ae^{-pt}}_{\text{natural response;}}$

and for Q(t) = const. $\underbrace{i_f = Q/P}_{\text{forced response}}$;

and for a complete response:

$$\boxed{i(t) = \frac{Q}{P} + Ae^{-pt}}$$

Complete response = natural response + forced response

Total solution = complementary solution + particular solution.

3.4.4 PROCEDURE TO FIND THE COMPLETE RESPONSE f(t) OF RL AND RC CIRCUITS WITH DC SOURCES

	RL	RC
1. Simplify the circuit by "killing" all independent sources and determine:	R_{eq}, L_{eq} (* $\tau = L_{eq}/R_{eq}$)	R_{eq}, C_{eq} (* $\tau = R_{eq}C_{eq}$)
2. Consider: ⟶ and use dc-analysis to find:	$L_{eq} \sim$ short circuit $i_L(0^-)$	$C_{eq} \sim$ open circuit $v_C(0^-)$
3. Repeat procedure 2 to find the forced response:	i.e., f(t) as $t \to \infty$ $f(\infty)$	
4. Obtain the total response as the sum of the natural and forced responses:	i.e., $f(t) = Ae^{-t/\tau} + f(\infty)$	

	RL	RC
5. Determine $f(0^+)$ by considering the conditions:	$i_L(0^+) =$ $i_L(0^-)$	$v_C(0^+) = v_C(0^-)$
6. Then:	$f(0^+) = A + f(\infty)$ and $f(t) = [f(0^+) - f(\infty)]e^{-t/\tau} + f(\infty)$	

Killing: setting them equal to zero

3.5 THE RLC CIRCUITS

3.5.1 PARALLEL RLC CIRCUIT (source free)

Circuit diagram:

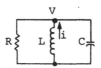

KCL equation for parallel RLC circuit:

$$\frac{v}{R} + \frac{1}{L}\int_{t_0}^{t} v\, dt - i(t_0) + C\frac{dv}{dt} = 0;$$

and the corresponding linear, second-order homogeneous differential equation is

$$C\frac{d^2v}{dt^2} + \frac{1}{R}\frac{dv}{dt} + \frac{v}{L} = 0.$$

General solution:

$$V = A_1 e^{S_1 t} + A_2 e^{S_2 t},$$

where

$$S_{1,2} = \frac{-1}{2RC} \pm \sqrt{\left(\frac{1}{2RC}\right)^2 - \frac{1}{LC}}$$

or

$$S_{1,2} = -\alpha \pm \sqrt{\alpha^2 - \omega_0^2}\,;$$

where α = exponential damping coefficient neper frequency

$$= \frac{1}{2RC}$$

and ω_0 = resonant frequency = $\dfrac{1}{\sqrt{LC}}$

3.5.2 SPECIAL RESPONSE OF PARALLEL RLC CIRCUIT

a) Overdamped	b) Critical damping	c) Underdamped
Condition 1) $\alpha > \omega_0$ or if $LC > 4R^2 C^2$	$\alpha = \omega_0$ or $LC = 4R^2 C^2$ or $L = 4R^2 C$	$\alpha < \omega_0$
2) S_1 and S_2 = negative real numbers, i.e., $\sqrt{\alpha^2 - \omega_0^2} < \alpha$ or $(-\alpha - \sqrt{\alpha^2 - \omega_0^2}) < (-\alpha + \sqrt{\alpha^2 - \omega_0^2}) < 0$	$S_1 = S_2 = \alpha$	S_1 and S_2 compose of real and complex quantities.
3) $v(t) \to A_1 e^{S_1 t} \to 0$, as $t \to \infty$	$v(t) = A_1 e^{S_1 t} + A_2 e^{S_2 t}$	$v(t) = e^{-\alpha t}(A_1 e^{j\omega_d t} + A_2 e^{j\omega_d t})$

a) Overdamped	b) Critical damping	c) Underdamped
		where $\omega_d =$ $\sqrt{\omega_0^2 - \alpha^2} =$ Natural Resonant Frequency or $v(t) = e^{-\alpha t} \left\{ (A_1 + A_2) \left[\dfrac{e^{j\omega_d t} + e^{-j\omega_d t}}{2} \right] + j(A_1 - A_2) \left[\dfrac{e^{j\omega_d t} - e^{-j\omega_d t}}{2j} \right] \right\}$ or $v(t) = e^{-\alpha t} \left[(A_1 + A_2)\cos \omega_d t + j(A_1 - A_2)\sin \omega_d t \right]$

Graphic representation:

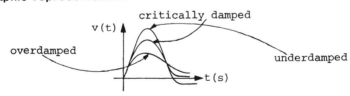

3.5.3 SERIES RLC CIRCUIT (source free)

Circuit diagram:

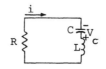

A-26

KVL equation for series RLC circuit:

$$Ri + \frac{1}{C}\int_{t_0}^{t} i\, dt + L\frac{di}{dt} - v_c(t_0) = 0$$

and the corresponding second-order differential equation in terms of i:

$$L\frac{d^2 i}{dt^2} + R\frac{di}{dt} + \frac{i}{C} = 0$$

or in terms of v:

$$LC\frac{d^2 v_c}{dt^2} + RC\frac{dv_c}{dt} - v_c = 0.$$

Special response of series RLC circuit:

a) Overdamped	b) Critical damping	c) Underdamped
$S_1, S_2 = \frac{-R}{2L} \pm \sqrt{\left(\frac{R}{2L}\right)^2 - \frac{1}{LC}}$ or $= -\alpha \pm \sqrt{\alpha^2 - \omega_0^2}$ where $\alpha = R/2L$, $\omega_0 = 1/\sqrt{LC}$, $i(t) = A_1 e^{S_1 t} + A_2 e^{S_2 t}$	$S_1 = S_2 = \alpha$ $i(t) = e^{-\alpha t}(A_1 + A_2)$	$S_{1,2} = -\alpha \pm j\omega_d$ $\omega_d = \sqrt{\omega_0^2 - \alpha^2}$ $i(t) = e^{-\alpha t}(B_1 \cos\omega_d t + B_2 \sin\omega_d t)$

3.5.4 COMPLETE RESPONSE OF RLC CIRCUIT

The general equation of a complete response of a second-order system in terms of voltage for an RLC circuit is given by,

$$\boxed{v(t) = \underbrace{V_f}_{} + \underbrace{Ae^{S_1 t} + Be^{S_2 t}}}$$

　　　　　　　forced response　natural response

(i.e., constant for
DC excitation)

Note: A and B can be obtained by

1) substituting v at $t = 0^+$

2) taking the derivative of the response, i.e.,

$$\frac{dv}{dt} = 0 + S_1 A\, e^{S_1 t} + S_2 B\, e^{S_2 t}, \text{ where}$$

$\dfrac{dv}{dt}$ at $t = 0^+$ is known.

CHAPTER 4

NETWORK THEOREMS

4.1 LINEARITY AND SUPERPOSITION

4.1.1 LINEARITY

1) A linear element: A passive element that can be represented by a linear voltage-current relationship.

2) A linear dependent source: A dependent current or voltage source whose output current or voltage is proportional only to the first power of some current or voltage variable in the circuit or to the sum of such quantities.

3) A linear circuit: A circuit composed entirely of independent sources, linear dependent sources and linear elements.

4.1.2 SUPERPOSITION THEOREM

The network response in any linear resistive network with zero initial conditions can be obtained by summing all the individual voltages or currents caused by each independent source acting alone. With all other independent sources set equal to zero, i.e., independent voltage sources are replaced by short circuits and independent current sources are replaced by open circuits.

A-29

4.2 THEVENIN'S AND NORTON'S THEOREMS

4.2.1 THEVENIN'S THEOREM

In any linear network, it is possible to replace everything except the load resistor by an equivalent circuit containing only a single voltage source in series with a resistor (R_{th} Thevenin resistance), where the response measured at the load resistor will not be affected.

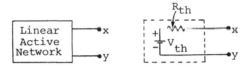

Procedures to find Thevenin equivalent:

1) Solve for the open circuit voltage v_{oc} across the output terminals. $V_{oc} = V_{th}$

2) Place this voltage v_{oc} in series with the Thevenin resistance which is the resistance across the terminals found by setting all independent voltage and current sources to zero. (i.e., short circuits and open circuits, respectively.)

4.2.2 NORTON'S THEOREM

Given any linear circuit, the passive and active components can be converted into an equivalent two-terminal network consisting of a single current source in parallel with a resistor (R_N - Norton resistance).

Procedures to find Norton equivalent:

1) Setting all sources to zero (i.e., voltage sources → short circuits, and current sources → open circuits). Then find resulting resistance R_N between the output terminals.

2) I_N is the current through a short circuit applied to the two terminals of the given network.

The Norton's equivalent is obtained by connecting current source I_N and R_N in parallel.

4.3 MAXIMUM POWER TRANSFER THEOREM

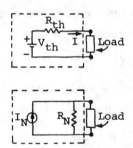

Maximum energy transfer occurs between the driving source and the load, if the following condition is satisfied:

$$R_L = R_{th} \quad \text{(for Thevenin equivalent circuit),}$$

or

$$R_L = R_N \quad \text{(for Norton equivalent circuit).}$$

4.4 MILLMAN'S THEOREM

The theorem states that voltage sources connected in parallel can be reduced to one equivalent circuit.

General Procedures: The application of Millman's theorem)

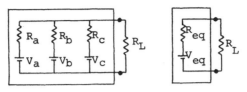

Millman's theorem

1) Given a similar circuit as above, convert all voltage sources to current sources and resistance to conductance into an equivalent parallel circuit as shown below:

2) Combine all parallel current sources and all conductance in parallel.

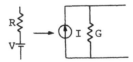

3) Convert the resulting equivalent current source to a voltage source and the equivalent conductance to an equivalent resistance.

(Note: $R = \frac{1}{G}$ and $V = \frac{I}{G}$.)

4.5 SUBSTITUTION (COMPENSATION) THEOREM

Any network branch voltage or branch current can be replaced by a current or a voltage source which will maintain the same voltage across the chosen branch, and the same current through the branch.

4.6 RECIPROCITY THEOREM

In a linear, single-source network, the current I produced in any branch of the network because of the single voltage source is interchangeable with the location of the voltage source without a change in current.

CHAPTER 5

USEFUL TECHNIQUES OF CIRCUIT ANALYSIS

5.1 MATRICES

5.1.1 A GENERAL FORM OF A RECTANGULAR MATRIX R

$$R = \begin{bmatrix} a_{11} & a_{12} & \cdots & a_{1N} \\ a_{21} & a_{22} & \cdots & a_{2N} \\ a_{M1} & a_{M2} & \cdots & a_{MN} \end{bmatrix}$$

a_{ij} = elements of a matrix

$i \Rightarrow$ row

$j \Rightarrow$ column

(Special case: $M = N$, is a square matrix.)

5.1.2 ADDITION, SUBTRACTION AND MULTIPLICATION OF MATRICES

Consider matrices $X = [x_{ij}]$ and $Y = [y_{ij}]$.

$m \times \ell$ matrix $\quad \ell \times N$ matrix

Addition and Subtraction:

$$X \pm Y = [x_{ij} \pm y_{ij}]$$

Multiplication:

$$XY = \sum_{k=1}^{\ell} x_{ik} y_{kj},$$

$$\text{where } i = 1,\ldots,M$$
$$\text{and } j = 1,\ldots,N$$

Note: $(X+Y)+Z = X+(Y+Z)$

$(X+Y)Z = XZ + YZ$

$(XY)Z = X(YZ)$

$XY \neq YZ$ (in general)

5.1.3 PROPERTIES OF MATRICES

1) Diagonal matrix: $D = \begin{bmatrix} a_{11} & & & \\ & a_{22} & & \\ & & \ddots & \\ & & & a_{NN} \end{bmatrix}$

2) Unit or identity matrix:

$$I = \begin{bmatrix} 1 & & & \\ & 1 & & \\ & & \ddots & \\ & & & 1 \end{bmatrix}$$

I is also a square matrix.

3) Transpose of a matrix:

$$(X^T)^T = X$$

$$(X+Y)^T = X^T + Y^T$$
$$(kX)^T = kX^T \quad \text{(where K is a scalar multiple)}$$
$$(XY)^T = Y^T X^T$$

4) Power of matrix:

$$X^n = \underbrace{X \; X \; \ldots \; X}_{n \text{ times}}$$

5) Inverse of a matrix:

If X is a square matrix, its inverse is X^{-1} such that

$$X \; X^{-1} = I.$$

Note: $X^{-1}X = I = XX^{-1}$

$$(XY)^{-1} = Y^{-1}X^{-1}$$

5.2 DETERMINANTS

5.2.1 THE DETERMINANT

In general, given a matrix X i.e.,

$$X = \begin{bmatrix} x_{11} & x_{12} & \ldots & x_{1N} \\ x_{21} & x_{22} & \ldots & x_{2N} \\ x_{N1} & x_{N2} & \ldots & x_{NN} \end{bmatrix},$$

the determinant of X (Δ_x) can be expressed in terms of minor (Δ_{jk}) as follows:

$$\Delta_x = x_{j_1}(-1)^{j+1}\Delta_{j_1} + x_{j_2}(-1)^{j+2}\Delta_{j_2} + \ldots + x_{jN}(-1)^{j+N}\Delta_{jN}$$

$$= \sum_{n=1}^{N} x_{jn}(-1)^{j+n}\Delta_{jn} \quad \text{— along row j.}$$

$$\Delta_x = x_{1K}(-1)^{1+K}\Delta_{1K} + x_{2K}(-1)^{2+K}\Delta_{2K} + \ldots + x_{NK}(-1)^{N+K}\Delta_{NK}$$

$$= \sum_{n=1}^{N} x_{nk}(-1)^{n+k}\Delta_{nk} \quad \text{— along column k.}$$

Also, in terms of the cofactor C_{jk} (i.e., $C_{jk} = (-1)^{j+k} \times \Delta_{jk}$),

$$\Delta_x = \sum_{n=1}^{N} x_{jn}C_{jn} = \sum_{n=1}^{N} x_{jk}C_{jk}$$

where j and k are positive integers $\leq N$.

Cramer's Rule:

In any system of n linear equations of n unknowns, for the kth variable v_k,

$$v_k = \frac{\Delta_k}{\Delta_G}.$$

(Note: Δ_G is the determinant of a conductance matrix.)

5.2.2 SECOND AND THIRD-ORDER DETERMINANT CALCULATIONS

1) Second-order determinant:

$$\Delta_x = \begin{vmatrix} x_{11} & x_{12} \\ x_{21} & x_{22} \end{vmatrix} = x_{11}x_{22} - x_{12}x_{21}$$

2) Third-order determinant:

$$\Delta_x = \begin{vmatrix} x_{11} & x_{12} & x_{13} \\ x_{21} & x_{22} & x_{23} \\ x_{31} & x_{32} & x_{33} \end{vmatrix} = \begin{array}{l} x_{11} x_{22} x_{33} + x_{12} x_{23} x_{31} + x_{13} x_{21} x_{32} \\ - x_{13} x_{22} x_{31} - x_{11} x_{23} x_{32} - x_{12} x_{21} x_{33} \end{array}$$

$$= x_{11}[x_{22}x_{33} - x_{23}x_{32}]$$
$$+ x_{12}(x_{23}x_{31} - x_{21}x_{33}) + x_{13}(x_{21}x_{32} - x_{22}x_{31})$$

5.3 SINGLE-LOOP AND SINGLE-NODE-PAIR CIRCUIT ANALYSIS

5.3.1 SINGLE-LOOP ANALYSIS
Analytical Procedure:

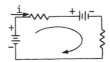

1) Use an arrow to represent the direction of the unknown current and represent it by i in the circuit.

2) For each resistor in the circuit, select a voltage reference.

3) Select a clockwise or counterclockwise direction of movement while applying KVL to the single closed path. Write down positive voltage when a positive terminal is encountered or negative voltage when a negative terminal is encountered.

4) For the resistive elements in the circuit, Ohm's law is applied.

5.3.2 NODE-PAIR ANALYSIS

Analytical Procedure:

1) Assume a voltage across any element, assigning an arbitrary reference polarity. (Note: Elements connected in parallel have a common voltage across them.)
2) Assign and label the direction of current flow for each resistor.
3) Apply KCL to either of the nodes in the circuit. Note: Apply KCL to the node at which the positive voltage reference has a preferred location.
4) Express the current in each resistor in terms of v and the conductance of the resistor by Ohm's law.

5.4 SOURCE TRANSFORMATIONS

Characteristics of Practical Voltage and Current Sources:

A practical voltage source	A practical current source
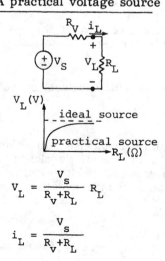	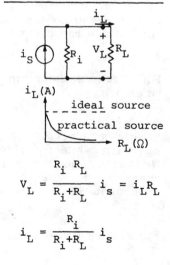

$$V_L = \frac{V_S}{R_V + R_L} R_L$$

$$i_L = \frac{V_S}{R_V + R_L}$$

$$V_L = \frac{R_i R_L}{R_i + R_L} i_S = i_L R_L$$

$$i_L = \frac{R_i}{R_i + R_L} i_S$$

A-39

Conditions for equivalence of practical voltage and current source:

1) Each source must produce identical current and identical voltage in any load that is placed across its terminals.

2) $i_L = \dfrac{V_s}{R_v + R_L} = \dfrac{R_i i_s}{R_i + R_L}$

Hence,

$$R_v = R_i = R_s \text{ and } V_s = R_s i_s,$$

where R_s = internal resistance of either practical source.

5.5 NODAL ANALYSIS

5.5.1 NODAL ANALYSIS - FORMAT APPROACH

1) Assign a reference node for the k nodes circuit.

2) For the circuit containing only voltage sources:

   Replace each voltage source by a short circuit without changing the assigned node voltages. Apply KCL at each of the nodes in this modified circuit by using the assigned node to reference voltages. Finally, correspond each source voltage to the variables $v_1, \ldots, v_{k-1}$.

3) For the circuit containing only current sources:

   Apply KCL at each non-reference node. For circuits containing only independent current sources, the conductance matrix can be obtained by using the equation: total current leaving each node (through all conductances) = total source current entering that node. (Note: Put the terms in order from $(v_1, \ldots, v_{n-1})$.)

   Finally, for the dependent current source, correspond the source current and the controlling quantity to the variables $v_1, \ldots, v_{k-1}$.

5.5.2 NODAL ANALYSIS - GENERAL APPROACH

1) Convert all voltage sources to current sources.
2) In each network, determine the number of nodes.
3) Choose a reference node and assign voltages to the remaining node.
4) Apply KCL at each node, except at the reference node.
5) Solve the unknown equations for nodal voltages.

5.5.3 MESH ANALYSIS

General Approach:

1) Assign closed loops of current called mesh currents, clockwise, to each loop of the circuit.
2) Apply KVL around each closed loop.
3) Solve resulting equations for the assumed loop currents.

Note: Mesh analysis is only applicable to planar network. By definition, a planar network is a circuit diagram on a plane surface such that no branch passes over or under any other branch.

5.6 FREE AND GENERAL NODAL ANALYSIS

A general procedure for writing a set of independent and sufficient nodal equations using the concepts "tree" and "cotree".

V_x = control voltage

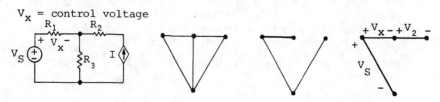

1) Draw a graph of the network given and indicate a tree. By definition, a tree is any set of branches which does not contain any loops. It makes connections between nodes, but not necessarily directly.
2) Label and place all voltage sources in the tree.
3) Label and place all current sources in the cotree. By definition, cotrees are those branches which do not belong to the tree.
4) If possible, place and label all control-voltage and control-current in the tree.
5) Complete the tree and assign voltage across each tree branch. (See Fig. for sample illustration)

Concept of link: Any branch belonging to the cotree is a link.

L = number of links = B(number of branches)
$\qquad\qquad\qquad$ - (N(number of nodes)- 1)
$\qquad\qquad$ = B - N + 1

CHAPTER 6

SINUSOIDAL ANALYSIS

6.1 SINUSOIDAL CURRENT, VOLTAGE AND PHASE ANGLE

6.1.1 SINUSOIDAL FORCING FUNCTION - GENERAL FORM

$$v(t) = V_m \sin(\omega t + \theta)$$

where V_m = maximum value

f = frequency = $\frac{1}{T}$ = $\frac{\text{cycles}}{\text{sec}}$ = hertz (Hz)

T = period (time duration of 1 cycle = sec)

ω = angular frequency = $2\pi f$ = $\frac{2\pi}{T}$ = $\frac{\text{radians}}{\text{sec}}$

6.1.2 LEAD AND LAG CONCEPT OF PHASOR ANGLE θ

Let $A = V_m \sin \omega t$ and $B = V_m \sin(\omega t + \theta)$, then

1) B leads A by θ rad.
2) A lags B by θ rad.
3) A leads B by $-\theta$ rad.
4) A leads $V_m \sin(\omega t - \theta)$ by θ rad.

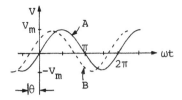

6.1.3 SINUSOIDAL CURRENTS AND VOLTAGES

Voltage across resistance (R), inductance (L) or capacitance (C) if current is indicated as follows:

Element	voltage	$i = I_m \sin \omega t$	$i = I_m \cos \omega t$
R	V_R =	$R I_m \sin \omega t$	$V_R = R I_m \cos \omega t$
L	V_L =	$\omega L I_m \cos \omega t$	$V_L = \omega L I_m (-\sin \omega t)$
C	V_C =	$\dfrac{I_m}{\omega C} (-\cos \omega t)$	$V_C = \dfrac{I_m}{\omega C} \sin \omega t$

Current in R, L or C if voltage is indicated as follows:

Element	current	$v = V_m \sin \omega t$	$v = V_m \cos \omega t$
R	i_R =	$\dfrac{V_m}{R} \sin \omega t$	$i_R = \dfrac{V_m}{R} \cos \omega t$
L	i_L =	$\dfrac{V_m}{\omega L} (-\cos \omega t)$	$i_L = \dfrac{V_m}{\omega L} \sin \omega t$
C	i_C =	$\omega C V_m \cos \omega t$	$i_C = \omega C V_m (-\sin \omega t)$

Note:

1) $i_R(t) = \dfrac{V_R(t)}{R}$, $V_R(t) = i_R(t) R$

2) $i_L(t) = \dfrac{1}{L} \displaystyle\int_{-\infty}^{t} V_L(\tau) d\tau$, $V_L(t) = L \dfrac{di_L(t)}{dt}$

3) $i_C(t) = C \dfrac{dV_C(t)}{dt}$, $V_C(t) = \dfrac{1}{C} \displaystyle\int_{-\infty}^{t} i_C(\tau) d\tau$

6.1.4 CHARACTERISTICS OF PHASE ANGLE IN PURE ELEMENT

Element	Current and voltage phase angle relationship.	Impedance magnitude	Diagram
R	Current and voltage in phase.	R	
L	Current lags the voltage by 90° or $\pi/2$ rad.	ωL	
C	Current leads the voltage by 90° or $\pi/2$ rad.	$1/\omega C$	
Series RL	Current lags the voltage by $\tan^{-1}(\omega L/R)$.	$\sqrt{R^2+(\omega L)^2}$	
Series RC	Current leads the voltage by $\tan^{-1}(1/\omega CR)$.	$\sqrt{R^2+(1/\omega C)^2}$	

6.2 CONCEPT OF PHASOR

6.2.1 PHASOR NOTATION

In general, the phasor form of a sinusoidal voltage or current is

$$V = V_m \underline{/\theta} \text{ and } I = I_m \underline{/\theta}.$$

Thus, for the voltage source $v(t) = V_m \cos\omega t$, the corresponding phasor form is $V_m = \underline{/\theta^0}$. For the current response $i(t) = I_m \cos(\omega t + \theta)$, the corresponding phasor form is $I_m = \underline{/\theta^0}$.

6.2.2 TIME DOMAIN TO FREQUENCY DOMAIN TRANSFORMATION, AND VICE VERSA

Time domain → frequency domain:

i.e., $v(t) = V_m \cos(\omega t + \theta) \to V = V_m \underline{/\theta}$

1) Assume a sinusoidal function i(t) in the time domain is given. Express i(t) as a cosine wave with a phase angle.

2) Using Euler's identity -- $e^{j\theta} = \cos\theta + j\sin\theta$ -- express the cosine wave as the real part of a complex quantity.

3) Drop the Re and the term $e^{j\omega t}$ to obtain the final phasor form (the frequency domain form).

Frequency domain → time domain:

1) Given a phasor current or voltage in polar form in the frequency domain, express the complex expression in exponential form.

2) Multiply the factor $e^{j\omega t}$ to the obtained exponential form.

3) Apply the Euler's identity and take the real part of the complex expression to obtain the time-domain representation.

6.2.3 TIME-DOMAIN AND FREQUENCY-DOMAIN RELATIONSHIPS OF VOLTAGE AND CURRENT FOR ELEMENT R, L AND C

Element	Voltage & Current Relationship	
voltage, current	Time domain	Frequency domain
R	$v = Ri$	$V = RI$
L	$v = L\,di/dt$	$V = (j\omega L)I$
C	$v = \dfrac{1}{C}\int i\,dt$	$V = \left(\dfrac{1}{j\omega C}\right)I$

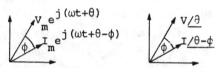

time domain plot   frequency domain plot

(Note: V_m and I_m are $1/\sqrt{2}$ times V and I.)

6.3 COMPLEX NUMBERS

6.3.1 IMAGINARY NUMBERS

$$j = \sqrt{-1}$$

Commonly used forms: $j^2 = -1$, $j^3 = j^2 j = -j$, $j^4 = (j^2)^2 = 1$, $j^5 = j$, etc.

6.3.2 COMPLEX NUMBERS

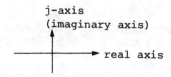

Rectangular form:	Polar form:	Exponential form:
$Z = x + jy$ ↑ ↑ real imaginary $Z^* = x - jy$ (complex conjugate of Z)	$Z = r(\cos\theta + j\sin\theta)$ where $x = r\cos\theta$ $y = r\sin\theta$ $r = \sqrt{x^2+y^2}$ and $\theta = \tan^{-1} y/x$	$Z = re^{j\theta}$ (By Euler's identity, i.e., $e^{j\theta} = (\cos\theta + j\sin\theta)$) and $Z^* = re^{-j\theta}$
Note: For x=0, pure imaginary, a point on j-axis. y=0, real number, a point on real axis	$Z = r\underline{/\theta} =$ $r(\cos\theta + j\sin\theta)$ $Z^* = r\underline{/-\theta} =$ $r(\cos\theta - j\sin\theta)$	

6.3.3 COMPLEX NUMBERS - MULTIPLICATION AND DIVISION

	Multiplication $Z_1 Z_2$	Division Z_1/Z_2
Exponential ($Z=re^{j\theta}$)	$r_1 r_2 e^{j(\theta_1+\theta_2)}$	$\dfrac{r_1}{r_2} e^{j(\theta_1-\theta_2)}$
Polar ($z=r\underline{/\theta}$)	$r_1 r_2 \underline{/\theta_1+\theta_2}$	$\dfrac{r_1}{r_2} \underline{/\theta_1-\theta_2}$
Rectangular ($Z=x+jy$)	$x_1 x_2 + jx_1 y_2$ $+ jy_1 x_2 + j^2 y_1 y_2$	$\dfrac{(x_1 x_2 + y_1 y_2) + j(y_1 x_2 - y_2 x_1)}{x_2^2 + y_2^2}$
[Note: $Z_1 = r_1 e^{j\theta_1}$ $Z_2 = r_2 e^{j\theta_2}$]		

6.3.4 POWER OF COMPLEX NUMBER

Given a complex number $Z = x+jy$, where $Z = re^{j\theta}$ ($r\cos\theta = x$ and $r\sin\theta = y$), then

$$Z^n = (re^{j\theta})^n = r^n e^{j\theta n} = r^n(\cos n\theta + j\sin n\theta),$$

A-48

where

$$\theta = \tan^{-1}\left(\frac{y}{x}\right) \text{ and } r = \sqrt{x^2 + y^2}.$$

6.3.5 ROOTS OF A COMPLEX NUMBER

Given a complex number $Z = x+jy$,
then
$$\sqrt[n]{Z} = \sqrt[n]{r}\,[e^{j\theta+2k\pi}]^{1/n}$$
$$= \sqrt[n]{r}\,e^{j(\frac{\theta+2k\pi}{n})}$$
$$= \sqrt[n]{r}\,[\cos\left(\frac{\theta+2k\pi}{n}\right) + j\sin\left(\frac{\theta+2k\pi}{n}\right)],$$

where $k = 0, 1, 2, \ldots, n-1$.

Note: For any complex number,

$$\theta = \theta + 2k\pi, \quad k = 0, \pm 1, \ldots.$$

6.3.6 COMMONLY USED FUNCTIONS OF A COMPLEX NUMBER OF THE FORM: $Z = x + jy$

a) $\sin z = \sin(x+jy) = \sin x \cos jy + \cos x \sin jy$
$\qquad = \sin x \cosh y + j\cos x \sinh y$

b) $\log_a z = \log_a(x+jy) = \log_a r + j\theta \log_a e$

(Note: $x+jy = re^{j\theta}$)

c) $e^z = e^{x+jy} = e^x e^{jy} = e^x(\cos y + j\sin y)$

6.3.7 EULER'S THEOREM

The Taylor expansion of $e^{j\theta}$ is given by

$$e^{j\theta} = 1 + j\theta + \frac{j^2\theta^2}{2!} + \ldots + \frac{j^n\theta^n}{n!} + \ldots$$
$$= \underbrace{[1 - \frac{\theta^2}{2!} + \frac{\theta^4}{4!} - \ldots]}_{\cos\theta} + j\underbrace{[\theta - \frac{\theta^3}{3!} + \frac{\theta^5}{5!} - \ldots]}_{\sin\theta}$$

Other forms:

1) $\cos\theta = \dfrac{e^{j\theta} + e^{j\theta}}{2}$, $\sin\theta = \dfrac{e^{j\theta} - e^{j\theta}}{2j}$

2) $\cos j\theta = \dfrac{e^{j(j\theta)} + e^{-j(j\theta)}}{2} = \cosh\theta$, $\sin j\theta = \dfrac{e^{j(j\theta)} - e^{-j(j\theta)}}{2j}$
$= j\sinh\theta$

6.4 IMPEDANCE AND ADMITTANCE

6.4.1 IMPEDANCE

Definition:

$$\text{impedance } (Z) = \frac{\text{phasor voltage}}{\text{phasor current}} \;\; [\text{ohms}].$$

Note: Impedance is a complex quantity.

6.4.2 TWO GENERAL FORMS OF AN IMPEDANCE

1) $\quad$ Polar form: $Z = |Z| \underline{/\theta}$

A-50

2)
$$\boxed{\text{Rectangular form: } Z = R \pm jx}$$

R = Resistive component
x = Reactive component

Impedance for elements R, L and C in the frequency domain are expressed as follows:

1) $Z_R = \dfrac{V}{I} = R$

2) $Z_L = \dfrac{V}{I} = j\omega L$

3) $Z_C = \dfrac{V}{I} = \dfrac{1}{j\omega C}$

Note: $+jx \to$ inductive reactance, $X_L = \omega L$

$-jx \to$ capacitive reactance, $X_C = 1/\omega C$

6.4.3 ADMITTANCE

Definition:

$$\boxed{\text{Admittance } (Y) = \dfrac{1}{Z} = \dfrac{\text{Phasor current}}{\text{Phasor voltage}} \quad [\text{mhos}]}$$

General form of Admittance:

$$\boxed{Y = G \pm jB}$$

Note: positive sign (i.e., +jB) $\to$ capacitive susceptance

negative sign (i.e., -jB) $\to$ inductive susceptance

6.4.4 REPRESENTATION OF IMPEDANCE AND ADMITTANCE IN TERMS OF PHASOR VOLTAGE AND CURRENT

Phaser diagram	Impedance	Admittance
$V = V\underline{/\theta}$ $i = I\underline{/0}$	$Z = \dfrac{V\underline{/\theta}}{I\underline{/\theta}} = Z\underline{/0°} = R$	$Y = \dfrac{I\underline{/\theta}}{V\underline{/\theta}} = Y\underline{/0°} = G$
$V = V\underline{/\theta}$ $i = I\underline{/\theta-\phi}$	$Z = \dfrac{V\underline{/\theta}}{I\underline{/\theta-\phi}} = Z\underline{/\phi}$ $= R + jX_L$	$Y = \dfrac{I\underline{/\theta-\phi}}{V\underline{/\theta}} = Y\underline{/-\phi}$ $= G - jB_L$

Phaser diagram	Impedance	Admittance
$V = V\underline{/\theta}$ $i = I\underline{/\theta+\phi}$	$Z = \dfrac{V\underline{/\theta}}{I\underline{/\theta+\phi}} = Z\underline{/-\phi}$ $= R - jX_C$	$Y = \dfrac{I\underline{/\theta+\phi}}{V\underline{/\theta}} = Z\underline{/-\phi}$ $= G + jB_C$

A-52

6.4.5 CONVERSION OF Z TO Y AND VICE VERSA IN POLAR FORM

1) $Z = \dfrac{1}{Y} = R \pm jx = \dfrac{G}{G^2+B^2} \pm j\dfrac{-B}{G^2+B^2}$

2) $Y = \dfrac{1}{Z} = G \pm jB = \dfrac{R}{R^2+x^2} \pm j\dfrac{-x}{R^2+x^2}$

6.5 AC ANALYSIS

Procedures similar to DC analysis and theorems are used for AC analysis except that they are in terms of phasor voltages and current (V and I) and impedance (Z).

Note: In the case of source conversions, the general format is as follows:

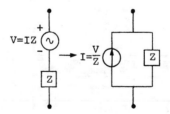

6.6 AVERAGE POWER AND rms VALUES

6.6.1 INSTANTANEOUS POWER

$p = vi$

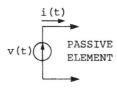

Note: p = +'ve, energy transfer from source to network.

p = -'ve, energy transfer from network to source.

In a resistive circuit, $p = i^2 R = v^2/R$.

In a inductive circuit, $p = Li\dfrac{di}{dt} = \dfrac{1}{L} v \int_{-\infty}^{t} vdt$.

In a capacitive circuit, $p = Cv\dfrac{dv}{dt} = \dfrac{1}{C} i \int_{-\infty}^{t} idt$.

6.6.2 AVERAGE POWER

$$\boxed{\text{Average power (P)} = \tfrac{1}{2} V_m I_m \cos\theta}$$

$$= \underbrace{V_{rms} I_{rms}}_{\text{apparent power}} \cos\theta, \text{ where } V_{rms} = V_m/\sqrt{2}$$
$$I_{rms} = I_m/\sqrt{2}$$

(Note: Rms=effective values)

Power factor (pf) = $\dfrac{\text{average power (P)}}{\text{apparent power }(V_{rms} I_{rms})} = \cos\theta$

Note: The unit of apparent power is voltamperes (VA). Since cos θ has maximum value of 1, the magnitude of the apparent power must be greater than the magnitude of the real power.

6.6.3 SPECIAL CASES OF pf

1) In a sinusoidal case, pf = $\cos\theta$, where θ = angle by which the voltage leads the current = pf angle.

2) For a purely resistive load, voltage and current are in phase, i.e., $\theta = 0$ and pf = 1. Hence, apparent power = average power.

3) For a purely reactive load, the phase difference between the voltage and current is either $+90°$ or $-90°$. Hence, pf = 0.

4) In general networks, 0 < pf < 1.

Summary:

Reactive power = $V_{rms} I_{rms} \sin\theta$ [voltamperes · reactive (VAR)] = Q

Apparent power = $V_{rms} I_{rms}$ [voltampere(VA)] = S

Average power = $V_{rms} I_{rms} \cos\theta$ = P

6.6.4 POWER TRIANGLE FOR INDUCTIVE AND CAPACITIVE LOAD

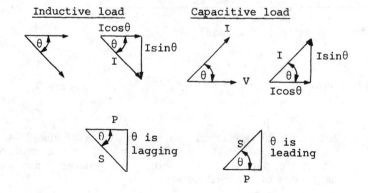

6.6.5 COMPLEX POWER

Complex power $(P) = V_{rms} I_{rms}^*$

$$= V_{rms} I_{rms} e^{j(\theta_v - \theta_i)}$$

$$= V_{rms} I_{rms} \cos\theta - jV_{rms} I_{rms} \sin\theta$$

$$= P - jQ$$

    real average power   reactive power

Complex Representation of P, Q, S and pf:

$P = \text{Re} V_{rms} I_{rms}^*$

$Q = I_m V_{rms} I_{rms}^*$

$S = |V_{rms} I_{rms}^*|$

$pf = P/S = \cos\theta$

6.6.6 RMS OR EFFECTIVE VALUE

Effective value can be obtained as follows:

1) Square the time function.

2) Take the average value of the squared function over a period.

3) Take the square root of the average of the squared function.

(Note: Effective value = square root of the mean square = rms value, i.e.,

$$I_{eff} = I_{rms} = \sqrt{\frac{1}{T} \int_0^T [i(t)]^2 \, dt}$$

Special cases:

1) Effective value of $a\sin\omega t$ and $a\cos\omega t = a/\sqrt{2}$

2) I_{eff} for sinusoidal current $i(t)$ equals $I_m \cos(\omega t - \theta)$, with $T = 2\pi/\omega = I_m/\sqrt{2} = 0.707\, I_m$.

CHAPTER 7

POLYPHASE SYSTEMS

7.1 SINGLE-PHASE, 2-PHASE AND 3-PHASE SYSTEMS

7.1.1 SINGLE-PHASE, THREE-WIRE SYSTEM

The representation of a general single-phase, three-wire system is:

Since $v_{an} = v_{nb} = v$, $v_{ab} = 2v_{an} = 2v_{nb}$.

7.1.2 TWO-PHASE SYSTEM

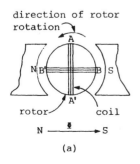

(a)

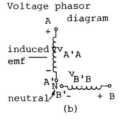

(b)

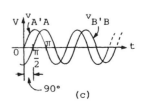

(c)

A-58

$$V_{BN} = V_{coil} \; \underline{/0^0}$$

$$V_{AN} = V_{coil} \; \underline{/90^0}$$

$$V_{AB} = V_{AN} + V_{NB} = V_{coil} \; \underline{/90^0} + V_{coil} \; \underline{/180^0} = \sqrt{2} \; V_{coil} \; \underline{/135^0}$$

7.1.3 THREE-PHASE SYSTEM

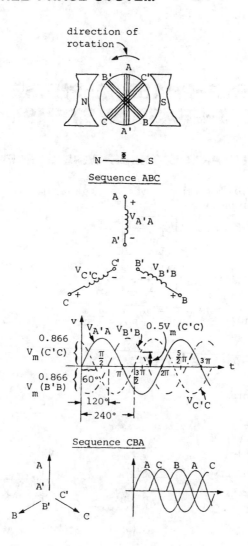

A-59

Characteristics

At any instant of time, the summation of all three phase voltages is zero, i.e.,

$$\Sigma (V_{A'A} + V_{B'B} + V_{C'C}) = 0.$$

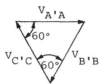

Note: In a three-phase system, the three coils on the rotor are placed $120°$ apart. (Assume each coil has an equal number of turns.)

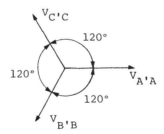

7.1.4 THREE-PHASE SYSTEM VOLTAGES

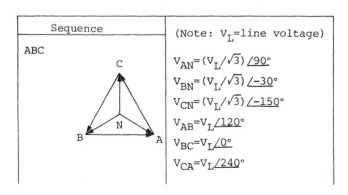

Sequence ABC	(Note: V_L=line voltage)
	$V_{AN}=(V_L/\sqrt{3})\underline{/90°}$
	$V_{BN}=(V_L/\sqrt{3})\underline{/-30°}$
	$V_{CN}=(V_L/\sqrt{3})\underline{/-150°}$
	$V_{AB}=V_L\underline{/120°}$
	$V_{BC}=V_L\underline{/0°}$
	$V_{CA}=V_L\underline{/240°}$

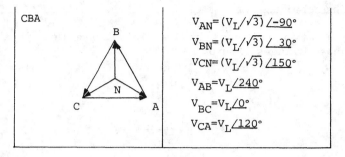

$V_{AN} = (V_L/\sqrt{3}) \angle -90°$
$V_{BN} = (V_L/\sqrt{3}) \angle 30°$
$V_{CN} = (V_L/\sqrt{3}) \angle 150°$
$V_{AB} = V_L \angle 240°$
$V_{BC} = V_L \angle 0°$
$V_{CA} = V_L \angle 120°$

7.2 THE WYE (Y) AND DELTA (Δ) CONNECTIONS

7.2.1 WYE (Y) AND DELTA (Δ) ALTERNATORS

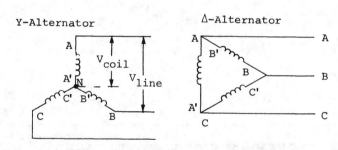

$I_{coil} = I_{line}$ $I_{coil} = \dfrac{1}{\sqrt{3}} I_{line}$

$V_{line} = \sqrt{3} \, V_{coil}$ $V_{line} = V_{coil}$

I_{coil} is more commonly referred to as I_{phase}.

7.2.2 THREE-PHASE Y-Y CONNECTION

Ideal voltage sources connected in Y (three-phase):

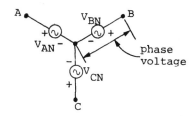

7.2.3 CHARACTERISTICS OF BALANCED THREE-PHASE SOURCES

1. $|V_{AN}| = |V_{BN}| = |V_{CN}|$ and $V_{AN} + V_{BN} + V_{CN} = 0$

2. If $V_{AN} = V_p \angle 0°$ is the reference where V_p = rms is the magnitude of any of the phase voltages, then

$V_{BN} = V_p \angle -120°$ and $V_{CN} = V_p \angle -240°$ (positive phase or sequence ABC),

or

$V_{BN} = V_p \angle 120°$ and $V_{CN} = V_p \angle 240°$ (negative phase or sequence CBA).

Phasor diagrams of:

a positive sequence   a negative sequence

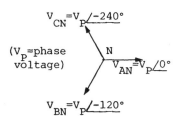
(V_p=phase voltage)

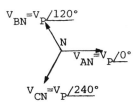

3. Phasor diagram of a line and phasor voltage relationship

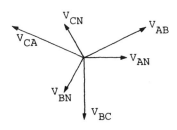

A-62

$$V_{line} = \sqrt{3}\, V_{phase}$$

$$V_{AB} = \sqrt{3}\, V_P \; \underline{/30^0}$$

$$V_{BC} = \sqrt{3}\, V_P \; \underline{/-90^0}$$

$$V_{CA} = \sqrt{3}\, V_P \; \underline{/-210^0}$$

7.2.4 DELTA CONNECTIONS

A balanced Δ-connected load with Y-connected source:

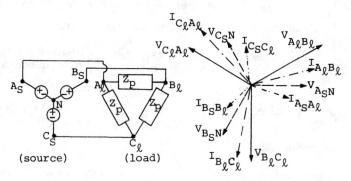

Phasor Diagram

Given that $V_{phase} = |V_{A_sN}| = |V_{B_sN}| = |V_{C_sN}|$,

assume $V_{line} = |V_{A_sB_s}| = |V_{A_sC_s}| = |V_{C_sA_s}|$

where $V_L = \sqrt{3}\, V_A$ and $V_{A_sB_s} = \sqrt{3}\, V_{A_sN} \; \underline{/30^0}$.

Then the phase currents are

$$I_{A_\ell B_\ell} = \frac{V_{A_sB_s}}{Z_p}, \quad I_{B_\ell C_\ell} = \frac{V_{B_sC_s}}{Z_p} \quad \text{and} \quad I_{C_\ell A_\ell} = \frac{V_{C_sA_s}}{Z_p}$$

and the line currents are

$$I_{A_\ell B_\ell} - I_{C_\ell A_\ell} = I_{A_s A_\ell}, \text{ etc.}$$

Note: The three-phase currents are equal in magnitude,

i.e., $I_P = |I_{A_s B_s}| = |I_{B_s C_s}| = |I_{C_s A_s}|$,

$I_L = |I_{A_s A_\ell}| = |I_{B_s B_\ell}| = |I_{C_s C_\ell}|$

and $I_L = \sqrt{3}\, I_P$.

7.3 POWER IN Y- AND Δ- CONNECTED LOADS

7.3.1 Y-CONNECTED LOAD

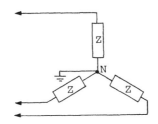

For a Y-connected load, the phase power with pf angle $\theta = P_P = V_{phase} I_{line} \cos\theta$ (Note: $V_P I_P = V_P I_L = \dfrac{V_L I_L}{\sqrt{3}}$

The total power = $P_t = 3P_p$, or $P_t = \sqrt{3}\, V_L I_L \cos\theta$ where $V_L = \sqrt{3}\, V_P$.

Δ-connected load

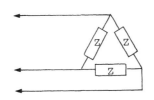

A-64

For a Δ-connected load, the phase power = P_p = $V_L I_P \cos\theta$ (Note: $V_P I_P = V_L I_P = V_L \frac{IL}{\sqrt{3}}$).

The total power = $P_t = 3P_p$ or $P_t = \sqrt{3} V_L I_L \cos\theta$.

Notice that the total power for any balanced three-phase load is equal to $\sqrt{3} V_L I_L \cos\theta$, where S_T (apparent power) = $\sqrt{3} V_L I_L$ and θ_T (reactive power) = $\sqrt{3} V_L I_L \sin\theta$.

CHAPTER 8

FREQUENCY DOMAIN ANALYSIS

8.1 COMPLEX FREQUENCIES

8.1.1 COMPLEX FREQUENCY

In general, the complex frequency s has the form $s = \delta + j\omega$ which describes an exponentially varying sinusoid. The complex frequency s consists of two parts:

1) The real part, δ, the neper frequency in nepers/sec.

2) The imaginary part, $j\omega$, where ω is the radian frequency in radians/sec.

The real part is related to the exponential variation; the imaginary part is related to the sinusoidal variation.

In general, a function $f(t)$ can be expressed in terms of the complex frequency s as

$$f(t) = k\, e^{st}$$

where k is a complex constant.

The characteristics of the function f(t) relate to the complex frequency s as follows:

s	f(t)
+	increases
−	decreases as t increases
0	constant sinusoidal amplitude

Note: Increasing the magnitude of the real part of s will increase the rate of the exponential increase or decrease. Increasing the magnitude of the imaginary part of s will increase the time function changing rate.

8.2 COMPLEX FREQUENCY IMPEDANCES AND ADMITTANCES

Table of complex frequency impedances and admittances for elements R, L and C:

element	impedance Z(s)	Admittance Y(s)
R	R	$\frac{1}{R}$
L	sL	1/sL
C	1/sC	sC

Note: $Z(s) = \frac{V}{I} = \frac{1}{Y(s)}$

8.3 THE S-PLANE (COMPLEX-FREQUENCY PLANE)

8.3.1 THE S-PLANE

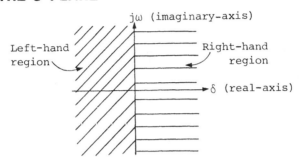

1) A point at the origin. ⟶ corresponds to a DC quantity
2) Points on δ -axis ⟶ a) $\delta > 0$ ~exponential functions decaying
   b) $\delta < 0$ ~exponential functions increasing
3) Points on $j\omega$-axis ⟶ purely sinusoidal functions.
4) Points in ▨ (left-hand-region) ⟶ describe frequencies of exponentially-decreasing sinusoids.
5) Points in ▤ (right-hand-region) ⟶ describe frequencies of exponentially-increasing sinusoids. (i.e., frequencies of positive real parts, time-domain quantities).

8.4 POLES AND ZEROS

Consider a rational function of the form

$$H(s) = \frac{b_m s^m + b_{m-1} s^{m-1} + \ldots + b_1 s + b_0}{a_n s^n + a_{n-1} s^{n-1} + \ldots + a_1 s + a_0}.$$

It can be expressed as

$$H(s) = k\frac{(s-z_1)\cdots(s-z_m)}{(s-p_1)\cdots(s-p_n)}$$

where the zeros of $H(s)$ (i.e., $z_1, z_2, \ldots, z_m$) can be obtained by setting the numerator of $H(s)$ equal to zero.

The poles of $H(s)$ (i.e., $p_1, p_2, \ldots, p_n$) can be obtained by setting the denominator of $H(s)$ equal to zero.

Poles and zeros are indicated in the S-plane (complex-frequency plane) as follows:

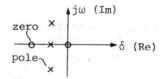

Procedures for graphical determination of magnitude and angular variation of frequency-domain function

Step 1: Find all poles and zeros of the frequency-domain function. Indicate them in the S-plane and, for the function to be determined, assign a test point on the S-plane corresponding to the frequency.

Step 2: Draw a corresponding arrow from each pole and zero to the test point.

Step 3: Calculate the length and angle of each pole arrow and zero arrow.

Step 4: Determine the magnitude of the frequency-domain function for the assumed frequency of the test point by the following ratio:

$$\frac{\text{product of the zero-arrow lengths}}{\text{product of the pole-arrow lengths}}$$

Step 5: Finally, use the formula

[Sum of zero-arrow angles]−[sum of pole-arrow angles]

to obtain the angular variation of the frequency-domain function evaluated at the test point.

8.5 RESONANCE (SERIES AND PARALLEL)

8.5.1 RESONANCE

In a network, when the voltage and the current at the input terminals are in phase, the network is in resonance.

Note: In resonance, power factor (pf) is unity.

8.5.2 PARALLEL RESONANCE

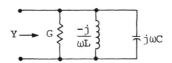

Characteristics

1) The complex admittance Y is

$$Y = G + j(\omega C - 1/\omega L).$$

2) Resonance condition for the circuit:

$$\omega C - \frac{1}{\omega L} = 0$$

$$\omega C = \frac{1}{\omega L}$$

$$\omega = \frac{1}{\sqrt{LC}} = \omega_0$$

where ω_0 = resonant frequency

or $f_0 = \dfrac{1}{2\pi\sqrt{LC}} \dfrac{\text{cycles}}{\text{sec}}$

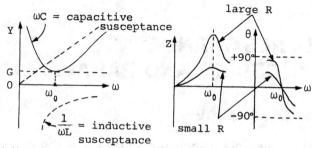

(a) Parallel circuit Y as a function of ω.

(b) Z and θ as a function of ω.

Note: At $\omega < \omega_0$, inductive susceptance > capacitive susceptance and Y is negative.

$\omega > \omega_0$, $\measuredangle$ Z is negative.

$\omega \to 0$, $\measuredangle$ Z is $+90°$

4) Pole-zero representation of Y

$$Y(s) = K \dfrac{(s+\alpha - j\omega_d)(s+\alpha - j\omega_d)}{s}$$

where α = exponential damping ratio = $\dfrac{1}{2RC}$,

ω_d = natural resonant frequency = $\sqrt{\omega_0^2 - \alpha^2}$

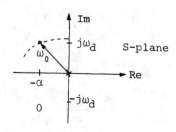

8.5.3 SERIES RESONANCE

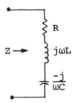

Characteristics

1) The complex impedance Z is

$$Z = R + j(\omega L - \frac{1}{\omega C}).$$

2) Resonance condition for the circuit:

$$\omega L - \frac{1}{\omega C} = 0$$

or

$$\omega = \frac{1}{\sqrt{LC}} = \omega_0$$

Hence, the resonant frequency $(f_0) = \dfrac{1}{2\pi \sqrt{LC}} \dfrac{\text{cycles}}{\text{sec}}$

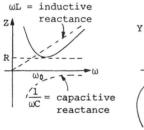
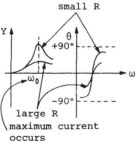

Z, Y, and θ as a function of ω.

When $\omega < \omega_0$, capacitive reactance > inductive reactance and ∢ Z is negative.

When $\omega > \omega_0$, inductive reactance > capacitive reactance and ∢ Z is positive approaching +90°.

When $\omega \to 0$, ∢ Z is -90°.

A-72

8.6 QUALITY FACTOR - Q

8.6.1 QUALITY FACTOR

By definition, the quality power Q is expressed as

$$Q = 2\pi \times \frac{\text{maximum energy stored}}{\text{total energy lost or dissipated per period}}$$

8.6.2 QUALITY FACTQR Q
FOR SOME GENERAL CIRCUITS

CIRCUIT		QUALITY FACTOR, Q
1. RC series		$Q = 2\pi \times \dfrac{\frac{1}{2}I^2_{max}/\omega^2 C}{(I^2_{max}/2) \times R \times T} = \dfrac{1}{\omega CR}$
2. RL series		$Q = 2\pi \times \dfrac{\frac{1}{2}LI^2_{max}}{(I^2_{max}/2) \times R \times T} = \dfrac{2\pi fL}{R} = \dfrac{\omega L}{R}$ $T = \dfrac{1}{f} = \dfrac{2\pi}{\omega}$
3. RLC series		At resonance: $Q_0 = \dfrac{\omega_0 L}{R} = \dfrac{1}{\omega_0 CR}$ or $Q_0 = \dfrac{\omega_0}{\omega_2 - \omega_1} = \dfrac{f_0}{f_2 - f_1} = \dfrac{f_0}{BW \text{(Bandwidth)}}$ Note: $\frac{1}{2}CV^2_{max} = \frac{1}{2}LI^2_{max}$ (at maximum)
4. RLC parallel		At resonance: $Q_0 = \dfrac{R}{\omega_0 L} = \omega_0 CR$

Note: Both circuits in (3) and (4) store a constant amount of energy at resonance.

8.7 SCALING

The method of scaling is used to ease the numerical calculations during networks analysis. There are basically two types of scaling: magnitude scaling and frequency scaling.

Magnitude scaling

A factor of K_m is increased for the impedance (z) of a two-terminal network with the frequency remaining constant, i.e.,
$R \to K_m R$,
$C \to C/K_m$ and
$L \to K_m L$.

Frequency scaling

A factor of K_f is increased for the frequency at any impedance, i.e.,
$R \to R$,
$C \to C/K_f$ and
$L \to L/K_f$.

CHAPTER 9

STATE - VARIABLE ANALYSIS

9.1 STATE - VARIABLE METHOD

Given a circuit with energy-storing elements (i.e., a capacitor and inductor), the circuit can be analyzed by using the state-variable method. A hybrid set of variables is selected (including capacitor voltages and inductor currents) to describe the energy state of the system.

By using a set of state variables, a set of n first-order, simultaneous differential equations can be obtained from the given nth-order differential equations (i.e., state equations).

9.1.1 CONDITIONS FOR THE STATE-VARIABLE METHOD

1) The state equation must be expressed in normal form, i.e., the derivative of each state variable must be expressed in terms of a linear combination of all the state variables and forcing functions.

2) The equations describing the derivatives must be of the same order as the state variables appearing in each equation.

9.2 STATE EQUATIONS FOR n-th ORDER CIRCUITS

9.2.1 STATE EQUATIONS FOR THE FIRST AND SECOND ORDER CIRCUITS

1) First-order circuit:

State equation

$$\dot{x}(t) = \frac{1}{RC}[v(t)-x(t)]$$

1) $x(t)$ represents the state variable $v_c(t)$.

2) $v(t) = Ri(t) + v_c(t)$,

$$i(t) = ic(t) = C\frac{dv_c}{dt}$$

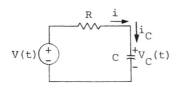

2) Second-order circuit

$$\dot{x}_1(t) = -\frac{R}{L}x_1(t) - \frac{1}{L}x_2(t) + \frac{1}{L}v(t)$$

$$\dot{x}_2(t) = \frac{1}{C}x_1(t)$$

$$x_1(t) = i_L(t)$$

$$x_2(t) = v_C(t)$$

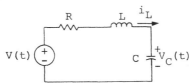

9.2.2 GENERAL STEPS TO OBTAIN THE STATE EQUATIONS FOR A LINEAR TIME-INVARIANT NETWORK

General steps to obtain the state equations for a linear time-invariant network

Method 1

Step 1: Assign state variables for the voltage across each capacitor and for the current through each inductor.

Step 2: Apply KVL and KCL to obtain a set of linear independent equations for each capacitor and inductor.

Step 3: Rearrange the equations obtained in step 2 so that all other variables in the network are in terms of the chosen state variables.

Step 4: Consider all the equations obtained in step 2 and 3. Simplify them so that the equations are expressed only in terms of the state variables and their corresponding derivatives. Therefore, all network variables not chosen as state variables are eliminated.

Step 5: Rearrange the equations obtained in step 4 in the compact form

$$\dot{x}(t) = Ax(t) + bu(t) \text{ (the normal-form equation)},$$

where $\dot{x}(t)$ = derivative of $x(t) = \dfrac{dx(t)}{dt}$,

$x(t)$ = n-column vector of all the state variables $(x_1, x_2, \ldots x_n)$ chosen in step 1 (i.e., $x(t) \triangleq [x_1(t) + x_2(t) + \ldots + x_n(t)]^T$,

A = constant n × n square matrix = system matrix (i.e., $A \triangleq (a_{ij})$),

A-77

b = an n-column vector, and bu(t) = the forcing function vector due to the independent sources (i.e., b $\triangleq$ [$b_1, b_2, \ldots, b_n$]).

Therefore,

$$\begin{bmatrix} \dot{x}_1(t) \\ \dot{x}_2(t) \\ \vdots \\ \dot{x}_n(t) \end{bmatrix} = \begin{bmatrix} A_{11} & A_{12} \ldots A_{1n} \\ \vdots & \vdots \\ A_{n1} & \ldots \ldots A_{nn} \end{bmatrix} \begin{bmatrix} x_1(t) \\ x_2(t) \\ \vdots \\ x_n(t) \end{bmatrix} + \begin{bmatrix} b_{11} \\ b_{21} \\ \vdots \\ b_{n1} \end{bmatrix} u(t)$$

9.3 NORMAL - FORM EQUATIONS

9.3.1 GENERAL PROCEDURES TO OBTAIN A SET OF NORMAL-FORM EQUATIONS

Method 2

Step 1: Obtain a normal tree for the given network using the nodal analysis as outlined in Chapter 5.

Step 2: Assign state variables for the voltage across each capacitor and the current through each inductor correspondingly, i.e.,

$$\overset{C}{\bullet\!\!-\!\!|\!\vdash\!\!-\!\!\bullet} \Rightarrow \underset{\text{free branch}}{\bullet\!\!-\!\!\overset{+\ \ v_C\ \ -}{}\!\!-\!\!\bullet}$$

$$\overset{L}{\bullet\!\!-\!\!\text{mm}\!\!-\!\!\bullet} \Rightarrow \underset{\text{link}}{\bullet\!\!-\!\!\overset{i_L}{\underset{\longrightarrow}{\text{-----}}}\!\!-\!\!\bullet}$$

or, for the resistive tree branches or links, indicated by using a new voltage or current variable (if necessary).

Step 3: a) For each capacitor, apply KCL (as outlined in Chapter 2) to write a set of equations.

b) For each inductor, repeat part (a) but use KVL instead.

c) If any new voltage and current variables were assigned to the resistors, write the equation for R by using KCL and KVL. Then express v_R and i_R in terms of the state variables and source quantities. Otherwise, skip this step.

Step 4: Put all the equations obtained in step 3 in order and rearrange to obtain the normal-form equations.

9.4 STATE TRANSITION MATRIX - e^{At}

9.4.1 STATE TRANSITION - e^{At}

Let us represent the state equations in the normal form

$$\dot{x}(t) = Ax(t) + bu(t) \quad (At\ t=t_0,\ x(t_0) = x_0.) \quad (1)$$

The solution of the matrix state equation (in (1)) is given by

$$\boxed{x(t) = e^{A(t-t_0)} x_0 + \int_{t_0}^{t} e^{A(t-\tau)} bu(\tau) d\tau\ ,}$$

where $e^{At} \triangleq$ the state transition matrix, which describes the change of state of the system from zero to the state at time t where $e^{A(t-t_0)}$ is e^{At} evaluated at $t = t - t_0$.
(Note: $x(t)$ and $e^{A(t-t_0)} bu(\tau)$ are n-column vectors.)

To determine the state transition matrix e^{At}, the Cayley-Hamilton theorem is applied.

9.4.2 CAYLEY-HAMILTON THEOREM

1) By definition,

$$e^{At} \triangleq \mu_0(t)I + \mu_1(t)A + \mu_2(t)A^2 + \ldots + \mu_{n-1}(t)A^{n-1} \quad (2)$$

where A is an n × n square matrix and $\mu_0(t)\ldots\mu_n(t)$ are scalar functions of time.

2) For equation (2) to hold, the following conditions must be satisfied:

a) I = unity matrix.

b) The characteristic equation of the matrix A equals

$Det[A - sI] = 0$,

where S_i, $i = 1,2,\ldots,n$ of A are the roots of the characteristic nth-order polynomial equation and are called the eigenvalues of A.

9.4.3 GENERAL PROCEDURES TO OBTAIN THE STATE TRANSITION MATRIX e^{At}, GIVEN MATRIX A

Step 1: Obtain a matrix in the form $A - sI$.

Step 2: Equate: $Det[A - sI] = 0$ and solve for the roots (i.e., S_i, $i = 1,2,\ldots,n$) of the characteristic equation.

Step 3: Express each root in n equations of the form $e^{tS_i} = \mu_0 + \mu_1 S_i + \ldots + \mu_{n-1} S_i^{n-1}$, and solve the scalar time functions: $\mu_0, \ldots, \mu_{n-1}$.

Step 4: Obtain the state transition matrix by substituting the time functions obtained in step 3 into Equation (2).

CHAPTER 10

FOURIER ANALYSIS

10.1 TRIGONOMETRIC FOURIER SERIES

10.1.1 DIRICHLET CONDITIONS FOR THE EXISTENCE OF A FOURIER SERIES

If $f(t)$ is a bounded periodic function of period T (i.e., $f(t+T) = f(t)$) and if $f(t)$ satisfies these Dirichlet conditions:

1) $f(t)$, if discontinuous, has a finite number of discontinuities in any period T;
2) $f(t)$ has a finite average value over period T;
3) The number of maxima and minima of $f(t)$ in any period T is finite

then $f(t)$ may be represented by a trigonometric Fourier series as described below:

10.1.2 TRIGONOMETRIC FORM OF A FOURIER SERIES

1) The function $f(t)$ is expressed over any interval $(t_0, t_0 + 2\pi/\omega_0)$ as

$$f(t) = a_0 + a_1 \cos \omega_0 t + a_2 \cos 2\omega_0 t + \ldots + b_1 \sin \omega_0 t + b_2 \sin 2\omega_0 t + \ldots$$

$$= a_0 + \sum_{n=1}^{\infty} (a_n \cos n\omega_0 t + b_n \sin n\omega_0 t) \quad (t_0 < t < t_0 + 2\pi/\omega_0)$$

where ω_0 = fundamental frequency = $2\pi/T$

and
$$a_0 = \frac{1}{T} \int_{t_0}^{(t_0+T)} f(t)\,dt \quad \text{(where } t_0 \text{ is assumed to be zero generally)}$$

$$a_n = \frac{2}{T} \int_{t_0}^{(t_0+T)} f(t) \cos n\omega_0 t\,dt$$

$$b_n = \frac{2}{T} \int_{t_0}^{(t_0+T)} f(t) \sin n\omega_0 t\,dt$$

10.1.3 SPECIAL INTEGRATION PROPERTIES FOR SINE AND COSINE

1) $$\int_0^T \sin^2 n\omega_0 t\,dt = \frac{T}{2}$$

$$\int_0^T \cos^2 n\omega_0 t\,dt = \frac{T}{2}$$

2) $$\int_0^T \sin n\omega_0 t\,dt = \int_0^T \cos n\omega_0 t\,dt = 0$$

3) $$\int_0^T \sin k\omega_0 t \cos n\omega_0 t\,dt = \int_0^T \sin k\omega_0 t \sin n\omega_0 t\,dt$$

$$= \int_0^T \cos k\omega_0 t \cos n\omega_0 t = 0$$

for $k \neq n$.

10.2 EXPONENTIAL FOURIER SERIES

A given function $g(t)$ can be expressed as a linear combination of exponential functions over the period t_0, $t_0 + 2\pi/\omega_0$, as follows:

$$g(t) = \ldots + G_{-n} e^{-jn\omega_0 t} + \ldots + G_{-2} e^{-j2\omega_0 t} + G_{-1} e^{-j\omega_0 t}$$

$$+ G_0 + G_1 e^{j\omega_0 t} + G_2 e^{j2\omega_0 t} + \ldots + G_n e^{jn\omega_0 t} + \ldots$$

$$= \sum_{n=-\infty}^{\infty} G_n e^{jn\omega_0 t} \quad (t_0 < t < t_0 + 2\pi/\omega_0),$$

it is assumed that $t_0 = 0$.

(Note: $T = 2\pi/\omega_0$.)

Therefore,

$$G_n = \frac{1}{T} \int_0^T g(t) e^{-jn\omega_0 t}\, dt$$

and

$$G_0 = \frac{1}{T} \int_0^T g(t)\, dt$$

Relationships between trigonometric and exponential Fourier series

Trigonometric series		Exponential series
a_0	=	G_0
a_n	=	$G_n + G_{-n}$
b_n	=	$j(G_n - G_{-n})$
$\frac{1}{2}(a_n - jb_n)$	=	G_n

10.3 COMPLEX FORM OF A FOURIER SERIES

The complex form of a Fourier series is given as

$$g(t) = \sum_{n=-\infty}^{\infty} C_n e^{jn\omega_0 t}$$

where C_0 is a complex constant $= \dfrac{1}{T} \displaystyle\int_{-\frac{T}{2}}^{\frac{T}{2}} g(t) e^{-jn\omega_0 t} \, dt,$

for $n = 0, \pm 1, \pm 2, \pm 3, \ldots$.

Note: $|C_n| = |C_{-n}|$, since $C_{-n} = C_n^*$.

10.3.1 SPECIAL CASE

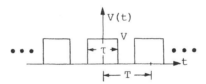

Given a train of rectangular pulses with period T, i.e.,

$$v(t) = \begin{cases} V & -\tau/2 < t < \tau/2 \\ 0 & \tau/2 < t < T - \tau/2, \end{cases}$$

then

$$C_n = \frac{1}{T} \int_{-\frac{\tau}{2}}^{\frac{\tau}{2}} A\, e^{-jn\omega_0 t} \, dt$$

$$= \frac{2A}{n\omega_0 T} \frac{(e^{jn\omega_0 \tau/2} - e^{-jn\omega_0 \tau/2})}{2j}$$

$$= \frac{A\tau}{T} \left[\frac{\sin(\tfrac{1}{2}n\omega_0 \tau)}{(\tfrac{1}{2}n\omega_0 \tau)} \right] = \frac{A\tau}{T} \, \text{Sa}(\tfrac{1}{2}n\omega_0 \tau),$$

where $\frac{\sin x}{x}$ = sampling function = Sa(x).

Thus, $v(t) = \frac{A\tau}{T} \sum_{n=-\infty}^{\infty} Sa(\tfrac{1}{2}n\omega_0\tau)e^{jn\omega_0 t}$.

10.4 WAVEFORM SYMMETRY PROPERTIES

Waveform symmetry	Properties
1) Even symmetry (i.e., cosine function)	a) $g(t) = g(-t)$ b) $b_n = 0$ c) $a_n = \frac{4}{T} \int_0^{T/2} g(t)\cos n\omega_0 t\, dt$
2) Odd symmetry (i.e., sine function)	a) $g(t) = -g(-t)$ b) $a_0 = a_n = 0$ c) $b_n = \frac{4}{T} \int_0^{T/2} g(t)\sin n\omega_0 t\, dt$
3) Half wave symmetry	a) $g(t) = -g(t+T/2)$ where T = period b) For n = odd, $a_n = \frac{4}{T} \int_0^{T/2} g(t)\cos n\omega_0 t\, dt$

Waveform symmetry	Properties
3) Half wave symmetry	$b_n = \dfrac{4}{T} \displaystyle\int_0^{T/2} g(t)\sin n\omega_0 t\, dt$ $n = \text{even}, \quad a_n = b_n = 0$
4) Half wave and even symmetry	a) For $n = \text{odd}$, $a_n =$ $\dfrac{8}{T} \displaystyle\int_0^{T/4} g(t)\cos n\omega_0 t\, dt$ $b_n = 0,\ n = \text{even},\ a_n = b_n = 0$
5) Half wave and odd symmetry	a) For $n = \text{odd}$, $a_n = 0$ and $b_n = \dfrac{8}{T} \displaystyle\int_0^{T/4} g(t)\sin n\omega_0 t\, dt$ $n = \text{even},\ b_n = a_n = 0$

10.5 THE FOURIER TRANSFORM

By definition, the Fourier transform of $g(t)$ is

$$\mathcal{F}\{g(t)\} = G(\omega) = \int_{-\infty}^{\infty} g(t)e^{-j\omega t}\, dt$$

The inverse Fourier transform of $F(\omega)$ is

$$\mathcal{F}^{-1}\{G(\omega)\} = g(t) = \frac{1}{2\pi} \int_{-\infty}^{\infty} G(\omega)e^{j\omega t}\, d\omega,$$

as long as $\int_{-\infty}^{\infty} |f(t)| \, dt$ converges (i.e., $\int_{-\infty}^{\infty} |f(t)| \, dt < \infty$).

Thus, $g(t)$ and $G(\omega)$ are called the Fourier transform pair.

Note: The Fourier transform of $g(t)$ can also be expressed in terms of sine and cosine by using the Euler's identity:

$$e^{-j\omega t} = \cos \omega t - j\sin \omega t.$$

Hence,
$$G(\omega) = \underbrace{\int_{-\infty}^{\infty} g(t)\cos\omega t \, dt}_{R(\omega)} - j \underbrace{\int_{-\infty}^{\infty} g(t)\sin\omega t \, dt}_{I(\omega)}$$

$$= |G(\omega)| \underline{/\theta(\omega)}$$

where
$$|G(\omega)| = [R^2(\omega) + I^2(\omega)]^{\frac{1}{2}}$$

and
$$\theta(\omega) = \tan^{-1}\left[\frac{I(\omega)}{R(\omega)}\right]$$

10.5.1 SOME USEFUL FOURIER TRANSFORM PAIRS

$$g(t) \iff G(\omega) = F\{g(t)\}$$

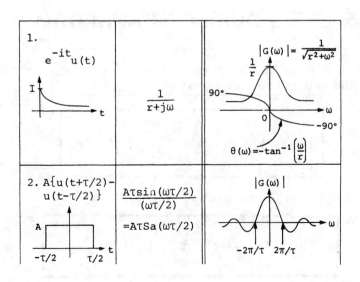

3. $\delta(t)$	1	$G(\omega)$, 1
4. 1	$2\pi\delta(\omega)$	$2\pi\delta(\omega)$
5. $u(t)$	$\pi\delta(\omega) + \dfrac{1}{j\omega}$	$\|G(\omega)\|$
6. $\text{sgn}(t)$	$\dfrac{2}{j\omega}$	$\|G(\omega)\|$
7. $\cos\omega_0 t$	$\pi\{\delta(\omega-\omega_0) + \delta(\omega+\omega_0)\}$	$G(\omega)$
8. $\sin\omega_0 t$	$j\pi\{\delta(\omega+\omega_0) - \delta(\omega-\omega_0)\}$	$jG(\omega)$

10.6 PARSEVAL'S IDENTITY

If $G(\omega) = F\{g(t)\}$, then

$$\int_{-\infty}^{\infty} |g(t)|^2 dt = \frac{1}{2\pi} \int_{-\infty}^{\infty} |G(\omega)|^2 d\omega.$$

In general, if $G(\omega) = F\{g(t)\}$ and $H(\omega) = F\{h(t)\}$, then

$$\int_{-\infty}^{\infty} g(t)h^*(t)dt = \frac{1}{2\pi} \int_{-\infty}^{\infty} G(\omega)H^*(\omega)d\omega,$$

where * denotes the complex conjugate.

Note: $G(-\omega) = G^*(\omega)$

10.7 CONVOLUTION THEOREM FOR THE FOURIER TRANSFORMS

Let $F(\omega) = F\{f(t)\}$ and $G(\omega) = F\{g(t)\}$. Then the convolution of f and g (i.e., f * g) is defined as

$$f * g = \int_{-\infty}^{\infty} f(\tau)g(t-\tau)d\tau = \frac{1}{2\pi} \int_{-\infty}^{\infty} F(\omega)G(\omega)e^{j\omega t} d\omega$$

Hence,

$$\boxed{F\{f * g\} = F(\omega)G(\omega) = F\{f\}\, F\{g\}.}$$

CHAPTER 11

LAPLACE TRANSFORMATION

11.1 DEFINITION OF LAPLACE TRANSFORM

By definition,

$$L\{g(t)\} = G(s) = \int_0^\infty e^{-st} g(t) dt$$

where $g(t)$ is a function of the real variable t, and s is a complex variable defined as $s = \delta + j\omega$. The function $g(t)$ is called the original function and the function $G(s)$ is called the image function.

The transformation of a time domain function into a complex frequency domain function is the operation $L\{g(t)\}$.

In order for the Laplace transform to be valid, the following conditions must be satisfied:

1) If the integral in eq.(1) converges for a real $s = s_0$, i.e.,

$$\lim_{\substack{A \to 0 \\ B \to \infty}} \int_A^B e^{-s_0 t} g(t) dt \quad \text{exists,}$$

then it converges for all s with Re(s) > s_0, and the image function is a single valued analytic function of s in the half-plane Re(s) > s_0.

2) g(t) is a piecewise continuous function.

1) It should be noted that in specifying the Laplace transform of a signal, both the algebraic expression and the range of values of s for which this expression is valid is required.

2) The range of values s for which the integral defining the Laplace transform converges is referred to as the region of convergence.

11.2 DEFINITION OF THE INVERSE LAPLACE TRANSFORM

If $L\{g(t)\} = G(s)$, then $g(t) = L^{-1}\{G(s)\}$ is the inverse Laplace transform of G(s). L^{-1} is called the inverse Laplace transform operator.

11.3 COMPLEX INVERSION FORMULA

The inverse Laplace transform of G(s) can be found directly by methods of complex variable theory. The result is

$$g(t) = \frac{1}{2\pi j} \int_{\delta-j\infty}^{\delta+j\infty} e^{st} G(s) ds = \frac{1}{2\pi j} \lim_{T \to \infty} \int_{\delta-jT}^{\delta+jT} e^{st} G(s) ds$$

where δ is chosen such that all the singular points of $G(s)$ lie to the left of the line $Re(s) = \delta$ in the complex s-plane.

11.4 GENERAL LAPLACE TRANSFORM PAIRS

SIGNAL	TRANSFORM	REGION OF CONVERGENCE
$\delta(t)$	1	All s
$u(t)$	$1/s$	$Re\{s\} > 0$
$-u(-t)$	$1/s$	$Re\{s\} < 0$
$\dfrac{t^{n-1}}{(n-1)!} u(t)$	$1/s^n$	$Re\{s\} > 0$
$-\dfrac{t^{n-1}}{(n-1)!} u(-t)$	$1/s^n$	$Re\{s\} < 0$
$e^{-\alpha t} u(t)$	$\dfrac{1}{s+\alpha}$	$Re\{s\} > \alpha$
$-e^{-\alpha t} u(-t)$	$\dfrac{1}{s+\alpha}$	$Re\{s\} < -\alpha$
$\dfrac{t^{n-1}}{(n-1)!} e^{-\alpha t} u(t)$	$\dfrac{1}{(s+\alpha)^n}$	$Re\{s\} > -\alpha$
$-\dfrac{t^{n-1}}{(n-1)!} e^{-\alpha t} u(-t)$	$\dfrac{1}{(s+\alpha)^n}$	$Re\{s\} < -\alpha$
$\delta(t-T)$	e^{-sT}	All s
$[\cos \omega_0 t] u(t)$	$\dfrac{s}{s^2+\omega_0^2}$	$Re\{s\} > 0$
$[\sin \omega_0 t] u(t)$	$\dfrac{\omega_0}{s^2+\omega_0^2}$	$Re\{s\} > 0$
$[e^{-\alpha t} \cos \omega_0 t] u(t)$	$\dfrac{s+\alpha}{(s+\alpha)^2+\omega_0^2}$	$Re\{s\} > -\alpha$
$[e^{-\alpha t} \sin \omega_0 t] u(t)$	$\dfrac{\omega_0}{(s+\alpha)^2+\omega_0^2}$	$Re\{s\} > -\alpha$

11.5 OPERATIONS FOR THE LAPLACE TRANSFORM

1) Linearity of the Laplace Transform:

If $x_1(t) \overset{L}{\longleftrightarrow} X_1(s)$ with region of convergence R_1

and $x_2(t) \longleftrightarrow X_2(s)$ with region of convergence R_2

then

$$ax_1(t) + bx_2(t) \overset{L}{\longleftrightarrow} aX_1(s) + bX_2(s)$$

with region of convergence containing $R_1 \cap R_2$

2) Time Shifting:

If $x(t) \overset{L}{\longleftrightarrow} X(s)$ with region of convergence (ROC) = R, then,

$$x(t-t_0) \overset{L}{\longleftrightarrow} e^{-st_0} X(s) \text{ with ROC} = R$$

3) Shifting in the s–Domain:

If $x(t) \overset{L}{\longleftrightarrow} X(s)$ ROC = R,

then,

$$e^{s_0 t} x(t) \overset{L}{\longleftrightarrow} X(s-s_0) \text{ with ROC } R_1 = R + \text{Re}\{s_0\}$$

Note: The ROC associated with $X(s-s_0)$ is that of $X(s)$, shifted by $\text{Re}\{s_0\}$. Thus, for any value s that is in R, the value $s + \text{Re}\{s_0\}$ will be in R_1.

For example

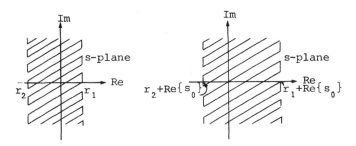

4) Time Scaling:

If $x(t) \overset{L}{\longleftrightarrow} X(s)$ ROC = R, then

$$x(at) \overset{L}{\longleftrightarrow} \frac{1}{|a|} X\left(\frac{s}{a}\right) \text{ with ROC } R_1 = \frac{R}{a}$$

5) Convolution Property:

If $x(t) \overset{L}{\longleftrightarrow} X_1(s)$ ROC = R_1

(and)

$x_2(t) \overset{L}{\longleftrightarrow} X_2(s)$ ROC = R_2,

then

$$x_1(t) * x_2(t) \overset{L}{\longleftrightarrow} X_1(s)X_2(s) \text{ with ROC containing } R_1 \cap R_2$$

6) Differentiation in the Time Domain

If $x(t) \overset{L}{\longleftrightarrow} X(s)$ ROC = R, then

$$\frac{dx(t)}{dt} \overset{L}{\longleftrightarrow} sX(s) \text{ with ROC containing R}$$

7) Differentiation in the s-domain:

Given:
$$X(s) = \int_{-\infty}^{+\infty} x(t)e^{-st}\,dt$$

Differentiating both sides:
$$\frac{dX(s)}{ds} = \int_{-\infty}^{+\infty} (-t)x(t)e^{-st}\,dt$$

Hence,

$$\boxed{-t\,x(t) \overset{L}{\longleftrightarrow} \frac{dX(s)}{ds} \qquad \text{ROC} = R.}$$

8) Integration in the Time Domain:

If $x(t) \overset{L}{\longleftrightarrow} X(s)$ ROC = R, then

$$\boxed{\int_{-\infty}^{t} x(\tau)\,d\tau \overset{L}{\longleftrightarrow} \frac{X(s)}{s} \qquad \text{ROC contains } R \cap \{\text{Re}\{s\} > 0\}}$$

11.6 HEAVISIDE EXPANSION THEOREM

By the theorem,

$$L^{-1}\left\{\frac{p(s)}{q(s)}\right\}, \text{ where } q(s)=(s-a_1)(s-a_2)\ldots(s-a_m) \text{ and } p(s) = \text{a polynomial of degree} < m.$$

$$= \sum_{n=1}^{m} \frac{p(a_n)}{q'(a_n)} e^{a_n t} \qquad \text{(i.e., Heaviside Expansion Formula)}$$

11.7 FINAL AND INITIAL VALUE THEOREM

Initial value theorem:

$$g(0^+) = \lim_{s \to \infty} \{sG(s)\}$$

Final value theorem:

$$g(\infty) = \lim_{s \to 0} sG(s).$$

Note: All poles of sG(s) lie in the left-hand side of the complex s-plane.

CHAPTER 12

TWO PORT NETWORK PARAMETERS

12.1 Z - PARAMETERS

12.1.1 IMPEDANCE OR Z-PARAMETERS

Impedance parameters are defined by the following two sets of equations.

$$v_1 = z_{11}i_1 + z_{12}i_2$$
$$v_2 = z_{21}i_1 + z_{22}i_2$$

where v_1 and v_2 are acting as independent variables.

(Note: A general linear two-port network is being considered.)

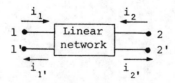

By setting i_1 and i_2 equal to zero, four impedance parameters are defined as follows:

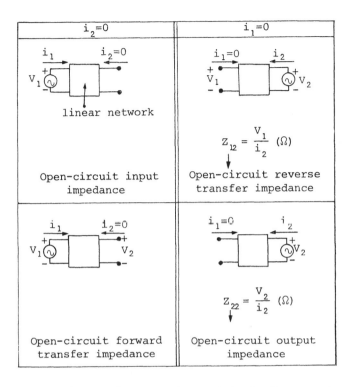

(Note: Since i_1 and i_2 are set equal to 0, Z-parameters are also called open-circuit impedance parameters.)

12.2 HYBRID - PARAMETERS

12.2.1 HYBRID OR H-PARAMETERS

The usage of hybrid parameters is for the analysis of transistor circuits. The hybrid parameters are defined by two sets of equations as follows:

$$v_1 = h_{11}I_1 + h_{12}v_2$$

$$I_2 = h_{21}I_1 + h_{22}v_2$$

where v_1 and I_2 are acting as independent variables.

The hybrid parameters are determined by the use of short-circuit and open-circuit conditions.

Hence,

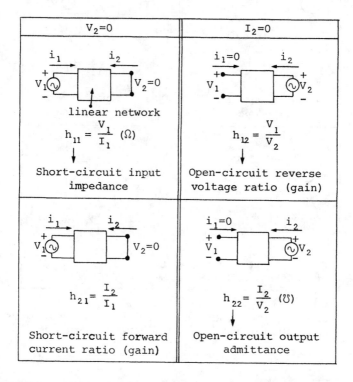

The hybrid parameters in a two-port network are defined as shown in the network below.

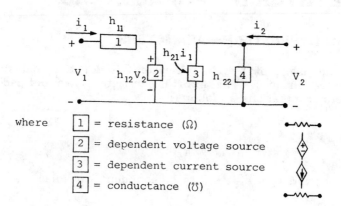

12.3 ADMITTANCE PARAMETERS

12.3.1 ADMITTANCE OR Y-PARAMETERS

Admittance parameters are described by the following two sets of equations:

$$i_1 = y_{11}v_1 + y_{12}v_2$$

$$i_2 = y_{21}v_1 + y_{22}v_2$$

Then each parameter is defined by setting v_1 or v_2 equal to zero. Hence, Y-parameters are also called the short-circuit admittance parameters.

Thus, by setting:

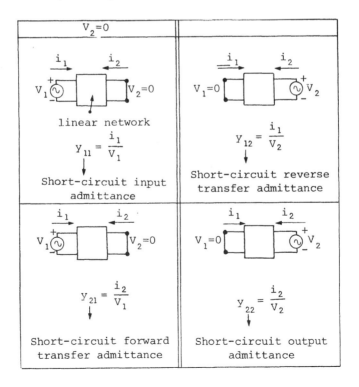

12.4 Z, Y, AND H PARAMETERS AND RELATIONSHIPS

Let us define the following matrices:

$$[Z] = \begin{bmatrix} z_{11} & z_{12} \\ z_{21} & z_{22} \end{bmatrix}, \quad [Y] = \begin{bmatrix} y_{11} & y_{12} \\ y_{21} & y_{22} \end{bmatrix}, \quad [H] = \begin{bmatrix} h_{11} & h_{12} \\ h_{21} & h_{22} \end{bmatrix}$$

then, $\Delta z = z_{11} z_{22} - z_{12} z_{21}$, $\Delta y = y_{11} y_{22} - y_{12} y_{21}$, $\Delta h = h_{11} h_{22} - h_{12} h_{21}$

are the determinants of [Z], [Y] and [H] respectively.

Now, the conversion between parameters are defined as follows:

(1A) $Z \rightarrow Y \implies \begin{bmatrix} \dfrac{z_{22}}{\Delta z} & \dfrac{-z_{12}}{\Delta z} \\ \dfrac{-z_{21}}{\Delta z} & \dfrac{z_{11}}{\Delta z} \end{bmatrix}$ (1B) $Z \rightarrow H \implies \begin{bmatrix} \dfrac{\Delta z}{z_{22}} & \dfrac{z_{12}}{z_{22}} \\ \dfrac{-z_{21}}{z_{22}} & \dfrac{1}{z_{22}} \end{bmatrix}$

(2A) $Y \rightarrow Z \implies \begin{bmatrix} \dfrac{y_{22}}{\Delta y} & \dfrac{-y_{12}}{\Delta y} \\ \dfrac{-y_{21}}{\Delta y} & \dfrac{y_{11}}{\Delta y} \end{bmatrix}$ (2B) $Y \rightarrow H \implies \begin{bmatrix} \dfrac{1}{y_{11}} & \dfrac{-y_{12}}{y_{11}} \\ \dfrac{y_{21}}{y_{11}} & \dfrac{\Delta y}{y_{11}} \end{bmatrix}$

(3A) $H \rightarrow Z \implies \begin{bmatrix} \dfrac{\Delta h}{h_{22}} & \dfrac{h_{12}}{h_{22}} \\ \dfrac{-h_{21}}{h_{22}} & \dfrac{1}{h_{22}} \end{bmatrix}$ (3B) $H \rightarrow Y \implies \begin{bmatrix} \dfrac{1}{h_{11}} & \dfrac{-h_{12}}{h_{11}} \\ \dfrac{h_{21}}{h_{11}} & \dfrac{\Delta h}{h_{11}} \end{bmatrix}$

CHAPTER 13

DISCRETE SYSTEMS AND Z - TRANSFORMS

13.1 DISCRETE - TIME SYSTEMS

13.1.1 A DISCRETE-TIME SYSTEM

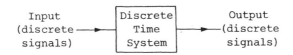

Since discrete signals may be represented by a sequence of numbers, knowing the characteristics of such a sequence is essential.

The characteristics of some general sequences are listed below:

1. Kronecker Delta Sequence:

$$\delta(n) = \begin{cases} 1, & n=0 \\ 0, & n=\pm 1, \pm 2, \ldots \end{cases}$$

$$\delta(n-i) = \begin{cases} 1, & n=i \\ 0, & \text{elsewhere} \end{cases}$$

(Note: i is an arbitrary integer.)

2. Unit Step Sequence:

$$u(n) = \begin{cases} 0, & n=-1,-2,-3\ldots \\ 1, & n=0,1,2,\ldots \end{cases}$$

3. Unit Alternating Sequence:

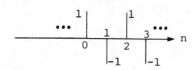

$$u(n) = \begin{cases} 0, & n=-1,-2,-3\ldots \\ (-1)^n, & n = 0,1,2,\ldots \end{cases}$$

4. Unit Ramp Sequence:

$$u(n) = \begin{cases} 0, & n=-1,-2,\ldots \\ n, & n=0,1,2,\ldots \end{cases}$$

13.2 FIRST – ORDER LINEAR DISCRETE SYSTEM

A first-order linear discrete system is represented by the linear first-order difference equation as follows:

$$y(n) + A_1 y(n-1) = B_0 u(n) + B_1 u(n-1) \qquad (1)$$

where u and y are denoted as the input and output of the system, respectively.

If the input signal is applied at n = 0, then eq.(1) becomes

$$y(0) = B_0 u(0) + B_1 u(-1) - A_1 y(-1)$$

where $y(-1)$ is the initial condition of the system.

(Note: $u(-1) = 0$)

In general, where an input signal is applied to n = j, then the response of the system is represented as follows:

$$y(j) = B_0 u(j) + B_1 u(j-1) - A_1 y(j-1)$$

Since $B_1 u(j-1) = 0$

hence, $y(j) = B_0 u(j) - A_1 y(j-1)$

(Note: The initial condition is defined by $y(j-1)$.)

Summary:

In general, a linear discrete system is described by the relationship as follows:

$$y(n) = B_0 u(n) + B_1 u(n-1) + \ldots + B_i u(n-i) - A_1 y(n-1)$$
$$- A_2 y(n-1) - \ldots - A_N y(n-N)$$

where $\begin{array}{c} B_0 \ldots B_i \\ A_1 \ldots A_N \end{array}$ = constants

and i and N = fixed non-negative integers.

If an input signal is applied at $n = j$, then

$$y(j) = B_0 u(j) - A_1 y(j-1) - A_2 y(j-2) - \ldots - A_N y(j-N)$$

(Note: $u(j-i) = 0$ where $i = 1, 2, 3, \ldots$)

13.3 CLOSED-FORM IDENTITIES

Closed-form identity is useful in expressing the response of a linear system.

Some generally used closed-form identities are given below:

1) $\sum_{m=0}^{N} r^m = \dfrac{1 - r^{N+1}}{1 - r}$ where $r \neq 1$

2) $\sum_{m=0}^{N} m r^m = \dfrac{r}{(1-r)^2} [1 - r^m - m r^m + m r^{m+1}]$ where $r \neq 1$

A-104

3) $\sum_{m=0}^{N} m^2 r^m = \frac{r}{(1-r)^3}[(1+r)(1-r^N)-2(1-r)Nr^N-(1-r)^2N^2r^N]$

where $r \neq 1$

13.4 THE Z - TRANSFORM

When a continuous function of time g(t) is sampled at regular intervals of period T, the usual Laplace transform techniques are modified.

The diagrammatic form of a simple sampler together with its associated input-output waveforms is shown below.

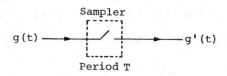

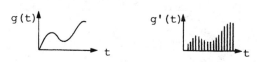

Note: The sampling frequency $\equiv f_s = \frac{1}{T}$

Defining the set of impulse function $\delta_T(t)$ by

$$\delta_T(t) \equiv \sum_{n=0}^{\infty} \delta(t-nT)$$

the input-output relationship of the sampler becomes

$$g'(t) = g(t) \cdot \delta_T(t)$$

$$= \sum_{n=0}^{\infty} g(nT) \cdot \delta(t-nT).$$

Note: For a given g(t) and T, the function g'(t) is unique. However the converse is not true.

The variable 'z' is introduced by means of the transformation:

$$z = e^{Ts}$$

and since any function of s can now be replaced by a corresponding function of z, we have

$$\boxed{G(z) = \sum_{n=0}^{\infty} g(nT) \cdot z^{-n}}$$

where $G'(s) \equiv G(z)$

and $s = \dfrac{1}{T} \ln z$

The z operator can now be defined in terms of the Laplace operator by the relationship

$$Z\{g(t)\} \equiv L\{g'(t)\}$$

or

$$Z\{g(t)\} = \Sigma \text{ residues of } \left[\left(\frac{1}{1-e^{Tx}z^{-1}}\right) G(z)\right]$$

The inverse z transform is

$$Z^{-1}\{G(z)\} \equiv g'(t)$$
$$= \frac{1}{2\pi j} \oint G(z) \cdot z^{n-1} dz$$

where the contour of integration encloses all the singularities of the integrand.

A-106

13.5 PROPERTIES OF Z - TRANSFORM

$g(t)$	$Z\{g(t)\} = G(z)$
1. Linearity: $Af(t)+Bg(t)$	$AF(z)+BG(z)$
2. Left shifting: $g(t+mT)$	$Z^m G(z) - \sum_{r=0}^{m-1} Z^{m-1} g(rT)$ $= Z^m G(z)$ when $g(rT) = 0$, $0 \le r \le m-1$
3. Right shifting: $g(t-mT)$	$Z^{-m} G(z)$
4. Summation: $\sum_{m=0}^{T/t} g(mT)$	$\left(\dfrac{z}{z-1}\right) G(z)$
5. Differentiating: $tg(t)$	$-Tz \dfrac{d}{dz} G(z)$
6. Integrating: $t^{-1}g(t)$	$-\dfrac{1}{T} \displaystyle\int_0^Z \dfrac{G(z)}{z} dz$
7. Convolution: $\sum_{r=0}^{t} g_1(t-r)g_2(r)$	$G_1(z)G_2(z)$

8. Initial value Theorem

$$g(0) = \lim_{|Z| \to \infty} G(z)$$

9. Final value Theorem:

$$g(\infty) = \lim_{z \to 1} (z-1)G(z)$$

if $(z-1)G(z)$ is analytic for $|z| \ge 1$.

13.6 METHODS OF EVALUATING INVERSE Z - TRANSFORMS

1) Cauchy's residue theorem;

   For $t = nT$,
   $$g(nT) = \sum_{\text{all } z_k} [\text{residues of } G(z)z^{n-1} \text{ at } z_k]$$

   where z_k defines all of the poles of $G(z)z^{n-1}$

2) Partial fractions:

   Expand $\frac{G(z)}{z}$ into partial fractions. The product of z with each of the partial fractions will then be recognizable from the standard forms in the table of z transforms. Note however that the continuous functions obtained are only valid at the sampling instants.

3) Power series expansion by long division using detached coefficients:

   $G(z)$ is expanded into a power series in z^{-1} and the coefficient of the term in z^{-n} is the value of $g(nT)$. i.e., the value of $g(t)$ at the nth sampling instant.

13.6.1 THE Z TRANSFORM AS A MEANS OF DETERMINING APPROXIMATELY THE INVERSE LAPLACE TRANSFORM

Since $Z = e^{Ts}$

$$S^{-1} = \frac{T}{2} \left[\frac{1}{v} - \frac{v}{3} - \frac{4v^3}{45} - \frac{44v^5}{945} - \cdots \right]$$

where
$$v \equiv \frac{1 - z^{-1}}{1 + z^{-1}},$$

the series being very rapid in its convergence. Given $G(s)$ to find its inverse Laplace transform, the following operations are carried out:

1) Divide the numerator and denominator of G(s) by the highest power of s, yielding as an alternate form for G(s) (the quotient of two polynomials in s^{-1}).

2) Choose as a numerical value of T, which makes $2\pi/T$ much larger than the imaginary part of the poles of G(s).

3) Substitute the alternative form for G(s) obtained in (1) above; the expansion for s^{-n} can be determined from the following short table of approximations.

   Do not, at this stage, insert the numerical value for T because tabulations with different intervals may be required.

4) Divide by T.

5) Insert the chosen value for T and divide the numerator by the denominator.

6) The coefficient of z^{-n} is the required value of the function at $t = nT$.

s^{-n}	Z-transform (approximation)
s^{-1}	$\dfrac{T}{2}\left[\dfrac{1 + Z^{-1}}{1 - Z^{-1}}\right]$
s^{-2}	$\dfrac{T^2}{12}\left[\dfrac{1 + 10Z^{-1} + Z^{-2}}{(1 - Z^{-1})^2}\right]$
s^{-3}	$\dfrac{T^3}{3}\left[\dfrac{Z^{-1} + Z^{-2}}{(1 - Z^{-1})^3}\right]$
s^{-4}	$\dfrac{T^4}{144}\left[\dfrac{1 + 20Z^{-1} + 102Z^{-2} + 20Z^{-3} + Z^{-4}}{(1 - Z^{-1})^4}\right]$
s^{-5}	$\dfrac{T^5}{124}\left[\dfrac{Z^{-1} + 11Z^{-2} + 11Z^{-3} + Z^{-4}}{(1 - Z^{-1})^4}\right]$
s^{-6}	$\dfrac{T^6}{4}\left[\dfrac{Z^{-2} + 2Z^{-3} + Z^{-4}}{(1 - Z^{-1})^6}\right]$
s^{-7}	$\dfrac{T^7}{8}\left[\dfrac{Z^{-2} + 3Z^{-3} + 3Z^{-4} + Z^{-5}}{(1 - Z^{-1})^7}\right]$

13.7 Z - TRANSFORM PAIRS

Transform Pair Signal	Transform	ROC
$\delta[n]$	1	All z
$u[n]$	$\dfrac{1}{1 - Z^{-1}}$	$|Z| > 1$
$u[-n - 1]$	$\dfrac{1}{1 - Z^{-1}}$	$|Z| < 1$
$\delta[n - m]$	Z^{-m}	All Z except 0 (if $m > 0$) or ∞ (if $m < 0$)
$\alpha^n u[n]$	$\dfrac{1}{1 - \alpha Z^{-1}}$	$|Z| > |\alpha|$
$-\alpha^n u[-n - 1]$	$\dfrac{1}{1 - \alpha Z^{-1}}$	$|Z| < |\alpha|$
$n\alpha^n u[n]$	$\dfrac{\alpha Z^{-1}}{(1 - \alpha Z^{-1})^2}$	$|Z| > |\alpha|$
$-n\alpha^n u[-n - 1]$	$\dfrac{\alpha Z^{-1}}{(1 - \alpha Z^{-1})^2}$	$|Z| < |\alpha|$
$[\cos\omega_0 n]\, u[n]$	$\dfrac{1 - [\cos\omega_0] Z^{-1}}{1 - [2\cos\omega_0] Z^{-1} + Z^{-2}}$	$|Z| > 1$
$[\sin\omega_0 n]\, u[n]$	$\dfrac{[\sin\omega_0] Z^{-1}}{1 - [2\cos\omega_0] Z^{-1} + Z^{-2}}$	$|Z| > 1$

CHAPTER 14

TOPOLOGICAL ANALYSIS

14.1 INCIDENT MATRIX

By definition, an augmented incident matrix Aa, is an n(nodes) × b(branches) matrix of a directed graph of any planar network, i.e.,

$$Aa = [a_{ij}]_{n \times b}$$

where

$$a_{ij} = \begin{cases} 1 & \text{when branch bj is incident to node } n_i \text{ and the reference current, } i_j, \text{ leaves the node.} \\ -1 & \text{when branch bj is incident to node } n_j \text{ and the reference current, } i_j, \text{ enters the node.} \\ 0 & \text{when branch bj is not incident to node } n_i. \end{cases}$$

The incident matrix Aa can be represented as:

$$Aa = \begin{matrix} & b_1 \ b_2 \ b_3 \ \ldots \ b_j \\ n_1 \\ n_2 \\ n_3 \\ \vdots \\ n_i \end{matrix} \begin{bmatrix} \\ \\ \\ \\ \\ \end{bmatrix}$$

A-111

Note: An incidence submatrix can be obtained by taking out any one of the rows of the incidence matrix Aa.

14.2 THE CIRCUIT (LOOP) MATRIX

By definition, an augmented circuit matrix, B_a, is an $\ell \times b$ matrix where ℓ = loops and b = branches.

Hence, $Ba = [b_{ij}]_{\ell \times b}$

where

$$b_{ij} = \begin{cases} 1 & \text{when branch bj is in loop } \ell_i \text{ and is oriented in the same direction.} \\ -1 & \text{when branch bj is in loop } \ell_i \text{ and is oriented in the opposite direction.} \\ 0 & \text{when branch bj is not in loop } \ell_i. \end{cases}$$

The circuit matrix can be represented as:

$$Ba = \begin{array}{c} \\ \ell_1 \\ \ell_2 \\ \ell_3 \\ \vdots \\ \ell_i \end{array} \begin{array}{c} b_1 \, b_2 \, b_3 \ldots b_j \\ \left[\begin{array}{cccc} & & & \\ & & & \\ & & & \\ & & & \\ & & & \end{array} \right] \end{array}$$

14.3 FUNDAMENTAL (LOOP) MATRIX

The fundamental loop (circuit) matrix, B_f, is defined as a $[b-(n-1)] \times b$ matrix where b = branches and n = nodes. i.e.,

$$B_f = [b_{ij}]_{[b-(n-1)] \times b}$$

where

$$b_{ij} = \begin{cases} 1 & \text{when branch bj is in the fundamental loop } \ell_i \text{ and is oriented in the same direction.} \\ -1 & \text{when branch bj is in the fundamental loop } \ell_i \text{ and is oriented in the opposite direction.} \\ 0 & \text{when branch bj is not in the fundamental loop } \ell_i. \end{cases}$$

Note: A fundamental loop cannot contain more than one chord; a chord is any branch of a cotree and a cotree is the set of all branches not in a tree. (A tree is defined in Chapter 5.)

The matrix representation of B_f is as follows:

$$B_f = \begin{array}{c} \\ \ell_1 \\ \ell_2 \\ \ell_3 \\ \vdots \\ \ell_i \\ \vdots \\ \ell_{b-(n-1)} \end{array} \overset{b_1 \; b_2 \; b_3 \; \ldots \; b_j \; \ldots \; b_b}{\begin{bmatrix} & & & & & \\ & & & & & \\ & & & & & \\ & & & & & \\ & & & & & \\ & & & & & \\ & & & & & \end{bmatrix}}$$

CHAPTER 15

NUMERICAL METHODS

15.1 NEWTON'S METHOD

Newton's method is used to find the roots of a polynomial equation.

Consider the equation:

$$F(s) = k_4 s^4 + k_3 s^3 + k_2 s^2 + k_1 s^1 + k_0 = 0.$$

In order to determine the real value of $s = s'$ such that $F(s) = 0$, the Newton's method is used as follows:

Since all the coefficients of $F(s)$ are of the same sign and real, the complex roots of $F(s)$ are complex conjugate pairs and the real roots, if any, are negative.

Now, by inspection, we begin with a guess, $s = s_0$, for the root. Also let $s_1 = s_0 - h_0$ where

s_1 = a closer approximation of the root obtained from s_0 and

$$h_0 = s_0 - s_1 = \frac{-F(s_0)}{F'(s_0)} \quad \text{and} \quad F'(s_0) = \frac{d}{ds} F(s) \bigg|_{s=s_0}$$

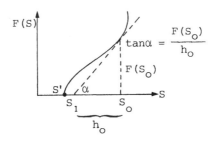

A-114

Then, in general,

$$S_{i+1} = S_i - \frac{F(s_i)}{F'(s_i)} = (i+1) = \text{iteration of the initial approximation } s_0.$$

i.e., S_i = the previous approximation (or guess) of the root.

and S_{i+1} = the new approximation of $F'(S_i)$, which is the derivative of $F(S)$ evaluated at $S = S_i$.

Finally, the iteration is stopped when S_{i+1} is approximately equal to S_i, indicating that the value of S_i is the root s'.

15.2 SIMPSON'S RULE

Simpson's rule states that

$$\int_{x_0=a}^{x=b} y(x)dx \cong \frac{h}{3}(y_0 + 4y_1 + 2y_2 + 4y_3 + 2y_4 + \ldots + 2y_{n-2} + 4y_{n-1} + y_n)$$

where $h = \frac{b-a}{n}$ (note: n is even.)

and $y_0 = y(a)$, $y_1 = y(a+h)$, $y_2 = y(a+2h)$, $y_n = y(a+nb) = y(b)$

Note: The more values between the limits of integration taken, (the larger n is), the more accurate the result will be.

Simpson's rule is a simple and reasonably accurate method which can be used to write programs for digital computers or programmable calculators. The flow chart shown below is for Simpson's rule of integration.

Handbook of Electrical Engineering

SECTION B
Electronics

CHAPTER 1

FUNDAMENTALS OF SEMICONDUCTOR DEVICES

1.1 CHARGED PARTICLES AND THE ENERGY GAP CONCEPT

The electron:

A) Negative charge = 1.60×10^{-19} coulomb

B) Mass = 9.11×10^{-31} kg

Hole - In a semiconductor, two electrons are shared by each pair of ionic neighbors through a covalent bond. When an electron is missing from this bond, it leaves a "hole" in the bond, creating a positive charge of 1.6×10^{-19} C.

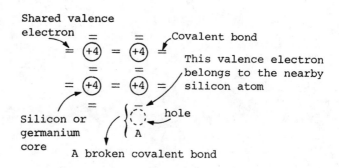

The energy gap concept and classification of materials

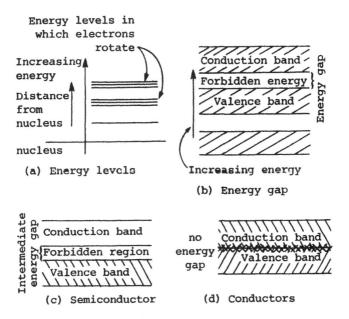

Drift and the Diffusion Current:

A) The diffusion current - The movement of charged particles due to a non-uniform concentration gradient.

B) The drift current - The movement of charges under the influence of an electric field.

1.2 FIELD INTENSITY, POTENTIAL AND ENERGY

Potential - The work done against an electric field in taking a unit of positive charge from point A to B.

$$v = -\int_{x_0 \text{ (point A)}}^{x \text{ (point B)}} E \cdot dx$$

Electric field intensity E:

$$E = \frac{-dv}{dx}$$

Potential energy U (in joules) equals the potential multiplied by the charge q.

Potential-energy barrier concept:

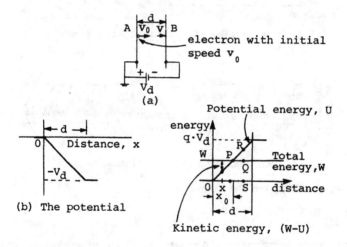

A) The kinetic energy is at its maximum when the electron leaves electrode A.

B) At P, no kinetic energy exists; the electron can therefore travel up to a distance x_0 from plate A.

The ev unit of energy: 1 ev = 1.60 × 10^{-19} J

1.3 MOBILITY AND CONDUCTIVITY

Mobility - When a metal is subjected to a constant E, a steady state is reached where the average value of the drift speed ν is attained. ν is proportional to E and is found by $\nu = \mu E$, where μ is the mobility of the electrons and where the electric field has small values.

Current density: $J = \frac{I}{A} = \frac{N \cdot q \cdot \nu}{L A} = n \cdot q \cdot \nu = \rho \cdot \nu$

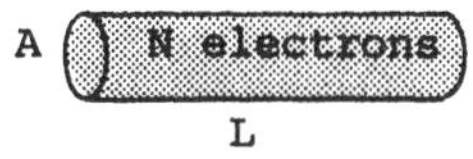
A
N electrons
L

$\rho \equiv n \cdot q$ is the charge density; ν is the drift speed of the electrons.

Conductivity: $J = nq \cdot \nu = n \cdot q \cdot \mu \cdot E$

$J = \sigma E$, where $\sigma = nq \cdot \mu$ = conductivity of the metal

1.4 ELECTRONS AND HOLES IN AN INTRINSIC SEMICONDUCTOR

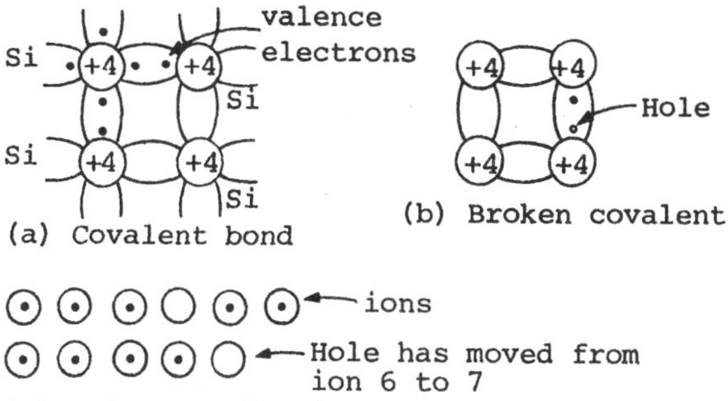
valence
electrons
Si
+4
Si
Si
Si
Hole
ions
Hole has moved from
ion 6 to 7

(a) Covalent bond (b) Broken covalent bond

(c) Hole contributing to conductivity.

In an intrinsic semiconductor:

$$n = p = n_i = \text{density of the intrinsic carriers.}$$

1.5 DONOR AND ACCEPTOR IMPURITIES

Donor impurity:

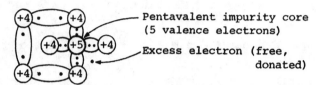
Pentavalent impurity core (5 valence electrons)
Excess electron (free, donated)

Acceptor impurity:

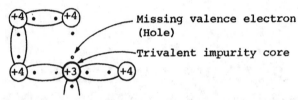
Missing valence electron (Hole)
Trivalent impurity core

The mass-action law: $n \cdot p = n_i^2$

1.6 CHARGE DENSITIES IN A SEMICONDUCTOR

$N_D + p = N_A + n$, where N_D = positive charges contributed by the donor per meter^3, and N_A = negative charges contributed by the acceptor per meter^3.

$n \approx N_D$ (In an n-type material the free-electron concentration is equal to the density of the donor atoms.)

$p = \dfrac{n_i^2}{N_D}$ = Concentration of holes in the n type semiconductor

$p \approx N_A$ and $n = \dfrac{n_i^2}{N_A}$ (in a p-type material)

Generation and recombination of charges - On average, a hole or electron will exist for τ_p seconds or τ_n seconds, respectively, before recombination. This time interval is known as the "Mean Lifetime" of the hole or electron.

1.7 DIFFUSION AND THE POTENTIAL VARIATION WITHIN A GRADED SEMICONDUCTOR

Diffusion - The diffusion hole-current density J_p is, proportional to the concentration gradient.

$$J_p = -q \cdot D_p \cdot \frac{dp}{dx}$$

The Einstein relationship:

A) The Einstein equation: $\dfrac{D_p}{\mu_p} = \dfrac{D_n}{\mu_n} = v_T$

(B) $\boxed{V_T = \dfrac{\overline{K}T}{q}}$ = "Volt-equivalent of temperature"

$= \dfrac{T}{11,600}$ (T = temperature in °K)

C) At room temperature, $V_T = 0.0259V$ and $\mu = 38.6D$

Total current:

$$J_p = q \cdot \mu_p \cdot p \cdot E - q \cdot D_p \cdot \frac{dp}{dx}$$

$$J_n = q \cdot \mu_n \cdot n \cdot E + q \cdot D_n \cdot \frac{dn}{dx}$$

The potential variation in a graded semiconductor:

(A) $\boxed{E = \dfrac{V_T}{p} \cdot \dfrac{dp}{dx}}$ (E is the built-in field)

B) $E = \dfrac{-dv}{dx}$, hence $\boxed{dv = -V_T \cdot \dfrac{dp}{p}}$

C)

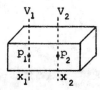 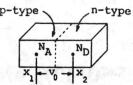

(a) A graded semi-conductor, p(x) is not constant

(b) A step-graded jn

D) $V_{21} = V_2 - V_1 = V_T \cdot \ln \dfrac{p_1}{p_2}$

E) $\boxed{p_1 = p_2 \cdot e^{V_{21}/V_T}}$ This is the "Boltzmann relationship of kinetic gas theory".

F) Mass-action law: $n_1 = n_2 \cdot e^{-V_{21}/V_T}$ = Boltzmann equation for electrons
$n_1 p_1 = n_2 \cdot p_2$

G) An open-circuited, step-graded junction

a) $V_0 = V_{21} = V_T \cdot \ln \dfrac{p_{p_0}}{p_{n_0}}$, $p_1 = p_{p_0}$ = thermal-equilibrium hole concentration in p-side and $p_2 = p_{n_0}$

= thermal-equilibrium hole in n-side

b) $p_{p_0} = N_A$ and $p_{n_0} = \dfrac{n_i^2}{N_D}$, such that

$$\boxed{V_0 = V_T \cdot \ln\left[\dfrac{N_A \cdot N_D}{n_i^2}\right]}$$

H) Analysis of p-n junction in thermal equilibrium.

Approximate doping profile for a p-n step junction:

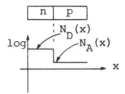

I) Unbiased p-n junction:

The equilibrium carrier concentration and potential as a function of distance in a p-n junction:

a) The region $x_n < x < x_p$ is called the "space-charge region" or "depletion region".

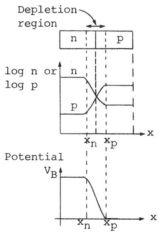

(a) Equilibrium carrier construction and potential

b) $$V_B = \text{built-in-voltage} = \frac{K \cdot T}{q} [\ln n_n - \ln n_p]$$

where

n_n = Electron concentration on n-side

n_p = Electron concentration on p-side

B-8

Depletion region width:

P	N		P	N
(a) Heavy doping		(b) Light doping

J) Forward bias junction

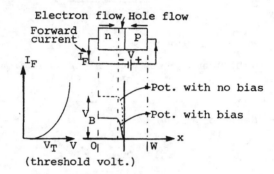

The potential distribution in a p-n junction

n_w (electron concentration at $x = w$) = n_{p0} = n_i^2/N_A

n_{p0} = Equilibrium electron concentration in p-region

$x = 0$ is the right-hand edge of the depletion region

n_A = Acceptor concentration

n_T = Electron concentration at $x = 0$

$$\boxed{n_T = n_n \exp\left\{\frac{-q(V_B - V)}{KT}\right\}}$$

$n_T = n_{p0} \exp\left[\dfrac{q \cdot V}{KT}\right]$, n_{p0} = electron concentration

Current density: $J_e = q \cdot D_e \cdot \dfrac{dn}{dx} = q \cdot D_e \dfrac{(n_w - n_T)}{w}$

$$= -q \cdot \frac{D_e n_i^2}{N_A w}\left(\exp\left[\frac{qV}{KT}\right] - 1\right)$$

The hole current is small compared to the electron current.

Reverse bias:

The potential distribution and electron concentration profile:

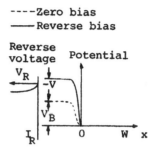

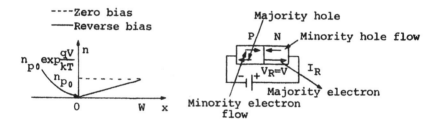

The total current flowing into the p-regin is:

$$I = I_s (\exp \frac{q \cdot v}{KT} - 1) \text{ where } I_s = \frac{q \cdot D_e \cdot n_i^2 \cdot A}{w \cdot N_A}$$

CHAPTER 2

JUNCTION DIODE: THEORY AND SIMPLE CIRCUIT ANALYSIS

2.1 THE OPEN-CIRCUITED P-N JUNCTION

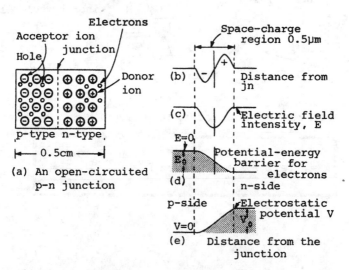

Fig. 1 A schematic diagram of a p-n junction. Since potential energy = potential x charge, the curve in (d) is proportional to the energy for a hole & curve in (e) is proportional to the -ve of that in (d). (It is assumed that the diode dimensions are large compared with the space charge region.)

2.2 THE P-N JUNCTION AS A RECTIFIER

Reverse bias - The polarity is such that it causes both the holes in p-type and the electrons in n-type to move away from the junction. This cannot continue for long because the holes must be supplied across the junction from the n-type material.

A zero current would result, except that thermal energy present will create a small current, known as the "reverse saturation current," I_0. This current increases with increasing temperature.

Forward bias - In this case, the resultant current crossing the junction is the sum of the hole and the electron minority currents.

2.3 V-I CHARACTERISTICS

For a p-n junction, $I = I_0(e^{V/\eta \cdot V_T} - 1)$,

$V_T = \dfrac{T}{11600}$ = Volt-equivalent of temperature

The characteristics:

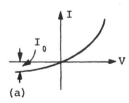

(a)

The reverse-bias voltage V_R is the voltage below which the current is very small (<1% of the maximum rated value).

2.3.1 TEMPERATURE DEPENDENCE OF THE V-I CHARACTERISTICS

$$I_0(T) = I_{01} \times 2^{(T-T_1)/10} \quad (I_0 = I_{01} \text{ at } T = T_1)$$

$$\boxed{\frac{dV}{dT} \approx -2.5 \text{ mv/°C}}$$

2.4 DIODE RESISTANCE, TRANSITION AND DIFFUSION CAPACITANCE

Diode resistance:

The static resistance R is the ratio V/I.

For a small-signal operation, the "Dynamic-resistance" r is:

$r = \frac{dV}{dI}$ (depending on the oeprating voltage)

$$g = \frac{1}{r} = \frac{dI}{dV}$$
$$= I_0 \frac{\exp\left[\frac{V}{\eta V_T}\right]}{\eta V_T}$$
$$= (I + I_0)/\eta \cdot V_T$$

For $I \gg I_0$,
$$\boxed{\gamma \approx \frac{\eta \cdot V_T}{I}}$$

Piecewise linear characterization of a semiconductor diode

The break-point is not at the origin but at a point V_y units from the origin. V_y is called the offset voltage.

R_f is the forward resistance.

Transition and diffusion capacitance:

In the reverse bias region, the transition capacitance predominates, while in the forward bias region, the diffusion or storage capacitance predominates.

Transition capacitance, C_T:

$$C_T = \left| \frac{dQ}{dV} \right|$$

Note: C_T depends on the magnitude of the reverse voltage, since the magnitude of the reverse voltage determines the depletion width.

Diffusion capacitance, C_D:

C_D is introduced when the p-n junction is forward biased due to the additional injected charge redistribution in the n-region.

This type of capacitance limits switching speed in logic circuits used as junction devices.

Static derivation of C_D:

$$C_D = \frac{dQ}{dV} = \tau \cdot \frac{dI}{dV} = \tau \cdot g = \frac{\tau}{r}$$

$$C_D = \frac{\tau \cdot I}{\eta \cdot V_T}$$

For reverse bias, g is very small and $C_D \ll C_T$.

For a forward current, $C_D \gg C_T$.

$$r \cdot C_D = \tau$$

The diode time constant equals the mean lifetime of minority carriers.

2.5 ANALYSIS OF SIMPLE DIODE CIRCUITS: CONCEPT OF DYNAMIC RESISTANCE AND A.C. LOAD LINE

2.5.1 THE D.C. LOAD LINE

The behavior of the diode at low frequency is marked by V-I characteristics.

Other elements of a circuit beyond the region bounded by the diode and its terminals can be replaced by a thevenin equivalent circuit.

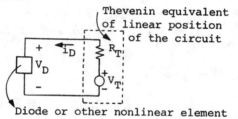

For the non-linear element $i_D = f(v_0)$, the thevenin equivalent is given as $v_D = v_T - i_D \cdot R_T$.

The problem is solved by plotting these equations on the same set of axes.

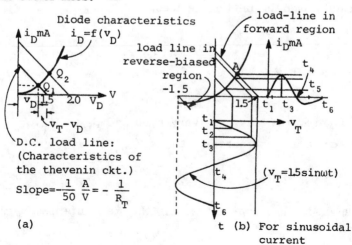

(a)   (b) For sinusoidal current

B-15

The straight-line characteristics of the thevenin circuit is "D.C. load line." As long as R_T remains constant, any change in v_T is accounted for by a horizontal shift of the load line.

If v_T is sinusoidal, the corresponding current i_D can be found as shown in (b).

Small-signal analysis, dynamic resistance:

Small-signal - When the total peak-to-peak swing of the signal is a small fraction of its D.C. component.

$v_T = V_{dc} + v_i = V_{dc} + V_{im} \sin \omega t$

(where V_{dc} = bias voltage, and $V_{im} << V_{Dc}$).

The operating point for $v_T = V_{dc}$ is called "the quiescent point". It is found as follows:

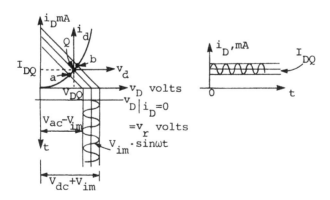

A new set of axes $i_d - v_D$ is constructed at Q (as shown below);

$$i_d = i_D - I_{DQ}, \quad v_d = v_D - V_{DQ}$$

B-16

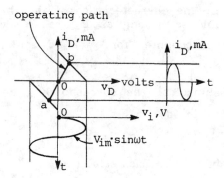

The operating path is "ab". The dynamic resistance r_d of the diode is equal to the inverse of the slope of line "ab".

$$r_d = \text{dynamic resistance} = \left.\frac{\Delta v_D}{i_D}\right|_{Q \text{ pt.}}$$

If r_d is found, circuit variables are obtained by using Ohm's law.

The original circuit has two parts:

(a) For calculating Q   (b) For calculating small-signal a.c. component

Calculation of r_d:

$$r_d = \left.\frac{dv_d}{di_D}\right|_{Q \text{ point}} \cong \frac{V_T}{I_{DQ}} = \frac{25\text{mV}}{I_{DQ}} \quad (\text{at } T=300°K)$$

Reactive elements:

$$I_{dm} = (\text{the peak current}) = V_{im}/|r_i + r_d + z_i|$$

B-17

2.5.2 THE A.C. LOAD LINE

For the circuit shown below, the D.C. load line and Q point are:

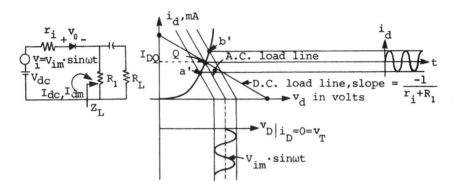

When the A.C. signal is present, the effective resistance seen by the diode is $r_i + (R_1 \| R_L)$. The A.C. load line is drawn through Q with slope $= \dfrac{-1}{[r_i+(R_1 \| R_L)]}$

The equations for the D.C. and A.C. load lines are:

$V_{dc} = I_{DQ}(r_i+R_1) + V_{DQ}$ D.C. load line

$v_i = i_d(r_i+[R_1 \| R_L]) + r_d$ A.C. load line

2.6 SCHOTTKY AND ZENER DIODES

Schottky diode:

It is formed by bonding a metal (platinum) to n-type silicon. A Schottky diode has negligible charge storage and is often used in high-speed switching applications.

B-18

Platinum acts as an acceptor material for electrons when bonded to n-type silicon.

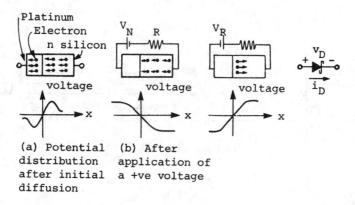

(a) Potential distribution after initial diffusion

(b) After application of a +ve voltage

Zener diode:

Unlike regular diodes, when you apply a high, reverse voltage across a Zener diode, you produce an almost constant-voltage region on the characteristic curve of the diode. One application of this special property of the Zener diode is voltage regulation.

The change of the Zener voltage V_Z as a result of a change of temperature is proportional to the Zener voltage as well as change in temperature.

$$T_c = \text{Temperature coefficient} = \frac{\Delta V_Z / V_Z}{\Delta T} \times 100\%/°C$$

2.7 DIODE LOGIC CIRCUITS

2.7.1 A DIODE "AND" GATE

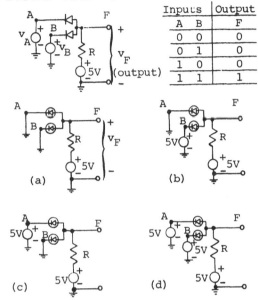

Inputs		Output
A	B	F
0	0	0
0	1	0
1	0	0
1	1	1

The diode AND gate redrawn for each possible combination of inputs. The diodes are assumed to be perfect rectifiers.

2.7.2 A DIODE "OR" GATE

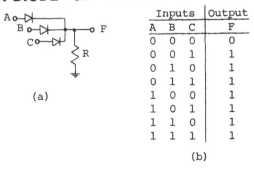

Inputs			Output
A	B	C	F
0	0	0	0
0	0	1	1
0	1	0	1
0	1	1	1
1	0	0	1
1	0	1	1
1	1	0	1
1	1	1	1

(b)

A diode OR gate. (a) The circuit. (b) The truth table. The output is 1 if any of the inputs is 1.

CHAPTER 3

THE BIPOLAR JUNCTION TRANSISTOR

3.1 THE JUNCTION TRANSISTOR THEORY

3.1.1 OPEN-CIRCUITED TRANSISTOR

Under this condition, the minority concentration is constant within each section and is equal to its thermal-equilibrium value n_{p_0} in p region and p_{n_0} in n region. The potential barriers at the junctions adjust to the contact difference of potential V_0, such that no free carriers cross a junction.

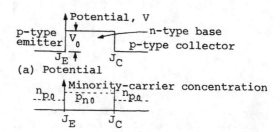

B-21

3.1.2 TRANSISTOR BIASED IN THE ACTIVE REGION

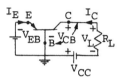

(a) p-n-p transistor biased in the active region.

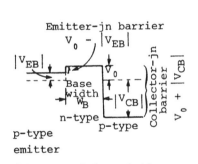

(b) Potential variation through the transistor.

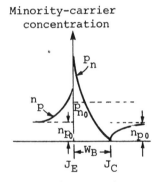

(c) Minority-carrier concentration

The dashed curve applies to the case before the application of external biasing voltages.

The forward biasing of the emitter junction lowers the emitter-base potential barrier by $|V_{EB}|$, permitting minority-carrier injection. Holes are thus injected into the base and electrons into the emitter region.

Excess holes diffuse across the n-type base, the holes which reach J_c fall down the potential barrier and are collected at the collector.

3.1.3 TRANSISTOR CURRENT COMPONENTS

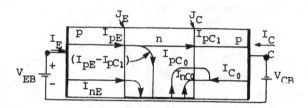

Transistor current compoments for a forward-biased emitter junction and a reverse-biased collector junction.

I_{pE} - Current due to holes crossing from emitter into base is the forward injection.

I_{nE} - Electrons crossing from base into the emitter.

$I_E = I_{pE} + I_{nE}$ (I_{pE} has a magnitude proportional to the slope at J_E of the p_n curve. Similarly, I_{nE} has a magnitude proportional to the slope at J_E of the n_p curve.)

$-I_{co} = I_{nco} + I_{pco}$ (I_{nco} consists of electrons moving from the p to the n region across J_c, and I_{pco} results from holes across J_c from n to p.)

I_c = Complete collector current = $I_{co} - I_{pc}$

$$\boxed{I_c = I_{co} - \alpha \cdot I_E}$$

Large-signal current gain α: This is the ratio of the negative of the collector-current increment from cutoff ($I_c = I_{co}$), to the emiiter-current change from cutoff ($I_E = 0$), e.g.,

$$\boxed{\alpha = \frac{-(I_c - I_{co})}{I_E - 0}}$$ The large-signal current gain of a C-B transistor.

α is not a constant, but vaires with emitter current I_E, V_{CB} and temperature.

A generalized transistor equation gives an expression for I_c in terms of any V_c and I_E:

$$I_c = -\alpha I_E + I_{co}(1 - e^{V_c/V_T})$$

A transistor as an amplifier, and parameter α':

A small voltage change ΔV_i between emitter and base causes a relatively large emitter-current change ΔI_E. $\Delta I_c = \alpha' \cdot \Delta I_E$, $\Delta V_L = -R_L \Delta I_c = -\alpha' \cdot R_L \cdot \Delta I_E$.

If the dynamic resistance of the emitter junction is r_e, then

$$\Delta V_i = r_e \cdot \Delta I_E, \text{ and } A = \alpha' \cdot R_L(\Delta I_E/r_e)\Delta I_E = \frac{-\alpha' \cdot R_L}{r_e}$$

The parameter α':

α' = The negative of the small-signal Short-circuit current transfer ratio $\equiv \left.\frac{\Delta I_c}{\Delta I_E}\right|_{V_{CB}}$

$\alpha' = -\alpha$ (assuming α is independent of I_E)

3.1.4 TRANSISTOR CONFIGURATION

The CB configuration:

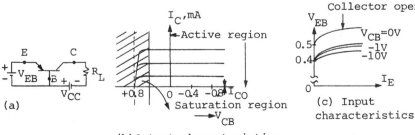

(a)

(b) Output characteristics

(c) Input characteristics

The common-emitter configuration:

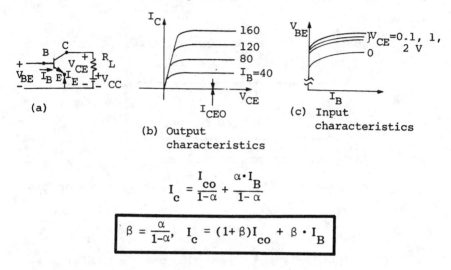

(a)

(b) Output characteristics

(c) Input characteristics

$$I_c = \frac{I_{co}}{1-\alpha} + \frac{\alpha \cdot I_B}{1-\alpha}$$

$$\boxed{\beta = \frac{\alpha}{1-\alpha}, \quad I_c = (1+\beta)I_{co} + \beta \cdot I_B}$$

CE cutoff currents - A transistor is in cutoff if $I_E = 0$ and $I_c = I_{co}$. It is not in cutoff if the base is open-circuited.

Common-emitter current gain:

Large-signal current-gain $\beta = \dfrac{I_c - I_{CBo}}{I_B - (-I_{CBo})}$

DC current gain $h_{FE} = \beta_{dc} = \dfrac{I_C}{I_B}$

Small-signal current gain $h_{fe} = \beta' = \left.\dfrac{\Delta I_c}{\Delta I_B}\right|_{V_{CE}}$

3.2 THE JUNCTION TRANSISTOR: SMALL-SIGNAL MODELS

3.2.1 THE HYBRID-Pi MODEL

It is useful to predict high-frequency performance.

Model:

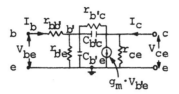

b,e,c: external terminals

$C_{b'c}$ is the depletion-region capacitance of the collector-base junction. C_{be} is an equivalent capacitance that accounts for a reduction in gain and in increase in phase shift at higher frequency.

Simplified circuit valid at low frequencies:

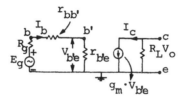

$$\frac{V_0}{V_{b'e}} = -g_m \cdot R_L \text{ (which is equivalent to } A_v \text{ if } r_{bb'} < r_{b'e})$$

Beta cutoff frequency f_β : At this frequency the magnitude of h_{fe} has decreased 3dB from its mid-frequency value.

$$h_{fe} = \frac{I_c}{I_b} = \frac{g_m \cdot r_{be}}{1 + j\omega/\omega\beta}$$

$$\omega_\beta = 1/r_{b'e}(c_{b'e} + c_{b'c})$$

$$f_\beta = \frac{1}{2\pi \cdot r_{b'e} \cdot c_{b'e}} \quad \text{(since } c_{b'e} \gg c_{b'c}\text{)}$$

$$f_T = h_{fe} \cdot f_\beta = \frac{g_m}{2\pi \cdot c_{b'e}}$$

3.2.2 h-PARAMETERS

h-Parameters for Each Transistor Configuration in Terms of the Other Two.

		Common base	Common collector
Common emitter	h_{ie}	$\dfrac{h_{ib}}{h_{fb} + 1}$	h_{ic}
	h_{fe}	$-\dfrac{h_{fb}}{h_{fb} + 1}$	$-(h_{fc} + 1)$
	h_{oe}	$\dfrac{h_{ob}}{h_{fb} + 1}$	h_{oc}
	h_{re}	$\dfrac{h_{ib} h_{ob}}{1 + h_{fb}} - h_{rb}$	$1 - h_{rc}$

		Common emitter	Common collector
Common base	h_{ib}	$\dfrac{h_{ie}}{h_{fe} + 1}$	$-\dfrac{h_{ic}}{h_{fc}}$
	h_{fb}	$\dfrac{-h_{fe}}{h_{fe} + 1}$	$-\dfrac{h_{fc} + 1}{h_{fc}}$
	h_{ob}	$\dfrac{h_{oe}}{h_{fe} + 1}$	$-\dfrac{h_{oc}}{h_{fc}}$
	h_{rb}	$\dfrac{h_{ie} h_{oe}}{h_{fe} + 1} - h_{re}$	$\dfrac{-h_{ic} h_{oc}}{h_{fc}} + h_{rc} - 1$

		Common emitter	Common base
Common collector	h_{ic}	h_{ie}	$\dfrac{h_{ib}}{1+h_{fb}}$
	h_{fc}	$-(1+h_{fe})$	$\dfrac{-1}{1+h_{fb}}$
	h_{oc}	h_{oe}	$\dfrac{h_{ob}}{1+h_{fb}}$
	h_{rc}	$1-h_{re} \cong 1 \text{(since } h_{re}\ll 1)$	$\dfrac{1-h_{ib}h_{ob}}{1+h_{fb}}$

3.2.3 THE CONCEPT OF BIAS STABILITY

Q-point variation due to uncertainty in β :

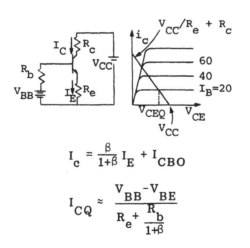

$$I_c = \frac{\beta}{1+\beta} I_E + I_{CBO}$$

$$I_{CQ} \approx \frac{V_{BB}-V_{BE}}{R_e + \dfrac{R_b}{1+\beta}}$$

$$\boxed{\begin{array}{c} I_{CQ} \approx \dfrac{V_{BB}-V_{BE}}{R_e} \approx \dfrac{V_{BB}-0.7}{R_e} \\[6pt] \text{iff} \quad R_b \ll \beta R_e \end{array}}$$

The effect of temperature on the Q-point:

$$\Delta I_{CQ} = \frac{K \cdot \Delta T}{R_e} + \left(1 + \frac{R_b}{R_e}\right) I_{CBO1}(e^{K \cdot \Delta T} - 1)$$

Stability factor:

$$S_I = \frac{\Delta I_{CQ}}{\Delta I_{CBO}}, \quad S_V = \frac{\Delta I_{CQ}}{\Delta V_{BE}}, \quad S_\beta = \frac{\Delta I_{CQ}}{\Delta \beta}$$

for the common-emitter amplifier:

$$S_I \cong 1 + \left(\frac{R_b}{R_e}\right), \quad S_V \cong -1/R_e, \quad \text{and} \quad S_\beta = \frac{I_{CQ1}}{\beta_1}\left(\frac{R_b + R_e}{R_b + (1+\beta_2)R_e}\right)$$

where $\beta_1, \beta_2, I_{CQ1}$ and I_{CQ2} are the lower and upper limits.

Temperature compensation using diode biasing:

Single diode compensations:

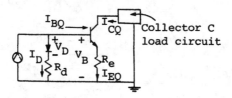

This compensation reduces the base-emitter voltage variation.

The diode is such that: $\dfrac{\Delta V_D}{\Delta T} = \dfrac{\Delta V_{BE}}{\Delta T}$.

$$I_{BB} = I_D + \frac{I_{EQ}}{1+\beta} = \text{constant}$$

$$V_B = V_D + I_D \cdot R_d = V_{BEQ} + I_{EQ} \cdot R_e$$

B-29

$$I_{EQ} = \frac{V_D - V_{BEQ} + I_D R_d}{R_e + R_d/(1+\beta)} \quad \text{and} \quad \frac{\Delta I_{EQ}}{\Delta T} = 0$$

Two-diode compensation:

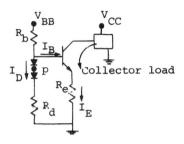

The quiescent emitter current is independent of variations of temperature if $R_b = R_d$.

$$I_{EQ} = \frac{(V_{BB} \cdot R_d + 2V_D \cdot R_b)/(R_b + R_d) - V_{BEQ}}{R_e}$$

B-30

CHAPTER 4

POWER SUPPLIES

4.1 DIODE RECTIFIERS

4.1.1 HALF-WAVE RECTIFIER

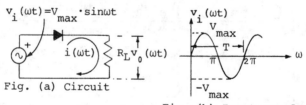

Fig. (a) Circuit

Fig. (b) Input waveform

Fig. (c) Output Current

Fig. (d) Output voltage

Fig. (e) Voltage across the diode

$$I_{dc} = \frac{I_{max}}{\pi} = 0.318\, I_{max}$$

B-31

$$v_{dc} = 0.318\, I_{max} \cdot R_L = 0.318\, V_{max}$$

4.1.2 FULL-WAVE RECTIFIER

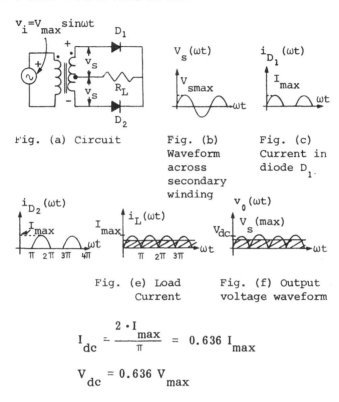

Fig. (a) Circuit

Fig. (b) Waveform across secondary winding

Fig. (c) Current in diode D_1.

Fig. (e) Load Current

Fig. (f) Output voltage waveform

$$I_{dc} = \frac{2 \cdot I_{max}}{\pi} = 0.636\, I_{max}$$

$$V_{dc} = 0.636\, V_{max}$$

4.1.3 PIV RATING

The PIV rating indicates the voltage which a rectifier diode can withstand in a reverse bias condition.

$PIV = 2 \cdot V_{max}$ - for conventional full-wave rectifier

$ = V_{max}$ - for the bridge rectifier

4.1.4 RIPPLE FACTOR

$$\text{Ripple factor (RF)} = \frac{\text{rms value of the ac component}}{\text{dc value of the waveform}}$$

$$i_{ac} = i_L - i_{dc} \quad (i_L = \text{the rectified load current})$$

$$I_{ac(rms)} = [I^2_{L(rms)} - I^2_{dc}]^{\frac{1}{2}}$$

$$RF = \left[\left(\frac{V_{L\,rms}}{V_{dc}}\right)^2 - 1\right]^{\frac{1}{2}}$$

$$RF_{Halfwave} = 1.21 \quad (V_{Lrms} = \frac{V_{max}}{2} \text{ and } V_{dc} = \frac{V_{max}}{\pi})$$

$$RF_{Fullwave} = 0.482 \quad \left\{V_{L(max)} = \frac{V_{max}}{\sqrt{2}} = 0.707\, V_{max}\right\}$$

$$RF_{Bridge-rectifier} = 0.482 \quad \left\{I_{L(max)} = \frac{I_{max}}{\sqrt{2}} = 0.707\, I_{max}\right\}$$

4.1.5 RECTIFIER EFFICIENCY

$$\eta_R = \frac{P_{L(dc)}}{P_{i(dc)}}$$

For the half-wave rectifier:

$$P_{L(dc)} = I_{dc} \cdot V_{dc} = \left(\frac{I_{max}}{\pi}\right)^2 \cdot R_L$$

$$P_{i(ac)} = I^2_{rms} \cdot R_L = \left(\frac{I_{max}}{2}\right)^2 \cdot R_L$$

$$\eta_R = 0.406$$

For conventional and bridge rectifier:

$$\eta_R = 0.812$$

4.1.6 COMPARISON OF RECTIFIERS

Rectifier	V_{dc}	RF	f_{output}	η_R	PIV.
Half-wave	$0.318 V_{s(max)}$	1.21	f_{input}	40.6%	$V_{s(max)}$
Full-wave	$0.636 V_{s(max)}$	0.482	$2f_{input}$	81.2%	$2V_{s(max)}$
Bridge	$0.636 V_{s(max)}$	0.482	$2f_{input}$	81.2%	$V_{s(max)}$

% load regulation:

$$\%LR = \frac{V_{NL} - V_{FL}}{V_{FL}} \times 100\%$$

V_{NL} = No-load dc voltage

V_{FL} = Full-load dc voltage

A high % LR is poor regulation.

4.2 FILTERS

4.2.1 SHUNT-CAPACITANCE FILTER

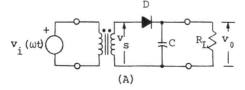

(A)

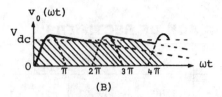

Half-wave rectifier with shunt-capacitance filter. (A) Circuit. (B) Output waveform.

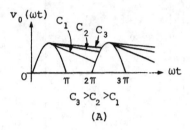

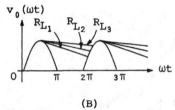

Output of Half-wave rectifier with (A) C and (B) R_L vary.

Half-Wave Rectifier

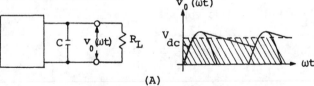

Full-Wave Rectifier

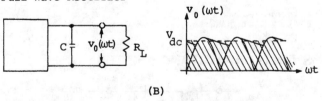

The same value of capacitance used with a full-wave rectifier in (B) produces a higher dc output than when used with a half-wave rectifier in (A).

The dc load voltage of a full-wave rectifier with a shunt-capacitance filter is given as:

$$V_{dc} = \frac{V_{s(max)}}{1 + \frac{1}{4f\,R_L\,C}}$$

$$RF = \frac{1}{4\sqrt{3}\cdot f\cdot C\cdot R_L}$$

4.2.2 PI-FILTER

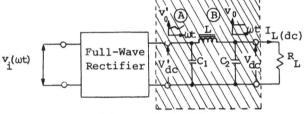

Full-wave rectifier and pi filter.

$$V_{dc} = V'_{dc} - I_{dc}\cdot r_L = \frac{V'_{dc}\cdot R_L}{R_L + r_L}$$

$$RF = \sqrt{2}/\,\omega^3\cdot R_L\cdot C_1\cdot C_2\cdot L$$

4.2.3 RC FILTER

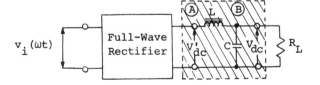

Full-wave rectifier and L-section (choke-input) filter.

$$V_{dc} = \frac{V'_{dc}\cdot R_L}{R_L + R}$$

B-36

4.2.4 L-SECTION FILTER

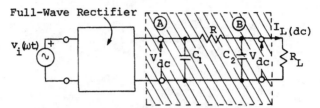

Full-wave rectifier and RC filter.

This filter is used with high-load current circuits.

4.2.5 VOLTAGE MULTIPLIER

The voltage multiplier is used when a high dc voltage with extremely light loading is required.

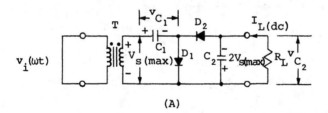

(A)

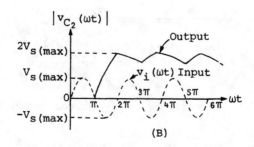

(B)

$$V_{C_2} = 2V_{s(max)}$$

4.2.6 COMPARISON OF FILTERS

Filter	V_{dc}	RF
Shunt-Capacitance Rectifier FW	$\dfrac{V_{s(max)}}{1 + 1/(4fR_LC)}$ For 60 Hz: $\dfrac{V_{s(max)}}{0.00417/R_LC}$	$\dfrac{1}{4\sqrt{3}fR_LC}$ For 60 Hz: $\dfrac{2.41 \times 10^{-3}}{R_LC}$
Pi Rectifier FW	$\dfrac{V_{s(max)} R_L/(R_L+r_L)}{1 + 1/(4fR_LC_1)}$ For 60 Hz: $\dfrac{V_{s(max)} R_L/(R_L+r_L)}{1 + 0.00417/R_LC}$	$\dfrac{\sqrt{2}}{\omega^3 C_1 C_2 L R_L}$ For 60 Hz: $\dfrac{0.026}{C_1 C_2 R_L L} \times 10^{-6}$
RC Rectifier FW	$\dfrac{V_{s(max)} R_L/(R_L+R)}{1 + 1/(4fR_LC_1)}$ For 60 Hz: $\dfrac{V_{s(max)} R_L/(R_L+R)}{1 + 0.00417/R_LC}$	$\dfrac{\sqrt{2}}{\omega^2 C_1 C_2 R_L R}$ For 60 Hz: $\dfrac{9.95}{C_1 C_2 R_L R} \times 10^{-6}$
L-Section Rectifier FW	$0.636 V_{s(max)}$ For 60 Hz: $0.636 V_{s(max)}$	$\dfrac{0.118}{(\omega^2 LC)}$ For 60 Hz: $\dfrac{0.83}{LC} \times 10^{-6}$

CHAPTER 5

MULTITRANSISTOR CIRCUITS

5.1 THE DIFFERENCE AMPLIFIER

5.1.1 BASIC DIFFERENCE AMPLIFIER

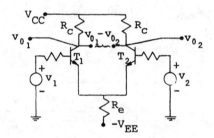

The Differential-mode or difference-mode input voltage $v_d = v_2 - v_1$.

The common-mode input voltage $V_a = \frac{V_2 + V_1}{2}$.

$v_2 = v_a + \frac{v_d}{2}$

$v_1 = v_a - \frac{v_d}{2}$

$v_a(t) = \frac{v_1(t) + v_2(t)}{2}$

V_d is the desired signal and is amplified while V_a is rejected.

5.1.2 Q-POINT ANALYSIS

The differential-mode input is assumed to be zero. $(V_1 = V_2)$

The equivalent circuit for T_1 or T_2 when $v_1 = v_2 = v_a$:

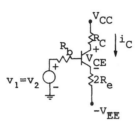

$$v_{E_1} = v_{E_2} = (i_{E_1} + i_{E_2})R_e - V_{EE}$$

$$= i_E(2R_e) - V_{EE} \quad (\text{when } i_{E_1} = i_{E_2} = i_E)$$

The load-line equation for $v_a = v_1 = v_2$ is

$$v_{CE} = V_{CC} - i_C R_C - i_E(2R_e) + V_{EE}$$

$$\approx V_{CC} + V_{EE} - i_C(R_C + 2R_e)$$

$$i_C \approx \frac{v_a + V_{EE} - 0.7}{2R_e + [R_b \div (h_{FE}+1)]}, \quad (V_{BE} = 0.7))$$

5.1.3 COMMON-MODE LOAD LINE

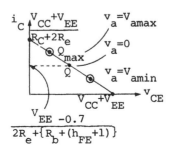

The common-mode input $v_a = 0$, for Q.

$Q = Q_{max}$ when $v_a = v_{amax}$;

$Q = Q_{min}$ when $v_a = v_{amin}$.

In both cases, $v_d = 0$.

The individual collector voltages v_{01} and v_{02} will vary with variations in v_a.

Difference-mode load-line equations

These equations determine the effect of a non-zero difference-mode input. ($v_2 = -v_1 = \frac{v_d}{2}$, $v_a = 0$ and Q is as shown in the previous figure.)

$\Delta v_{CE_1} = -R_c \cdot \Delta i_{c_1}$, using small-signal notation

$$\boxed{v_{ce_1} = -R_c \cdot i_{c_1}}, \quad \Delta v_{CE_2} = -R_c \cdot \Delta i_{c_2},$$

$$\text{or} \quad \boxed{v_{ce_2} = -R_c \cdot i_{c_2}}$$

These are "Difference-mode load line" equations, such that $v_{E_1} = v_{E_2}$.

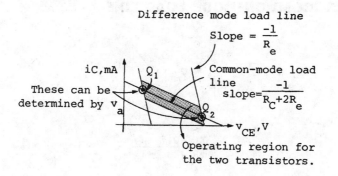

B-41

Small-signal analysis:

Equivalent circuit with all components reflected into emmiter.

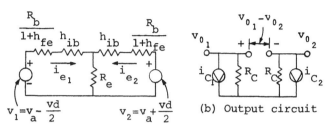

(a) Input circuit

(b) Output circuit

Circuit used to calculate i_a and i_d:

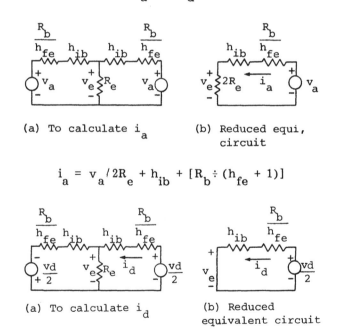

(a) To calculate i_a

(b) Reduced equi, circuit

$$i_a = v_a / 2R_e + h_{ib} + [R_b \div (h_{fe} + 1)]$$

(a) To calculate i_d

(b) Reduced equivalent circuit

$$i_d = \frac{v_d/2}{h_{ib} + [R_b \div (1+h_{fe})]}$$

Assuming $i_c \approx i_e$, and $\quad v_{01} - v_{02} = \dfrac{R_c}{h_{ib} + [R_b \div (1+h_{fe})]} \cdot v_d$

Common-mode Rejection Ratio:

A_d = The difference-mode gain = $\dfrac{R_c/2}{h_{ib}+[R_b \div (1+h_{fe})]}$

A_a = The common-mode gain = $\dfrac{R_c}{2R_e+h_{ib}+[R_b \div (1+h_{fe})]}$

$v_{01} = A_d \cdot V_d - A_a \cdot V_a$ and $v_{02} = -A_d \cdot v_d - A_a \cdot v_a$.

Common-mode rejection ratio:

$$\boxed{\text{"CMRR"} = \dfrac{A_d}{A_a} \cong \dfrac{R_e}{h_{ib}+[R_b/h_{fe}]}}$$

If "CMRR" $\gg \dfrac{V_a}{V_d}$, then the output voltage is proportional to V_d.

5.1.4 DIFFERENCE AMPLIFIER WITH CONSTANT CURRENT SOURCE

$\text{CMRR} = \dfrac{R_e}{(V_T/I_{EQ})+(R_b/h_{fe})}$, if $\dfrac{R_b}{h_{fe}}$ is small,

$$\text{CMRR} < \dfrac{R_e \cdot I_{EQ}}{V_T}$$

CMRR can be increased only by increasing R_e.

In the previous circuit, R_e is replaced by another transistor which is a constant-current source.

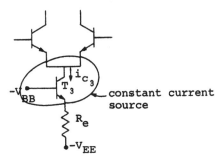
constant current source

Quiescent operation:

$$I_{CQ_3} \approx \frac{V_{EE} - V_{BB} - 0.7}{R_e'}$$ (I_{CQ_3} is constant as long as T_3 does not saturate.)

The condition for keeping T_3 in the linear region is:

$$v_{CE} > V_T \left[2.2 + \ln\left(\frac{h_{fe}}{h_{fe}}\right) \right]$$

Load-lines that describe circuit operation:

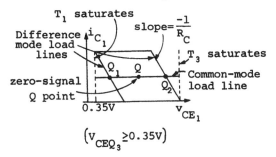

$$\left(v_{CEQ_3} \geq 0.35V \right)$$

Small-signal operation:

Equivalent circuit looking into the bases

The input impedance R_i between the bases of T_1 and T_2 is:

$$R_i = 2 \cdot h_{ie}.$$

B-44

5.1.5 DIFFERENCE AMPLIFIER WITH EMITTER RESISTORS FOR BALANCE

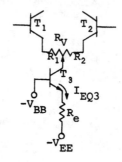

Difference amplifier with balance control R_U.

To compensate for different h_{fe1} and h_{fe2}, R_v is used when T_1 and T_2 have different characteristics.

The condition which ensures that the emitter currents of T_1 and T_2 are the same is:

$$R_2 - R_1 = R_b \left(\frac{1}{h_{fe1}} - \frac{1}{h_{fe2}} \right).$$

R_v results in symmetrical operation but it causes a loss in the current gain.

$$A_d = \frac{R_c}{R_b \left(\frac{1}{h_{fe1}} + \frac{1}{h_{fe2}} \right) + 2h_{ib} + Rv},$$

if

$$(R_e)_{eff} \approx \frac{1}{h_{ob3}} \gg h_{ib1} + \frac{R_b}{h_{fe1}} + R_1$$

B-45

5.2 THE DARLINGTON AMPLIFIER

The Darlington amplifier is used to provide increased input impedance and a very high current gain ($h_{fe_1} \cdot h_{fe_2}$).

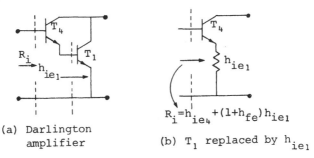

Basic Darlington amplifier

Current gain $A_i = \alpha(1+h_{fe})^2 + \alpha(1+h_{fe}) = \alpha(1+h_{fe})(h_{fe}+2)$

$\approx h_{fe}^2$ (assuming identical transistors)

$R_i = h_{ie_4} + (1+h_{fe})h_{ie_1}$

$= \dfrac{(1+h_{fe})V_T}{I_{EQ_4}} + \dfrac{(1+h_{fe})^2 \cdot V_T}{I_{EQ_1}}$

$= 2(1+h_{fe})h_{ie_1} = 2 \cdot h_{ie_4}$

Input impedance:

(a) Darlington amplifier

(b) T_1 replaced by h_{ie_1}

$R_i = h_{ie_4} + (1+h_{fe})h_{ie_1}$

In this case, emitter current in T_4 can be adjusted by setting

$$R_b, \quad I_{EQ_4} = I_{BQ_1} + \dfrac{0.7}{R_b}$$

$$R_i = h_{ie_4} + (1+h_{fe})(R_b \| h_{ie_1})$$

$$i_0 = h_{fe} \cdot i_{b_4} + h_{fe}(1 + h_{fe})i_{b_4} \cdot \frac{R_b}{R_b + h_{ie_1}}$$

$$A_i = h_{fe}^2 \frac{R_b}{R_b + h_{ie}} + h_{fe} \cdot \left(1 + \frac{R_b}{R_b + h_{ie_1}}\right)$$

Darlington amplifier with bias resistor.

5.3 THE CASCADE AMPLIFIER

The Amplifier:

The cascade amplifier is used as a dc level shifter when

the voltage of interest consists of a small-signal ac component v_i and a fixed dc level V_i. v_i from the level shifter should have a dc level different from V_i. Typically this final output level is to be 0V.

DC analysis:

T_1 is the emitter follower and T_2 acts as a constant current source. DC component of the output voltage, V_L:

$$V_i - \frac{R_i I_{E2}}{1+h_{fe}} - 0.7 - R_c \cdot I_{E2}$$

Small signal analysis:

v_L (small-signal component of the output voltage) $\cong \dfrac{v_i}{1 + \dfrac{(R_i \div h_{fe})+h_{ib}+R_c}{1/h_{ob2}}}$

Since $\dfrac{1}{h_{ob2}}$ is much larger than $\dfrac{R_i}{h_{fe}} + h_{ib} + R_c$, the load voltage $v_L \approx v_i$, while a negligible attenuation of the signal has resulted from the shift in dc level.

5.4 THE OP AMPLIFIER

Typical configuration:

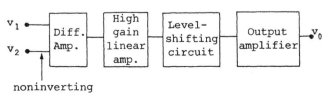

B-48

CHAPTER 6

SMALL-SIGNAL, LOW-FREQUENCY ANALYSIS AND DESIGN

6.1 HYBRID PARAMETERS

6.1.1 GENERAL TWO-PORT NETWORK

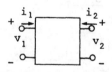

(a) Two-part network

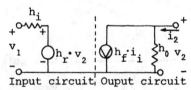

Input circuit | Ouput circuit

(b) Equivalent circuit of two-part network.

6.1.2 HYBRID EQUATIONS

$$v_1 = h_i \cdot i_1 + h_r \cdot v_2 \qquad v_1 = h_{11} \cdot i_1 + h_{12} \cdot v_2$$

$$i_2 = h_f \cdot i_1 + h_0 \cdot v_2 \qquad i_2 = h_{21} \cdot i_1 + h_{22} \cdot v_2$$

B-49

6.1.3 TERMINAL DEFINITIONS FOR H-PARAMETERS

$h_i = \dfrac{v_1}{i_1}\bigg|_{v_2=0}$ = short-circuit input impedance

$h_r = \dfrac{v_1}{v_2}\bigg|_{i_1=0}$ = open-circuit reverse voltage gain

$h_f = \dfrac{i_2}{i_1}\bigg|_{v_2=0}$ = short-circuit forward current gain

$h_o = \dfrac{i_2}{v_2}\bigg|_{i_1=0}$ = open-circuit output admittance

6.2 THE C-E CONFIGURATION

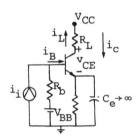

(a) Complete circuit

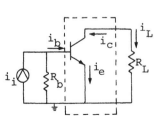

(b) Small-signal circuit

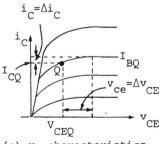

(c) v_i characteristics

B-50

$$h_{oe} = \left.\frac{i_c}{v_{ce}}\right|_{i_b=0} = \left.\frac{\Delta i_c}{\Delta v_{ce}}\right|_{Q\ point}$$

where i_c and v_{ce} are small variations about nominal operating point.

$$h_{oe} = \frac{h_{FE}}{h_{FC}} \cdot \frac{I_{CQ}}{V_T} \left[\frac{e^{v_{CE}/V_T}}{[e^{v_{CE}/V_T} + h_{FE}/h_{FC}]^2} \right]$$

$$\approx \frac{h_{FE}}{h_{FC}} \frac{I_{CQ}}{V_T} e^{-v_{CE}/V_T}$$

$$h_{fe} = \left.\frac{i_c}{i_b}\right|_{Q\ point} = \left.\frac{\Delta i_C}{\Delta i_B}\right|_{Q\ point}$$

$$h_{ie} = \left.\frac{v_{be}}{i_b}\right|_{v_{CE}=0} = \left.\frac{v_{be}}{i_b}\right|_{Q\ point}$$

$$= \frac{V_T}{I_{BQ}} \approx h_{fe} \cdot \frac{V_T}{I_{CQ}} \approx h_{fe} \cdot \frac{V_T}{I_{EQ}}$$

B-51

Equivalent circuit:

[Circuit diagram showing hybrid-parameter model with Base, Collector, and emitter terminals, including h_{ie}, $h_{re} \cdot v_{ce}$, $h_{fe} \cdot i_b$, h_{oe}, v_{be}, and v_{ce}]

This diagram can be simplified by ignoring h_{oe} and h_{re}.

C-E amplifier equivalent circuit:

[Circuit diagram showing C-E amplifier with i_i, R_b, v_{be}, h_{ie}, $h_{fe} \cdot i_b$, R_L, input Z_i and output Z_o, with currents i_b, i_c and voltage v_{ce}]

$$\frac{i_b}{i_i} = \frac{R_b}{R_b + h_{ie}}$$

$$A_i = \frac{i_L}{i_i} = \frac{-h_{fe}}{1 + h_{fe}[(25 \times 10^{-3})/I_{EQ}R_b]}$$

Requirement of high gain and stability:

$$h_{ie} = h_{fe} \cdot \frac{V_T}{I_{EQ}} \ll R_b \ll h_{fe} \cdot R_e$$

$$Z_i = \frac{R_b h_{ie}}{R_b + h_{ie}} \approx h_{ie} \quad (\text{if } R_b \gg h_{ie})$$

$$Z_o = \left.\frac{v_{ce}}{i_c}\right|_{i_i=0} = 1/h_{oe}$$

6.3 THE C-B CONFIGURATION

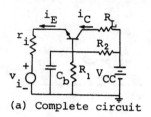

(a) Complete circuit

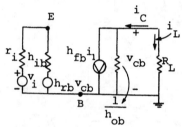

(b) Hybrid model

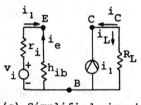

(c) Simplified circuit

For the hybrid model:

$$V_{eb} = h_{ib}i_1 + h_{rb} \cdot v_{cb} = h_{ib}(-i_e) + h_{rb} \cdot v_{cb}$$

$$i_c = h_{fb} \cdot i_1 + h_{ob} \cdot v_{cb} = h_{fb}(-i_e) + h_{ob} \cdot v_{cb}$$

$$h_{ib} = \left. \frac{v_{eb}}{-i_e} \right|_{v_{cb}=0} = \frac{V_T}{I_{EQ}} \approx \frac{h_{ie}}{1+h_{fe}}$$

$$h_{fb} = \left. \frac{i_c}{i_1} \right|_{v_{cb}=0} = \left. \frac{i_c}{-i_e} \right|_{v_{cb}=0} = -\alpha$$

$$h_{ob} = \left. \frac{i_c}{v_{cb}} \right|_{i_e=i_1=0}$$

Simplified equivalent circuit:

$$i_e = (1 + h_{fe})i_b = (i + h_{fe})\left(\frac{-v_{eb}}{h_{ie}}\right)$$

$$h_{ib} = \left.\frac{-v_{eb}}{i_e}\right|_{v_{cb}=0} = \frac{h_{ie}}{1+h_{fe}} \approx \frac{h_{ie}}{h_{fe}}$$

$$h_{fb} \text{ (short-circuit current gain)} = \left.\frac{i_c}{-i_e}\right|_{v_{cb}=0} = \frac{-h_{fe}}{1+h_{fe}} \doteq -1$$

$$h_{ob} \approx \frac{h_{oe}}{h_{fe}}$$

Modified circuits:

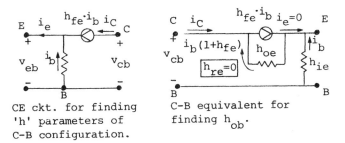

CE ckt. for finding 'h' parameters of C-B configuration.

C-B equivalent for finding h_{ob}.

To find the CB parameters h_{ob}, h_{fb}, and h_{ib}, divide the corresponding CE parameters by $1 + h_{fe}$.

6.4 THE C-C (EMITTER-FOLLOWER) CONFIGURATION

Characteristics:

A) A voltage gain slightly less than unity;
B) A high input impedance, and
C) A low output impedance.

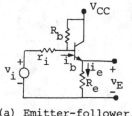

(a) Emitter-follower.

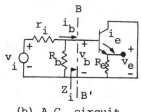

(b) A.C. circuit (c) Equivalent circuit

$$A_v = \frac{v_e}{v_i} = \frac{R_b}{r_i+R_b}\left[\frac{1}{1+[h_{ie}+(r_i\|R_b)]/[(1+h_{fe})R_e]}\right]$$

$$Z_i = h_{ie} + (1+h_{fe})R_e$$

$$Z_0 = h_{ib} + \frac{r_i}{1+h_{fe}}$$

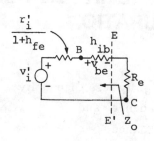

6.5 SIGNIFICANT PARAMETERS

	Configuration		
	CE	EF (CC)	CB
Gain	$A_i \approx -h_{fe}$	$A_v \approx 1$	$A_i \approx -h_{fb} = \dfrac{-h_{fe}}{1+h_{fe}}$
Input impedance	$h_{ie} = \dfrac{(25 \times 10^{-3}) h_{fe}}{I_{EQ}}$	$Z_i = h_{ie} + (h_{fe}+1) R_e$	$h_{ib} = \dfrac{h_{ie}}{1+h_{fe}}$
Output impedance	$\dfrac{1}{h_{oe}} > 10^4 \Omega$	$Z_o \approx h_{ib} + \dfrac{r_i'}{h_{fe}+1}$	$\dfrac{1}{h_{ob}} = \dfrac{1+h_{fe}}{h_{oe}}$
Simplest equivalent circuit	(see figure)	(see figure)	(see figure)

6.6 SMALL-SIGNAL EQUIVALENT CIRCUIT OF THE FET

6.6.1 EQUIVALENT CIRCUIT

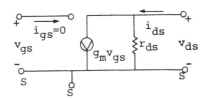

6.6.2 TRANSCONDUCTANCE

$$g_m = \left. \dfrac{\partial i_{DS}}{\partial v_{GS}} \right|_{Q \text{ point}}$$

For the MOSFET:

$$i_{DS} = k_n(V_{GS}-V_T)^2, \text{ and } g_m = 2k_n(V_{GS}-V_T)\Big|_{V_{GSQ}}$$

$$= 2\sqrt{k_n \cdot I_{DSQ}}$$

g_m of a FET is analogous to $1/h_{ib}$ in BJT:

$$\left(\frac{1}{h_{ib}}\right)_{BJT} \gg (g_m)_{FET}$$

6.6.3 DRAIN SOURCE RESISTANCE

$$r_{ds} = \frac{\partial V_{DS}}{\partial i_{DS}}\Big|_{Q\text{ point}} \quad \alpha \quad \frac{1}{I_{DQ}}$$

The drain source resistance is analogous to h_{oe} of the transistor.

6.6.4 AMPLIFICATION FACTOR

$$\mu = \frac{-\partial v_{DS}}{\partial v_{GS}}\Big|_{Q\text{ point}} = g_m \cdot r_{ds}$$

Equivalent Model:

6.7 THE COMMON-SOURCE VOLTAGE AMPLIFIER

$$A_v = \frac{V_L}{V_i} = -g_m(R_L \| Z_0)\left[\frac{1}{1+[r_i \div (R_3+(R_1\|R_2))]}\right]$$

with $r_i \ll R_3 + (R_1 \| R_2)$ and $R_L \ll Z_0$, $A_V \doteq -g_m R_L$.

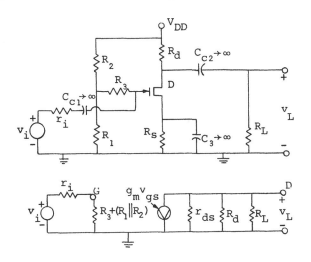

6.8 THE COMMON-DRAIN VOLTAGE AMPLIFIER (THE SOURCE FOLLOWER)

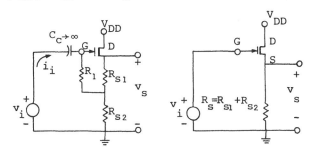

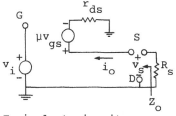

Equivalent circuit

$$Z_0 = \left.\frac{v_s}{i}\right|_{v_i=0} \text{ (as seen by } R_s) = \frac{r_{ds}}{1+\mu} \approx 1/g_m$$

$$\text{(when } \mu = g_m \cdot r_{ds} \gg 1)$$

A'_v (open-circuit voltage gain) $= \frac{\mu}{1+\mu} \cong 1$, when $\mu \gg 1$

$$A_v = \frac{v_s}{v_i} = \frac{\mu}{1+\mu}\left[\frac{g_m \cdot R_s}{1+g_m \cdot R_s}\right]$$

$$Z_i = \frac{v_g}{i_i} \cong \frac{R_1}{1 - \left(\frac{v_s}{v_g}\right)\left[\frac{R_{s2}}{R_{s1}+R_{s2}}\right]}$$

$$Z_{02} \text{ (looking into the source)} \cong \frac{R_s(r_{ds}+R_s)}{r_{ds}+(2+\mu)R_s}$$

$$\cong \frac{r_{ds}+R_s}{\mu}$$

Equivalent circuit for Z_i:

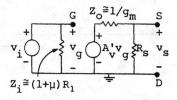

Assuming $R_{s2} \gg R_{s1}$,

$$Z_i \cong (1+\mu)R_1$$

6.9 THE PHASE-SPLITTING CIRCUIT

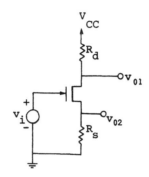

$$v_{01} = \frac{-\mu \cdot R_d}{(1+\mu)R_s + r_{ds} + R_d} v_i$$

$$v_{02} = \frac{(\mu/1+\mu)R_S}{R_s + \left(\frac{r_{ds}}{1+\mu}\right) + \frac{R_d}{1+\mu}} v_i$$

Z_{01} (looking into the source) $\cong \dfrac{R_s[r_{ds}+(1+\mu)R_s]}{r_{ds}+(2+\mu)R_s}$

$$\cong R_s$$

CHAPTER 7

AUDIO-FREQUENCY LINEAR POWER AMPLIFIERS

7.1 THE CLASS A COMMON-EMITTER POWER AMPLIFIER

The power amplifiers are classified according to the portion of the input sine wave cycle during which the load current flows.

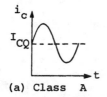
(a) Class A

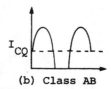
(b) Class AB

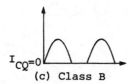
(c) Class B

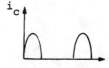
(d) Class C current flows for less than one-half cycle

Q-point placement:

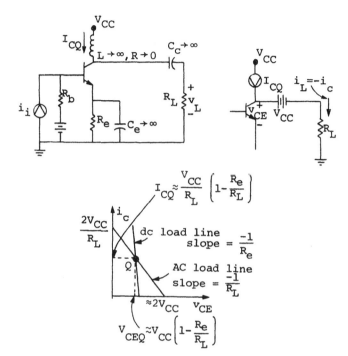

DC-load-line equation:
$$V_{CC} = v_{CE} + i_C \cdot R_e$$

AC-load-line equation:
$$v_{ce} = -i_c R_L = i_L \cdot R_L$$

or
$$i_C - I_{CQ} = \frac{-1}{R_L}(v_{CE} - V_{CEQ})$$

To place "Q" for the maximum symmetrical swing
$$I_{CQ} = \frac{V_{CC}}{R_{ac} + R_{dc}}$$
$$\approx V_{CC}/R_L$$

$$V_{CEQ} = V_{CC} \frac{R_L}{R_L + R_e} = \frac{V_{CC}}{1 + (R_e/R_L)}$$

$$\approx V_{CC} \quad (\text{because } R_L \gg R_e)$$

Power calculations:

$$i_C = I_{CQ} + i_c = \frac{V_{CC}}{R_L} + i_c$$

$$i_L = -i_c$$

$$i_{supply} = i_L + i_C = I_{CQ} = \frac{V_{CC}}{R_L}$$

$$v_{CE} = V_{CC} - i_C R_L, \quad v_L = i_L R_L = -i_c R_L$$

For a sinusoidal signal current:

$$i_i = I_{im} \cdot \sin \omega t, \quad i_c = I_{cm} \cdot \sin \omega t$$

Supplied power $P_{CC} = V_{CC} \cdot I_{CQ} \approx \dfrac{V_{CC}^2}{R_L}$

Power transferred to load:

$$P_L = \frac{I_{LM}^2 \cdot R_L}{2} = \frac{I_{cm}^2 \cdot R_L}{2}$$

(since $i_L = -i_c$, $I_{LM} = -I_{cm}$).

$$P_{Lmax} = \frac{I_{CQ}^2 \cdot R_L}{2} = \frac{V_{CC}^2}{2R_L} \quad (\text{when } I_{cm} = I_{CQ}).$$

Collector dissipation:

$$P_C = P_{CC} - P_L = \frac{V_{CC}^2}{R_L} - \frac{I_{cm}^2 \cdot R_L}{2}$$

$$P_{Cmin} = V_{CC}^2 / 2R_L \quad \text{and} \quad P_{Cmax} = \frac{V_{CC}^2}{R_L} = V_{CEQ} I_{CQ}$$

Efficiency:

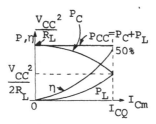

Variation of power and efficiency with collector current

$$\eta = \frac{P_L}{P_{CC}} = \frac{I_{cm}^2 (R_L/2)}{V_{CC} \cdot I_{CQ}} = \frac{1}{2}\left(\frac{I_{cm}}{I_{CQ}}\right)^2$$

$\eta_{max} = 50\%$

Figure of merit $= \dfrac{P_{Cmax}}{P_{L\,max}} = 2.$

The maximum-dissipation hyperbola:

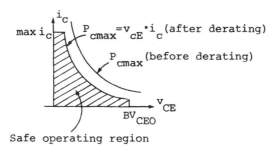

Safe operating region

For the safe operation, the Q-point must lie on or below the hyperbola $v_{CE} \cdot i_c = P_{c_1 max}$.

This hyperbola represents the locus of all operating points at which the collector dissipation is exactly P_{cmax}.

The ac load line, with slope $\dfrac{-1}{R_L}$, must pass through the Q-point, intersect the v_{CE} at a voltage less than BV_{CEO} and intersect the i_c axis at a current less than max i_c, i.e.,

B-64

$$2V_{CC} \leq BV_{CEO},$$

$$2I_{CQ} \leq i_{cmax}.$$

$$\text{Slope of hyperbola} \Big|_{Q \text{ point}} = \frac{\partial i_c}{\partial v_{CE}} = \frac{-I_{CQ}}{V_{CEQ}} = \frac{-1}{R_L}$$

7.2 THE TRANSFORMER-COUPLED AMPLIFIER

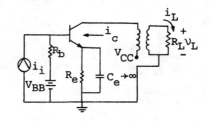

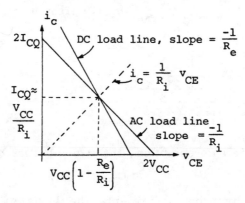

Load lines:

DC Load line: $V_{CC} = v_{CE} + i_E \cdot R_e \approx v_{CE} + i_c R_e$

AC load line: Slope $= \dfrac{i_c}{v_{ce}} = \dfrac{-1}{R_L'}$ (where $R_L' = N^2 R_L$).

B-65

$$I_{CQ} \approx \frac{V_{CC}}{R_L'}\left[1 - \frac{R_e}{R_L'}\right], \quad V_{CEQ} \approx V_{CC}\left[1 - \frac{R_e}{R_L'}\right]$$

Power calculations:

The signal i_L is sinusoidal; thus,

$$i_c = I_{cm} \cdot \sin \omega t.$$

Supplied power: $P_{CC} = V_{CC} \cdot I_{CQ} = \dfrac{V_{CC}^2}{R_L'}$

Load power:

$$P_L = \frac{I_{LM}^2}{2} R_L = \frac{I_{cm}^2}{2} R_L', \quad (I_{LM} = N \cdot I_{CM})$$

$$P_{Lmax} = \frac{V_{CC}^2}{2R_L'}$$

Collector dissipation:

$$P_C = \frac{V_{CC}^2}{R_L'} - \frac{I_{cm}^2}{2} R_L'$$

$$P_{cmax} = \frac{V_{CC}^2}{R_L'} = V_{CEQ} \cdot I_{CQ}$$

Efficiency:

$$\eta = \frac{1}{2}\left(\frac{I_{cm}}{I_{CQ}}\right)^2$$

$\eta_{max} = 50\%$

Fig. of merit $= \dfrac{P_{cmax}}{P_{Lmax}} = 2.$

7.3 CLASS B PUSH-PULL POWER AMPLIFIERS

Circuit:

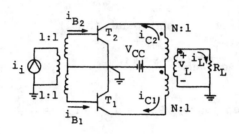

Waveforms:

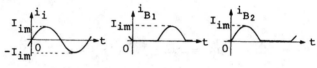

(a) Input current (b) Base current in T_1 (c) in T_2

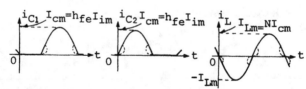

(d) Collector current in T_1 (e) Collector current in T_2 (f) Load current

Load-line determination:

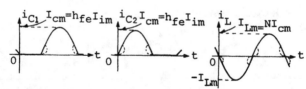

DC load line: $v_{CE} = V_{CC}$

AC load line:
$$\text{Slope} = \frac{i_c}{CE} = \frac{-1}{R'_L}$$

The maximum value of both i_{c_1} and i_{c_2} is $I_{cm} = \frac{V_{CC}}{R'_L}$.

Power calculations:
$$i_i = I_{im} \cdot \sin \omega t$$

Supplied power = $P_{CC} = \frac{2}{\pi} \cdot V_{CC} \cdot I_{cm}$

$$P_{CC(max)} = \frac{2}{\pi} V_{CC} \cdot \frac{V_{CC}}{R'_L} = \frac{2 V_{CC}^2}{\pi \cdot R'_L}$$

$$P_L = \frac{1}{2} I_{cm}^2 \cdot R'_L, \quad P_{L,max} = \frac{V_{CC}^2}{2R'_L}$$

Power dissipated in the collector:
$$2P_C = \frac{2}{\pi} V_{CC} \cdot I_{cm} - \frac{R'_L \cdot I_{cm}^2}{2} = P_{CC} - P_L$$

$I_{cm} = \frac{2}{\pi} \cdot \frac{V_{CC}}{R'_L}$ (The collector current at which the collector dissipation is maximum)

$$2P_{cmax} = \frac{2}{\pi^2} \cdot \frac{V_{CC}^2}{R'_L}$$

Efficiency: $\eta = \frac{P_L}{P_{CC}} = \frac{\pi}{4} \frac{I_{cm}}{V_{CC}/R'_L}$

$$\boxed{\eta_{max} = \frac{\pi}{4} = 78.5\%}$$

Figure of merit: $\frac{P_{cmax}}{P_{Lmax}} = \frac{2}{\pi^2} \approx \frac{1}{5}$

Power and efficiency variation:

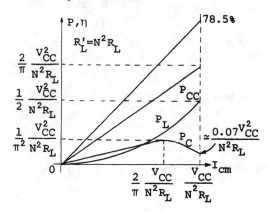

7.4 AMPLIFIERS USING COMPLEMENTARY SYMMETRY

The circuit:

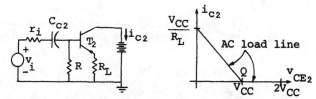

Circuit and load line for T_2 of complementary-symmetry emitter-follower

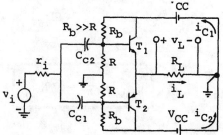

Complementary-symmetry amplifier

Each transistor is essentially a class "B" emitter follower.

$$i_L = i_{C_1} - i_{C_2}$$

$$I_{cm} = \frac{V_{CC}}{R_L} \quad \text{(If } v_i \text{ is sinusoidal, } i_L \text{ is also sinusoidal.)}$$

$$i_L = I_{CM} \cdot \sin \omega t$$

$$= \frac{V_{CC}}{R_L} \cdot \sin \omega t.$$

$$P_{L,max} = \frac{V_{CC}^2}{2R_L}$$

CHAPTER 8

FEEDBACK AMPLIFIERS

8.1 CLASSIFICATION OF AMPLIFIERS

8.1.1 VOLTAGE AMPLIFIER

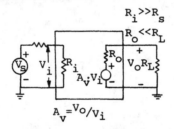

The voltage amplifier provides a voltage output proportional to the voltage input. The proportionality factor is independent of the source and load impedances. For the ideal amplifier:

$$R_i = \infty, \quad R_o = 0, \quad A_v = \frac{V_o}{V_s}$$

8.1.2 CURRENT AMPLIFIER

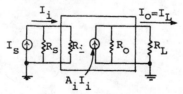

Norton's equivalent circuit of a current amplifier.

$$A_i = \frac{I_L}{I_i} \quad \text{(with } R_i = 0 \text{ representing the short-circuit current amplification.)}$$

$$R_i << R_s \quad \text{and} \quad R_o >> R_L$$

8.1.3 TRANSCONDUCTANCE AMPLIFIER

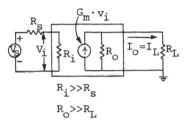

$R_i >> R_s$

$R_o >> R_L$

Thevenin's equivalent in its input circuit and a Norton's equivalent in its output circuit.

Output current is proportional to the signal voltage and is independent of R_S and R_L.

G_m (the short circuit mutual or transfer conductance)

$$= \frac{I_o}{v_i} \text{ for } R_L = 0$$

8.1.4 TRANSRESISTANCE AMPLIFIER

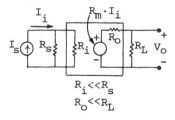

$R_i << R_s$

$R_o << R_L$

V_o is proportional to the signal current I_s and is independent of R_S and R_L.

$$R_m = \frac{V_o}{I_i} \text{ with } R_L = \infty \quad (R_m = \text{open circuit mutual resistance})$$

8.1.5 IDEAL AMPLIFIER CHARACTERISTICS

Parameter	Amplifier type			
	Voltage	Current	Transconductance	Transresistance
R_i	∞	0	∞	0
R_o	0	∞	∞	0
Transfer Characteristics	$V_o = A_v V_s$	$I_L = A_i I_s$	$I_L = G_m V_s$	$V_o = R_m I_s$

8.2 THE CONCEPT OF "FEEDBACK"

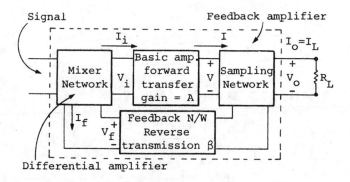

8.2.1 SAMPLING NETWORK

The output voltage is sampled by connecting the feedback network in shunt across the output voltage or node sampling.

Current or loop sampling - The feedback network is connected in series with the output.

8.2.2 TRANSFER RATIO OR GAIN

Each of the four quantities A_V, A_I, G_M and R_M is a different type of transfer gain of the basic amplifier without feedback (A).

A_f = The transfer gain with feedback. = A_{Vf} or A_{If} or G_{Mf} or R_{Mf}

$$A_{Vf} = \frac{V_o}{V_s}, \quad \frac{I_o}{I_s} = A_{If}, \quad \frac{I_o}{V_s} = G_{Mf}, \quad \frac{V_o}{I_s} = R_{Mf}$$

8.3 TRANSFER GAIN WITH FEEDBACK

8.3.1 SINGLE-LOOP FEEDBACK AMPLIFIER

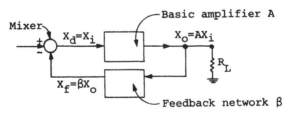

$X_d = X_s - X_f = X_i$ = The difference signal.

8.3.2 FEEDBACK AMPLIFIER TOPOLOGIES

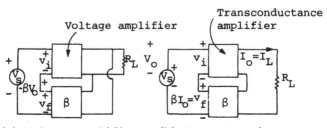

(a) Voltage amplifier with moltage-series feedback

(b) Current-series feedback

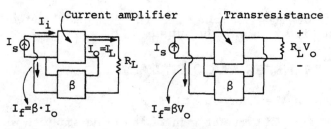

(c) Current-shunt feedback

(d) Voltage-shunt feedback

$$\beta \equiv \frac{X_f}{X_o}$$

β is a complex function of the signal frequency and is often a positive or negative real number.

$$A = \frac{X_o}{X_i}$$

$$\boxed{A_f = \frac{X_o}{X_s} = \frac{A}{1+\beta A}}$$

If $|A_f| < |A|$, the feedback is termed negative; if $|A_f| > |A|$, the feedback is positive, or regenerative.

In the case of negative feedback, the gain of the ideal amplifier is divided by $|1+\beta A|$, which exceeds unity.

Loop gain:

The loop gain $= -A \cdot \beta =$ The return ratio.

The return difference $D = 1 + A\beta$

$$N = \text{dB of feedback} = 20 \log \left| \frac{A_f}{A} \right| = 20 \log \left| \frac{1}{1+A\beta} \right|$$

8.4 FEATURES OF THE NEGATIVE FEEDBACK AMPLIFIER

Sensitivity – The fractional change in amplification with feedback divided by the fractional change without feedback is the sensitivity of the transfer gain.

$$\frac{|dA_f/A_f|}{|\frac{dA}{A}|} = \frac{1}{|1+\beta A|}$$

$$S = \text{Sensitivity} = \frac{1}{|1+\beta A|}$$

Desensitivity = $D = 1 + \beta A$.

$A_f = A/D$

if $|\beta A| \gg 1$, $A_f = \frac{A}{1+\beta A} \quad \frac{A}{\beta \cdot A} = \frac{1}{\beta}$

Frequency distortion - If the feedback network does not contain ractive elements, the overall gain is not a function of frequency.

Reduction of noise - The noise introduced in an amplifier is divided by the factor D if feedback is introduced.

8.5 INPUT RESISTANCE

Voltage-series feedback:

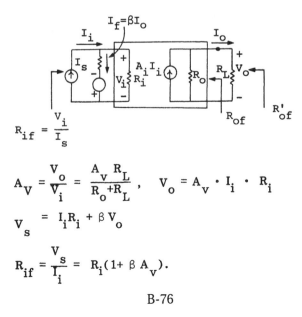

$R_{if} = \frac{V_i}{I_s}$

$A_V = \frac{V_o}{V_i} = \frac{A_v R_L}{R_o + R_L}$, $\quad V_o = A_v \cdot I_i \cdot R_i$

$V_s = I_i R_i + \beta V_o$

$R_{if} = \frac{V_s}{I_i} = R_i(1 + \beta A_v)$.

$$A_v = \lim_{R_L \to \infty} A_v$$

Current-series feedback:

$$R_{if} = R_i(1+ \beta G_M), \quad G_M = \lim_{R_L \to 0} G_M$$

$$G_M = \frac{I_o}{V_i} = \frac{G_M \cdot R_o}{R_o + R_L}$$

Current-shunt feedback:

$$A_I = \frac{I_o}{I_i} = A_i \cdot R_o/(R_o + R_L)$$

$$I_s = (1 + \beta A_I)I_i$$

$$R_{if} = \frac{V_i}{(1+ \beta A_I)I_i} = \frac{R_i}{1 + \beta A_I}$$

$$A_i = \lim_{R_L \to 0} A_I$$

Voltage-shunt feedback:

$$R_{if} = \frac{R_i}{1 + \beta \cdot R_M}, \quad R_M = \frac{V_o}{V_i} = \frac{R_m \cdot R_L}{R_o + R_L}$$

$$R_m = \lim_{R_L \to \infty} R_M$$

8.6 OUTPUT RESISTANCE

For voltage sampling, $R_{of} < R_o$; whereas, for current sampling, $R_{of} > R_o$.

Voltage series feedback

$$R_{of} = \frac{V}{I} = \frac{R_o}{1 + \beta A_V}, \quad \text{where } D = 1 + \beta A_V$$

$$R'_{of} = R'_o/1 + \beta A_v, \quad \text{where } R'_o = R_o \| R_L.$$

Voltage-shunt feedback:

$$R_{of} = \frac{R_o}{1 + \beta R_m}, \quad R'_{of} = R'_o/1 + \beta R_M$$

Current-shunt feedback:

$$R_{of} = \frac{V}{I} = R_o(1+\beta A_i)$$

$$R'_{of} = R'_o \frac{1+\beta A_i}{1+\beta A_I}$$

$$= R_{of} \text{ (for } R_L = \infty, A_I = 0 \text{ and } R'_o = R_o).$$

Current-series feedback:
$$R_{of} = R_o(1+\beta G_m) \text{ and } R'_{of} = R'_o \frac{1 + \beta G_m}{1 + \beta G_M}$$

8.7 FEEDBACK AMPLIFIER ANALYSIS

Topology Characteristic	Voltage shunt	Current shunt	Current series	Voltage series
Feedback signal X_f	Current	Current	Voltage	Voltage
Sample signal X_o	Voltage	Current	Current	Voltage
Input circuit: set†	$V_o = 0$	$I_o = 0$	$I_o = 0$	$V_o = 0$
Output circuit: set†	$V_i = 0$	$V_i = 0$	$I_i = 0$	$I_i = 0$
Signal source	Norton	Norton	Thévenin	Thévenin
$\beta = X_f/X_o$	I_f/V_o	I_f/I_o	V_f/I_o	V_f/V_o
$A = X_o/X_i$	$R_M = V_o/I_i$	$A_I = I_o/I_i$	$G_M = I_o/V_i$	$A_v = V_o/V_i$
$D = 1 + \beta A$	$1 + \beta R_M$	$1 + \beta A_I$	$1 + \beta G_M$	$1 + \beta A_v$
A_f	R_M/D	A_I/D	G_M/D	A_v/D
R_{if}	R_i/D	R_i/D	R_iD	R_iD
R_{of}	$\dfrac{R_o}{1 + \beta R_m}$	$R_o(1 + \beta A_i)$	$R_o(1 + \beta G_m)$	$\dfrac{R_o}{1 + \beta A_v}$
$R'_{of} = R_{of}\|R_L$	$\dfrac{R'_o}{D}$	$R'_o \dfrac{1 + \beta A_i}{D}$	$R'_o \dfrac{1 + \beta G_m}{D}$	$\dfrac{R'_o}{D}$

Analysis of a feedback amplifier:

Identify the topology - The input loop is defined as the mesh containing the applied signal voltage V_s and either (a) the base-to-emitter region of the first BJT, or (b) the gate-to-source region of the first FET, or (c) the section between the two inputs of a differential amplifier.

Mixing - There is a series mixing in the input circuit if there is a circuit compoment y in series with V_s and if y is connected to the output. If this is true, the voltage across y is the feedback signal $X_f = V_f$.

If this is not true, test for shunt comparison. Shunt mixing is present if the feedback signal subtracts from the applied excitation as a current at the input node.

Type of sampling:

A) Set $V_o = 0$ (i.e., set $R_L = 0$). If X_f becomes zero, the original system exhibited voltage sampling.

B) Set $I_o = 0$. If X_f becomes zero, current sampling was present in the original amplifier.

The amplifier without feedback - The amplifier configuration without providing feedback but with taking the loading of β network into account is obtained by applying following rules.

To find the input circuit:

A) Set $V_o = 0$ for voltage sampling. (i.e., short-circuit the output node)

B) Set $I_o = 0$ for current sampling. (i.e., open-circuit the output loop)

To find the output circuit :

A) Set $V_i = 0$ for shunt comparison.

B) Set $I_i = 0$ for series comparison. In other words, open-circuit the input loop

Outline of analysis - To find A_f, R_{if} and R_{of} the following steps are carried out:

A) Find out the topology first. It will determine whether X_f is a voltage or a current. The same applies to X_o.
B) Using the above rules, draw the basic amplifier circuit without feedback.
C) If X_f is a voltage, use a Thevenin's source; use Norton's source if X_f is a current.
D) Substitute the proper model for each active device (h-parameter model).
E) Indicate X_f and X_o on the circuit obtained by steps B, C, and D. Calculate $\beta = \dfrac{X_f}{X_o}$
F) Apply KVL and KCL to the equivalent circuit obtained after step D, to find A.
G) Calculate D, A_f, R_{if}, R_{of} and R'_{of} from A and B.

CHAPTER 9

FREQUENCY RESPONSE OF AMPLIFIERS

9.1 FREQUENCY DISTORTION

A plot of gain (phase) versus frequency of an amplifier is called the amplitude (phase) frequency-response characteristic.

If the amplification A is independent of frequency, and if the phase shift θ is proportional to frequency (or is zero), then the amplifier will preserve the form of the input signal, although the signal will be shifted in time by an amount θ/ω.

Frequency-response characteristics:

A) Low-frequency region: The amplifier behaves like a simple high-pass circuit.

B) Mid-band frequency: Amplification and delay is quite constant. The gain is normalized to unity.

C) High-frequency region: The circuit behaves like a low-pass network.

Low and high-frequency response:

Low-frequency response:

The low-frequency region is like a simple high-pass circuit.

$$V_o = \frac{R_1}{R_1 + \frac{1}{SC_1}}, \quad V_i = \frac{S}{S + \frac{1}{R_1 C_1}} V_i$$

$$|A_L(f)| = \frac{1}{1-j(f_L/f)}, \quad f_L = \frac{1}{2\pi R_1 C_1}, \quad \theta_L = \arctan \frac{f_L}{f}$$

The low-3dB-frequency: At this frequency, the gain has fallen to 0.707 times its mid-band value A_o; $f = f_L$ is the low-3dB frequency.

High-frequency response:

$$V_o = \frac{\frac{1}{SC_2}}{R_2 + \frac{1}{SC_2}} \quad V_i = \frac{1}{1 + SR_2 C_2} V_i$$

$$|A_H(f)| = 1/\sqrt{1+(f/f_H)^2} \quad \theta_H = -\arctan \frac{f}{f_H}$$

$$f_H = 1/2\pi R_2 C_2$$

The high-3dB frequency: At $f = f_H$, the gain is reduced to $1/\sqrt{2}$ times its mid-band value, this is called the high-3dB frequency.

Bandwidth: B.W. = frequency range from f_L to f_H.

9.2 THE EFFECT OF COUPLING AND EMITTER BYPASS CAPACITORS ON LOW-FREQUENCY RESPONSE

An emitter resistance R_e is used for self-bias in an

amplifier and to avoid degeneration; C_Z is used to bypass R'_e.

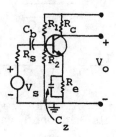

$$V_o = -I_b h_{fe} R_c = \frac{-V_s \cdot h_{fe} \cdot R_c}{R_s + h_{ie} + Z_b + Z'_e}$$

$$Z_b = \frac{1}{j\omega c_b} \quad \text{and} \quad Z'_e = (1+h_{fe}) \frac{R_e}{1+j\omega C_z \cdot R_e}$$

$$j\omega(Z_b + Z'_e) = \frac{1}{C_b} + \frac{1+h_{fe}}{C_Z} = \frac{1}{C_1}$$

$$C_1 = C_b + \frac{C_Z}{1+h_{fe}}$$

$$A_{vs} = \frac{V_o}{V_s} = \frac{-h_{fe} R_c}{R_s + h_{ie}} \cdot \frac{1}{1 - j/\omega c_1 (R_s + h_{ie})}$$

$$A_o = \text{mid-band gain} = \frac{-h_{fe} \cdot R_c}{R_s + h_{ie}}, \quad \frac{A_{vs}}{A_o} = \frac{1}{1-j(f_L/f)}$$

Low-3dB freq. $f_L = 1/2\pi C_1 (R_s + h_{ie})$ (at f_L, $R_s + h_{ie} = R_1$)

9.3 THE HYBRID-π TRANSISTOR MODEL AT HIGH FREQUENCIES

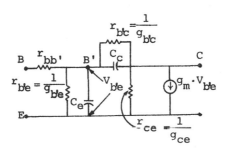

$C_c = C_{b'c}$ is the measured C_B ouptut capacitance with the input open ($I_E = 0$). (C_c is the transition capacitance, which varies as V_{CB}^{-n}, where n is $\frac{1}{2}$ or $\frac{1}{3}$). $C_e = C_{De} + C_{Te}$, where C_{De} is the emitter diffusion capacitance and C_{Te} is emitter-junction capacitance.

The diffusion capacitance:

$$C_{De} = g_m \frac{w^2}{2D_B} \quad (D_B = \text{the diffusion constant for minority carriers})$$

Simplified model:

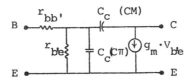

B-84

Variations of Hybrid-π barameters:

Dependence of parameters upon current, voltage and temperature

Parameter	Variation with increasing:		
	$\|I_C\|$	$\|V_{CE}\|$	T
g_m	$\|I_C\|$	Independent	$1/T$
$r_{bb'}$	Decreases		Increases
$r_{b'e}$	$1/\|I_C\|$	Increases	Increases
C_e	$\|I_C\|$	Decreases	
C_c	Independent	Decreases	Independent
h_{fe}		Increases	Increases
h_{ie}	$1/\|I_C\|$	Increases	Increases

9.4 THE C-E SHORT-CIRCUIT CURRENT GAIN

Approximate equivalent circuit:

$$I_L = -g_m V_{b'e}$$

$$V_{b'e} = I_i / g_{b'e} + j\omega(C_e + C_c)$$

$$A_i = \frac{I_L}{I_i} = \frac{-g_m}{(g_{b'e} + j\omega[C_e + C_c])} = \frac{-h_{fe}}{1+j(f/f_\beta)}$$

B-85

$$f_\beta = \frac{g_{b'e}}{2\pi(C_e+C_c)} = \frac{1}{h_{fe}} \frac{g_m}{2\pi(C_e+C_c)}$$

At $f = f_\beta$, $|A_i|$ is equal to $1/\sqrt{2} = 0.707$ of its low-frequency value, h_{fe}.

B.W. = The frequency range up to f_β.

The parameter f_T:

f_T is the frequency at which the short-circuit common-emitter current gain attains unit magnitude.

$$f_T \approx h_{fe} \cdot f_\beta = \frac{g_m}{2\pi(C_e+C_c)} \approx \frac{g_m}{2\pi \cdot C_e}$$

Variation of f_T on collector current:

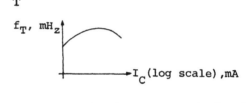

f_T also represents the short-circuit current gain-bandwidth product.

The plot of the short-circuit CE current gain versus the frequency:

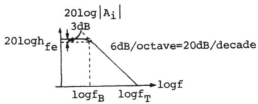

At f_T (the highest frequency of interest),

$$\frac{\omega C_c}{g_m} = \frac{2\pi \cdot f_T \, C_c}{g_m} = \frac{C_c}{C_e+C_c} \cong 0.03$$

9.5 THE COMMON-DRAIN AMPLIFIER AT HIGH FREQUENCIES

The source-follower and its small-signal high-frequency equivalent circuit:

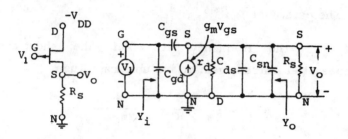

$$A_v = \text{The voltage gain} = \frac{(g_m + j\omega C_{gs})R_s}{1+(g_m+g_d+j\omega C_T)R_s}$$

$$C_T = C_{gs} + C_{ds} + C_{sn}$$

$$A_v(\text{at low frequency}) \approx \frac{g_m \cdot R_s}{1+(g_m+g_d)R_s} \approx \frac{g_m}{g_m+g_d}$$

$$= \frac{\mu}{1+\mu}$$

Input and output admittance:

$$y_i = j\omega C_{gd} + j\omega C_{gs}(1-A_v) \approx j\omega C_{gd}$$

$$y_o = g_m + g_d + j\omega C_T$$

$$R_o \text{ (at low frequency)} = 1/g_m + g_d \approx 1/g_m$$

9.6 THE COMMON-SOURCE AMPLIFIER AT HIGH FREQUENCIES

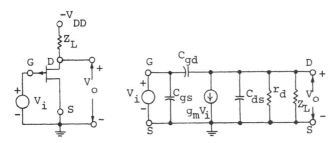

$$I = -g_m \cdot V_i + V_i \cdot y_{gd}$$

The amplification is $A_v = \dfrac{V_o}{V_i} = \dfrac{I \cdot Z}{V_i} = \dfrac{-g_m + y_{gd}}{y_L + g_d + y_{ds} + y_{gd}}$

$$A_v \text{ (at low frequencies)} = \dfrac{-g_m}{y_L + g_d} = \dfrac{-g_m \cdot r_d \cdot Z_L}{r_d + Z_L} = -g_m \cdot Z'_L$$

The input admittance is

$$y_i = y_{gs} + (1-A_v)y_{gd} = G_i + j\omega C_i \approx j\omega C_{gd}$$

9.7 THE EMITTER-FOLLOWER AMPLIFIER AT HIGH FREQUENCIES

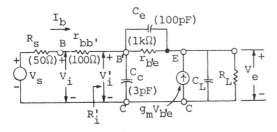 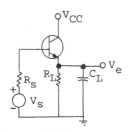

High-frequency equivalent circuit   Emitter-follower

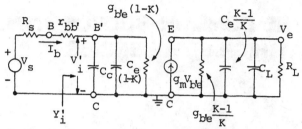

Fig: Equivalent circuit using Miller's theorem

The nodal equations at the nodes of B' and E are:

$$G'_s V_s = [G'_s + g_{b'e} + s(C_c + C_e)]v'_i - (g_{b'e} + sC_e)V_e$$

$$0 = -(g + sC_e)v'_i + [g + \frac{1}{R_L} + s(C_e + C_L)]V_e$$

$$G'_s = 1/R_s + r_{bb'} \quad \text{and} \quad g = g_m + g_{b'e}$$

Single pole solution:

$$K = V_e / V'_i$$

$$V_e = \frac{g_m \cdot V_{b'e}}{\frac{1}{R_L} + j\omega C_L} = \frac{g_m \cdot R_L (V'_i - V_e)}{1 + j\omega C_L R_L} \quad \text{(for K=1)}$$

$$K = \frac{K_o}{1 + jf/f_H}, \quad K_o = \frac{g_m \cdot R_L}{1 + g_m \cdot R_L} \approx 1.$$

$$f_H = \frac{1 + g_m \cdot R_L}{2\pi \cdot C_L \cdot R_L} \approx \frac{g_m}{2\pi C_L} = \frac{f_T \cdot C_e}{C_L}$$

Input admittance:

$$Y'_i = \frac{I_b}{V'_i} = j\omega [C_c + (1-K)C_e] + (1-K)g_{b'e}$$

$$Y'_i = j2\pi f \cdot C_c + (g_{b'e} + j2\pi \cdot f \cdot C_e) \frac{1 - K_o + jf/f_H}{1 + jf/f_H}$$

$$Y'_i = j2\pi f \cdot C_c + j \, g_{b'e} \cdot f/f_H - 2\pi \cdot f^2 \frac{C_e}{f_H}$$

9.8 SINGLE-STAGE CE TRANSISTOR AMPLIFIER RESPONSE

Equivalent circuit:

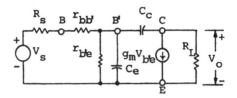

The transfer function:

$$\frac{V_o}{V_s} = \frac{-G'_s \cdot R_L(g_m - sC_c)}{s^2 C_e \cdot C_c \cdot R_L + s[C_e + C_c + C_c R_L(g_m + g_{b'e} + G'_s)] + G'_s}$$

The transfer function is of the form

$$A_{vs} = \frac{V_o}{V_s} = \frac{K_1(s-s_0)}{(s-s_1)(s-s_2)}$$

Approximate analysis:

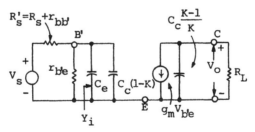

After application of Miller's theorem

$K = \frac{V_{ce}}{V_{b'e}}$, for $|K| \gg 1$. Also the output capacitance is C_c and the output time constant is $C_c \cdot R_L$.

$K = -g_m \cdot R_L$ (neglecting C_c)

B-90

Input capacitance $C = C_e + C_c(1 + g_m \cdot R_L)$

Input loop resistance $= R = R_s' \parallel r_{b'e}$

$$A_{vs} = v_o/v_s = (-g_m \cdot R_L \cdot G_s')/(G_s' + g_{b'e} + sC)$$

$$A_{vs} = \frac{A_{vso}}{1 + j\,f/f_H}$$

The high 3-dB frequency $= f_H = \dfrac{G_s' + g_{b'e}}{2\pi C} = \dfrac{1}{2\pi R \cdot C}$

$$|A_{vs}| = \frac{|A_{vso}|}{[1 + (f/f_H)^2]^{\frac{1}{2}}}, \quad \theta_1 = -\pi - \arctan f/f_H$$

CHAPTER 10

OPERATIONAL AMPLIFIERS

10.1 THE BASIC OPERATIONAL AMPLIFIER

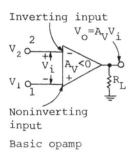

Basic opamp

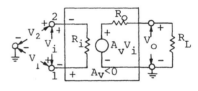

Low-frequency model of opamp

Ideal op amp:

A) $R_i = \infty$

B) $R_o = 0$

C) $A_v = -\infty$

D) B.W. $= \infty$

E) $V_o = 0$ when $V_1 = V_2$, independent of the magnitude of V_1

F) no drift of characteristics

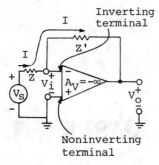
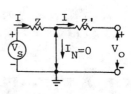

Ideal op amp with feedback impedances

This is the basic inverting circuit. This topology represents voltage-shunt feedback.

$$A_{vf} = \text{voltage gain with feedback} = \frac{-Z'}{Z}$$

Inverting operational amplifier:

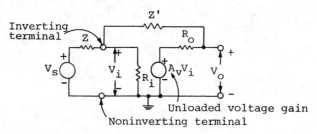

Small signal model

For a small-signal model, $|A_v| \neq \infty$, $R_i \neq \infty$ and $R_o \neq 0$.

$$A_{vf} = -y/y' - \left(\frac{1}{A_v}\right)(y'+y+y_i)$$

(where the y's are the admittances)

$$-A_v = \frac{V_o}{V_i} \text{ (with } Z'\text{)} = \frac{A_v + R_o y'}{1 + R_o y'}$$

Non-inverting op amp:

B-93

The configuration is that of a voltage-series feedback amplifier, with the feedback voltage, v_f, equal to v_2.

The feedback factor $\beta = \dfrac{V_2}{V_o} = \dfrac{Z}{Z+Z'}$ ($I_2 = 0$)

If $A_v \beta \gg 1$, then

$$A_{vf} \approx 1/\beta = \dfrac{Z+Z'}{Z} = 1 + \dfrac{Z'}{Z}$$

Configuration:

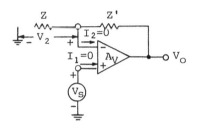

10.2 OFFSET ERROR VOLTAGES AND CURRENTS

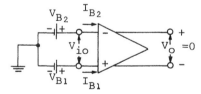

Input bias currents I_{B_1} and I_{B_2} and offset voltage V_{io}.

The input offset current is the difference between the input currents entering the input terminals of a balanced amplifier. In the figure above,

$$I_{io} = I_{B_1} - I_{B_2} \quad \text{(when } V_o = 0\text{)}$$

B-94

Input offset current drift - This drift is described by the ratio $\frac{\Delta V_{io}}{\Delta T}$, where

ΔV_{io} = the change of input offset voltage

and ΔT = change in temperature.

Input offset voltage - When applied to the input terminals, this voltage will balance the amplifier.

Input offset voltage drift:

$$\frac{\Delta V_{io}}{\Delta T}$$

ΔV_{io} = change of input offset voltage.

Output offset voltage - This voltage marks the difference between the dc voltages measured at the output terminals on grounding the two input terminals.

Input common mode range - This is the range of the common mode input signal for which the differential amplifier remains linear.

Input differential range - This range is the maximum difference signal that can safely be applied to the op amp input terminals.

Output voltage range - This is the maximum output swing that can be obtained without significant distortion at a specified load resistance.

Full-power bandwidth - This bandwidth is the maximum frequency at which a sinusoid whose size is the output voltage range is obtained.

Slew rate - This is the time rate of change of the closed-loop amplifier output voltage under large signal conditions.

The model of an op amp and balancing techniques:

Model:

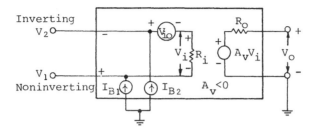

Universal balancing technique:

It is necessary to apply a small dc voltage in the input to bring the d output voltage to zero.

The following circuit supplies a small voltage in a series where, the non-inverting terminal is in the range $\pm V\left[\dfrac{R_2}{R_2+R_3}\right]$ = ±15mV, if ±15 V supplies are used.

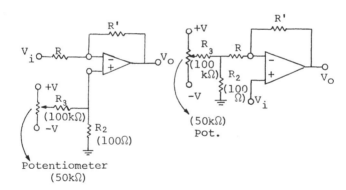

10.3 MEASUREMENT OF OPERATIONAL AMPLIFIER PARAMETERS

Input offset voltage V_{io}:

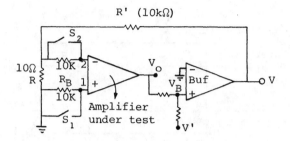

Set $V' = 0$ to get $V_o = 0$. The close s_1 and s_2.

If $V_o = 0$, then $V_i = 0$, and V_{io} appears between the inverting and non-inverting terminals.

$$V = \frac{V_{io}}{R}(R+R') = 1001\, V_{io} \approx 10^3 V_{io} \equiv V_3.$$

From the meter reading V_3 in volts, we get V_{io} in mv.

Power supply rejection ratio $= \dfrac{\Delta V_{io}}{\Delta V_{ce}}$ (ΔV_{io} and ΔV_{ce} the difference in the two input offset voltages)

Input bias current:

S_1 and S_2 are open and closed, respectively, and $V' = 0$.

Voltage across $R = V_{io} - R_B \cdot I_{B_1}$ and $V = \dfrac{R+R'}{R}(V_{io} - R_B \cdot I_{B_1})$

$$\approx 10^3(V_{io} - 10^4 I_{B_1}) \equiv V_4$$

$-I_{B_1} = (V_4 - V_3) 10^{-7} A = 100(V_4 - V_3)\, mA$

Open s_2 and close s_1; $V' = 0$ and we get I_{B_2}.

Bias current $I_B = \frac{1}{2}(I_{B_1} + I_{B_2})$ and I_{io}(offset) $= I_{B_1} - I_{B_2}$.

Open-loop differential voltage gain $A_v = A_d$:

S_1 and S_2 are closed, and V' is set to the output voltage $= -10V$ then, $V_o = -V' = 10v$.

$$V = \frac{R+R'}{R}(V_{io}+V_i) \approx 10^3 \left(V_{io} + \frac{V_o}{A_v}\right) \equiv V_5$$

$$A_v = \frac{10^3 \cdot V_o}{V_5 - V_3} = 10^4/V_5 - V_3$$

If we want to know the voltage gain A_v when there is a load, it is necessary to place R_L between V_o and the ground.

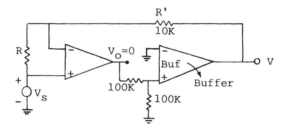

Close S_1 and S_2; $V' = 0$; apply signal v_s.

$$V_o = A_d \cdot V_d + A_c \cdot V_c = 0$$

$$V_1 = V_s \text{ and } V_2 = V_s \cdot \frac{R'}{R+R'} + V \cdot \frac{R}{R+R'} \cong V_s + \frac{V \cdot R}{R'}$$

$$V_d = V_1 - V_2 - V_{io} = \frac{-R}{R'}(V+V_3)$$

$$V_c = V_s + (VR/2R')$$

$$-A_d \frac{R}{R'}(V+V_3) + A_c \left[V_s + \frac{V \cdot R}{2R'}\right] = 0$$

If the measured value of V is V_6, then

$$\rho \cdot \frac{R}{R'}(V_6+V_3) = v_s$$

Slew rate - The slew rate is the maximum rate of change of the output voltage when supplying the rated output.

B-98

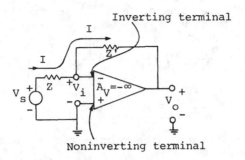

For a single-ended input amplifier, adjust $Z = R = 1K$ and $Z' = R' = 10K\Omega$.

V_s is a high-frequency square-wave. Its slopes are measured with respect to the lines of the leading and trailing edges of the output signal.

The slower of the two is the slew rate.

Frequency response of the op amp:

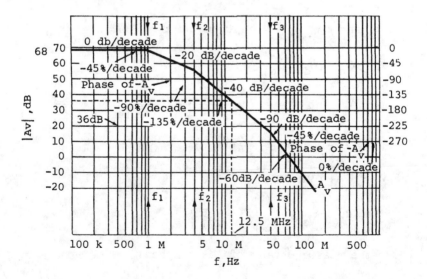

CHAPTER 11

OPERATIONAL AMPLIFIER SYSTEMS

11.1 BASIC OPERATIONAL AMPLIFIER APPLICATIONS

11.1.1 SIGN CHANGER, OR INVERTER

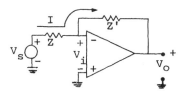

If $Z = Z'$, then $A_{vf} = \dfrac{V_o}{V_s} = \dfrac{-Z'}{Z} = -1$.

The sign of the input signal is changed at the output.

The circuit acts as a phase inverter.

11.1.2 SCALE CHANGER

If the ratio $\dfrac{Z'}{Z} = K$ (a real constant), then $A_{vf} = -K$.
The scale has been multiplied by the factor $-K$.

B-100

11.1.3 PHASE SHIFTER

If Z and Z' are equal in magnitude and differ in angle, then the op amp shifts the phase of a sinusoidal input voltage.

11.1.4 ADDER

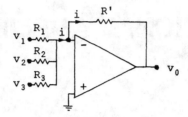

$$v_o = \frac{-R'}{R}(v_1 + v_2 + v_3), \text{ if } R_1 = R_2 = R_3 = R$$

11.1.5 NON-INVERTING ADDER

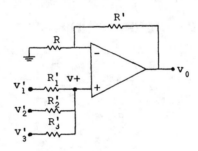

$$v_o = \left(1 + \frac{R'}{R}\right)v_+$$

For n equal to resistors, each of value R'_2

$$\frac{R'_{p_2}}{R'_2 + R'_{p_2}} = \frac{R'_2 \div (n-1)}{R'_2 + [R'_2 \div (n-1)]} = \frac{1}{n}$$

$$v_+ = \frac{1}{n}(v'_1 + v'_2 \ldots).$$

$$R'_{p_2} = R'_1 \parallel R'_3 \parallel R'_4 \ldots \parallel R'_n$$

B-101

11.1.6 VOLTAGE-TO-CURRENT CONVERTER (TRANSCONDUCTANCE AMPLIFIER)

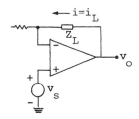

For a floating load

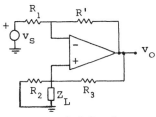

A grounded load

Floating load (neither side is grounded):

The current in $Z_L = i_L = \dfrac{v_{s(t)}}{R_1}$.

Grounded load: $\qquad i_L(t) = -\dfrac{v_{s(t)}}{R_2}$

11.1.7 CURRENT-TO-VOLTAGE CONVERTER

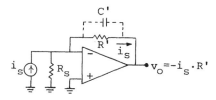

C' is used to reduce the high-frequency noise as well as the possibility of oscillations.

$$v_o = -i_s R'$$

The circuit acts like an ammeter with zero voltage across the meter.

B-102

11.1.8 D.C. VOLTAGE FOLLOWER

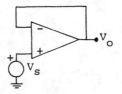

$V_o = V_s$ (because the inputs are tied (virtually) together)

The follower has a high input resistance and a low output resistance.

11.2 AC-COUPLED AMPLIFIER

This is used for amplifying an ac signal, while any dc signal is to be blocked.

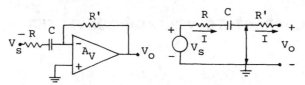

Equivalent circuit

$$V_o = -IR' = \frac{-V_s}{R + \frac{1}{sC}} R'$$

$$A_{vf} = \frac{V_o}{V_s} = \frac{-R'}{R} \cdot \frac{s}{s + \frac{1}{RC}}$$

f_L = low 3-dB frequency = $1/2\pi RC$

AC voltage follower:

This follower is used to connect a signal source with

B-103

high internal source resistance to a load of low impedance, which may be capacitive.

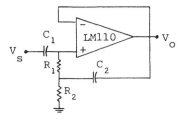

Analog integration and differentiation:

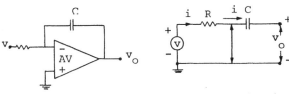

Integrator   Equivalent circuit

$$v_o = \frac{-1}{C} \int i\, dt = \frac{-1}{RC} \int v\, dt$$

If the input voltage is constant, $v = V$, then the output will be a ramp, $v_o = \frac{-Vt}{RC}$.

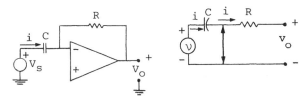

Differentiator   Equivalent circuit

$$v_o = -R_i = -RC\frac{dv}{dt}$$

If the input signal is $v = \sin \omega t$, then the output will be $v_o = -RC\omega \cos \omega t$. This results in amplification of the high-frequency components of amplifier noise, and the noise output may completely mask the differentiated signal.

11.3 ACTIVE FILTERS

11.3.1 IDEAL FILTERS

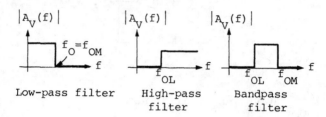

Low-pass filter High-pass filter Bandpass filter

An approximation for an ideal low-pass filter is:

$$\frac{A_v(s)}{A_{vo}} = \frac{1}{P_n(s)}$$

where $P_n(s)$ is a polynomial in the variable s with zeros in the left-hand plane.

11.3.2 BUTTERWORTH FILTER

$P_n(s) = B_n(s)$, known as the "Butterworth polynomial"

$$|B_n(\omega)| = \left[1 + \left(\frac{\omega}{\omega_o}\right)^{2n}\right]^{\frac{1}{2}}$$

Butterworth low-pass filter response:

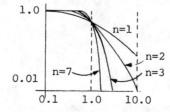

The transfer function is

$$\frac{A_v(s)}{A_{vo}} = \frac{1}{(s/\omega_o)^2 + 2K(s/\omega_o) + 1}$$

where $\omega_o = 2\pi f_o$ = High-frequency 3-dB point.

This represents a second-order filter.

For the first-order filter, $\dfrac{A_v(s)}{A_{vo}} = \dfrac{1}{\dfrac{s}{\omega_o} + 1}$

Circuit

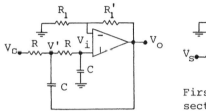

Second-order low-pass section

First-order low-pass section

Mid-band gain of op amp,

$$A_{vo} = \dfrac{V_o}{V_i} = \dfrac{R_1 + R_1'}{R_1}$$

$$\dfrac{A_v(s)}{A_{vo}} = 1/[(RC_s)^2 + (3 - A_{vo})RCs + 1]$$

obtained by applying KCL to node V'.

$$\omega_o = \dfrac{1}{RC} \quad \text{and} \quad 2K = 3 - A_{vo} \ldots$$

by comparing the coefficients of s^2 in $\dfrac{A_v(s)}{A_{vo}}$ and typical second-order transfer function.

Even-order Butterworth filters are synthesized by cascading second-orders prototypes such as those shown above and choosing A_{vo} of each op amp such that $A_{vo} = 3 - 2K$.

Normalized Butterworth polynomials:

n	Factors of $B_n(s)$
1	$(s+1)$
2	$(s^2+1.414s+1)$
3	$(s+1)(s^2+s+1)$
4	$(s^2+0.765s+1)(s^2+1.848s+1)$
5	$(s+1)(s^2+0.618s+1)(s^2+1.618s+1)$

Odd-order filters - Cascade the first-order filter with the second order.

CHAPTER 12

FEEDBACK AND FREQUENCY COMPENSATION OF OP AMPS

12.1 BASIC CONCEPTS OF FEEDBACK

Standard inverting configuration:

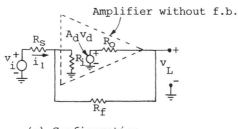

(a) Configuration

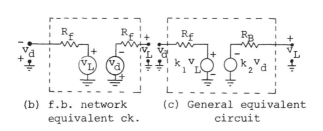

(b) f.b. network equivalent ck.

(c) General equivalent circuit

B-108

Current-differencing negative feedback circuit:

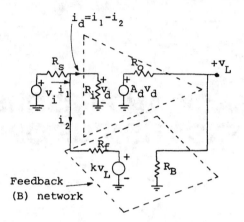

(a) Current-differencing -ve - feedback ckt.

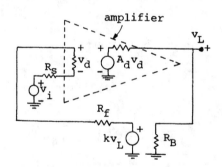

(b) Voltage-differencing -ve f.b. circuit

Gain of a feedback amplifier with current differencing:

$$i_1 = \frac{v_i + v_d}{R_s}, \quad i_2 = \frac{-v_d + K \cdot v_L}{R_f}, \quad i_d = \frac{-v_d}{R_i}$$

$$v_L \cong -(v_i A_d \div R_s)(R_s \| R_f) \div [1 + (A_d \cdot K/R_f)(R_s \| R_f)]$$

$$\cong -[R_f \div (K \cdot R_s)]v_i \quad \ldots \quad \text{(For inverting configuration } K = 1\text{)}$$

B-109

Loop-gain T of the amplifier:

$$\frac{-A_d K}{R_f}(R_s // R_f)$$

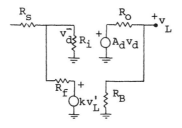

Circuit used to calculate T and A_o:

$$T = \left.\frac{v_L}{v_i'}\right|_{v_i=0} = \frac{v_L}{v_d} \cdot \frac{v_d}{i_d} \cdot \frac{i_d}{v_L'}$$

Open loop gain, or gain without feedback A_o:

$$A_o = \left.\frac{v_L}{v_i}\right|_{v_L'=0} = \frac{-A_d}{R_s}(R_s \| R_f)$$

Overall gain:

$$A_v = \frac{v_L}{v_i} = \frac{A_o}{1-T}$$

For voltage-differencing circuit:

$$T \cong -A_d \cdot K$$

$$A_o \approx A_d$$

$$v_L \approx v_i A_d / (1 + A_d K)$$

The loop gain T:

   The loop gain controls the "amount" of feedback present in a circuit.

B-110

If $T = 0$ ($K=0$), there is no feedback.

When $T \gg 1$, the gain of an amplifier with feedback approaches $A_v = v_L/v_i = -(R_f/KR_s)$ for the current-differencing configuration, and $A_v = \frac{1}{K}$ for the voltage-differencing configuration. (These amplifiers are shown in the diagram above.) We note that in both cases the gain with feedback is more or less independent of the amplifer gain, A_d. Consequently, the amplifier employing feedback is much more stable against variations in temperature and other parameters as the loop gain increases.

The feedback also has the effect of decreasing the gain from A_o (without feedback) to $\frac{A_o}{(1-T)}$.

Feedback amplifiers and the sensitivity function:

$$\text{Sensitivity function:} \quad \int_{A_o}^{A_v} = \frac{dA_v/A_v}{dA_o/A_o}$$

$$\int_{A_o}^{A_v} = 1/(1-T)$$

12.2 FREQUENCY RESPONSE OF A FEEDBACK AMPLIFIER

Bandwidth and Gain-bandwidth Product:

For current-differencing negative feedback Amplifier:

$$A_v = \frac{A_o}{1-T}$$

B-111

Single-pole amplifier:

$$A_d = \frac{A_{dm}}{1+(s/\omega_1)} \quad \cdots \quad A_{dm} = \text{gain at low frequency}$$

$$T = \frac{-T_m}{1 + \left(\frac{s}{\omega_1}\right)}, \quad T_m = \frac{A_{dm} \cdot K(R_s \| R_f)}{R_f}, \quad A_o = \frac{-A_{om}}{1 + \left(\frac{s}{\omega_1}\right)},$$

$$A_{om} = \frac{A_{dm}(R_s \| R_f)}{R_s}$$

$$A_v = \frac{-A_{om}}{1+T_m} \left\{ \frac{1}{1 + \frac{s}{\omega_1}(1+T_m)} \right\}$$

A pole is located at $s = -\omega_1(1 + T_m)$

f_n (upper 3-dB frequency) $= f_1(1 + T_m)$

$G \times (B.W.) = \left[A_{v(f=0)} \right] \cdot f_n = A_{om} \cdot f_1$ (A constant, independent of feedback)

Characteristic:

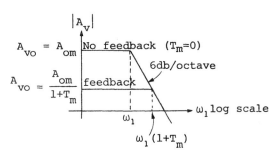

(a) locus of pole motion

(b) Gain versus frequency.

Double-pole amplifier:

$$A_d(s) = \frac{A_{dm}}{\left(1 + \frac{s}{\omega_1}\right)^2}, \quad T = \frac{-T_m}{\left(1 + \frac{s}{\omega_1}\right)^2}$$

$$A_o = \frac{-A_{om}}{\left(1 + \frac{s}{\omega_1}\right)^2}, \quad A_v(s) = \frac{-A_{om}}{1 + T_m + (2s/\omega_1) + (s^2/\omega_1^2)}$$

Poles are located at $s = -\omega_n (\xi \pm \sqrt{\xi^2 - 1})$, $\quad \xi = \dfrac{1}{\sqrt{1 + T_m}}$

Characteristics:

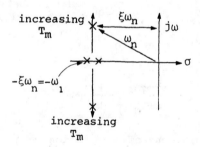

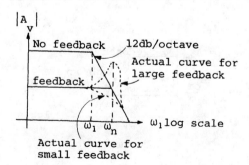

B-113

12.3 STABILIZING NETWORKS

No frequency compensation:

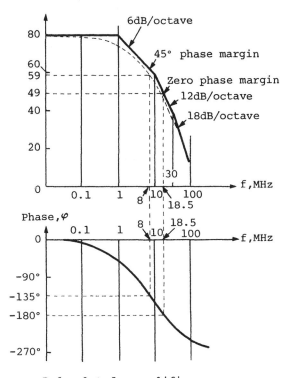

Bode plot for amplifier

The amplifier's open-loop gain is

$$A_o = \frac{-10^4}{\left(1 + \frac{s}{2\pi 10^6}\right)\left(1 + \frac{s}{2\pi \times 10^7}\right)\left(1 + \frac{s}{2\pi \cdot 30 \times 10^6}\right)}$$

If sufficient feedback is applied to make $|T| = 1$ at 18.5 MHZ, the amplifier is said to be marginally stable.

If sufficient feedback is applied to make $|T| = 1$ at a frequency < 18.5MHz, the actual phase at $|T| = 1$ is < 180°. The difference between 180° and the actual phase at $|T| = 1$ is called "Phase-Margin".

The frequency at which $|T| = 1$ is called the gain crossover frequency.

$$T(s) = A_o(s) \cdot K/(R_f/R_s), \quad \beta = \frac{K}{(R_f \div R_s)}$$

$$A_v(s) = \frac{A_o(s)}{1 - \beta \cdot A_o(s)} \approx \frac{-1}{\beta} \quad (\text{if } |T(s)| \gg 1)$$

For a 45° phase margin,

$|T(8\text{MHz})| = 1 = \beta |A_o(8\text{MHz})| \approx \beta(936), \quad \beta \approx -59\text{dB}$

At low frequencies, $T_m = \beta \cdot A_{om} = 21\text{dB}$

For a 45° phase margin:

$\beta = 1 / |A_o(\omega \text{ corresponding to } 135°)|$

$|T_{max}| = \beta |A_{om}| = A_{om} \div |A_o(\omega \text{ corresponding to } 135°)|$

Simple lag compensation:

Simple lag compensation is designed to introduce an additional negative real pole in the transfer function of the open-loop amplifier gain, A_o.

$$A_{o1}(s) = A_o(s) \div \left(1 + \frac{s}{\alpha}\right) \ldots$$

The pole α is adjusted so that $|T|$ drops to 0dB at a frequency where the poles of A_o contribute negligible phase shift.

This pole increases T but decreases the cross-over frequency.

Effect of lag compensation:

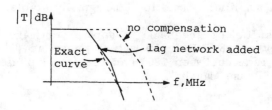

Lag compensating network:

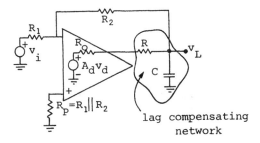

The break frequency of this lag filter:

$$\frac{\alpha}{2\pi} \approx \frac{1}{2\pi(R+R_o)C} \quad \ldots \quad (\text{assuming } R+R_o \ll R_2).$$

Lead compensation:

The transfer function $H(s) = \dfrac{s+\delta_1}{s+\delta_2}$

($\delta_2 > \delta_1$, i.e., the pole is located at a higher frequency than the zero.)

Bode diagram for $H(s)$ and the network:

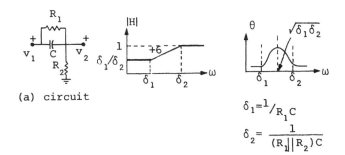

(a) circuit

$\delta_1 = 1/R_1 C$

$\delta_2 = \dfrac{1}{(R_1 \| R_2) C}$

This network introduces a low-frequency attenuation:

$$H(0) = R_2 \div (R_1+R_2) = \delta_1/\delta_2$$

B-116

CHAPTER 13

MULTIVIBRATORS

13.1 COLLECTOR-COUPLED MONOSTABLE MULTIVIBRATORS

Stable-state:

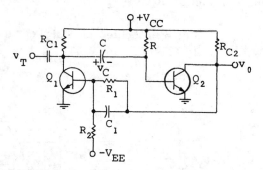

Collector-coupled monostable

In this state Q_1 is off, Q_2 is saturated and v_o is low.

$v_{c_1} \approx V_{cc}$ and $v_{c_2} \approx 0$, the current in R must be sufficient to saturate Q_2, i.e.,

$$I_{B_2} = I_R = \frac{(V_{cc} - V_{BES_2})}{R} \gg I_{BS_2} = \frac{I_{cs_2}}{\beta}$$

or

$$\frac{V_{cc} - V_{BES_2}}{R} \geq \frac{V_{cc} - V_{CES_2}}{\beta_2 \cdot R_{c_2}}$$

B-117

C_1 is a speed-up capacitor to couple change in v_o to v_{B_1}, turning Q_1 on rapidly.

Circuit operation:

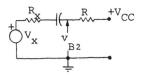

Voltages for the monostable shown in the figure in steady-state and just after triggering into the Quasi-Stable by a trigger pulse at $t = 0$.

	v_{C_1}	v_{B_2}	v_o	v_{B_1}	Q_1	Q_2
Stable state (t=0)	$\approx V_{cc}$	$\approx 0V$	$\approx 0V$	$< 0V$	off	on
Quasi-stable state (t=0$^+$)	$\approx 0V$	$\approx -V_{cc}$	$\approx V_{cc}$	$\approx 0V$	on	off

The initial trigger pulse need not be of amplitude V_{cc}. It only starts Q_2 off; the circuit's regenerative action itself drives v_{C_1} to $\approx 0v$ and v_{B_2} to $\approx -V_{cc}$.

Output pulse width:

The equivalent-circuit at the collector of Q_1 at $t = 0^+$ when Q_1 has just turned on and Q_2 has just turned off.

τ (Time constant with which v_{B_2} moves toward V_{cc})

$= (R+R_x) \cdot C \approx R \cdot C$, $(R \gg r_{sat})$

$v_{B_2}(t) = V_F - (V_F - V_I)e^{-t/\tau}$

(where $V_I = V_{BE_2} - V_{cc} + V_{CES_1}$

= initial value of v_{B_2}, and

$V_F = V_{cc}$ = the final value of V_{B_2})

B-118

T = The duration of the output pulse

$$= \tau \cdot \ln \frac{2V_{cc} - V_{BES_2} - V_{CES_1}}{V_{cc} - V_{BET_2}}$$

For a typical transistor, $V_{BES} + V_{CES} \approx 2V_{BET}$

$T \approx \tau \ln 2 \approx 0.69\tau \approx 0.69 R_c$

Recovery time:

This is the time one must wait before the monostable multivibrator should be triggered.

When the quasi-stable state ends at $t = T$, the output returns low although the circuit has not returned to its stable state.

At $t = T^-$, $v_{c_1} = V_{CES_1}$, $v_{B_2} = V_{BET_2}$ and the voltage across C is

$$v_c = V_{CES_1} - V_{BET_2} \approx 0.$$

The v_{c_1} increases exponentially from $v_I = V_{CES}$ towards its final value, $V_f = V_{cc}$, with

$$\tau_R \text{ (Recovery time constant)} = (R_{c_1} + (R \| R_{i_2})) \cdot C$$

$$\approx (R_{c_1} + R_{i_2}) \cdot C$$

V_{cc}' (the output-volt with $v_{EE} = 0$) = $\dfrac{V_{cc} \cdot R_1 + V_{BES_1} \cdot R_{c_2}}{R_1 + R_{c_2}}$

Waveforms:

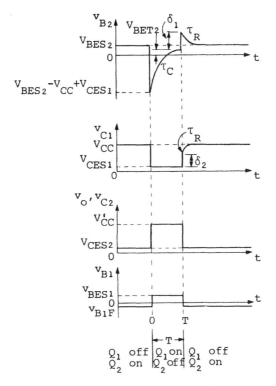

Waveforms for the collector-coupled monostable

13.2 EMITTER-COUPLED MONOSTABLE MULTIVIBRATORS

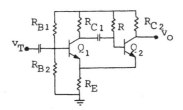

Circuit operation:

$(R+R_E)I_{B2} + R_E \cdot I_{C2} = V_{cc} - V_{BES2}$ (KVL equation for Q_2 with Q_1 off and Q_2 on)

$R_E \cdot I_{B2} + (R_E + R_{c2})I_{c2} = V_{cc} - V_{CE2}$

$v(0) = V_{cc} - R_{c2}(V_{cc} - V_{CES2})/(R_{c2} + R_E)$

$= R_E(V_{cc} - V_{CES2})/(R_{c2} + R_E) + V_{CES2}$

$v_E(0^-) = v_o(0^-) - V_{CES2}$

$v_{B2}(\bar{0}) = v_E(0) + V_{BES2}$

$v_{B1}(\bar{0}) < v_E(\bar{0}) + V_{BET1}$ (To maintain Q_1 off)

$v_{c1}(\bar{0}) = v_{cc}$ (with Q_1 off)

$v_{B1}(0^+) = V_{B1} = \dfrac{V_{cc} \cdot R_{B2}}{R_{B1} + R_{B2}} = v_{B1}(\bar{0})$

$v_E(0^+) = V_{B1} - V_{BES1}$

$i_{c1}(0^+) \approx i_{E1}(0^+) = v_E(\bar{0})/R_E, \quad v_{c1}(0^+) = V_{cc} - i_{c1}(0^+)R_{c1}$

The condition for Q_1 in a quasi-stable state:

$$\dfrac{R_E + R_{c1}}{R_E} > \dfrac{R_{B1} + R_{B2}}{R_{B2}}$$

Voltages with a Narrow Trigger Pulse Applied at t = 0

Parameter	Stable state t = 0− (Q_1 off, Q_2 saturated)	Quasi-stable state t = 0+ (Q_1 saturated, Q_2 off)
v_0	$V_{CC} - (V_{CC} - V_{CES_2}) \dfrac{R_{C_2}}{R_E + R_{C_2}}$ $\cong V_{CC}\left(1 - \dfrac{R_{C_2}}{R_E + R_{C_2}}\right)$	V_{CC}
v_E	$v_0(0^-) - V_{CES_2}$ $= (V_{CC} - V_{CES_2})\left(1 - \dfrac{R_{C_2}}{R_E + R_{C_2}}\right)$ $\cong V_{CC}\left(1 - \dfrac{R_{C_2}}{R_E + R_{C_2}}\right)$	$V_{B_1} - V_{BES_1}$ $\cong V_{CC} \dfrac{R_{B_2}}{R_{B_1} + R_{B_2}}$
v_{B_2}	$v_E(0^-) + V_{BES_2}$ $= (V_{CC} - V_{CES_2})\left(1 - \dfrac{R_{C_2}}{R_E + R_{C_2}}\right)$ $+ V_{BES_2}$ $\cong V_{CC}\left(1 - \dfrac{R_{C_2}}{R_E + R_{C_2}}\right)$	$v_{B_2}(0^-) + \left[v_{C_1}(0^+) - v_{C_1}(0^-)\right]$ $\cong V_{CC}\left(\dfrac{R_{B_2}}{R_{B_1} + R_{B_2}} - \dfrac{R_{C_2}}{R_E + R_{C_2}}\right)$
v_{B_1}	$\cong V_{CC} \dfrac{R_{B_2}}{R_{B_1} + R_{B_2}}$	$\cong V_{CC} \dfrac{R_{B_2}}{R_{B_1} + R_{B_2}}$
v_{C_1}	V_{CC}	$v_E(0^+) + V_{CES_1}$ $\cong \dfrac{V_{CC} R_{B_2}}{R_{B_1} + R_{B_2}}$

Condition for operation:

Conditions for Operation of the Monostable

1) $v_{B_1}(0^-) < v_E(0^-) + V_{BET_1}$

2) $i_{B_2}(0^-) > i_{C_2}(0^-)/\beta$

3) $v_{C_1}(0^+) < v_{B_1}(0^+)$

4) $v_{B_2}(0^+) < v_E(0^+) + V_{BET_2}$

Waveforms - With Q_1 saturated and Q_2 off in the quasi-stable state and Q_1 off and Q_2 saturated in the stable state.

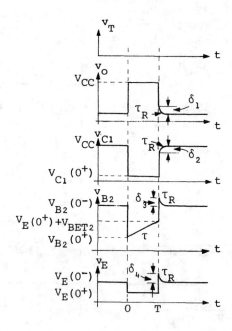

Pulse width and recovery:

$$T = \tau \cdot \ln\left[\frac{V_F - V_I}{V_F - (V_E(0^+) + V_{BET_2})}\right]$$

$$\approx R \cdot C \ln \left[\frac{1 + \dfrac{R_{C_2}}{R_E + R_{C_2}} - \dfrac{R_{B_2}}{R_{B_1} + R_{B_2}}}{1 - \dfrac{R_{B_2}}{R_{B_1} + R_{B_2}}} \right]$$

τ_R = (Recovery time constant)

$= (R_{C_1} + \lambda_{\pi_2} + R_E \parallel R_{C_2})C$

τ_R determines the circuits retrigger rate and the rate of decay of the overshoot.

Model of an emitter-coupled monostable multivibrator during the recovery time t > T.

The recovery time for the emitter-coupled monostable multivibrator is greater than in the collector-coupled monostable multivibrator.

13.3 COLLECTOR-COUPLED ASTABLE MULTIVIBRATORS

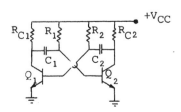

Typical waveforms at each collector and base:

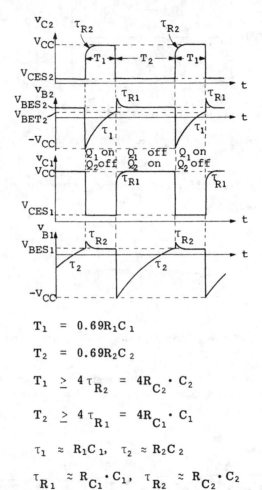

$T_1 = 0.69 R_1 C_1$

$T_2 = 0.69 R_2 C_2$

$T_1 \geq 4\tau_{R_2} = 4R_{C_2} \cdot C_2$

$T_2 \geq 4\tau_{R_1} = 4R_{C_1} \cdot C_1$

$\tau_1 \approx R_1 C_1, \quad \tau_2 \approx R_2 C_2$

$\tau_{R_1} \approx R_{C_1} \cdot C_1, \quad \tau_{R_2} \approx R_{C_2} \cdot C_2$

13.4 EMITTER-COUPLED ASTABLE MULTIVIBRATORS

When the bias levels in the emitter-coupled monostable multivibrator are properly adjusted, the circuit becomes astable. Only one capacitor controls the timing of Q_1 and Q_2.

To keep Q_2 active with c open, require

$$\beta \cdot i_{B_2} < i_{CS_2} \quad \text{or} \quad \beta \cdot \frac{(V_{CC}-V_{B_2})}{R} < \frac{V_{CC}-V_{C_2}}{R_{C_2}},$$

which, with $v_{B_2} = v_{C_2}$, reduces to $\beta \cdot R_{C_2} < R$.

To keep Q_1 active with c open, V_{B_1} should fall inside the following interval:

$$V_{BE} + \frac{R_E(1+\beta)(V_{CC}-V_{BE})}{R+(1+\beta)R_E} < V_{B_1}$$

$$< \frac{V_{CC} \cdot R_E[R+(1+\beta)R_{C_1}]+V_{BE} \cdot R \cdot R_{C_1}}{R(R_{C_1}+R_E)+(1+\beta)R_E \cdot R_{C_1}}$$

$$T_1 = RC \ln\left[\frac{V_{CC}-v_{B_2}(0^+)}{V_{CC}-V_{B_1}}\right]$$

$$\tau_{R_1} = (R_{C_1}+R\|R_{C_2}\|R_E)C \approx (R_{C_1}+R_{C_2}\|R_E)C$$

$$\tau_{R_2} = [(1+\beta)R_E\|R]C$$

Waveforms:

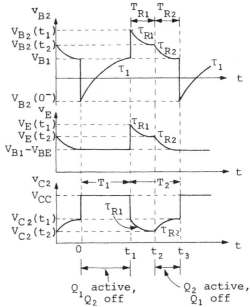

B-126

Approximate Voltages

	$t = 0^-$ Q_1 off Q_2 active	$t = 0^+$ Q_1 active Q_2 off
v_{C_1}	V_{CC}	$V_{CC} - v_E(0^+)R_{C_1}/R_E$
v_{B_2}	V_{B_1}	$V_{B_1} - v_E(0^+)R_{C_1}/R_E$
v_E	$V_{B_1} - V_{BE_2}$	$V_{B_1} - V_{BE_1}$

CHAPTER 14

LOGIC GATES AND FAMILIES

14.1 LOGIC LEVEL CONCEPTS

"1" and "0" limits must be separated by some voltage range to ensure that the state of a line can be determined even in the presence of noise, etc.

Voltages	Ge V	Si V
V_{BET}	0.1	0.5
V_{BES}	0.3	0.7
V_{CES}	1.0	0.1
V_{BE} (active)	0.2	0.6

V_{BET}: The v_{BE} threshold voltage below which little conduction occurs.

V_{BES} and V_{CES}: The v_{BE} and v_{CE} values in saturation

A high or "1" level corresponding to a voltage above some minimum upper level and a low or "0" to a voltage below some maximum lower level.

Graphical interpretation of V_{BET}, V_{BES} and V_{CES}:

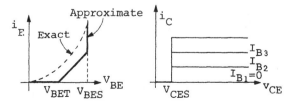

Demonstration of the logic level concept:

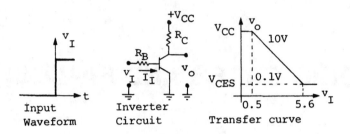

Input Waveform   Inverter Circuit   Transfer curve

$I_{CS} = (V_{CC} - V_{CES}) \div R_C$ ....In saturation, with no load

$I_{BS} = I_{CS}/\beta$ ....Corresponding base current of the edge of saturation.

$V_I = [R_B(V_{CC} - V_{CES}) \div (\beta RC)] + V_{BES}$ ....Voltage to saturate a transistor

The circuit will saturate for any

$V_I \geq [R_B(V_{CC} - V_{CES}) \div (\beta R_C)] + V_{BES} = 5.65$

...[Assuming $V_{CC} = +10V$, $R_C = 1K$, $R_B = 10K$]

With v_I satisfying the above equation, $V_o = V_{CES} = 0.1V$.

Inverter logic levels for Fig. 2.2:

v_I, Volts	v_o, Volts	V	Level
≥ 5.65	0.1	5.65 to 10V	"1"
≤ 0.5	10	-2 to 0.5V	"0"

Voltages above 5.65V are designated as "high" or "1" levels; and voltages below 0.5V, as "low" or "0" levels.

The upper "1" limit and the lower "0" limit are arbitrary, depending on power supply and breakdown voltage.

Logic level notation:

Parameter	Definition
V_{OL}	Low or "0" level output voltage
V_{OH}	High or "1" level output voltage
V_{IL}	Low-level input voltage
V_{IH}	High-level input voltage
$\overline{V_{IL}}$	Maximum V_{IL} level
$\underline{V_{IH}}$	Minimum V_{IH} level
I_{IH}, I_{IL}	Input current for $V_I = V_{OH}$ and $V_I = V_{OL}$, respectively.
I_{OH}, I_{OL}	Maximum available output load current for $V_O = V_{OH}$ and $V_O = V_{OL}$, respectively.

14.2 BASIC PASSIVE LOGIC

14.2.1 RESISTOR LOGIC (RL)

AND function can be implemented with passive components only, such as resistors.

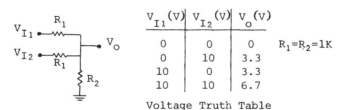

Voltage Truth Table

Fan-out (FO):

The number of equivalent gate inputs that can be driven from the output of a similar gate is the fan-out.

Noise Margin (NM):

This is a measure of the amount of noise that can be tolerated on signal lines before the voltage levels on these lines cease to look like "1s" or "0s".

Graphical Interpretation of NM and "1" and "0":

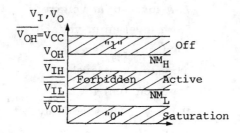

14.2.2 DIODE LOGIC CIRCUITS

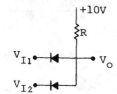

V_{I1}	V_{I2}	V_O
0	0	0.7
0	5	0.7
5	0	0.7
5	5	5.7

Basic Circuit

14.2.3 LOGIC CIRCUIT CURRENT, VOLTAGE, AND PARAMETER DEFINITIONS AND NOTATION

Parameter	Definition and test condition
$\overline{V_{IH}}$	Minimum input voltage that will look like a "1" at the input of a gate in worst case.
$\overline{V_{IL}}$	Maximum input voltage that will still look like a "0" at the input of a gate in worst case.
$\overline{V_{OH}}$	Minimum output voltage which, when applied to $FO_H = N_H$ other gate inputs, will look like a "1" at the input of each driven gate with a NM_H.

$\overline{V_{OL}}$ Maximum output voltage which, when applied to $FO_L = N_L$ other gate inputs, will look like a "0" at the input of each driven gate with a NM_L.

I_{IH} Current required at a gate input with $\underline{V_{OH}}$ present at the input.

I_{IL} Current required at a gate input with $\overline{V_{OL}}$ present at the input.

I_{OH} Available output current when $V_o = \underline{V_{OH}}$.

I_{OL} Available output current when $V_o = \overline{V_{OL}}$.

$FO_H = |I_{OH}/I_{IH}|$ Fan-out for a high-level output.

$FO_L = |I_{OL}/I_{IL}|$ Fan-out for a low-level output.

FO   The smaller of FO_H and FO_L.

$NM_H = \underline{V_{OH}} - \underline{V_{IH}}$ Noise margin for a high-level output.

$NM_L = \overline{V_{IL}} - \overline{V_{OL}}$ Noise margin for a low-level output.

NM   The smaller of NM_H and NM_L.

14.3 BASIC ACTIVE LOGIC

14.3.1 RESISTOR-TRANSISTOR LOGIC (RTL)

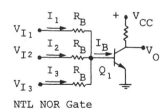

NTL NOR Gate

RTL parameter expressions:

1) $\underline{V_{IH}} = \dfrac{R_B}{\beta}\left[\dfrac{V_{CC}-V_{CES}}{R_C} + FO_L|I_{IL}|\right] + 3V_{BES} - 2\overline{V_{OL}}$

2) $\underline{V_{OH}} = NM_H + \underline{V_{IH}}$

3) $\underline{V_{OH}} = (V_{CC}R_B + FO_H V_{BES} R_C)/(R_B + FO_H R_C)$

4) $|I_{OH}| = (V_{CC} - \underline{V_{OH}})/R_C$

5) $I_{IH} = (\underline{V_{OH}} - V_{BES})/R_B$

6) $FO_H = |I_{OH}/I_{IH}|$

7) $\overline{V_{IL}} = V_{BET}$

8) $\overline{V_{OL}} = \overline{V_{IL}} - NM_L = V_{CES}$

9) $I_{IL} = (\overline{V_{OL}} - V_{BES})/R_B$

10) $I_{OL} = \beta[\underline{V_{IH}} - V_{BES} - 2(V_{BES} - \overline{V_{OL}})]/R_B - (V_{CC}-V_{CES})/R_C$

11) $FO_L = |I_{OL}/I_{IL}|$

14.3.2 DIODE-TRANSISTOR LOGIC (DTL)

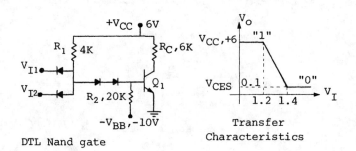

DTL Nand gate

Transfer Characteristics

B-133

DTL NAND gate characteristics:

Table 1	Table 2
$\underline{V_{OH}} = 6V$	$FO_L = FO = 9$
$\overline{V_{OL}} = .412V$	$NM_L = NM = .788V$
$\underline{V_{IH}} = 1.4V$	$I_{IL} = -0.83mA$
$\overline{V_{IL}} = 1.2V$	$I_{OL} = 12.2mA$

DTL NAND gate aprameters:

$$\underline{V_{IH}} = V_{DX} + V_{DY} + V_{BES} - V_{D_1}, \quad \overline{V_{IL}} = V_{DX} + V_{DY} + V_{BET} - V_{D_1}$$

$$|I_{IL}| = (V_{CC} - V_D - V_{CES})/R_1 - (V_{CES} + V_D - V_{DX} - V_{DY} + V_{BB})/R_2$$

$$I_{OL} = \frac{\beta(V_{CC} - 2.1)}{R_1} - \frac{\beta(V_{BES} + V_{BB})}{R_2} - \frac{V_{CC} - V_{CES}}{R_C}$$

$$V_{OL} = V_{CES} + \left[\frac{V_{CC} - V_{OL}}{R_C} + (FO_L)|I_{IL}| \right] r_{sat},$$

$$\overline{V_{OL}} = \overline{V_{IL}} - NM_L$$

14.4 ADVANCED ACTIVE LOGIC GATES

14.4.1 TRANSISTOR-TRANSISTOR LOGIC (TTL)

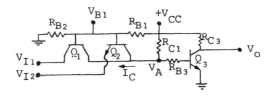

V_{CC}, R_{C_1} and R_{B_3} are chosen so that with Q_1 and Q_2 off, the current from V_{CC} through R_{C_1} is sufficient to saturate Q_3:

$$\frac{V_{CC} - V_{BES_3}}{R_{C_1} + R_{B_3}} = I_{B_3} \geq I_{BS_3}$$

$$= \frac{V_{CC} - V_{CES}}{R_{C_3}} + FO_L I_{IL}$$

Q_1 and Q_2 are off when both inputs V_{I_1}, and V_{I_2}, are low. Furthermore, Q_3 will be turned of if Q_1 or Q_2, or both are turned on when V_{I_1} or V_{I_2}, or both, are high. Hence, this circuit acts as a NAND gate.

Totem-pole output with phase-splitter driver:

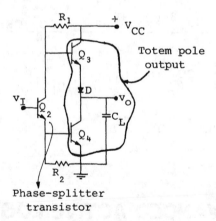

The circuit's response to a positive pulse at the base of Q_2: With v_I low, Q_2 is off, V_{C_2} is high and Q_3 is on, while Q_4 is off and v_o is high.

When v_I goes positive, Q_2 saturates and V_{C_2} droops, turning Q_3 off. V_{E_2} then rises and Q_4 saturates. $(\beta+1)I_{B_4}$ output current quickly discharges C_L and v_o rapidly falls to V_{CES}. When v_I drops from a high level to

0V, Q_2 turns off. V_{E_2} falls to 0V, turning Q_4 off; and V_{C_2} rises, turning on Q_3. Q_3 operates as an EF and quickly charges C_L with the large available current $(\beta+1)I_{B_3}$ and v_o rapidly rises.

The totem-pole output thus utilizes EF Q_3 to rapidly raise v_o and the CE transistor Q_4 to rapidly discharge v_o. The phase splitter Q_2 provides the proper phase drive for the totem-pole output transistors. Other topologies, such as DTL, can also use this output connection.

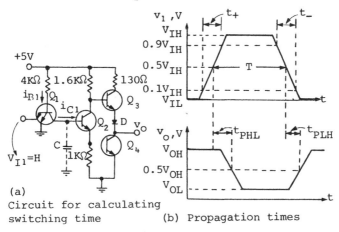

(a) Circuit for calculating switching time
(b) Propagation times

SWITCHING AND PROPAGATION TIMES

Characteristics:

Guaranteed Input and Output Voltage Levels for TTL

Parameter	Value, V	Interpretation for a TTL NAND gate
$\overline{V_{IL}}$	0.8	An input voltage ≤ 0.8V is guaranteed to turn on Q_1(E-B junction forward biased).
V_{IH}	2.0	An input voltage ≥ 2.0V is guaranteed to turn off Q_1(E-B junction reverse biased).
$\overline{V_{OL}}$	0.4	With $V_I \leq \overline{V_{IL}}$, the output is guaranteed to be ≤ 0.4V under full fan-out.
$\underline{V_{OH}}$	2.4	With $V_I \geq V_{IH}$, the output is guaranteed to be ≥ 2.4V under full fan-out.

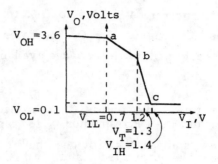

TTL Transfer curve

14.4.2 EMITTER-COUPLED LOGIC (ECL)

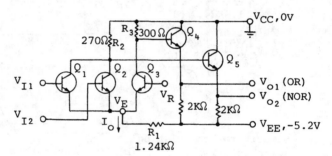

Basic ECL OR/NOR gate

CHAPTER 15

BOOLEAN ALGEBRA

15.1 LOGIC FUNCTIONS

15.1.1 NOT FUNCTION

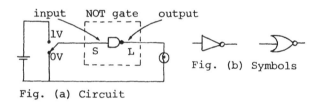

Fig. (a) Circuit

Fig. (b) Symbols

Logic equation and truth table:

$$L = \overline{S}$$

S	L=$\overline{S}$
1	0
0	1

15.1.2 AND FUNCTION

The AND functioning circuit:

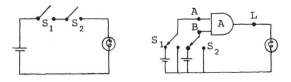

Logic equation and truth table:

L = A * B

A	B	L
0	0	0
0	1	0
1	0	0
1	1	1

15.1.3 OR FUNCTION

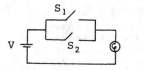
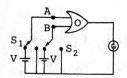

Logic equation and truth table:

L = A + B

A	B	L
0	0	0
0	1	1
1	0	1
1	1	1

15.2 BOOLEAN ALGEBRA

Boolean Theorems:

Theorem	Name
A + B = B + A A · B = B · A	Commutative law
(A+B)+C = A+(B+C) (A·B)·C = A·(B·C)	Associative law

$$A \cdot (B+C) = A \cdot B + A \cdot C$$
$$A + (B \cdot C) = (A+B) \cdot (A+C)$$
Distributive law

$$A + A = A$$
$$A \cdot A = A$$
Identity law

$$\overline{\overline{A}} = \overline{A}$$
$$\overline{\overline{A}} = A$$
Negation

$$A + A \cdot B = A$$
$$A \cdot (A+B) = A$$
Redundancy

$$0 + A = A$$
$$1 \cdot A = A$$
$$1 + A = 1$$
$$0 \cdot A = 0$$

$$\overline{A} + A = 1$$
$$\overline{A} \cdot A = 0$$

$$A + \overline{A} \cdot B = A + B$$
$$A \cdot (\overline{A} + B) = A \cdot B$$

$$\overline{A+B} = \overline{A} \cdot \overline{B}$$
$$\overline{A \cdot B} = \overline{A} + \overline{B}$$
De Morgan's laws

Proof of theorems:

Proof of theorem $(A+B) \cdot (A+C) = A + (B \cdot C)$

$$(A+B) \cdot (A+C) = AA + AB + AC + BC$$
$$= A + A(B+C) + BC$$
$$= A(1+B+C) + BC$$
$$= A + BC$$

Proof of theorem $A + A \cdot B = A$

A	B	A·B	A+A·B
0	0	0	0
0	1	0	0
1	0	0	1
1	1	1	1

Proof of theorem $A + \bar{A} \cdot B = A + B$

A	B	A+B	$\bar{A}$	$\bar{A}B$	$A+\bar{A}B$
0	0	0	1	0	0
0	1	1	1	1	1
1	0	1	0	0	1
1	1	1	0	0	1

Proof of theorem $\overline{A+B} = \bar{A} \cdot \bar{B}$

A	B	A+B	$\overline{A+B}$	$\bar{A}$	$\bar{B}$	$\bar{A} \cdot \bar{B}$
0	0	0	1	1	1	1
0	1	1	0	1	0	0
1	0	1	0	0	1	0
1	1	1	0	0	0	0

Manipulations of logic equations:

Simplify $\quad L = \bar{X}Y + XY + \bar{X}\,\bar{Y}$

$\qquad\qquad = Y(X+\bar{X}) + \bar{X}\,\bar{Y}$

$\qquad\qquad = Y(1) + \bar{X}\,\bar{Y} \ = Y + \bar{X}\,\bar{Y}$

$\qquad\qquad = Y + \bar{X}$

If $L = \bar{X}Y + X\bar{Y}$, find $\bar{L}$.

$\quad \bar{L} = \overline{\bar{X}Y + X\bar{Y}}$

$\qquad = \overline{(\bar{X}Y)}\,\overline{(X\bar{Y})}$

$\qquad = (\bar{\bar{X}}+\bar{Y})(\bar{X}+\bar{\bar{Y}})$

$\qquad = (X+\bar{Y})(\bar{X}+Y)$

$\qquad = X\bar{X} + XY + \bar{Y}\bar{X} + \bar{Y}Y$

$\qquad = XY + \bar{Y}\,\bar{X}$

15.3 THE NAND AND NOR FUNCTIONS

The NAND function:

Logic equation: $L = \overline{ABCD...}$

The NAND operation is commutative, i.e., $L = \overline{ABC} = \overline{BAC} = ...$, but not associative.

Truth table:

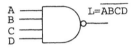

A	B	L	A·B
0	0	1	0
0	1	1	0
1	0	1	0
1	1	0	1

The NOR function:

The logic equation is $L = \overline{A+B}$

The NOR operation is commutative, i.e., $L = \overline{A+B+C...} = \overline{B+A+C...}$ but it is not associative.

Truth table:

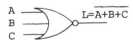

A	B	L	A+B
0	0	1	0
0	1	0	1
1	0	0	1
1	1	0	1

The exclusive OR function:

The logic equation is $L = A \oplus B = \overline{A}B + A\overline{B}$

The function is both commutative and associative.

In practice, these gates with more than two inputs are not available.

Truth table:

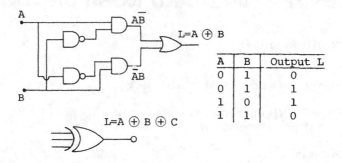

A	B	Output L
0	1	0
0	1	1
1	0	1
1	1	0

15.4 STANDARD FORMS FOR LOGIC FUNCTIONS

Sum of products (SP):

The logic function is written as a simple sum of terms, e.g.,

$$L = (\overline{W} + XY)(X + YZ)$$
$$= (\overline{W} + XY)X + (\overline{W} + XY)YZ$$
$$= \overline{W}X + XXY + \overline{W}YZ + XYYZ$$
$$= \overline{W}X + XY + \overline{W}YZ + XYZ \text{ (the desired form)}.$$

To find the logic equation for L from the truth table:

Select the rows for which L = 1, these are called "Minterms."

Find the logic expression for these in the product form.

i.e., Row #3 (Minterm): $\overline{X}Y\overline{Z} = \overline{0}\,1\,\overline{0} = 1$

B-143

Take the sum of all the minterms:

$$L = \overline{X}Y\overline{Z} + \overline{X}YZ + X\overline{Y}\overline{Z} + XY\overline{Z} + XYZ$$

Row #	X	Y	Z	L
1	0	0	0	0
2	0	0	1	0
3	0	1	0	1
4	0	1	1	1
5	1	0	0	1
6	1	0	1	0
7	1	1	0	1
8	1	1	1	1

Product of sums (PS):

This consists of a product of terms in which each term consist of a sum of all or part of the variables.

$$L = (\overline{W}+XY)(X+YZ) = (\overline{W}+X)(\overline{W}+Y)(X+Y)(X+Z)$$

To find a PS form from the truth table:

Pick-up rows for which $L = 0$, these are called the "Maxterms".

Express the logic expression for these in the sum form, i.e.,

For row #1: $X + Y + Z$

Take product of all these "Maxterms", i.e.,

$$L = (X+Y+Z)(X+Y+\overline{Z})(\overline{X}+Y+\overline{Z})$$

15.5 THE KARNAUGH MAP

It is a graphical technique for reducing logic equations to a minimal form.

Two variable Karnaugh map:

```
    A
B \  A=0  A=1
B=0 [   |    ]
B=1 [   |    ]
```

Two-variable map

A	B	Term in logic equation = to 1
0	0	$\bar{A}\bar{B}$
0	1	$\bar{A}B$
1	0	$A\bar{B}$
1	1	AB

```
  A
B \  0        1
0 [ ĀB̄  |  AB̄ ]
1 [ ĀB  |  AB ]
```

Correspondence with truth-table

Truth Table

A	B	L
0	0	0
0	1	1
1	0	1
1	1	1

```
  A
B \  0   1
0 [    | 1 ]
1 [ 1  | 1 ]
```

K map for
$L = \bar{A}B + A\bar{B} + AB$

An example

Set of rules for simplification:

    A group of two adjacent cells combines to yield a single variable.

    A single cell which can't be combined represents a two-variable term.

    It is permissible for groups to overlap because of the fact that in boolean algebra, $A+A = A$.

Three-variables map:

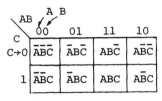

A primary map

The set of rules for simplification:

A group of 4 adjacent cells (in-line or square) combines to yield a single variable.

A group of two adjacent cells combines to yield a two-variable term.

A single cell which can't be combined represents a 3-variable term.

Use of these terms:

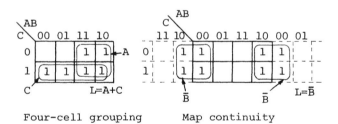

Four-cell grouping   Map continuity

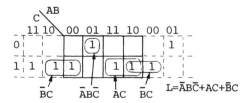

Two-cell groupings and single cell.

Four-variable map:

The rules are:

Eight adjacent cells yield a single variable. 4-adjacent cells yield a two-variable term. Two adjacent cells yield a 3-variable term. Individual cells represent 4-variable terms.

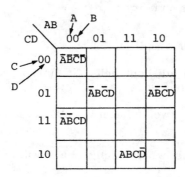

Primary map

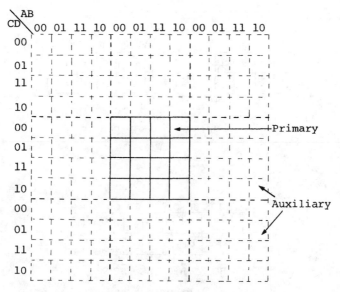

Continuity with auxiliary maps

CHAPTER 16

REGISTORS, COUNTERS AND ARITHMETIC UNITS

16.1 SHIFT REGISTERS

16.1.1 SERIAL-IN SHIFT REGISTERS

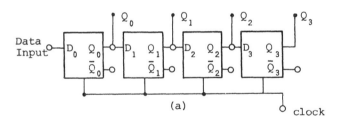

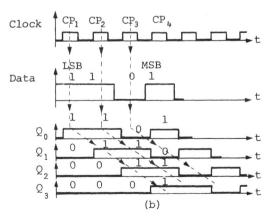

Four-bit shift registers: (a) logic diagram; (b) waveforms

B-148

Data can be taken from this register in either serial or parallel form. For serial removal, it is necessary to apply four additional clock pulses; the data will then appear on Q_3 in serial form.

To read data in parallel form, it is only necessary to enter the data serially. Once the data are stored, each bit appears on a separate output line, Q_0 to Q_3.

16.1.2 PARALLEL-IN SHIFT REGISTERS

The flip-flops have asynchronous preset and clear capability.

The unit has synchronous serial or asynchronous parallel-load capability and a clocked serial output.

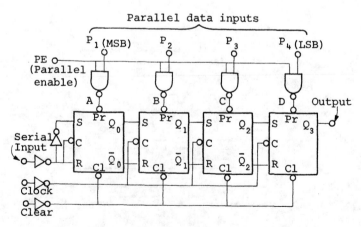

Simplified 54/7494 4-bit shift register.

B-149

16.1.3 UNIVERSAL SHIFT REGISTERS

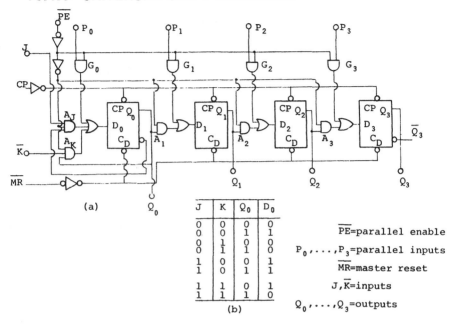

(a)

(b)

$\overline{PE}$=parallel enable
$P_0,\ldots,P_3$=parallel inputs
$\overline{MR}$=master reset
$J,\overline{K}$=inputs
$Q_0,\ldots,Q_3$=outputs

J	K	Q_0	D_0
0	0	0	0
0	0	1	1
0	1	0	0
0	1	1	0
1	0	0	1
1	0	1	1
1	1	0	1
1	1	1	0

This register contains four clocked master-slave flip-flops with D inputs.

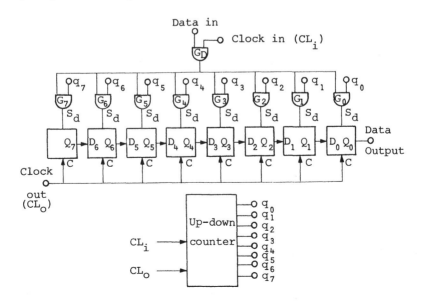

B-150

16.2 COUNTERS

16.2.1 THE RIPPLE COUNTER

This counter uses the maximum count capability of the three stages; hence, it is a mod-8 counter (the maximum modulus of an N-flipflop counter is 2^N).

Count $C = (Q_2 \times 2^2) + (Q_1 \times 2^1) + (Q_0 \times 2^0)$

To obtain the decimal output, a binary-to-decimal decoder as shown below is used.

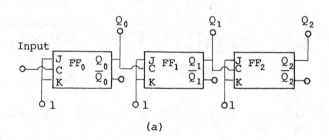

(a)

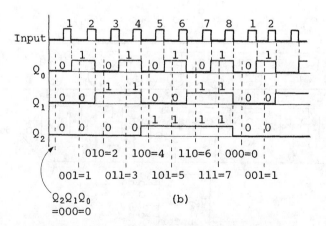

(b)

Mod-8 ripple counter: (a) logic diagram; (b) waveforms.

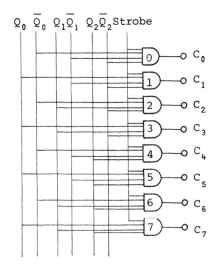

Binary-to-decimal decoder

16.2.2 THE SYNCHRONOUS COUNTER

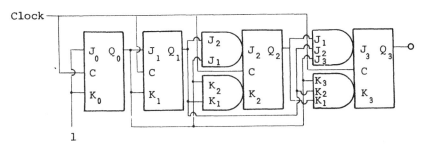

Circuit of synchronous parallel counter

The state table:

Counter state	State table	
	Q_1 (2^1)	Q_0 (2^0)
0	0	0
1	0	1
2	1	0
0	0	0
1	0	1
2	1	0
0	0	0
⋮	⋮	⋮

FF$_o$ must change state (toggle) with each clock pulse. This is done by connecting J$_o$ and K$_o$ to a high level.

FF$_1$ must change state whenever Q$_o$ = 1. This is achieved by connecting J$_1$ and K$_1$ directly to Q$_o$.

FF$_2$ changes state only when Q$_o$ = Q$_1$ = 1. Thus Q$_o$ and Q$_1$ are connected through AND gates to J$_2$ and K$_2$.

FF$_3$ changes state only when Q$_o$ = Q$_1$ = Q$_2$ = 1. This requires 3-input AND gates connecting Q$_o$, Q$_1$ and Q$_2$ to J$_3$ and K$_3$.

Each following stage requires an additional input to the AND gate.

The propagation delay of an F-F increases with the load and limits the speed attainable with the counter.

$$f_{max} \leq \frac{1}{t_{pd}(FF_3) + t_{pd}(AND) + t_s}$$

16.3 ARITHMETIC CIRCUITS

16.3.1 ADDITION OF TWO BINARY DIGITS, THE HALF ADDER

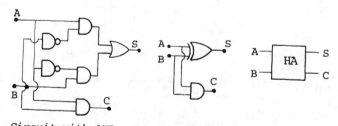

Circuit with AND, OR, & NOT gates

Truth Table:

A	B	S	C
0	0	0	0
0	1	1	0
1	0	1	0
1	1	0	1

The logic equations:

$C = AB$ and $S = \overline{A}B + A\overline{B}$

16.3.2 THE FULL ADDER

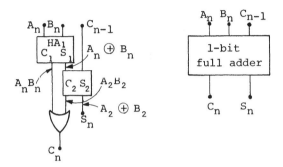

Truth table for adding A_n and B_n and a carry C_{n-1}.

A_n	B_n	C_{n-1}	S_n	C_n
0	0	0	0	0
0	1	0	1	0
1	0	0	1	0
1	1	0	0	1
0	0	1	1	0
0	1	1	0	1
1	0	1	0	1
1	1	1	1	1

Logic equations:

$$S_n = C_{n-1}(\overline{A}_n\overline{B}_n + A_nB_n) + \overline{C}_{n-1}(\overline{A}_nB_n + A_n\overline{B}_n)$$

$$= C_{n-1} \oplus (A_n \oplus B_n)$$

$$C_n = A_nB_n + C_{n-1}(A_n + B_n)$$

B-154

16.3.3 PARALLEL ADDITION

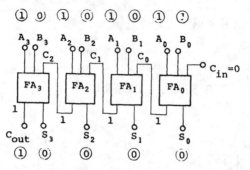

4-bit parallel adder.

If the propagation delay time $t_A = 2t_{pd}$ is the same for each adder, then the carry from FA_0 appears at FA_1 after time t_A, the carry from FA_1 appears at FA_2 after $2t_A$, and so on.

16.3.4 LOOK-AHEAD-CARRY ADDERS

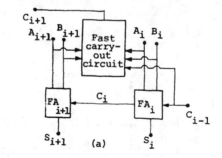

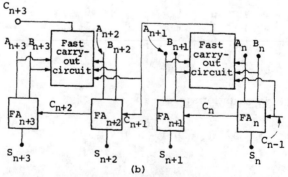

Carry-circuits: (a) 2-bit adder with a fast carry output; (b) 4-bit adder with fast carry output.

B-155

CHAPTER 17

OSCILLATORS

17.1 HARMONIC OSCILLATORS

A harmonic oscillator generates a sinusoidal output.

17.1.1 THE RC PHASE SHIFT OSCILLATOR

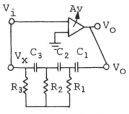

Amplifier with RC feedback network

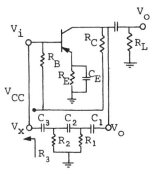

CE version of the circuit

For the feedback network:

$$R_3 C_3 = R_2 C_2 = R_1 C_1 = RC$$

$$\frac{V_i}{V_o} = \frac{j\omega RC}{1+j\omega RC}$$

The overall transfer function = $\dfrac{V_x}{V_o}$ = $\left(\dfrac{j\omega RC}{1+j\omega RC}\right)^3$

B-156

The circuit will oscillate at the frequency for which the phase shift from V_o to V_x is 180°.

$$f_o = 0.577/2\pi RC$$

$$\left|\frac{V_x}{V_o}\right| = \left(\frac{0.577}{1.58}\right)^3 = \frac{1}{8},$$

i.e., the voltage is attenuated by a factor of 8.

17.1.2 THE COLPITTS OSCILLATOR

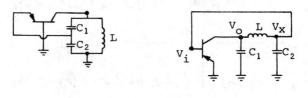

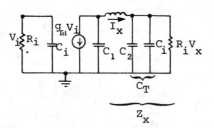

$$V_x = I_x \cdot Z_x, \quad I_x = -g_m \cdot V_i \cdot \frac{1/j\omega C_1}{(1/j\omega C_1)+j\omega L+Z_x}$$

$$V_x = V_i \text{ and } I_x - Z_x = \frac{I_x [1/j\omega C_1+j\omega L+Z_x]}{-g_m [1/j\omega C_1]}$$

$$Z_x = \frac{R_i [1/j\omega C_T]}{R_i+(1/j\omega C_T)}$$

$$1 + g_m R_i - \omega^2 LC_1 + j(\omega c_1 R_i + \omega C_T R_i - \omega^3 LC_1 C_T R_i) = 0$$

B-157

$$\omega^3 LC_1 C_T \cdot R_i = \omega C_1 R_i + \omega C_T \cdot R_i, \quad \omega^2 = \frac{C_1 + C_T}{L \cdot C_T \cdot C_1}$$

$$f_o = 1/2\sqrt{LC}, \quad \text{where } C = \frac{C_1 \cdot C_T}{C_1 + C_T}$$

Condition for sustained oscillations:

$$g_m R_i = C_1/C_T \quad (\text{since } R_i \approx r_{b'e}, \; g_m R_i \approx g_m r_{b'e} = h_{fe}),$$

$$h_{fe} = C_1/C_T$$

17.1.3 THE HARTLEY OSCILLATOR

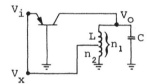

The feedback factor: $\quad \beta_v = \dfrac{V_x}{V_o} = \dfrac{h_2}{h_1}$

The Hartley oscillator and its ac and output equivalent circuit:

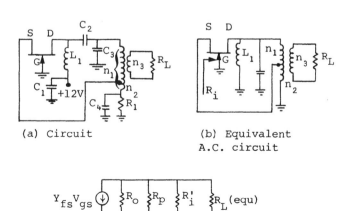

(a) Circuit (b) Equivalent A.C. circuit

(c) Output equivalent circuit

The voltage amplification: $A_v \approx R_L' \gamma_{fs}$

The Barkhausen's criterion: $\dfrac{n_2 \cdot R_L'}{n_1} \gamma_{fs} = 1\underline{/0°}$

$$\frac{n_2}{n_1} = \frac{1}{R_L' \gamma_{fs}}$$

17.1.4 THE CLAPP OSCILLATOR

$$\left| j\omega_o L - \frac{j}{\omega_o C_3} \right| = \left| \frac{-j}{\omega_o C} \right|$$

$$\omega_o L - \frac{1}{\omega_o C_3} = 1/\omega_o C$$

$$f_o = 1/2\pi \sqrt{L \; \frac{C_3 C}{C_3 + C}}$$

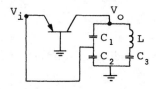

17.1.5 THE CRYSTAL OSCILLATOR

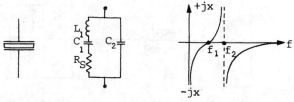

Resonant crystal, equivalent circuit and impedance characteristics

The series resonant frequency:

$$f_1 = \frac{1}{2\pi\sqrt{L_1 C_1}} \quad , \quad f_2 = \text{The parallel resonant}$$

$$\text{frequency} = \frac{1}{2\pi\sqrt{L_1 C_1 C_2/(C_1+C_2)}} = f_1\sqrt{1+\frac{C_1}{C_2}}$$

The Clapp and Colpitts crystal oscillators:

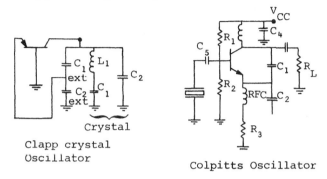

Clapp crystal Oscillator

Colpitts Oscillator

17.1.6 TUNNEL DIODE OSCILLATORS

These oscillators exhibit negative resistance when suitably biased.

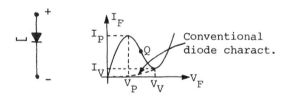

Dynamic resistance: $r = \Delta V_F / \Delta I_F$.

Tunnel diode dc biasing circuit:

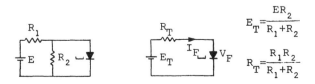

$$E_T = \frac{ER_2}{R_1+R_2}$$

$$R_T = \frac{R_1 R_2}{R_1+R_2}$$

B-160

Tunnel diode oscillator:

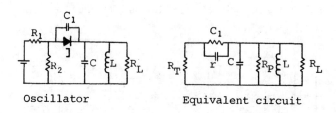

Oscillator   Equivalent circuit

AC equivalent circuit for deriving criteria for oscillations:

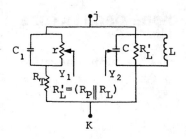

$Y_{jk} = Y_1 + Y_2, \quad Y_1 = 1/Z_1$

$Z_1 = R_T + \left(r \| \dfrac{1}{j\omega c_1} \right)$

$R_T = \dfrac{-r}{1+\omega^2 r^2 C_1^2}$ (The condition for oscillations at ω)

$f_{max}^2 = \dfrac{-(r+R_T)}{r^2 R_T C_1^2 (2\pi)^2}$ = The highest possible frequency.

f_o (The actual frequency of oscillations) $= \dfrac{1}{2\pi \sqrt{LC}} \cdot \dfrac{1}{\sqrt{1 + \dfrac{r \cdot C_1}{C(r+R_T)}}}$

17.2 RELAXATION OSCILLATORS

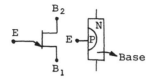

The unijunction transistor (UJT) oscillator:

Useful relationships:

$R_{BB} = R_{B1} + R_{B2}$ (R_{BB} is the interbase resistance)

n = The standoff ratio = $R_{B1}/R_{B1} + R_{B2}$

Equivalent circuit and characteristics:

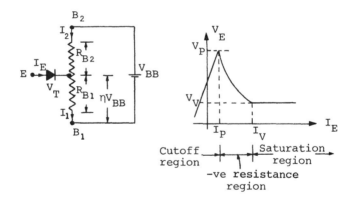

UJT oscillator:

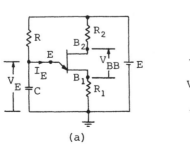

(a)   (b)

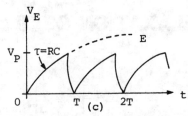

(a) Circuit of UJT oscillator; (b) AC equivalent circuit for (a); (c) emitter voltage waveform for the circuit of (a).

$$T = (R \cdot C)\ln \left| \frac{1}{1-n} \right|, \quad K = \ln \left| \frac{1}{1-n} \right|$$

f_o = The frequency of oscillations

   = 1/RCK

Plot of K as a function of n:

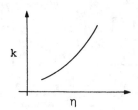

CHAPTER 18

RADIO-FREQUENCY CIRCUITS

18.1 NON-LINEAR CIRCUITS

A non-linear circuit is defined as any circuit that produces frequencies not present in the input signal.

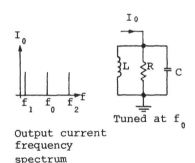

Output current frequency spectrum

Tuned at f_0

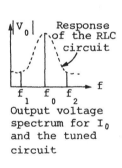

Output voltage spectrum for I_0 and the tuned circuit

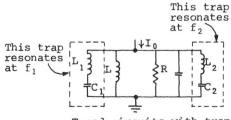

Tuned circuits with traps

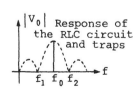

Response of the RLC circuit and traps

Class 'C' amplification:

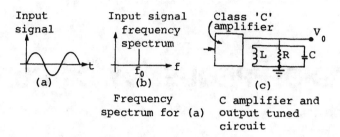

Frequency spectrum for (a)
C amplifier and output tuned circuit

Waveforms and frequency spectra for a class 'C' circuit:

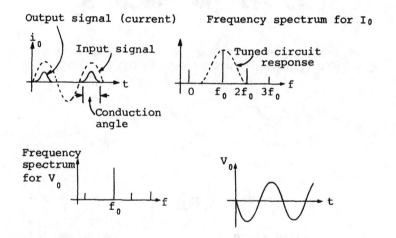

Frequency multiplication:

Diagram and signals for a frequency multiplier:

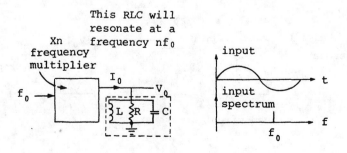

B-165

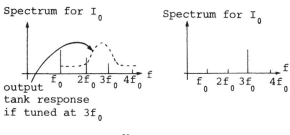

Mixing

Two signals are combined to yield an output signal with a frequency content which includes both the sum and difference of the two input frequencies.

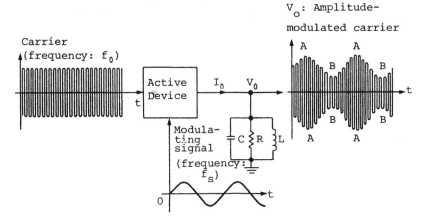

Block diagram of amplitude modulation system

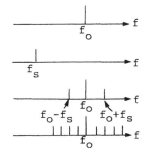

18.2 SMALL-SIGNAL RF AMPLIFIERS

Tuned amplifiers are mainly characterized by 1) center frequency f_o; 2) a bandwidth (B of 3dB 3) power gain at f_o: G_o and a noise figure at f_o, NF.

Basic circuits:

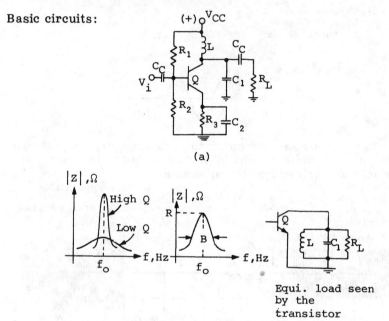

Equi. load seen by the transistor

Bandwidth is a function of the circuit Q. A high Q means a narrow bandwidth, while a lower Q results in a wider bandwidth.

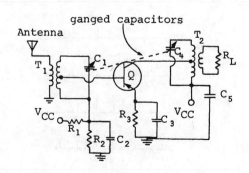

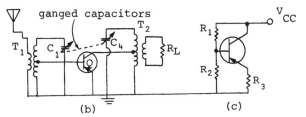

(a) RF amplifier with tuned input and output;
(b) AC circuit for (a); (c) DC circuit for (a)

Both the transformers are wound around an air core, that is, a non-magnetic core. The air core is used at high frequencies because the inductance required is small and the core losses are small, yielding a high Q.

18.3 TUNED CIRCUITS

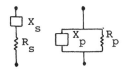

$$Q = \frac{X_s}{R_s}, \quad Z_s = R_s \pm jX_s, \quad Z_p = (R_p)(\pm jX_p)/R_p \pm jX_p.$$

$$Z_s = Z_p \rightarrow R_s = \frac{R_p X_p^2}{R_p^2 + X_p^2}, \quad X_s = \frac{X_p R_p^2}{R_p^2 + X_p^2}$$

$$Q = \frac{R_p}{X_p}$$

18.4 CIRCUITS EMPLOYING BIPOLAR TRANSISTORS

Figure:

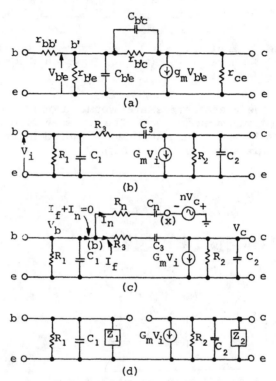

(a) Transistor hybrid-pi model; (b) transformed version of the circuit in (a); (c) unilateralized circuit; (d) unilateralized circuit including loading on input and output due to R_3, C_3, R_n, and C_n.

The circuit of Fig. (a) is transformed to (b). The parameters are:

$$R_1 = r_{bb'} + r_{b'e} \left[\frac{r_{b'e} + r_{bb'}}{r_{b'e} + r_{bb'} + \omega^2 (C_{b'e} + C_{b'c})^2 (r_{bb'} r_{b'e}^2)} \right]$$

$$C_1 = C_{b'e} + C_{b'c} \left/ \left[\left(1 + \frac{r_{bb'}}{r_{b'e}}\right)^2 + \omega^2 r_{bb'}^2 (C_{b'e} + C_{b'c})^2 \right] \right.$$

$$R_2 = \left[\frac{1}{r_{ce}} + \frac{1}{r_{b'c}} + g_m \frac{\frac{1}{r_{b'c}}\left(\frac{1}{r_{bb'}} + \frac{1}{r_{b'e}}\right) + \omega^2 C_{b'e} C_{b'c}}{\left(\frac{1}{r_{bb'}} + \frac{1}{r_{b'e}}\right)^2 + \omega^2 C_{b'e}^2} \right]^{-1}$$

$$C_2 = C_{b'c} \left[1 + \frac{g_m \left(\frac{1}{r_{bb'}} + \frac{1}{r_{b'e}}\right)}{\left(\frac{1}{r_{bb'}} + \frac{1}{r_{b'c}}\right)^2 + \omega^2 C_{b'e}^2} \right]$$

$$R_3 = r_{bb'} \left(1 + \frac{C_{b'e}}{C_{b'c}}\right) + \frac{1 + r_{bb'}/r_{b'e}}{r_{b'c} \cdot \omega^2 \cdot C_{b'c}^2}$$

$$C_3 = C_{b'c} \left/ \left[1 + \frac{r_{bb'}}{r_{b'e}} - \frac{r_{bb'}}{r_{b'c}} \frac{C_{b'e}}{C_{b'c}} \right] \right.$$

$$G_m = g_m \left/ \sqrt{\left(1 + \frac{r_{bb'}}{r_{b'e}}\right)^2 + [r_{bb'} \cdot \omega(C_{b'e} + C_{b'c})]^2} \right.$$

Ideally, the amplifier should be unilateral, i.e., it should be possible for a signal to be transmitted from the input to the output, but not vice versa.

A method for unilateralizing a transistor:

$I_f = -I_n$ (so that there is no net current leaving point b)

$$\frac{V_b - V_c}{R_3 + (1/j\omega C_3)} = -\frac{V_b - (-n \cdot v_c)}{R_n + (1/j\omega C_n)}$$

$$\frac{-V_c}{R_3 + (1/j\omega C_3)} \approx \frac{n \, v_c}{R_n + (1/j\omega C_n)}$$

$$R_n = n \cdot R_3 \quad \text{and} \quad C_n = \frac{C_3}{n}, \quad n = \frac{V_x}{V_c}$$

Once the circuit is unilateral, the input and output may be treated independently (Fig. d).

18.5 ANALYSIS USING ADMITTANCE PARAMETERS

The performance of an active device is predicted using Y-parameters, because these can be obtained at any specific frequency.

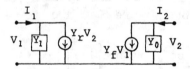

Using Y-parameters, because these can be obtained at any specific frequency.

$$Y_i = \frac{I_1}{V_1}\bigg|_{V_2=0}, \quad Y_f = \frac{I_2}{V_1}\bigg|_{V_2=0}$$

$$Y_r = \frac{I_1}{V_2}\bigg|_{V_1=0}, \quad Y_o = \frac{I_2}{V_2}\bigg|_{V_1=0}$$

A Y-parameter model with source and load added:

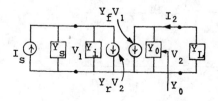

$$V_2 = -Y_f \cdot V_1/Y_o + Y_L, \quad I_1 = (V_i \cdot Y_i) + \frac{-Y_f \cdot V_1}{Y_o + Y_L} Y_r$$

$$Y_i = \frac{I_1}{V_1} = Y_i - \left(\frac{Y_f \cdot Y_r}{Y_o + Y_L} \right)$$

$$V_1 = -V_2 Y_r/Y_i + Y_s, \quad I_2 = \frac{-Y_f V_2 Y_r}{Y_i + Y_s} + (Y_o V_2)$$

$$Y_o = Y_o - (Y_f Y_r/Y_i + Y_s)$$

CHAPTER 19

FLIP-FLOPS

19.1 TYPES OF FLIP-FLOPS

19.1.1 THE BASIC FLIP-FLOP

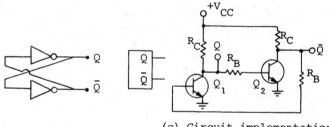

(c) Circuit implementation

The basic flip-flop simply consists of two RTL inverters.

19.1.2 R-S FLIP-FLOP

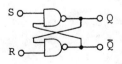

R	S	Q
1	1	unchanged
1	0	1
0	1	0
0	0	undefined

19.1.3 SYNCHRONOUS R-S FLIP-FLOP (CLOCKED R-S FLIP-FLOP)

The output state is set in synchronization with clock pulses.

NAND-gate implementation:

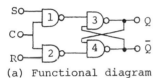

(a) Functional diagram

R_n	S_n	Q_{n+1}
0	0	Q_n
0	1	1
1	0	0
1	1	ND

(b) Truth table

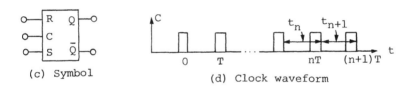

(c) Symbol   (d) Clock waveform

Gates 3 and 4 form an R-S latch with steering gates 1 and 2, used to input data to the device.

The clock is normally low, the outputs of gates 1 and 2 are then normally high and the state of the flip-flop cannot change. When C is high, the R and S inputs are steered to gates 3 and 4 and the flip-flop responds according to the truth table.

If R and S are low, the state of the flip-flop does not change, while Q in t_{n+1} is the same as it was in t_n, i.e., $Q_{n+1} = Q_n$.

19.1.4 PRESET AND CLEAR

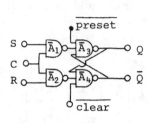

Synchronous		
R_n	S_n	Q_{n+1}
0	0	Q_n
0	1	1
1	0	0
1	1	ND
Asynchronous		
clear	preset	Q
0	1	0
1	0	1

(b) Truth table

General R-S Flip-Flop
Signal Polarities

If S is high, Q=1
If R is high, Q=0
If $\overline{preset}$ is low, Q=1
If $\overline{clear}$ is low, Q=0

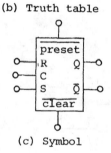

(c) Symbol

The two dc control lines are normally high.

The functional diagram: The two dc control lines do not require the clockpulse and, in fact, override the clocked inputs.

The asynchronous or dc truth table holds for the dc $\overline{clear}$ and $\overline{preset}$ inputs.

19.1.5 D-TYPE FLIP-FLOP

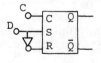

D_n	Q_{n+1}
1	1
0	0

(c) Symbol

B-175

With the D input connected to the S input and an inverter between R and S, the undefined output state that occurred in the R-S flip-flop is not possible here.

19.1.6 J-K FLIP-FLOP

(a) Symbol

J_n	K_n	Q_{n+1}
0	0	Q_n
0	1	0
1	0	1
1	1	$\bar{Q}_n$

(b) Truth table

Q_n	Q_{n+1}	J_n	K_n
0	0	0	X
0	1	1	X
1	0	X	1
1	1	X	0

(c) Excitation table, X = don't care

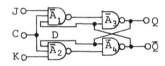

The J and K inputs listed in the truth table are the inputs present during time t_n. The output state is the flip-flop's state during t_{n+1} after a clock pulse at nT.

Excitation table: If the state Q_n of the flip-flop before clocking is known and the desired state Q_{n+1} after clocking is known, the necessary J and K inputs can be read from this table.

19.1.7 T-TYPE FLIP-FLOP

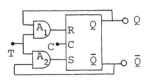

Functional Diagram

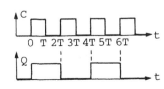

Waveforms when T=1

B-176

19.1.8 MASTER-SLAVE FLIP-FLOPS

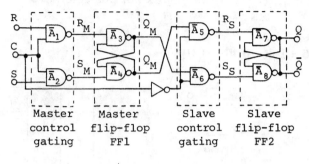

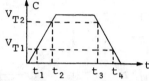

Time	Operation
t_1	disable slave from master
t_2	enable master (slave remains disabled)
t_3	disable master (slave remains disabled)
t_4	enable slave

Summary of Operations in the M-S Flip-Flop

19.2 FLIP-FLOP TIMING

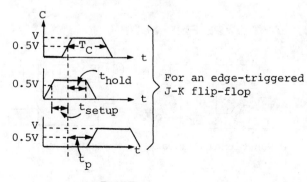

Hold, set-up and propagation time for a flip-flop:

The input data (J and K, R and S, or D) must be present and stable for some set-up time prior to the clock transition edge and for some hold time following the clock transition.

The clock transition edge is the rising edge of the clock pulse for a positive edge-triggered device and the falling edge of the clock pulse for a positive edge-triggered device.

Propagation time is measured from the 50% point on the clock transition edge to the 50% point on the output pulse edge.

19.3 COLLECTOR-COUPLED FLIP-FLOPS

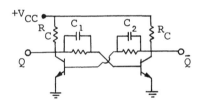

Collector-coupled ff

Test Conditions for a Flip-Flop

1) Do two stable states exist, in each of which at least one transistor is not active?

2) Is the incremental loop gain, with all transistors active, greater than 1?

With Q_1 off, $I_{B_2} = (V_{CC} - V_{BES_2})/(R_C + R_B)$. If Q_2 saturates, $I_{C_2} = (V_{CC} - V_{CES_2})/R_C$.

The condition for Q_2 to be saturated when Q_1 is off is:

$$I_{B_2} \approx (V_{CC} - V_{BES})/(R_C + R_B) \geq \frac{I_{C_2}}{\beta} = \frac{(V_{CC} - V_{CES})}{\beta \cdot R_C}$$

which is satisfied by any transistor with a minimum β,

$$\beta \geq \frac{V_{CC} - V_{CES}}{V_{CC} - V_{BES}} \cdot \frac{R_C + R_B}{R_C} \approx \frac{R_C + R_B}{R_C}$$

19.4 EMITTER-COUPLED FLIP-FLOPS

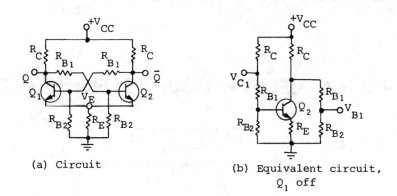

(a) Circuit

(b) Equivalent circuit, Q_1 off

With Q_1 off and Q_2 saturated, the emitter voltage is $V_E = (I_{B_2} + I_{C_2})R_E$. With Q_2 saturated, $V_{B_2} = V_E + V_{BES_2}$ and $V_{C_2} = V_E + V_{CES_2}$.

$$I_{C_2} = \left(\frac{V_{CC} - V_{C_2}}{R_C} + \frac{-V_{C_2}}{R_{B_1} + R_{B_2}} \right) \text{ mA}$$

$$I_{B_2} = \frac{V_{CC} - V_{B_2}}{R_C + R_{B_1}} - \frac{V_{B_2}}{R_{B_2}}$$

Q_2 will saturate when Q_1 is off, if and only if $\beta \geq I_{C_2}/I_{B_2}$.

19.5 SWITCHING SPEED OF A FLIP-FLOP

Transition time: This is the time required for conduction to occur between two transistors. The transition time is reduced by using speed-up capacitors.

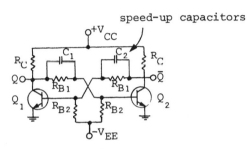

Fixed bias flip-flop

Without C_2 present, the input capacitance C_{i_1} is at the base of Q_1 will limit the time constant with which v_{B_1} can rise, and thus the rate at which Q_1 can turn on to:

$$[R_C \| r_{sat_2} + R_{B_1} + (R_{B_2} \| R_{i_1})]C_{i_1}$$

A model for Q_1 is turning on: The optimum value of C_2 for which v_{B_1} rises in zero time to its final value with no overshoot or undershoot is

$$C_2 = \frac{R_{B_2} \cdot C_{i_1}}{R_{B_1}}$$

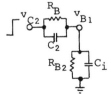

(a) Compensated attenuator

Equivalent circuit with Q_1 off and Q_2 saturated:

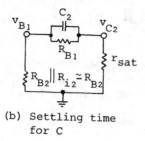

(b) Settling time for C

The voltage across C_2 changes at a rate

$$\tau_2 = [(R_{B_2} + r_{sat}) \| R_{B_1}] C_2,$$

$$\tau_2 \approx (R_{B_1} \| R_{B_2}) C_2 \text{ [with } r_{sat} \text{ very small]}$$

Equivalent circuit for determining the recharging rate for C_1:

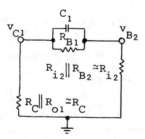

The value of the voltage on C_1 changes from its initial V_{C_2} value as soon as Q_1 has turned off. As Q_1 continues to be off, C_1 approaches its final steady-state value of

$$V_{C_2} = V_{C_1} - V_{B_2} \quad \text{at a rate} \quad \tau_1 = [(R_{i2} + R_c) \| R_{B_1}] \cdot C_1.$$

The settling time of the flip-flop: The time taken for C_1 and C_2 to recharge from their voltage levels ($V_{C_1} = V_{C_2} - V_{B_1}$ and $V_{C_2} = V_{C_1} - V_{B_2}$) to their new values ($V_{C_1} = V_{C_1} - V_{B_2}$ and $V_{C_2} = V_{C_2} - V_{B_1}$). (The values are interchanged.)

The resolution time is the sum of the transition time and the settling time. It is the total time the circuit requires to settle into its new steady state. It is a measure of the input trigger frequency that the circuit can resolve.

$$\frac{\text{The maximum frequency to}}{\text{which the flip-flop will respond}} = \frac{1}{\text{The resolution time}}$$

19.6 REGENERATIVE CIRCUITS

Schmitt trigger:

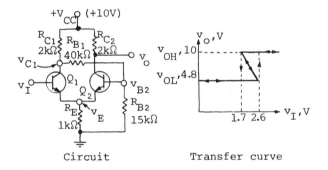

Circuit   Transfer curve

Transfer curve:

The lower trace is the output response from $V_I = 0$ to 2.6V.

The maximum voltage for which Q_1 will remain off

$$V_{IL} = V_{B_2} = \frac{V_{CC} \cdot R_{B_2}}{R_{C_1} + R_{B_1} + R_{B_2}} = 2.6\text{v}.$$

While $V_I < \overline{V_{IL}}$, the output voltage is

B-182

$$V_o = V_{CC} - \frac{V_{B2} \cdot R_{C2}}{R_E} = V_{OL} \left\{ \text{where } I_{C2} \approx I_{E2} = \frac{V_E}{R_E} \approx \frac{V_{B2}}{R_E} \right\}$$

$$= 4.8v.$$

When Q_1 is ON and Q_2 is OFF, the nodal equation at v_{C_1} is

$$\frac{V_{CC} - v_{C_1}}{R_{C_1}} = i_{C_1} + \frac{v_{C_1}}{R_{B_1} + R_{B_2}}$$

The minimum $V_I = V_{IH}$ for which Q_1 shuts off occurs when $V_I = v_E = v_{B_2}$. With Q_1 on and Q_2 off,

$$v_{B_2} = \frac{v_{C_1} \cdot R_{B_2}}{R_{B_1} + R_{B_2}}$$

Hence,

$$v_{IH} = v_{B_2} = \frac{V_{CC}}{R_{C_1}} \left[\frac{1}{R_E} + \frac{1}{R_{B_2}} + \frac{R_{B_1} + R_{B_2}}{R_{B_2} \cdot R_{C_1}} \right] = 1.7v$$

For $v_I > v_{IH} = 1.7$, $V_o = V_{CC}$.

For inputs in the range $\overline{v_{IL}} = 2.6V > V_I > 1.7 = v_{IH}$, the output will be either high or low, depending on its prior time history.

CHAPTER 20

WAVESHAPING AND WAVEFORM GENERATORS

20.1 COMMON WAVEFORMS

The step function: This function has an instantaneous change in level.

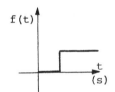 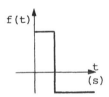

A ramp function:

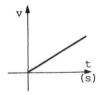 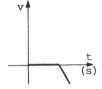

The exponential function:

 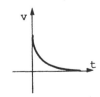

B-184

Pulse wave:

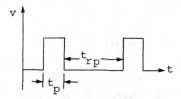

Pulse repetition time (PRT) = $t_p + t_{rp}$ = T.

Pulse repetition rate (PRR) = $\frac{1}{T}$.

% duty cycle = $\frac{t_p}{PRT} \times 100\%$

20.2 LINEAR WAVESHAPING CIRCUITS

RC low-pass circuit:

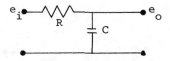

If the sinusoidal input e_i is applied, then

$$e_o = e_i \left(\frac{-jX_c}{R - jX_c} \right), \text{ where } X_c = \frac{1}{2\pi fc}$$

At low frequencies, $-jX_c \gg R$, then $e_o \cong e_i$.

At high frequencies, $-jX_c \ll R$, then $e_o \cong 0$

Frequency response:

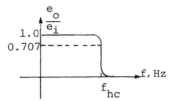

Cut-off frequency or half-power point frequency: The frequency at which the output (e_o) becomes 70.7 percent of the input c_i.

$$f_{nc} = \frac{1}{2\pi RC} = \frac{1}{2\pi\tau}, \quad \tau = \text{Time constant of the filter}$$

Step input to an RC low-pass:

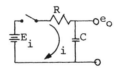

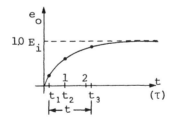

$$E_i = v_R + v_C = iR + v_C$$

$$= RC\frac{dv_c}{dt} + v_c$$

$$v_c = \text{The instantaneous voltage across C} = I_o R\left[1 - e^{-(t/\tau)}\right]$$

Curve characteristics:

$t_o = 0$ and $v_c = 0$ $t_3 = 2.3\tau$, $v_c = 0.9E_i$

$t_1 = 0.1\tau$, $v_c = 0.1E_i$ $t_4 = 5\tau$, $v_c \cong E_i$

$t_2 = \tau$, $v_c = 0.632E_i$

$t_r =$ The rise time $= t_3 - t_1 = 2.2\tau$

Pulse wave input to an RC low-pass:

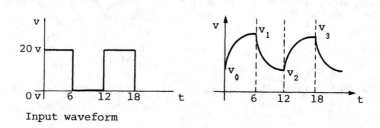

Input waveform

In steady state: $v_o = v_2$ and $v_1 = v_3$

$v_1 = E_i - (E_i - v_o)e^{-(t/\tau)}$

$v_2 = E_i - (E_i - v_1)e^{-(t/\tau)}$

Effects of the circuit τ on the output waveform:

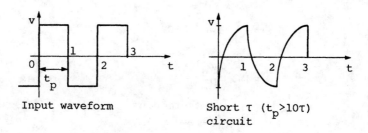

Input waveform   Short τ ($t_p > 10\tau$) circuit

B-187

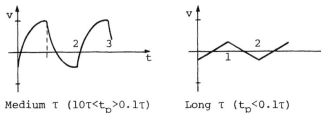

Medium τ ($10τ<t_p>0.1τ$) Long τ ($t_p<0.1τ$) circuit

RC high-pass circuit:

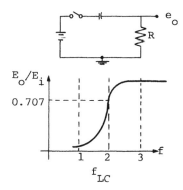

$$e_o = \left(\frac{R}{R-jx_c}\right)e_i, \quad x_c = 1/2\pi f_c$$

Lower-cutoff frequency $f_{\ell c} = 1/2\pi RC = \frac{1}{2\pi} \cdot \frac{1}{\tau}$

Pulse wave input:

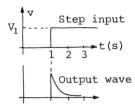

$$v_R = E_i - v_c = (E_i \pm V_{ci})e^{-(t/\tau)}$$

$$v_R = E_o = E_i e^{-t/\tau} \quad \text{(for } V_{ci} = 0\text{)}$$

Effects of τ on the output waveform:

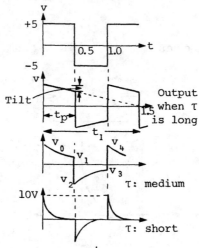

$$\text{Fractional tilt } F_t = \frac{t_p}{t_t}$$

(t_p = pulse width, t_t = Tilt time to reach 0 volt)

An R-C high pass circuit under short τ condition is called a differentiator.

RLC circuit:

Step input to a series RL circuit:

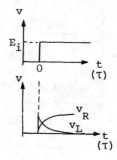

$$v_R = E_i \left(1 - e^{\frac{-tR}{L}}\right)$$

$$v_L = E_i \, e^{\frac{-tR}{L}}$$

Step input to a series RLC circuit:

Current in a underclamped circuit

Attenuators:

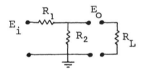

Simple attenuator

Attenuation factor: $\dfrac{E_i}{E_o} = \dfrac{R_2}{R_1 + R_2}$

$\dfrac{E_i}{E_o}$ (with R_L) = $\dfrac{R_2 \| R_L}{R_1 + (R_2 \| R_L)}$

Uncompensated and compensated attenuator:

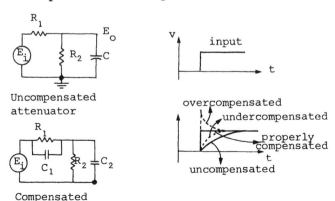

Uncompensated attenuator

Compensated attenuator

B-190

$$\frac{E_o}{E_i} = \text{Attenuation factor} = \frac{R_2 \| X_C}{R_1 + (R_2 \| X_C)}$$

$$\frac{E_o}{E_i} = \frac{Z_2}{Z_1 + Z_2}$$

$$Z_1 = R_1 \| X_{C1} \quad \text{and} \quad Z_2 = R_2 \| X_{C2}$$

For proper compensation:

$$R_1 C_1 = R_2 C_2$$

20.3 SWEEP GENERATORS

Voltage sweep principles:

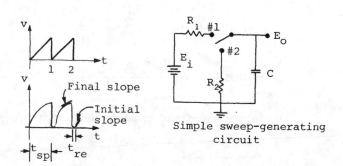

Simple sweep-generating circuit

t_{sp} = Sweep time, the time during which the function increases linearly.

t_{re} = Retrace or flyback time, the time the function takes to drop back down to the initial base voltage.

$$\% \text{ slope error} = \frac{\text{Initial slope} - \text{final slope}}{\text{initial slope}} \times 100\%$$

Astable sweep circuit:

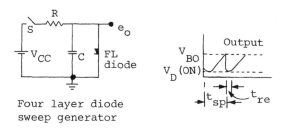

Four layer diode sweep generator

V_{BO} = (Breakdown voltage)
$= V_{CC} - [V_{CC} - V_D(ON)]e^{-(t_{sp}/\tau_c)}$

$t_{sp} = \tau_c \cdot \ln \dfrac{V_{CC} - V_D(ON)}{V_{CC} - V_{BO}}$ (s)

$V_D(ON)$ = Voltage across FL diode when it is on

$\tau_C = RC$

Transistor sweep generator:

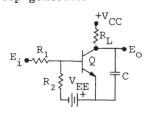

Input pulse waveform

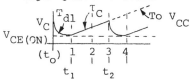

V_C = voltage across capacitor just before the transistor is ON

τ_C (charging time constant) = $R_L \cdot C$ seconds

τ_d (discharge time constant) = $R_T \cdot C$ seconds

R_T = Equivalent resistance of the ON transistor = $\dfrac{V_{CE}(ON)}{I_C(ON)}$ ohms

$t_{sp} = t_2 - t_1 = \tau_c \cdot \ln\left[\dfrac{V_{CC} - V_{CE}(ON)}{V_{CC} - V_C}\right]$ seconds

Slope error = $\dfrac{V_C - V_{CE}(ON)}{V_{CC} - V_{CE}(ON)} \times 100$

Miller sweep circuit:

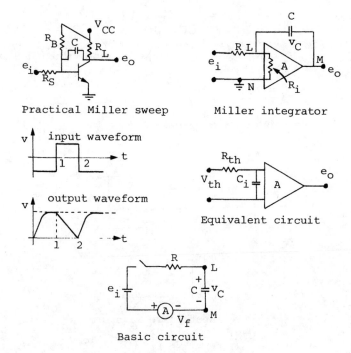

Practical Miller sweep

Miller integrator

input waveform

output waveform

Equivalent circuit

Basic circuit

Basic circuit: V_f is a fictitious voltage and is made such that its terminal voltage is always equal to the voltage across C but opposite in polarity. In such a case, a constant current E_i/R will flow n the circuit.

$$V_c = E_i \frac{t}{\tau} \text{ volts}$$

Miller integrator: This simulates the basic circuit.

Equivalent circuit:

$$C_i = C(1+A)$$

$$V_{th} = E_i(R_i/R+R_i)$$

$$R_{th} = R_i R/R_i + R$$

$$V_{ci} = V_{th} - V_{th}\, e^{-(t/R_{th} C_i)} \quad \text{... Capacitor charge equation.}$$

$$e_o = A \cdot e_i$$

$$= A \cdot V_{th} - A \cdot V_{th}\, e^{-(t/R_{th} C_i)}$$

The slope of the output for a high A:

$$A = \frac{e_o}{t} = \frac{E_i}{RC} = E_i \frac{t}{\tau_c}$$

B-194

Handbook of Electrical Engineering

SECTION C
Electromagnetics

CHAPTER 1

VECTOR ANALYSIS

1.1 VECTOR ALGEBRA

1.1.1 NOTATIONS AND UNIT VECTOR

A vector quantity of A is represented by $\bar{A}$. (Note: a bar or an arrow on top of A may be used as a symbol of a vector quantity.)

A scalar quantity is represented by A which defines a magnitude.

By definition, vector $\bar{A}$ can be expressed in component form as:

$$\bar{A} = \bar{a}_x A_x + \bar{a}_y A_y + \bar{a}_z A_z$$

The magnitude of $\bar{A}$ (i.e., $|\bar{A}|$) = $\sqrt{A_x^2 + A_y^2 + A_z^2}$ and the direction of $\bar{A}$ is indicated by the unit vectors $\bar{a}_x$, $\bar{a}_y$ and $\bar{a}_z$ in the cartesian coordinate. (Note: $|\bar{a}_x| = |\bar{a}_y| = |\bar{a}_z| = 1$) (see figures below)

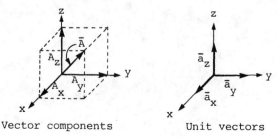

Vector components   Unit vectors

Also, notice that a unit vector in the direction of $\overline{A}$ is given by the equation:

$$\overline{a}_A = \frac{\overline{A}}{|\overline{A}|} = \frac{\overline{A}}{\sqrt{A_x^2 + A_y^2 + A_z^2}}$$

(Note: i, j and k are also used as unit vectors.)

1.1.2 VECTOR OPERATIONS

Vector Addition:

$$\overline{A} + \overline{B} = (A_x \overline{a}_x + A_y \overline{a}_y + A_z \overline{a}_z) + (B_x \overline{a}_x + B_y \overline{a}_y + B_z \overline{a}_z)$$

$$= (A_x + B_x)\overline{a}_x + (A_y + B_y)\overline{a}_y + (A_z + B_z)\overline{a}_z$$

Vector addition shown graphically:

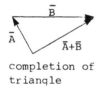
completion of triangle

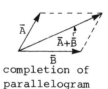
completion of parallelogram

Properties of vector addition:

A) Commutative - $\overline{A} + \overline{B} = \overline{B} + \overline{A}$

B) Associative - $\overline{A} + (\overline{B} + \overline{C}) = (\overline{A} + \overline{B}) + \overline{C}$

C) Distributive - $m(\overline{A} + \overline{B}) = m\overline{A} + m\overline{B}$

Vector Subtraction:

$$\overline{A} - \overline{B} = (A_x - B_x)\overline{a}_x + (A_y - B_y)\overline{a}_y + (A_z - B_z)\overline{a}_z$$

Vector subtraction show graphically: $(\overline{A} - \overline{B}) = \overline{A} + (-\overline{B})$

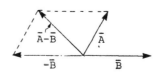

Vector multiplication by a scalar:

This obeys the associative and distributive laws.

$$(r + s)(\overline{A} + \overline{B}) = r(\overline{A} + \overline{B}) + s(\overline{A} + \overline{B})$$
$$= r\overline{A} + r\overline{B} + s\overline{A} + s\overline{B}$$

where r and s are scalar quantities.

Vector equality:

$\overline{A} = \overline{B}$ if and only if $\overline{A} - \overline{B} = 0$.

1.1.3 VECTOR FIELD

A field is mathematically a function of the vector connecting an arbitrary origin to a general point in space.

A vector field is defined as a vector function of a position vector.

1.1.4 VECTOR DOT AND CROSS PRODUCT

The dot (scalar) product:

$$\overline{A} \cdot \overline{B} = |\overline{A}| |\overline{B}| \cos \theta_{AB}$$

where θ_{AB} is the smaller angle between vectors $\overline{A}$ and $\overline{B}$.

Note: The dot product of a unit vector by itself equals one (i.e., $\overline{a}_x \cdot \overline{a}_x = \overline{a}_y \cdot \overline{a}_y = \overline{a}_z \cdot \overline{a}_z = 1 \cdot 1 \cos 0° = 1$), but by a different unit vector equals zero (i.e., $\overline{a}_x \cdot \overline{a}_y = \overline{a}_x \cdot \overline{a}_z = \overline{a}_z \cdot \overline{a}_y = 1 \cdot 1 \cos 90° = 0$)

The result of a dot (scalar) product must be a scalar.

If $\overline{A} = \overline{a}_x A_x + \overline{a}_y A_y + \overline{a}_z A_z$

and $\overline{B} = \overline{a}_x B_x + \overline{a}_y B_y + \overline{a}_z B_z$

then,

$$\bar{A} \cdot \bar{B} = A_x B_x + A_y B_y + A_z B_z$$

$$\bar{A} \cdot \bar{A} = |\bar{A}|^2$$

$$\bar{A} \cdot \bar{a} = |\bar{A}||\bar{a}| \cos \theta_{Aa} = |\bar{A}| \cos \theta_{Aa}$$

where $\bar{a}$ is a unit vector in any direction.

The dot product obeys the commutative and distributive laws,

i.e., $\bar{A} \cdot \bar{B} = \bar{B} \cdot \bar{A}$ and $\bar{A} \cdot (\bar{B}+\bar{C}) = \bar{A} \cdot \bar{B} + \bar{A} \cdot \bar{C}$

The cross (vector) product:

$$\bar{A} \times \bar{B} = (|\bar{A}||\bar{B}| \sin \theta_{AB}) \bar{a}_n$$

where vector $\bar{a}_n$ represents a unit normal vector.

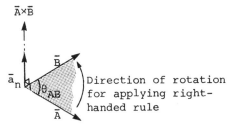

Direction of rotation for applying right-handed rule

Note: $\bar{a}_n$ is always perpendicular to the plane formed by vectors $\bar{A}$ and $\bar{B}$.

Note: $\bar{a}_n$ is always perpendicular to the plane formed by vector $\bar{A}$ and $\bar{B}$.

The cross product of $\bar{A}$ and $\bar{B}$ can also be expressed as

$$\bar{A} \times \bar{B} = (A_y B_z - A_z B_y)\bar{a}_x + (A_z B_x - A_x B_z)\bar{a}_y + (A_x B_y - A_y B_x)\bar{a}_z$$

or in determinant form,

$$\overline{A} \times \overline{B} = \begin{vmatrix} \overline{a}_x & \overline{a}_y & \overline{a}_z \\ A_x & A_y & A_z \\ B_x & B_y & B_z \end{vmatrix}$$

Note: The cross product is not commutative but,

$$\overline{A} \times \overline{B} = -\overline{B} \times \overline{A}$$

The cross product of a unit vector:

The cross product of a unit vector can be found by applying the following simple rule:

Observe in the figure shown, anticlockwise direction resulting in a positive vector and clockwise direction resulting in a negative vector.

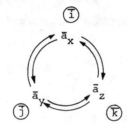

Thus,

$$\overline{a}_x \times \overline{a}_y = \overline{a}_z \quad \text{(Anticlockwise)}$$

$$\overline{a}_y \times \overline{a}_x = -\overline{a}_z \quad \text{(Clockwise)}$$

etc.

Table of unit vector cross products

$\overline{a}_x \times \overline{a}_x = 0$	$\overline{a}_x \times \overline{a}_y = \overline{a}_z$	$\overline{a}_x \times \overline{a}_z = -\overline{a}_y$
$\overline{a}_y \times \overline{a}_x = -\overline{a}_z$	$\overline{a}_y \times \overline{a}_y = 0$	$\overline{a}_y \times \overline{a}_z = \overline{a}_x$
$\overline{a}_z \times \overline{a}_x = \overline{a}_y$	$\overline{a}_z \times \overline{a}_y = -\overline{a}_x$	$\overline{a}_z \times \overline{a}_z = 0$

Note: The cross product of a unit vector by itself equals zero. i.e., $\overline{a}_x \times \overline{a}_x = 0$.

1.2 DIFFERENT TYPES OF COORDINATE SYSTEMS AND DIFFERENTIAL VOLUME

DIFFERENT TYPES OF COORDINATE SYSTEMS:

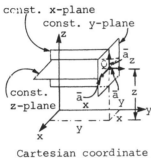

Cartesian coordinate system

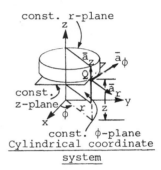

Cylindrical coordinate system

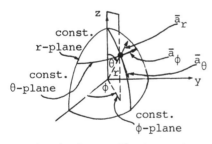

Spherical coordinate system

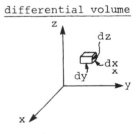

Cartesian:
dv=dxdydz

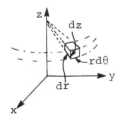

Cylindrical:
dv=rdrdϕdz

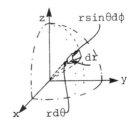

Spherical:
dv=r²sinθdrdθdϕ

CHAPTER 2

COULOMB'S LAW AND ELECTRIC FIELD

2.1 COULOMB'S LAW

By definition, the force between two point charges of arbitrary positive or negative strengths is given by the Coulomb's law as follows:

$$F = k \frac{Q_1 Q_2}{4 \pi \varepsilon_0 d^2}$$

where Q_1 and Q_2 = positive or negative charges on either object in coulombs.

d = distance separating the two point charges.

k = the constant of proportionality

= $(4\pi\varepsilon_0)^{-1}$ = 9×10^9 newton-meter^2/coul.

ε_0 = permittivity in free space

= 8.854×10^{-12} F/m

Note: $\varepsilon = \varepsilon_0 \varepsilon_r$ for media other than free space, where ε_r is the relative permittivity of the media.

The force F can be expressed in a vector form to indicate its direction as follows:

$$\overline{F} = \frac{Q_1 Q_2}{4 \pi \varepsilon_0 d^2} \overline{a}_d$$

where unit vector $\bar{a}_d$ is in the direction of d and $\bar{a}_d = \dfrac{\bar{d}}{|\bar{d}|} = \dfrac{\bar{d}}{d}$.

2.2 ELECTRIC FIELD INTENSITY

Assuming a fixed position point charge Q, then by Coulomb's law, the force on a test charge Q_t due to Q is

$$\bar{F} = \dfrac{Q\,Q_t}{4\pi\varepsilon_0 d_t^2}\,\bar{a}_{d_t}$$

Now, by definition, the electric field intensity $\bar{E}$ due to Q equals the force per unit charge in V/m or N/c.

Hence,

$$\bar{E} = \dfrac{\bar{F}_t}{Q_t} = \dfrac{Q}{4\pi\varepsilon_0 d_t^2}\,\bar{a}_{d_t}$$

Thus, in general $\bar{E} = \dfrac{Q}{4\pi\varepsilon_0 d^2}\bar{a}_d$ where d is the magnitude of the vector $\bar{d}$.

In general, if Q is located at $\bar{d}' = a\bar{a}_x + b\bar{a}_y + c\bar{a}_z$ and the field is at $\bar{d} = x\bar{a}_x + y\bar{a}_y + z\bar{a}_z$, then

$$\boxed{\bar{E} = \dfrac{Q(\bar{d} - \bar{d}')}{4\pi\varepsilon_0 |\bar{d} - \bar{d}'|^3}}$$

2.3 ELECTRIC FIELD INTENSITY DUE TO: POINT CHARGES, VOLUME CHARGE DISTRIBUTION,

LINE OF CHARGE AND SHEET OF CHARGE

Electric field intensity due to two point charges is illustrated in diagram below:

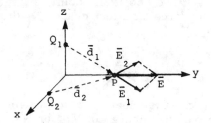

$$\overline{E} = \overline{E}_1 + \overline{E}_2 = \frac{Q_1}{4\pi\epsilon_0 d_1^2}\overline{a}_{d_1} + \frac{Q_2}{4\pi\epsilon_0 d_2^2}\overline{a}_{d_2}$$

In general, for k point charges, the electric field

$$\boxed{\overline{E} = \sum_{j=1}^{k} \frac{Q_j}{4\pi\epsilon_0 \cdot d_j^2} \cdot \overline{a}_{d_j}}$$

Field due to volume charge distribution:

Gauss's law for electric fields in free space is defined as:

$$\oint_S (\epsilon_0 \overline{E}) \cdot ds = \int_V \rho du \equiv Q$$

where Q is the total charge within a finite volume i.e.,

$$Q = \int_V \rho dv = \int_V dQ$$

and ρ is the volume charge density, i.e.,

$$\rho = \frac{dQ}{dv} \text{ in c/m}^3.$$

(see figure)

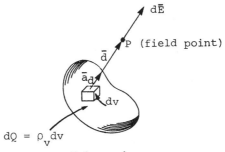

Volume charge

Thus the differential electric field $d\overline{E}$ due to dQ in the figure is

$$d\overline{E} = \frac{dQ}{4\pi t_0 d^2}\overline{a}_d = \frac{\rho\,dv}{4\pi\varepsilon_0 d^2}\overline{a}_d$$

and $\quad \overline{E} = \displaystyle\int_V \frac{\rho\,dv}{4\pi\varepsilon_0 d^2}\overline{a}_d$ = total electric field at P.

In general, the electric field of the summation of all the volume charge in a given region is:

Field of a Line Charge:

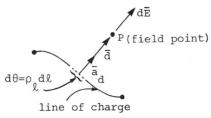

line of charge

$dQ = \rho_\ell\, d_\ell\quad$ where ρ_ℓ is the line charge density.

$$d\overline{E} = \frac{dQ}{4\pi\varepsilon d^2}\overline{a}_d$$

Hence,

$$d\overline{E} = \frac{\rho_\ell\, d\ell}{4\pi\varepsilon_0 d^2}\overline{a}_d$$

C-10

$$\bar{E} = \int_\ell \frac{\rho_\ell}{4\pi\varepsilon_0 d} \bar{a}_d \, d\ell$$

A Special Case:

For an infinitely long wire, the electric field intensity at a distance d from the line charge is

$$\bar{E} = \frac{\rho_\ell}{2\pi\varepsilon_0 d}$$

Field of a Sheet of Charge

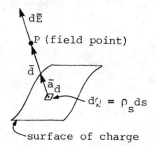

$$dQ = \rho_s \, ds$$

$$d\bar{E} = \frac{\rho_s \, ds}{4\pi\varepsilon_0 d^2} \bar{a}_d$$

Hence,

$$\bar{E} = \int_S \frac{\rho_s \, ds}{4\pi\varepsilon_0 d^2} \bar{a}_d$$

2.4 STREAMLINES AND SKETCHES OF FIELDS

Streamlines: Continuous lines from the test charge which are, at every point, tangential to $\bar{E}$ and are an indication of the direction of $\bar{E}$.

The equation of the streamlines is obtained by solving

$$\boxed{\dfrac{E_y}{E_x} = \dfrac{dy}{dx}}$$

(Assuming $E_z = 0$, thus the streamlines are confined to the XY-plane.)

An example:

Given: $\bar{E} = \dfrac{1}{r}\bar{a}_r$ (The field of the uniform line charge with $\rho_L = 2\pi\varepsilon_0$.)

Then, in cartesian coordinates:

$$\bar{E} = \dfrac{x}{x^2+y^2}\bar{a}_x + \dfrac{y}{x^2+y^2}\bar{a}_y$$

The equation for the streamlines is

$$\dfrac{dy}{dx} = \dfrac{E_y}{E_x} = \dfrac{1}{x},$$

then $\ln y = \ln x + \ln c$

and the streamlines equation is $y = cx$.

The sketch:

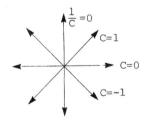

CHAPTER 3

ELECTRIC FLUX DENSITY, GAUSS'S LAW AND DIVERGENCE

3.1 ELECTRIC FLUX AND FLUX DENSITY

By definition, the electric flux, ψ (from Faraday's experiment), is given by

$$\psi = Q$$

where Q is the charge in coulombs.

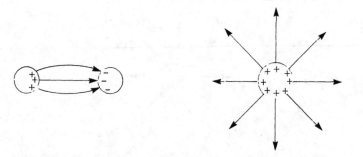

In other words, one coulomb of electric flux corresponds to one coulomb of electric charge.

Note: Electric flux goes from positive to negative charge and from positive charge to infinity if the negative charge is absented.

The electric flux density $\overline{D}$ is a vector quantity. In general, at a point M of any surface S (see figure), $Dds\cos\theta = d\psi$, where $d\psi$ is the differential flux through the differential surface ds of M and θ is the angle of $\overline{D}$ with respect to the normal vector from ds. (Note: $\overline{D} = \dfrac{d\psi}{ds}$ is the case where $\overline{D}$ is normal to ds and the direction and magnitude of the electric flux density varies along the surface.)

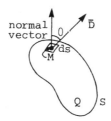

3.2 GAUSS'S LAW AND ITS APPLICATIONS

Gauss's law states that the net electric flux passing out of a closed surface is equal to the total charge within such surface.

Hence, since

$$d\psi = \overline{D} \cdot \overline{ds}$$

$$\psi = \int \overline{D} \cdot \overline{ds}$$

and by Gauss's law,

$$\psi_{net} = \oint_s \overline{D} \cdot \overline{ds} = Q_s$$

where Q_s is the total number of charges enclosed by the surface.

Application of Gauss's Law:

A spherical surface is shown in the figure below:

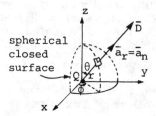

Applying Gauss's law, the charge Q_s enclosed by the spherical surface is

$$Q_s = \oint_S \overline{D} \cdot \overline{ds}$$

where ds in this case is equal to $4\pi r^2$ (Note: r is the radius of the sphere). Hence,

$$Q_s = D 4\pi r^2$$

and

$$\overline{D} = \frac{Q_s}{4\pi r^2} \cdot \overline{a}_r$$

Since electric field intensity $\overline{E}$ is equal to $\frac{Q}{4\pi\varepsilon_0 d^2} \overline{a}_d$ and d is equal to r in this case, then $\overline{D} = \varepsilon_0 \overline{E}$.

Some hints for choosing a special Gaussian surface:

A) The surface must be closed.
B) D remains constant through the surface and normal to the surface.
C) D is either tangential or normal to the surface at any point on the surface.

It is easier in solving a problem if we can choose a special Guassian surface.

Special Cases:

Case 1: Given an infinite line charge of uniform density $\rho_\ell =$

$\frac{Q}{\ell}$, the electric flux density can be obtained by enclosing the line of charge with an imaginary cylindrical Guassian surface of length ℓ and radius r, and then apply Gauss's law. (See figure)

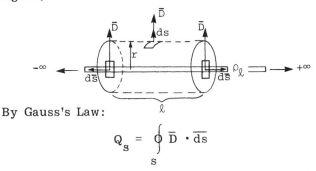

By Gauss's Law:
$$Q_s = \oint_s \bar{D} \cdot \overline{ds}$$

At the two ends of the cylindrical surface, $\bar{D} \cdot \overline{ds} = 0$ since $\bar{D}$ and $\overline{ds}$ are orthogonal or $\bar{D}$ are tangential to the surface.

Finally, $\bar{D}$ and $\overline{ds}$ are parallel in the rest of the surface and therefore,
$$Q_s = D \int ds$$
$$= D(2\pi\ell r)$$

For
$$Q_s = \rho_\ell \ell$$

Therefore,
$$\bar{D} = \frac{\rho_\ell}{2\pi r} \bar{a}_r$$

Case 2:

Given two circular cylindrical coaxial capacitors, with inner cylinder of radius $r = a$ and charge density ρ_s, and outer cylinder $r = b$, the electric flux density can be determined by using the Gaussian surface. (See figure)

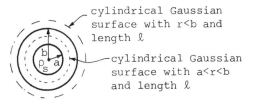

cylindrical Gaussian surface with r<b and length ℓ

cylindrical Gaussian surface with a<r<b and length ℓ

(cross-sectional view)

C-16

If the Gaussian surface is located at a distance $a<r<b$, then the electric flux density $\overline{D}$ is equal to the flux density for an infinite line of charge, i.e.,

$$\overline{D} = \frac{\rho_\ell}{2\pi r} \overline{a}_r \quad \text{(Note: } \rho_\ell = 2\pi a \rho_s\text{)}$$

If the Gaussian surface is located at $r > b$, the charges on the inner and outer conductors will cancel each other. Then the electric flux density is equal to zero.

3.3 DIVERGENCE

By definition, the divergence of a vector field $\overline{B}$ at a defined point is given as

$$\text{Div } \overline{B} = \lim_{\Delta v \to 0} \frac{\oint_s \overline{B} \cdot d\overline{s}}{\Delta v}$$

where $\oint_s \overline{B} \cdot d\overline{s}$ is the total flux flowing outward of an enclosed small surface of the chosen point divided by the enclosed small volume Δv as Δv approaches zero. (See figure)

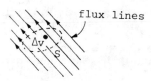
flux lines

The divergence is the characteristic of the variation of a vector field from different points in space and the unit is flux lines/m^3.

Cartesian Coordinates

In the figure below, a differential volume in the

cartesian coordinate system is shown. The vector field at the defined point $\bar{B}$ can be expressed as:

$$\bar{B} = B_x \bar{a}_x + B_y \bar{a}_y + B_z \bar{a}_z$$

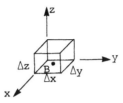

Now, apply the divergence formula.

Thus, $\oint_s \bar{B} \cdot \overline{ds}$ must be determined first.

This is equal to the summation of the integration of all six faces of the closed surface. Hence, since

$$\int_{front} + \int_{back} \simeq \partial \frac{B_x}{\partial x} \Delta x \Delta y \Delta z$$

$$\int_{left} + \int_{right} \simeq \partial \frac{B_y}{\partial y} \Delta x \Delta y \Delta z$$

and

$$\int_{top} + \int_{bottom} \simeq \partial \frac{B_z}{\partial z} \Delta x \Delta y \Delta z$$

Therefore,

$$\oint_s \bar{B} \cdot \overline{ds} \simeq \partial \frac{B_x}{\partial x} \Delta x \Delta y \Delta z + \partial \frac{B_y}{\partial y} \Delta x \Delta y \Delta z$$

$$\simeq \quad + \partial \frac{B_z}{\partial z} \Delta x \Delta y \Delta z$$

$$\simeq \Delta x \Delta y \Delta z \left(\partial \frac{B_x}{\partial x} + \partial \frac{B_y}{\partial y} + \partial \frac{B_z}{\partial z} \right)$$

$$\simeq \Delta v \left(\partial \frac{B_x}{\partial x} + \partial \frac{B_y}{\partial y} + \partial \frac{B_z}{\partial z} \right)$$

$$\lim_{\Delta v \to 0} \frac{\oint_s \overline{B} \cdot \overline{ds}}{\Delta v} = \frac{\partial B_x}{\partial x} + \frac{\partial B_y}{\partial y} + \frac{\partial B_z}{\partial z}$$

Thus,
$$\text{Div } \overline{B} \text{ (cartesian)} = \frac{\partial B_x}{\partial x} + \frac{\partial B_y}{\partial y} + \frac{\partial B_z}{\partial z}$$

In other coordinate systems, using the same approach,

$$\text{Div } \overline{B} \text{ (cylindrical)} = \frac{1}{r}\frac{\partial}{\partial r}(rB_r) + \frac{1}{r}\frac{\partial B_\phi}{\partial \phi} + \frac{\partial B_z}{\partial z}$$

$$\text{Div } \overline{B} \text{ (spherical)} = \frac{1}{r^2}\frac{\partial}{\partial r}(r^2 B_r) + \frac{1}{r\sin\theta}\frac{\partial}{\partial \theta}(\sin\theta B_\theta)$$
$$+ \frac{1}{r\sin\theta}\frac{\partial B_\phi}{\partial \phi}$$

Note: The result of the divergence operation is a scalar and the operand is a vector.

3.4 THE DEL OPERATOR

The del operator is a vector operator which is defined in the cartesian coordinate system as

$$\nabla \equiv \overline{a}_x \frac{\partial}{\partial x} + \overline{a}_y \frac{\partial}{\partial y} + \overline{a}_z \frac{\partial}{\partial z}$$

A Special Case

In the rectangular system of coordinates, the dot product of the operator ∇ with a vector $\overline{B}$ is equal to the divergence of $\overline{B}$.

$$\nabla \cdot \bar{B} = \left(\bar{a}_x \frac{\partial}{\partial x} + \bar{a}_y \frac{\partial}{\partial y} + \bar{a}_z \frac{\partial}{\partial z}\right) \cdot (B_x \bar{a}_x + B_y \bar{a}_y + B_z \bar{a}_z)$$

$$= \frac{\partial B_x}{\partial x} + \frac{\partial B_y}{\partial y} + \frac{\partial B_z}{\partial z}$$

$$= \text{Div } \bar{B}$$

(Note: The quantity of $\nabla \cdot \bar{B} = \text{Div } \bar{B}$ is a scalar.)

3.5 THE DIVERGENCE THEOREM AND MAXWELL'S FIRST EQUATION

The divergence of the flux density vector field $\bar{D}$ is

$$\text{Div } \bar{D} = \lim_{\Delta v \to 0} \frac{\oint_S \bar{D} \cdot d\bar{s}}{\Delta v}$$

Applying Gauss's law, namely $\oint_S \bar{D} \cdot d\bar{s} = Q_S$ to Div $\bar{D}$ results in the following relationships:

$$\text{Div } \bar{D} = \lim_{\Delta v \to 0} \frac{Q_S}{\Delta v} = \rho$$

This is known as Maxwell's first equation. It states that the net flux density crossing the surface of an enclosed small volume is equal to the charge density within the volume.

Now, the divergence theorem is obtained using the above relationships.

Since,

$$Q_S = \int_V \rho \, dv = \oint_S \bar{D} \cdot d\bar{s}$$

C-20

and $\rho = \text{Div } \overline{D}$

Thus,
$$\int_V \text{Div } \overline{D} \, dv = \oint_S \overline{D} \cdot d\overline{s}$$

or
$$\int_V \nabla \cdot \overline{D} \, dv = \oint_S \overline{D} \cdot d\overline{s}$$

(the divergence theorem). This theorem relates the volume to the surface integral. It states that the total flux flowing outward of an enclosed surface is equal to the integral of the divergence of the flux density throughout the volume enclosed by such surface.

Note: In general, $\int_V \text{Div } \overline{B} \, dv = \oint_S \overline{B} \cdot d\overline{s}$, which is true for any vector field.

CHAPTER 4

ENERGY AND POTENTIAL

4.1 WORK

In the figure shown below, a small test charge Q is presented in an electric field $\bar{E}$.

The force acting on Q due to $\bar{E}$ is given by $\bar{F} = Q\bar{E}$.

Thus, an opposite and equal force acting on Q against the direction of the electric field can be expressed as $\bar{F}' = -Q\bar{E}$.

This is the force applied externally and energy is required.

The work done, dw, or the energy needed to move a point charge through a distance dL, is expressed as

$$dw = \bar{F}' \cdot d\bar{L}$$
$$= -Q\bar{E} \cdot d\bar{L}$$

where dw is the differential work done by moving the point charge through a differential distance $d\bar{L}$.

Or

$$W = -Q \int_A^B \bar{E} \cdot d\bar{L}$$

where A and B specify the starting and ending positions of the travelling point charge.

C-22

Hence, from the above equation, work done is independent of the path taken.

Also note, by convention, the work done by the electric field is a negative quantity.

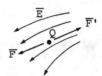

The following table gives the equation for dL in different coordinate systems:

Coordinate systems	$d\overline{L}$
Cartesian	$\overline{a}_x dx + \overline{a}_y dy + \overline{a}_z dz$
Cylindrical	$\overline{a}_r dr + \overline{a}_\phi r d\phi + \overline{a}_z dz$
Spherical	$\overline{a}_r dr + \overline{a}_\theta r d\theta + \overline{a}_\phi r \sin\theta \, d\phi$

4.2 POTENTIAL

The potential difference between two points p and p', symbolized as $v_{p'p}$ (or $\Phi p'p$) is defined as the work done in moving a unit positive charge by an external force from the initial point p to the final point p'.

i.e.,

$$v_{p'p} = -\int_p^{p'} \overline{E} \cdot d\overline{L} = v_{p'} - v_p$$

(Note: Zero or ground is chosen as the common reference point for $v_{p'}$ and v_p, and from the equation, the potential difference is independent of the path selected.)

The unit for potential difference is joule per coulomb or volt. Notice that when energy is applied in moving a positive point charge from point p to p', the potential at point p' is said to be of a higher potential than p or $v_{p'p}$ is positive.

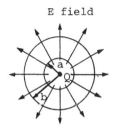

E field

In the figure, a point charge Q is located at the center of two equipotential surfaces and the direction of the electric field is shown. Assume that the equipotential surface of radius b is the potential reference. Then the potential difference between any points located at $r = a$ and $r = b$ is given by:

$$v_{ab} = \frac{Q}{4\pi\varepsilon_0}\left(\frac{1}{a} - \frac{1}{b}\right) = v_a - v_b$$

Now, if the potential reference is chosen at infinity, (i.e., as $b \to \infty$) then,

$$v_a = \frac{Q}{4\pi\varepsilon_0}\left(\frac{1}{a} - \frac{1}{\infty}\right) = \frac{Q}{4\pi\varepsilon_0 a}$$

or in general, the potential difference between any points of radius from Q is $V = \frac{Q}{4\pi\varepsilon_0 r}$.

Potential due to point charges:

The potential due to point charges in free space is defined as the summation of the potential contributions of each charge.

i.e.,

$$V = \sum_{j=1}^{k} \frac{Q_j}{4\pi\varepsilon_0 R_j} = \sum_{j=1}^{k} v_j$$

or

$$V = \int_V \frac{\rho \, dv}{4\pi\varepsilon_0 R}$$

(Note: $dQ = \rho dv$)

4.3 GRADIENT

The gradient of a scalar function is a vector quantity. It is defined as the maximum rate of change of a scalar field in free space together with the vector direction in which the rate of change occurs.

In different coordinate systems, the gradient of a function is defined as follows:

$$\text{grad } g \text{ (rectangular)} = \bar{a}_x \frac{\partial g}{\partial x} + \bar{a}_y \frac{\partial g}{\partial y} + \bar{a}_z \frac{\partial g}{\partial z}$$

$$\text{grad } g \text{ (cylindrical)} = \bar{a}_r \frac{\partial g}{\partial r} + \bar{a}_\phi \frac{1}{r}\frac{\partial g}{\partial \phi} + \bar{a}_z \frac{\partial g}{\partial z}$$

$$\text{grad } g \text{ (spherical)} = \bar{a}_r \frac{\partial g}{\partial r} + \bar{a}_\theta \frac{1}{r}\frac{\partial g}{\partial \theta} + \bar{a}_\phi \frac{1}{r\sin\theta}\frac{\partial g}{\partial \phi}$$

Since the del operator is defined as

$$\nabla \equiv \bar{a}_x \frac{\partial}{\partial x} + \bar{a}_y \frac{\partial}{\partial y} + \bar{a}_z \frac{\partial}{\partial z}$$

in rectangular coordinates, thus grad $g = \nabla g$ (this is interchangeable in all coordinate systems).

The relationship between v and E can be expressed in terms of gradient as follows:

$$\bar{E} = -\nabla v$$

4.4 ENERGY

The total energy stored in a region of continuous charge distribution due to k point charges is

$$\boxed{\begin{aligned} W_E &= \frac{1}{2} \sum_{j=1}^{k} Q_j V_j \\ &= \frac{1}{2} \int_v \rho V dv \end{aligned}}$$

where ρ is the charge density of the region.

By Maxwell's first equation, i.e. $\rho = \nabla \cdot \bar{D}$ and the vector identity
$$\nabla \cdot (V\bar{D}) \equiv V(\nabla \cdot \bar{D}) + \bar{D} \cdot (\nabla V)$$
we have,

$$W_E = \frac{1}{2} \int_v V(\nabla \cdot \bar{D}) dv$$

$$= \frac{1}{2} \int_v [\nabla \cdot (V\bar{D}) - \bar{D} \cdot (\nabla V)] dv$$

By the divergence theorem, i.e.,

$$\int_v \nabla \cdot (VD) dv = \oint_s (VD) \cdot ds$$

if we take the surface integral to be infinitely large then v and D will approach zero. Hence,

$$\oint_s (V\bar{D}) \cdot d\bar{s} = 0$$

Thus

$$W = -\frac{1}{2} \int_V \overline{D} \cdot (\nabla V) dv$$

$$= \frac{1}{2} \int_V \overline{D} \cdot \overline{E} \, dv \quad \text{since } E = -\nabla V$$

$$= \frac{1}{2} \int_V \varepsilon_0 E^2 dv \quad \text{in free space}$$

$$= \frac{1}{2} \int_V \frac{\overline{D}^2}{\varepsilon_0} dv$$

or

$$\frac{dW_E}{dv} = \frac{1}{2} \overline{D} \cdot \overline{E}$$

4.5 THE DIPOLE

Definition:

Two oppositely charged points having equal magnitude and separated by a small distance d compared to the distance to the point P at which the electric and potential fields are to be determined.

Configuration:

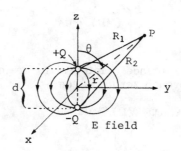
E field

C-27

The total potential due to the two point charges is

$$V = \frac{Q}{4\pi\epsilon_0} \cdot \frac{R_2 - R_1}{R_1 R_2}$$

$$= \frac{Q \cdot d \cdot \cos\theta}{4\pi\epsilon_0 \cdot r^2}$$

(Note: $R_1 \simeq R_2$ and $R_2 - R_1 \cong d\cos\theta$.)

Now, $\bar{E}$ is obtained by use of the gradient relationship:

$$\bar{E} = -\nabla V$$

$$= -\left(\bar{a}_r \frac{\partial V}{\partial r} + \bar{a}_\theta \frac{1}{r}\frac{\partial V}{\partial \theta} + \bar{a}_\phi \frac{1}{r\sin\theta}\frac{\partial V}{\partial \phi}\right)$$

$$\cong \frac{Qd}{4\pi\epsilon_0 r^3}(2\cos\theta\,\bar{a}_r + \sin\theta\,\bar{a}_\theta)$$

(Note: Spherical coordinates are used.)

Now, let's define the

$$\boxed{\text{Dipole moment } \bar{p} = Q\bar{d}}$$

then in general

$$\boxed{V = \frac{\bar{p} \cdot \bar{a}_R}{4\pi\epsilon_0 \cdot R^2}}$$

CHAPTER 5

CURRENT DENSITY, CAPACITANCE, DIELECTRICS, AND CONDUCTORS

5.1 CAPACITANCE

5.1.1 EXAMPLE OF CAPACITANCE

By definition the capacitance of two oppositely charged conductors in a uniform dielectric medium as shown in the figure is given as

$$\boxed{C = \frac{Q}{V_0}}$$

and is measured in farads (F).

$$F \equiv C/V$$

Where V_0 is the potential difference between the two conductors and Q is the total charge in either conductor.

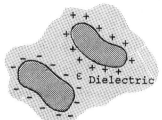

Two conductors surrounded
by a uniform dielectric

Capacitance of the parallel-plate capacitor:

$$\overline{E} = \frac{\rho_s}{\epsilon} \overline{a}_z$$

ϵ is the permitivity of the homogeneous dilectric

$$\overline{D} = \rho_s \cdot \overline{a}_z$$

On lower plate: $D_n = D_z = \rho_s$

D_n is the normal value of $\overline{D}$.

On upper plate: $D_n = -D_z$

$$V_0 = \text{The potential difference} = -\int_{\text{upper}}^{\text{lower}} \overline{E} \cdot d\overline{L}$$

$$= -\int_d^0 \frac{\rho_s}{\epsilon} dz = \frac{\rho_s d}{\epsilon}$$

$$\boxed{C = \frac{Q}{V_0} = \frac{\epsilon S}{d}} \quad Q = \rho_s S \text{ and } V_0 = \frac{\rho_s d}{\epsilon}$$

considering conductor planes of area S of linear dimensions much greater than d.

Total energy stored in the capacitor:

$$W_E = \frac{1}{2} \int_{vol} \varepsilon E^2 dv = \frac{1}{2} \int_0^S \int_0^d \frac{\varepsilon \rho_s}{\varepsilon^2} dz\, ds$$

$$W_E = \frac{1}{2} C V_0^2 = \frac{1}{2} Q V_0 = \frac{1}{2} \frac{Q^2}{C}$$

Examples of Capacitance:

Two concentric conducting spherical shells (Spherical capacitor):

$$C = \frac{Q}{V_{ab}} = \frac{4\pi\varepsilon}{\frac{1}{a} - \frac{1}{b}}$$

where radius is $a < b$, $b > a$.

and where $V_{ab} = \frac{Q}{4\pi\varepsilon} \left(\frac{1}{a} - \frac{1}{b} \right)$

An isolated spherical conductor:

$$C = 4\pi\varepsilon a$$

The capacitance of a coaxial cable (inner radius a and outer radius b, length L)

$$C = \frac{2\pi\varepsilon L}{\ln(b/a)}$$

Multiple dielectric capacitors:

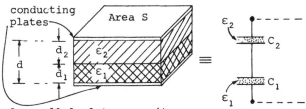

A parallel-plate capacitor containing two dielectrics with the dielectric interface parallel to the conducting plates;
$C = 1/\{(d_1/\varepsilon_1 S) + (d_2/\varepsilon_2 S)\}$.

$$C = \frac{1}{\left(\frac{1}{C_1}\right) + \left(\frac{1}{C_2}\right)} \quad \text{where} \quad C_1 = \frac{\varepsilon_1 S}{d_1} \quad \text{and} \quad C_2 = \frac{\varepsilon_2 S}{d_2}$$

V_0 = A potential difference between the plates

$\quad = E_1 d_1 + E_2 d_2$

$E_1 = V_0/d_1 + \left(\dfrac{\varepsilon_1}{\varepsilon_2}\right) d_2$

ρ_{S_1} = The surface charge density = $D_1 = \varepsilon_1 E_1$

$\quad = \dfrac{V_0}{\left(\dfrac{d_1}{\varepsilon_1}\right) + \left(\dfrac{d_2}{\varepsilon_2}\right)} = D_2$

$$C = \frac{Q}{V_0} = \frac{\rho_s \cdot S}{V_0} = \frac{1}{(d_1/\varepsilon_1 S) + (d_2/\varepsilon_2 S)} = \frac{1}{(1/C_1) + (1/C_2)}$$

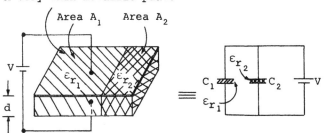

$$C_{eq} = C = C_1 + C_2 \quad \left(C_1 = \frac{\varepsilon r_1 A_1}{d} \text{ and } C_2 = \frac{\varepsilon r_2 A_2}{d}\right)$$

$$= \frac{\varepsilon r_1 A_1 + \varepsilon r_2 A_2}{d}$$

5.2 DIELECTRIC AND CURRENT DENSITY

Current Density:

The current density is a vector and it is defined in terms of the charge density and the drift velocity as

$$\overline{J} = \rho \overline{U} \quad A/m^2$$

since drift velocity can be expressed in terms of electric field intensity as

$$\overline{U} = \mu \overline{E}$$

where μ is the mobility of the electron in $m^2/v \cdot s$.

Thus, $\overline{J} = \rho \mu \overline{E} = \sigma \overline{E}$ where σ is the conductivity of the material.

Now, the total current flowing through a surface at which the current density $\overline{J}$ exists is

$$I = \int_S \overline{J} \cdot d\overline{s} \quad (A)$$

If $\overline{J}$ is uniform and, letting A represent a uniform cross-sectional area, then

$$I = JA$$

Since $V = EL$ (Note: $V_{pp'} = -\int_{p'}^{p} \overline{E} \cdot d\overline{L} = EL$),

For constant E field

$$J = \frac{I}{A} = \sigma E = \frac{\sigma V}{L}$$

Therefore, total current I is

$$I = \frac{\sigma V}{L} A$$

Now applying Ohm's law, V = IR, the resistance R can be expressed as

$$R = \frac{V}{I} = \frac{L}{\sigma A}$$

If the net charge density is to be considered within a closed surface, then

$$\nabla \cdot \overline{J} = -\frac{\partial \rho}{\partial t}$$

which is known as the continuity of a current equation.

5.3 BOUNDARY CONDITIONS

It is very important to determine the boundary conditions for any electromagnetics problems in which perfect dielectric materials and perfect conductors are being considered.

From the Maxwell's integral relation,

$$\oint_S \overline{D} \cdot d\overline{s} = \int_V \rho dv,$$

the general boundary condition is given as

$$D_{n_1} - D_{n_2} = \rho_s \ C/m^2$$

where ρ_s is the surface charge density and D_{n_1} and D_{n_2} are shown in the figure below.

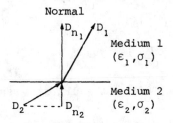

From the general boundary condition, there are two important cases to be considered.

Case 1

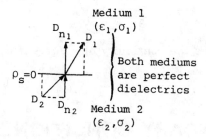

Properties: For perfect dielectric materials, $\sigma = 0$. Thus,

$$\sigma_1 = \sigma_2 = 0$$

Hence,

$$\rho_s = 0 \text{ and } D_{n_1} = D_{n_2}$$

Case 2

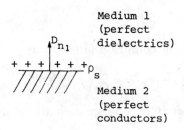

C-35

Properties: For perfect conductor materials, $\sigma \to \infty$ and $\overline{E} = \overline{B} = 0$. Thus,

$$\sigma_2 \to \infty \quad \text{and} \quad \overline{E}_1 = \overline{B}_2 = 0$$

Hence, $\quad D_{n_1} = \rho_s \quad \text{or} \quad \overline{n} \cdot \overline{D} = \rho_s$

5.4 CONDUCTORS IN ELECTROSTATIC FIELDS

A) There are only normal components of the static electric field intensity, i.e., $E_t = 0$.

B) The surface of the conductor is equipotential, i.e., there is no potential difference between any point within the conductor surface.

C) There is no static electric field inside the conductor.

CHAPTER 6

POISSON'S AND LAPLACE'S EQUATIONS

6.1 POISSON'S AND LAPLACE'S EQUATIONS

Derivation of Poisson's equation:

$$\nabla \cdot \overline{D} = \rho$$
$$\overline{D} = \varepsilon \overline{E}$$
$$\overline{E} = -\nabla V$$

By substitution

$$\nabla \cdot \overline{D} = \nabla \cdot (\varepsilon \overline{E}) = -\nabla \cdot (\varepsilon \nabla V) = \rho$$

Poisson's equation:

$$\boxed{\nabla \cdot \nabla V = \frac{-\rho}{\varepsilon}}$$

Poisson's equation can be used to find out the potential function.

Poisson's equation in cartesian coordinates:

$$\boxed{\nabla^2 V = \frac{\partial^2 V}{\partial x^2} + \frac{\partial^2 V}{\partial y^2} + \frac{\partial^2 V}{\partial z^2} = \frac{-\rho}{\varepsilon}}$$

Laplace's equation:

If the region of interest is charge-free (i.e., $\rho = 0$) then Poisson's equation becomes Laplace's equation.

$$\boxed{\nabla^2 V = 0}$$

The operation $\nabla^2 V$ is called the Laplacian of V.

6.2 LAPLACE'S EQUATION IN DIFFERENT COORDINATE SYSTEMS

$$\nabla^2 V = \frac{1}{r}\frac{\partial}{\partial r}\left(r\frac{\partial V}{\partial r}\right) + \frac{1}{r^2}\left(\frac{\partial^2 V}{\partial \phi^2}\right) + \frac{\partial^2 V}{\partial z^2} \quad \text{(cylindrical)}$$

and in spherical coordinates is

$$\nabla^2 V = \frac{1}{r^2}\frac{\partial}{\partial r}\left(r^2\frac{\partial V}{\partial r}\right) + \frac{1}{r^2 \sin\theta}\frac{\partial}{\partial \theta}\left(\sin\theta \frac{\partial V}{\partial \theta}\right)$$

$$+ \frac{1}{r^2 \sin^2\theta}\frac{\partial^2 V}{\partial \phi^2} \quad \text{(spherical)}$$

$$\nabla^2 V = \frac{\partial^2 V}{\partial x^2} + \frac{\partial^2 V}{\partial y^2} + \frac{\partial^2 V}{\partial z^2} = 0 \quad \text{(cartesian)}$$

6.3 UNIQUENESS THEOREM

If there exists any solution to Poisson's or Laplace's equation which also satisfies the boundary conditions, then the solutions must be identical (i.e., unique).

Proof:

Assume V_1 and V_2 are two solutions of Laplace's equation.

Thus, $\nabla^2 V_1 = \nabla^2 V_2 = 0$

and $\nabla^2 (V_1 - V_2) = 0$

Let V_B be the potential on the boundaries. For V_1, let V_{B_1} be the boundary value. For V_2, let V_{B_2} be the boundary value. Therefore,

$$V_{B_1} = V_{B_2} = V_B$$

By the vector identity,

$$\nabla \cdot (A\bar{B}) \equiv A(\nabla \cdot B) + B \cdot (\nabla A)$$

Substitute $(V_1 - V_2)$ for A. Substitute $\nabla(V_1 - V_2)$ for $\bar{B}$, we have

$$\nabla \cdot [(V_1 - V_2) \nabla(V_1 - V_2)]$$
$$\equiv (V_1 - V_2)[\nabla \cdot \nabla(V_1 - V_2)]$$
$$+ \nabla(V_1 - V_2) \cdot \nabla(V_1 - V_2)$$

Integrating on both sides throughout the volume enclosed by the boundary surfaces, we have

$$\int_V \nabla \cdot [(V_1 - V_2) \nabla(V_1 - V_2)] dv$$
$$\equiv \int_V (V_1 - V_2)[\nabla \cdot \nabla(V_1 - V_2)] dv$$
$$+ \int_V [\nabla(V_1 - V_2)]^2 dv$$

Apply the divergence theorem on the left side of the identity.

Therefore,
$$\int_V \nabla \cdot [(V_1 - V_2) \nabla (V_1 - V_2)] dv$$
$$= \oint_S [(V_{B_1} - V_{B_2}) \nabla (V_{B_1} - V_{B_2})] \cdot ds = 0$$

By hypothesis,
$$\nabla^2 (V_1 - V_2) = 0$$

therefore,
$$\int_V [\nabla (V_1 - V_2)]^2 dv = 0$$

Since $[\nabla(V_1 - V_2)]^2$ is always positive.

Thus,
$$[\nabla (V_1 - V_2)]^2 = 0$$
and
$$\nabla (V_1 - V_2) = 0$$

This implies $V_1 - V_2$ = constant.

Consider a point on the boundary; we have
$$V_1 - V_2 = V_{B_1} - V_{B_2} = 0$$

Thus, $V_1 = V_2$ and the theorem has been proved.

CHAPTER 7

THE STEADY MAGNETIC FIELD

7.1 BIOT-SAVART LAW

The Biot-Savart Law states that the differential magnetic field strength $d\vec{H}$ at any point P, produced by a differential element $d\vec{\ell}$ carrying the current I, is proportional to $Id\vec{\ell} \times \overline{a_R}$, where $\overline{a_R}$ is a unit vector leading from $d\vec{\ell}$ to the point P, and is also inversely proportional to the square of the distance from the differential element to the point P.

In mathematical form,

$$\boxed{d\vec{H} = \frac{Id\vec{\ell} \times \overline{a_R}}{4\pi R^2}} \quad \text{(Amperes/meter i.e. A/m)}$$

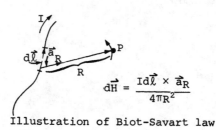

Illustration of Biot-Savart law

Integral form of Biot-Savart Law:

$$\bar{H} = \int_S \frac{\bar{K} \times \bar{a}_R}{4\pi R^2} ds$$

Other forms of the Biot-Savart Law:

$$\bar{H} = \oint \frac{Id\bar{\ell} \times \bar{a}_R}{4\pi R^2}$$

where K is the surface current density.

$$\bar{H} = \int_V \frac{\bar{J} \times \bar{a}_R}{4\pi R^2} dv$$

where J is the current density.

To illustrate the application of the Biot-Savart Law, take the following example:

Given an infinitely long straight filament carrying a current I, find the field $\bar{H}$ at point P.

$$d\bar{H} = \frac{Id_z \bar{a}_z \times (r\bar{a}_r - z\bar{a}_z)}{4\pi(r^2 + z^2)^{3/2}}$$

$$= \frac{Id_z \, r\bar{a}_\phi}{4\pi(r^2 + z^2)^{3/2}}$$

$$H = \frac{I}{4\pi} \int_{-\infty}^{\infty} \frac{rd_z \, \bar{a}_\phi}{(r^2 + z^2)^{3/2}}$$

$$= \frac{I}{2\pi r} \bar{a}_\phi \quad \text{(see figure on following page)}$$

C-42

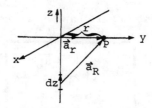

7.2 AMPERE'S CIRCUITAL LAW

Statement: The line integral of the tangential component of $\overline{H}$ about any closed path is exactly equal to the current enclosed by that path.

$$\oint \overline{H} \cdot d\overline{L} = I$$

Applications:

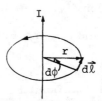

An infinitely long straight filament carrying a direct current I.

The path is a circle of radius r.

Then Ampere's circuital law becomes

$$\oint \overline{H} \cdot d\overline{L} = \int_0^{2\pi} H_\phi \, r \, d\phi = H_\phi \, r \int_0^{2\pi} d\phi$$

$$= H_\phi \cdot 2\pi r = I$$

$$\overline{H} = I/2\pi r \; \overline{a}_\phi$$

An example of a coaxial cable:

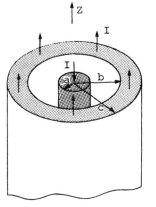

A coaxial cable with current I in the inner conductor and -I in the outer conductor.

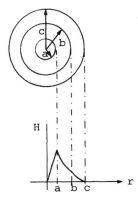

Relationship between $\vec{H}$ and r.

The following table shows H_ϕ as a function of r.

$r < a$	$\bar{H}_\phi = \dfrac{Ir}{2\pi a^2}$
$b > r > a$	$\bar{H}_\phi = \dfrac{I}{2\pi r}$
$c > r > b$	$\bar{H}_\phi = \dfrac{I}{2\pi r} \dfrac{c^2 - r^2}{c^2 - b^2}$
$r > 0$	$\bar{H}_\phi = 0$

7.3 CURL OF A VECTOR FIELD

The curl of any vector is defined as a vector where the direction is given by the right-hand rule and the magnitude is given by the limit of the quotient of the closed line integral and the area of the enclosed path as the area approaches 0.

$$\boxed{(\operatorname{curl} \overline{H})_n = \lim_{\Delta s_n \to 0} \frac{\oint \overline{H} \cdot d\overline{L}}{\Delta s_n}}$$

Δs_n is the area enclosed by the closed line integral, and n is any component; this is normal to the surface enclosed by the closed path.

Different forms:

$$\boxed{\operatorname{curl} \overline{H} = \left(\frac{\partial H_z}{\partial y} - \frac{\partial H_y}{\partial z}\right)\overline{a}_x + \left(\frac{\partial H_x}{\partial z} - \frac{\partial H_z}{\partial x}\right)\overline{a}_y + \left(\frac{\partial H_y}{\partial x} - \frac{\partial H_x}{\partial y}\right)\overline{a}_z}$$

In cartesian coordinates.

In determinant form

$$\boxed{\operatorname{curl} H = \begin{vmatrix} \overline{a}_x & \overline{a}_y & \overline{a}_z \\ \frac{\partial}{\partial x} & \frac{\partial}{\partial y} & \frac{\partial}{\partial z} \\ H_x & H_y & H_z \end{vmatrix} \\ \operatorname{curl} \overline{H} = \nabla \times \overline{H}}$$

Hence,

$$\boxed{\nabla \times \overline{H} = \left(\frac{\partial H_z}{\partial y} - \frac{\partial H_y}{\partial z}\right)\overline{a}_x + \left(\frac{\partial H_x}{\partial z} - \frac{\partial H_z}{\partial x}\right)\overline{a}_y + \left(\frac{\partial H_y}{\partial x} - \frac{\partial H_x}{\partial y}\right)\overline{a}_z}$$

(cartesian)

$$\nabla \times \bar{H} = \left[\frac{1}{r}\frac{\partial H_z}{\partial \phi} - \frac{\partial H_\phi}{\partial z}\right]\bar{a}_r + \left[\frac{\partial H_r}{\partial z} - \frac{\partial H_z}{\partial r}\right]\bar{a}_\phi$$

$$+ \left[\frac{1}{r}\frac{\partial(rH_\phi)}{\partial r} - \frac{1}{r}\frac{\partial H_r}{\partial \phi}\right]\bar{a}_z$$

(cylindrical)

$$\nabla \times \bar{H} = \frac{1}{r\sin\theta}\left[\frac{\partial(H_\phi \sin\theta)}{\partial \theta} - \frac{\partial H_\theta}{\partial \phi}\right]\bar{a}_r$$

$$+ \frac{1}{r}\left[\frac{1}{\sin\theta}\frac{\partial H_r}{\partial \phi} - \frac{\partial(rH_\phi)}{\partial r}\right]\bar{a}_\theta + \frac{1}{r}\left[\frac{\partial(rH_\theta)}{\partial r} - \frac{\partial H_r}{\partial \theta}\right]\bar{a}_\phi$$

(spherical)

7.4 STOKE'S THEOREM

$$\oint_\ell \bar{F} \cdot d\bar{\ell} = \int_S (\nabla \times \bar{F}) \cdot d\bar{s}$$

$\bar{F}$ is any vector field, s is a surface bounded by ℓ. It gives the relation between a closed line integral and surface integral.

By using the Divergence Theorem and Stoke's theorem we can derive a very important identity:

$\nabla \cdot \nabla \times \bar{A} \equiv 0$ $\bar{A}$ is any vector field.

7.5 MAGNETIC FLUX AND MAGNETIC FLUX DENSITY

$B = \mu_0 H$, B is the magnetic flux density in free space.

Unit of B is webers per square meter (wb/m²) or Tesla (T) a new unit.

$\mu_0 = 4\pi \times 10^{-7}$ H/m (permeability of free space)

H is in amperes per meter (A/m).

$$\oint_S \overline{B} \cdot d\overline{s} = 0$$

This is Gauss's law for the magnetic field.

$$\nabla \cdot \overline{B} = 0$$

After application of the divergence theorem. This is the fourth and last equation of Maxwell.

7.6 SCALAR AND VECTOR MAGNETIC POTENTIALS

The scalar magnetic potential:

$$\overline{H} = -\nabla V_m$$

V_m is the scalar magnetic potential.

Dimensions of V_m are amperes.

V_m also satisfies Laplace's equation:

$$\nabla^2 V_m = 0 \quad \text{for } J = 0$$

V_m satisfies Laplace's equation in homogeneous magnetic materials but it does not satisfy the equation in any region in which current density is present.

V_m is not a single valued function of position.

$$V_{m,ab} = \int_a^b \overline{H} \cdot d\overline{L} \quad \text{(specified path)}$$

The magnetic scalar potential V_m is not a conservative field.

The vector magnetic potential:

$$\overline{B} = \nabla \times \overline{A}$$

$\overline{A}$ is the vector magnetic potential.

$$\overline{H} = \frac{1}{\mu_0} \nabla \times \overline{A}$$

$$\nabla \times \overline{H} = \overline{J} = \frac{1}{\mu_0} \nabla \times \nabla \times \overline{A}$$

The units of $\overline{A}$ are webers per meter.

$\overline{A}$ can be determined from the differential current element:

$$\overline{A} = \oint \mu_0 I d\overline{L}/4\pi R \qquad d\overline{A} = \frac{\mu_0 I d\overline{L}}{4\pi R}$$

Direction of $d\overline{A}$ is the same as that of $Id\overline{L}$.

Alternative expressions for $\overline{A}$:

$$\overline{A} = \int_s \frac{\mu_0 \cdot \overline{K} \, ds}{4\pi R}$$

$$\overline{A} = \int_{vol} \frac{\mu_0 \overline{J} dv}{4\pi R}$$

CHAPTER 8

FORCES, TORQUES AND INDUCTANCE IN MAGNETIC FIELDS

8.1 MAGNETIC FORCE ON MOVING PARTICLES

$$\boxed{\overline{F} = Q\overline{E}}$$

The force on a charged particle in an electric field.

$$\boxed{\overline{F} = Q\overline{U} \times \overline{B}}$$

The force on a charged particle in a magnetic field. U is the velocity of the charged particle.

Lorentz force equation:

$$\boxed{\overline{F} = Q(\overline{E} + \overline{U} \times \overline{B})}$$

The force on a moving particle when both electric and magnetic fields are present.

8.2 MAGNETIC FORCE ON A DIFFERENTIAL CURRENT ELEMENT

$$\overline{F} = Q\overline{U} \times \overline{B} \quad \text{and} \quad d\overline{F} = dQ\overline{U} \times \overline{B}$$

$$\boxed{\begin{aligned} d\overline{F} &= \overline{J} \times \overline{B}\, dv \\ d\overline{F} &= \overline{K} \times \overline{B}\, ds \\ d\overline{F} &= Id\overline{L} \times \overline{B} \end{aligned}}$$

where $\overline{J} = \rho\overline{U}$

$\overline{J}dv = \overline{K}ds = Id\overline{L}$

$$\boxed{\begin{aligned} \overline{F} &= \int_{vol} \overline{J} \times \overline{B}\, dv \\ \overline{F} &= \int_{S} \overline{K} \times \overline{B}\, ds \\ \overline{F} &= \oint Id\overline{L} \times \overline{B} = -I \oint \overline{B} \times d\overline{L} \\ \overline{F} &= I\overline{L} \times \overline{B} \end{aligned}}$$

8.3 MAGNETIC FORCE BETWEEN DIFFERENTIAL CURRENT ELEMENT

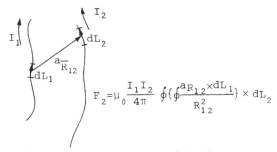

$$F_2 = \mu_0 \frac{I_1 I_2}{4\pi} \oint \{ \oint \frac{a_{R_{12}} \times dL_1}{R_{12}^2} \} \times dL_2$$

Force between two differntial elements.

C-50

The magnetic field at point 2 due to a current element at point 1 is:

$$d\bar{H}_2 = \frac{I_1 d\bar{L}_1 \times \bar{a}_{R_{12}}}{4\pi R_{12}^2}$$

The differential force on a differential current element is

$$d\bar{F} = I d\bar{L} \times \bar{B} \qquad d(d\bar{F}_2) = I_2 d\bar{L}_2 \times d\bar{B}_2$$

$$d(dF_2) = \mu_0 \cdot \frac{I_1 I_2}{4\pi R_{12}^2} d\bar{L}_2 \times (d\bar{L}_1 \times \bar{a}_{R_{12}})$$

since $d\bar{B}_2 = \mu_0 d\bar{H}_2$

Force between two differential current elements.

$$\boxed{\bar{F}_2 = \mu_0 \frac{I_1 I_2}{4\pi} \oint \left[\oint \frac{\bar{a}_{R_{12}} \times d\bar{L}_1}{R_{12}^2} \right] \times d\bar{L}_2}$$

Total force between two filamentary circuits.

8.4 FORCE AND TORQUE ON A CLOSED CIRCUIT

The force on a filamentary closed circuit is

$$\bar{F} = -I \oint \bar{B} \times d\bar{L}$$

for a uniform magnetic field

$$\bar{F} = -I\bar{B} \times \oint d\bar{L}$$

sine $\oint dL = 0$

therefore

$$\bar{F} = 0 \text{ in a closed circuit.}$$

Torque:

The torque or moment of a force is defined as the cross product of the lever arm and the force.

$$\boxed{\overline{T} = \overline{R} \times \overline{F}}$$

$\overline{R}$ is the vector lever arm.

If two forces $\overline{F}_1$ and $\overline{F}_2$ with lever arm $\overline{R}_1$ and $\overline{R}_2$ are applied to an object and there is no translation on that object, then

$$\boxed{\overline{T} = (\overline{R}_1 - \overline{R}_2) \times \overline{F}_1 = \overline{R}_{21} \times \overline{F}_1}$$

Torque on a Differential Current Loop:

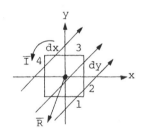

$d\overline{F}_1 = Idx(B_{oy}\overline{a}_z - B_{oz}\overline{a}_y)$ on side 1.

$d\overline{T}_1 = \overline{R}_1 \times d\overline{F}_1 = -\dfrac{1}{2} dxdy\, I\, B_{oy}\overline{a}_x = d\overline{T}_3$

$d\overline{T}_1 + d\overline{T}_3 = -dxdy IB_{oy}\overline{a}_x$, B_o is the value of the magnetic field at the center of the loop.

$d\overline{T}_2 + d\overline{T}_4 = dxdy IB_{ox}\overline{a}_y$

$d\overline{T} = Idxdy(B_{ox}\overline{a}_y - B_{oy}\overline{a}_x)$

$\qquad = Idxdy(\overline{a}_z \times \overline{B}_o)$

$$\boxed{d\overline{T} = Id\overline{A} \times \overline{B}}$$ $d\overline{A}$ is the vector area of the loop, i.e., dxdydz.

Differential magnetic dipole moment

$$d\bar{m} = Id\bar{A}$$

$$d\bar{T} = d\bar{m} \times \bar{B}$$

$$\bar{T} = \bar{m} \times \bar{B}$$

8.5 MAGNETIC MATERIALS

Magnetization and permeability:

The magnetic dipole moment

$$\bar{m} = Id\bar{S}$$

current I circulating about a path enclosing an area of ds.

If there are n number of identical magnetic dipole/unit volume then

Magnetization $\boxed{\bar{M} = nId\bar{s}}$ = the magnetic dipole moment per unit volume.

$$\bar{M} = \frac{1}{V} \sum_{i=1}^{nV} I_i \, d\bar{S}_i$$

unit of M are A/m.

Effect of Alignment of the magnetic dipoles due to application of a magnetic field:

$$dI_b = nI\, d\bar{S} \cdot d\bar{L} = \bar{M} \cdot d\bar{L}$$

$$I_b = \oint \bar{M} \cdot d\bar{L}$$

where I_b is the bound current.

$$\oint \bar{M} \cdot d\bar{L} = \int_s \bar{J}_b \cdot d\bar{s}$$

J_b is the bound current density.

$$\boxed{\nabla \times \bar{M} = \bar{J}_b}$$

Relationship between $\bar{B}$, $\bar{H}$ and $\bar{M}$:

$$\nabla \times \bar{H} = \bar{J}, \quad \nabla \times \bar{H} = \bar{J}_f + \bar{J}_B \quad \text{for free space}$$

$$\nabla \times \left(\frac{\bar{B}}{\mu_0} - \bar{M}\right) = \bar{J}_f, \quad \boxed{\nabla \times \bar{H} = \bar{J}_f}$$

$$\boxed{\bar{B} = \mu_0(\bar{H} + \bar{M})}$$

$$\boxed{\bar{M} = \chi_m \bar{H}}$$

χ_m is the magnetic susceptibility.

$$\boxed{\bar{B} = \mu \bar{H}}$$

where $\mu_0 = \mu_0 \mu_R$ = Permeability.

$$\boxed{\mu_R = 1 + \chi_m}$$

μ_R = Relative permeability

Tensor Permeability:

For an anisotropic magnetic material

$$B_x = \mu_{xx} H_x + \mu_{xy} H_y + \mu_{xz} H_z$$
$$B_y = \mu_{yx} H_x + \mu_{yy} H_y + \mu_{yz} H_z$$
$$B_z = \mu_{zx} H_x + \mu_{zy} H_y + \mu_{zz} H_z$$

where μ is a a tensor in $\overline{B} = \mu \overline{H}$.

8.6 MAGNETIC CIRCUIT

$$\boxed{\overline{H} = -\nabla V_m}$$

When dealing with magnetic circuits V_m is known as the magnetomotive force (mmf).

$$\boxed{V_{mAB} = \int_A^B \overline{H} \cdot d\overline{L}}$$

$$\boxed{\phi = \int_S \overline{B} \cdot d\overline{s}}$$

Total magnetic flux flowing through the cross section of a magnetic circuit.

$$\boxed{V_m = \phi R}$$

R: Reluctance is defined as the ratio of mmf to total flux.

$$R = \frac{L}{\mu S}$$

L = length of magnetic material

S = uniform cross section

$$\oint \overline{H} \cdot d\overline{L} = NI$$

I is the current flowing through an N-turn coil.

8.7 POTENTIAL ENERGY AND FORCES ON MAGNETIC MATERIALS

$$W_E = \frac{1}{2} \int_{vol} \overline{D} \cdot \overline{E} \, dv$$

General expression for energy in electrostatic field. Assuming a linear relationship between $\overline{D}$ and $\overline{E}$.

Total energy stored in a steady magnetic field. $\overline{B}$ is linearly related to $\overline{H}$ since $\overline{B} = \mu\overline{H}$.

$$W_H = \frac{1}{2} \int_{vol} \mu H^2 \, dv$$

$$W_H = \frac{1}{2} \int_{vol} \frac{B^2}{\mu} \, dv$$

8.8 INDUCTANCE AND MUTUAL INDUCTANCE

$$\boxed{L = \frac{N\phi}{I} = \frac{\text{Total flux linkage}}{\text{current linked}}}$$

Unit of inductance is H which is equivalent to wb/A.

Applications:

Inductance per meter length of a coaxial cable of inner radius a and outer radius b.

$$L = \frac{\mu_0}{2\pi} \ln \frac{b}{a} \qquad H/m.$$

A toroidal coil of N turns and IA,

$$L = \frac{\mu_0 N^2 s}{2\pi R}$$

R = Mean radius of the toroid.

Different expressions for inductance:

$$\boxed{L = \frac{2W_H}{I^2}} \qquad L = \frac{\int_{vol} \overline{B} \cdot \overline{H} \, dv}{I^2}$$

$$= \frac{1}{I^2} \int_{vol} \overline{H} \cdot (\nabla \times \overline{A}) dv$$

$$L = \frac{1}{I^2} \left[\int_{vol} \nabla \cdot (\overline{A} \times \overline{H}) dv + \int_{vol} \overline{A} \cdot (\nabla \times \overline{H}) dv \right]$$

$$L = \frac{1}{I^2} \int_{vol} \overline{A} \cdot \overline{J} \, dv$$

$$L = \frac{1}{I^2} \int_{vol} \left\{ \int_{vol} \frac{\mu \overline{J}}{4\pi R} \, dv \right\} \cdot \overline{J} \, dv$$

C-57

Mutual inductance between circuits 1 and 2, L_{12}

$$\boxed{L_{12} = \frac{N_2 \cdot \phi_{12}}{I_1}}$$

where N is the number of turns

$$= \frac{1}{I_1 I_2} \int_{vol} (\mu \overline{H}_1 \cdot \overline{H}_2) dv$$

$$\boxed{L_{12} = L_{21}}$$

8.9 BOUNDARY CONDITIONS

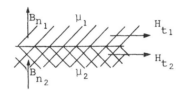

$B_{n_1} = B_{n_2}$ normal component of B is continuous

$H_{n_2} = \frac{\mu_1}{\mu_2} H_{n_1}$ normal component of H is discontinuous.

$M_{n_2} = \frac{\chi_{m_2} \mu_1}{\chi_{m_1} \mu_2} M_{n_1}$

$H_{t_1} - H_{t_2} = K$

K is the normal component of the surface current.

Hence,
$$\frac{B_{t_1}}{\mu_1} - \frac{B_{t_2}}{\mu_2} = K$$

and
$$M_{t_2} = \frac{\chi_{m_2}}{\chi_{m_1}} M_{t_1} - \chi_{m_2} K$$

CHAPTER 9

EMF IN A TIME-VARYING FIELD AND MAXWELL'S EQUATIONS

9.1 FARADAY'S LAW

Faraday's Law can be stated as follows:

$$\text{emf} = \oint \bar{E} \cdot d\bar{L} = -\frac{d\phi}{dt} \quad (v)$$

The minus sign is by Lenz's Law which indicates that the induced e.m.f. is always acting against the changing magnetic fields which produce that e.m.f.

Other forms of Faraday's Law which can be applied to different situations:

$$\text{emf} = -N \frac{d\phi}{dt}$$

N is the number of turns in a filamentary conductor.

$$\text{emf} = -\frac{d}{dt} \int_s \bar{B} \cdot d\bar{s}$$

or

$$\text{emf} = -\int_s \frac{\partial \overline{B}}{\partial t} \cdot d\overline{s} = \oint \overline{E} \cdot d\overline{L}$$

Finally we have

$$\nabla \times \overline{E} = -\frac{\partial \overline{B}}{\partial t}$$ Maxwell's equation

Different cases in Faraday's Law:

A) Conductors moving in a time-independent fields:

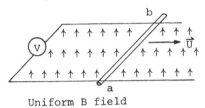

Uniform B field

$$\text{emf} = \int_b^a (\overline{U} \times \overline{B}) \cdot d\overline{\ell}$$

B) Conductors moving in a time-varying fields:

$$\text{emf} = -\int_s \frac{\partial \overline{B}}{\partial t} \cdot d\overline{s} + \oint (\overline{U} \times \overline{B}) \cdot d\overline{\ell}$$

9.2 DISPLACEMENT CURRENT

$$\nabla \times \overline{H} = \overline{J}_c$$

where J_c is defined as the conduction current density.

$$\nabla \cdot \overline{J}_c = \nabla \cdot (\nabla \times \overline{H}) = 0$$

But $\nabla \cdot \bar{J}_c = -\frac{\partial \rho}{\partial t}$ by the continuity equation.

Hence,
$$\nabla \times \bar{H} = \bar{J}_c + \bar{J}_D$$

where J_D is defined as the displacement current density.

Furthermore,
$$J_D = \frac{\partial \bar{D}}{\partial t} \quad \text{and} \quad I_c = \int_S \bar{J}_c \cdot d\bar{s}$$

So far there are three types of current density, namely conduction current density $\bar{J} = \sigma \bar{E}$, convection current density $\bar{J} = \rho \bar{U}$ and displacement current density $\bar{J}_d$.

Time-Varying version of Ampere's circuital law:

$$\oint \bar{H} \cdot d\bar{L} = I + I_d = I + \int_S \frac{\partial \bar{D}}{\partial t} \cdot d\bar{s}$$

I_d is the displacement current.

9.3 MAXWELL'S EQUATIONS

Maxwell's equation in differential form:

$$\nabla \times \bar{E} = -\frac{\partial \bar{B}}{\partial t}$$
$$\nabla \times \bar{H} = \bar{J} + \frac{\partial \bar{D}}{\partial t}$$
$$\nabla \cdot \bar{D} = \rho$$
$$\nabla \cdot \bar{B} = 0$$

Auxiliary equations relating $\bar{D}$ and $\bar{E}$:

$$\bar{D} = \varepsilon \bar{E} \qquad \bar{D} = \varepsilon_0 \bar{E} + \bar{P}$$
$$\bar{B} = \mu \bar{H} \qquad \bar{B} = \mu_0 (\bar{H} + \bar{M})$$
$$\bar{J} = \sigma \bar{E} \qquad \bar{J} = \rho \bar{U}$$

Lorentz force equation

$$\bar{F} = \rho (\bar{E} + \bar{U} \times \bar{B})$$

Maxwell's equations in integral form:

$$\oint \bar{E} \cdot d\bar{L} = - \int_S \frac{\partial \bar{B}}{\partial t} \cdot d\bar{s}$$

$$\oint \bar{H} \cdot d\bar{L} = I + \int_S \frac{\partial \bar{D}}{\partial t} \cdot d\bar{s}$$

$$\oint_S \bar{D} \cdot d\bar{s} = \int_{vol} \rho \, dv$$

$$\oint \bar{B} \cdot d\bar{s} = 0$$

These four integral equations enable us to find the boundary conditions on $\bar{B}$, $\bar{D}$, $\bar{H}$ and $\bar{E}$ which are necessary to evaluate the constants obtained in solving Maxwell's equations in partial differential form.

CHAPTER 10

UNIFORM PLANE WAVE AND POYNTING THEOREM

10.1 MAXWELL'S AND WAVE EQUATIONS

For convenience, Maxwell's equations in free space are written in terms of E and H only

$$\nabla \times \overline{H} = \varepsilon_0 \frac{\partial \overline{E}}{\partial t}$$

$$\nabla \times \overline{E} = -\mu_0 \frac{\partial \overline{H}}{\partial t}$$

$$\nabla \cdot \overline{E} = 0$$

$$\nabla \cdot \overline{H} = 0$$

Phasor notation:

For some component

$$\overline{E}_x = \overline{E}_{xyz} \cos(\omega t + \phi)$$

since $\quad e^{j\omega t} = \cos \omega t + j \sin \omega t$

Therefore $\bar{E}_x$ can be written as

$$\bar{E}_x = \text{Re}[E_{xyz}\, e^{j\phi} \cdot e^{j\omega t}]$$

where Re means the real part is to be taken.

$$E_{xs}\ (\text{phasor}) = E_{xyz}\, e^{j\omega}$$

Maxwell's Equations in Phasor Form:

$$\boxed{\begin{aligned}\nabla \times \bar{H}_s &= j\omega\varepsilon_0 \bar{E}_s \\ \nabla \times \bar{E}_s &= -j\omega\mu_0 \bar{H}_s \\ \nabla \cdot \bar{E}_s &= 0 \\ \nabla \cdot \bar{H}_s &= 0\end{aligned}}$$

Wave equations:

$$\nabla \times \nabla \times \bar{E}_s = \nabla(\nabla \cdot \bar{E}_s) - \nabla^2 \bar{E}_s = -j\omega\mu_0 \nabla \times \bar{H}_s$$

$$= \omega^2 \mu_0 \varepsilon_0 \bar{E}_s = -\nabla^2 \bar{E}_s$$

$$\boxed{\nabla^2 \bar{E}_s = -\omega^2 \mu_0 \varepsilon_0\, \bar{E}_s}$$

$$\nabla^2 E_{xs} = \frac{\partial^2 E_{xs}}{\partial x^2} + \frac{\partial^2 E_{xs}}{\partial y^2} + \frac{\partial^2 E_{xs}}{\partial z^2} = -\omega^2 \mu_0 \varepsilon_0\, E_{xs}$$

For

$$\frac{\partial^2 \bar{E}_{xs}}{\partial y^2} = \frac{\partial^2 \bar{E}_{xs}}{\partial z^2} = 0$$

i.e., E_{xs} independent of x and y.

C-64

This can be simplified to
$$\frac{d^2 E_{xs}}{dz^2} = -\omega^2 \mu_0 \epsilon_0 E_{xs}$$

$$E_x = E_{x_0} \cos[\omega(t - z\sqrt{\mu_0 \epsilon_0})]$$

and

$$E_{x'} = E_{x'_0} \cos[\omega(t + z\sqrt{\mu_0 \epsilon_0})]$$

E_{x_0} = value of E_x at $z = 0$, $t = 0$.

Velocity of the travelling wave:

To find the velocity U, let us keep the value of E_x to be constant, therefore

$$t - z\sqrt{\mu_0 \epsilon_0} = \text{constant}.$$

Take differentials; we have
$$dt - \frac{1}{U} dz = 0$$

$$\frac{dz}{dt} = U$$

in free space

velocity of light = $U = \dfrac{1}{\sqrt{\mu_0 \epsilon_0}} = 3 \times 10^8$ m/s

wave length = $\lambda = \dfrac{U}{f} = \dfrac{2\pi U}{\omega}$

The field is moving in the Z direction with velocity U. It is called a travelling wave.

Form of the $\overline{H}$ field:

If $\overline{E}_s$ is given, $\overline{H}_s$ can be obtained from

$$\nabla \times \overline{E}_s = -j\omega \mu_0 \epsilon_0 \overline{H}_s$$

$$\frac{\partial E_{xs}}{\partial z} = -j\omega \mu_0 H_{ys}$$

$$= E_{x_0}(-j\omega \sqrt{\mu_0 \epsilon_0}) e^{-j\omega \sqrt{\mu_0 \epsilon_0} z}$$

$$H_y = E_{x_0}\sqrt{\frac{\varepsilon_0}{\mu_0}} \cos\left[\omega\left(t - z\sqrt{\mu_0\varepsilon_0}\right)\right]$$

$$\frac{E_x}{H_y} = \sqrt{\frac{\mu_0}{\varepsilon_0}} \quad \text{is a constant}$$

$$\eta = \sqrt{\frac{\mu}{\varepsilon}}$$

η = The intrinsic impedance: It is the square root of the ratio of permeability to permittivity and is measured in Ω.

$$\eta_0 = \sqrt{\frac{\mu_0}{\varepsilon_0}} = 377\,\Omega$$

$\eta_0 = \eta$ of free space.

The term uniform plane wave is used because the H and E fields are uniform throughout any plane, Z = constant, and it is also called a transverse electromagnetic (TEM) wave since both the E and H fields are perpendicular to the direction of propagation.

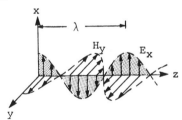

A uniform TEM wave.

10.2 WAVE MOTION IN PERFECT (LOSSLESS) DIELECTRIC

For an isotropic and homogeneous medium the wave equation is:

$$\nabla^2 \overline{E}_s = -\omega^2 \mu \varepsilon \overline{E}_s$$

For E_{xs},

$$\frac{\partial^2 E_{xs}}{\partial z^2} = -\omega^2 \mu \varepsilon E_{xs},$$

Assume

$$E_x = E_{x_0} e^{-\alpha z} \cos(\omega t - \beta z)$$

or

$$E_{xs} = E_{x_0} \cdot e^{-\alpha z} e^{-j\beta z}$$

α : attenuation constant and

β : the phase constant

$$\boxed{\gamma = \alpha + j\beta}$$

γ is the propagation constant, so

$$E_{xs} = E_{x_0} e^{-\gamma z}$$

and solving

$$\frac{\partial^2 E_{xs}}{\partial z^2} = -\omega^2 \mu \varepsilon E_{xs}$$

results

$$\boxed{\gamma = \pm j\omega\sqrt{\mu\varepsilon}}$$

and

$$\boxed{\alpha = 0 \text{ and } \beta = \omega\sqrt{\mu\varepsilon}}$$

$$E_x = E_{x_0} \cos[\omega(t - z\sqrt{\mu\varepsilon})]$$

is a wave travelling in the +z direction at velocity

C-67

$$U = \frac{U_0}{\sqrt{\mu_R \varepsilon_R}}$$

$$\boxed{\beta = \frac{2\pi}{\lambda}}$$

For the magnetic field intensity:

$$\boxed{H_y = \frac{E_{x0}}{\eta} \cos(\omega t - \beta z)}$$

where η is the intrinsic impedance.

$$\boxed{\eta = \sqrt{\frac{\mu}{\varepsilon}}}$$

Note: E_x and H_y are in phase and perpendicular to each other and also perpendicular to the direction of propagation.

10.3 WAVE MOTION IN LOSSY DIELECTRICS

All dielectric materials have some conductivity σ. Maxwell's curl equations are:

$$\nabla \times \overline{H}_s = (\sigma + j\omega\varepsilon)\overline{E}_s$$

$$\nabla \times \overline{E}_s = -j\omega\mu\overline{H}_s$$

The new value of the propagation constant is

$$\gamma = \pm\sqrt{(\sigma + j\omega\varepsilon)j\omega\mu} = j\omega\sqrt{\mu\varepsilon}\sqrt{1 - \frac{j\sigma}{\omega\varepsilon}}$$

E_{xs} (x component of the electric field intensity propagating in +z direction) $= E_{x_0} e^{-\alpha z} e^{-j\beta z}$

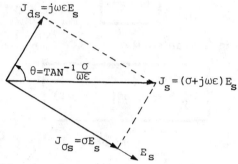

Relationship between J_{ds}, J_s, and $J_{\sigma s}$ in a vector form.

$$H_{ys} = \frac{E_{x0}}{\eta} e^{-\alpha z} e^{-j\beta z}$$

$$\eta = \frac{j\omega\mu}{\sigma + j\omega\varepsilon} = \frac{\mu}{\varepsilon} \frac{1}{\sqrt{1 - j(\sigma/\omega\varepsilon)}}$$

The electric and magnetic fields are no longer in time phase. The loss tangent $= \frac{\sigma}{\omega\varepsilon}$

$$\nabla \times \overline{H}_s = (\sigma + j\omega\varepsilon)\overline{E}_s = \overline{J}_{\sigma s} + \overline{J}_{ds}$$

$$\boxed{\frac{\overline{J}_{\sigma s}}{\overline{J}_{ds}} = \frac{\sigma}{j\omega\varepsilon}}$$

the ratio of conduction current density to displacement current density.

10.4 THE POYNTING THEOREM AND ELECTROMAGNETIC POWER

The Poynting theorem is a power theorem for the electromagnetic field and is helpful in finding the power in a uniform plane wave.

The Poynting Vector is defined as:

$$\bar{P} = \bar{E} \times \bar{H} \quad VA/m^2 \quad \text{or} \quad W/m^2$$

It is the instantaneous power density.

Taking the divergence of P, we have

$$\nabla \cdot \bar{P} = \nabla \cdot (\bar{E} \times \bar{H}) = \bar{H} \cdot \nabla \times \bar{E} - \bar{E} \cdot \nabla \times \bar{H}$$

Applying Maxwell's equation,

$$\nabla \times \bar{E} = -\frac{\partial \bar{B}}{\partial t} \quad \text{and} \quad \nabla \times \bar{H} = \bar{J} + \frac{\partial \bar{D}}{\partial t}$$

we have,

$$\nabla \cdot \bar{P} = -\bar{H} \cdot \frac{\partial \bar{B}}{\partial t} - E \cdot \frac{\partial \bar{D}}{\partial t} - J \cdot \bar{E}$$

For μ and ε are time constant.

Then,

$$\bar{H} \cdot \frac{\partial \bar{B}}{\partial t} = \frac{\partial}{\partial t}\left(\frac{\bar{H} \cdot \bar{B}}{2}\right) \quad \text{and} \quad \bar{E} \cdot \frac{\partial \bar{D}}{\partial t} = \frac{\partial}{\partial t}\left(\frac{\bar{E} \cdot \bar{D}}{2}\right)$$

Therefore, by substitution,

$$\nabla \cdot \bar{P} = -\frac{\partial}{\partial t}\left[\frac{\bar{H} \cdot \bar{B}}{2} + \frac{\bar{E} \cdot \bar{D}}{2}\right] - \bar{J} \cdot \bar{E}$$

Take the integration throughout an arbitrary volume V

$$\int_V \nabla \cdot \bar{P} \, dv = -\frac{\partial}{\partial t} \int_V \left[\frac{\bar{H} \cdot \bar{B}}{2} + \frac{\bar{E} \cdot \bar{D}}{2}\right] dv - \int_V \bar{J} \cdot \bar{E} \, dv$$

By the divergence theorem

$$-\oint_S \bar{P} \cdot d\bar{s} = \underbrace{\frac{\partial}{\partial t} \int_V \left[\frac{\bar{H} \cdot \bar{B}}{2} + \frac{\bar{E} \cdot \bar{D}}{2} \right] dv}_{B} + \underbrace{\int_V \bar{J} \cdot \bar{E} \, dv}_{C}$$

$\underbrace{\phantom{-\oint_S \bar{P} \cdot d\bar{s}}}_{A}$

A) Ingoing power-flux over the enclosed surface S.

B) Time rate of increase of total electromagnetic energy in volume V enclosed by S.

C) Ohmic power loss or total energy dissipated.

Time–Average Poynting Vector and Power

$$P_{av} = \int_S \bar{P}_{av} \cdot d\bar{s} \quad (w)$$

$$= \frac{1}{2\pi} \int_0^{2\pi} P(t) \, d(\omega t) \quad (w)$$

An example:

In a perfect dielectric $\bar{E}$ and $\bar{H}$ are given by:

$$E_x = E_{x0} \cos(\omega t - \beta z), \quad H_y = \frac{E_{x0}}{\eta} \cos(\omega t - \beta t)$$

$$P_z = \frac{E_{x0}^2}{\eta} \cos^2(\omega t - \beta z)$$

$$P_{z,av} = \frac{1}{T} \int_0^T \frac{E_{x0}^2}{\eta} \cos^2(\omega t - \beta z) \, dt$$

$$\boxed{P_{z,av} = \frac{1}{2} \cdot \frac{E_{x0}^2}{\eta}}$$

If rms values are used instead of peak amplitudes, the factor ½ will not be present.

The average power flowing through any area S normal to the Z axis is:

$$P_{z,av} = \frac{1}{2} \frac{E_{x0}^2}{\eta} S$$

For a lossy dielectric, E_x and H_y are not in time phase

$$P_{z,av} = \frac{1}{2} \frac{E_{x0}^2}{\eta_m} e^{-2\alpha z} \cdot \cos\theta_\eta$$

where

$$\eta = \eta_m \angle \theta_\eta$$

CHAPTER 11

PROPAGATION OF ELECTROMAGNETIC WAVES

11.1 PROPAGATION IN CONDUCTIVE MEDIUM

Some properties:

A) All time-varying fields attenuate very rapidly within a good conductor.

B) Propagating energy will decrease as it propagates because of ohmic losses.

C) In general, we take $\sigma/\omega\varepsilon \gg 1$ as good conductors in order to find α, β and η.

In general, $\gamma = j\omega\sqrt{\mu\varepsilon}\sqrt{1 - j\dfrac{\sigma}{\omega\varepsilon}}$

For a good conductor,

$$\gamma = j\sqrt{-j\,\omega\mu\sigma} \qquad \left(1 - j\dfrac{\sigma}{\omega\varepsilon} \simeq -j\dfrac{\sigma}{\omega\varepsilon}\right)$$

Hence,

$$\alpha = \beta = \sqrt{\pi f \mu \sigma} = \sqrt{\dfrac{\omega\mu\sigma}{2}}$$

$$\eta = \sqrt{\dfrac{\omega\mu}{\sigma}}\;\underline{/45°}$$

D) H and E fields are out of phase by 45° at any fixed location of the conductor.

$$\overline{E} = E_0 e^{-\alpha z} e^{j(\omega t - \beta z)} \overline{a}_x$$

$$\overline{H} = \frac{E_0}{|\eta|} e^{-\alpha z} e^{j(\omega t - \beta z - \pi/4)} \overline{a}_y$$

conduction current density (J):

$$\overline{J} = \sigma \overline{E}$$

$$= \sigma E_0 e^{-\alpha z} e^{j(\omega t - \beta z)} \overline{a}_x$$

11.2 SKIN DEPTH

The skin depth or depth of penetration is defined as the distance by which $|E|$ decreases to $e^{-1} = 0.368$. Therefore,

$$\text{(skin depth) } \delta = \frac{1}{\sqrt{\pi f \mu \sigma}}$$

$$= \frac{1}{\alpha} = \frac{1}{\beta}$$

Some equations in terms of δ:

$$U = \frac{\omega}{\beta} = \sqrt{\frac{2\omega}{\mu \sigma}} = \omega \delta$$

$$\lambda = \frac{2\pi}{\beta} = \frac{2}{\sqrt{\pi f \mu \sigma}} = 2\pi \delta$$

11.3 POWER CONSIDERATION

Time average Poynting Vector:

$$\overline{P}_{av} = \frac{1}{4} \sigma \delta E_0^2 e^{-2z/\delta} \overline{a}_z$$

Note: the power density attenuation of one skin depth is

$$e^{-2} = 0.135$$

To find the power loss, P_L, crossing a surface of width

$$0 < y < a$$

and length

$$0 < x < b$$

$$P_L = \int_0^a \int_0^b \left[\frac{1}{4} \sigma \, \delta E_0^2 \, e^{-2z/\delta} \right]_{z=0} dx\,dy$$

$$= \frac{1}{4} \sigma\delta \, ab E_0^2$$

or in terms of current density J_0

$$P_L = \frac{1}{4} \frac{1}{\sigma} \delta \, ab J_0^2 \quad \text{for} \quad J_0 = \sigma E_0$$

11.4 REFLECTION OF UNIFORM PLANE WAVES

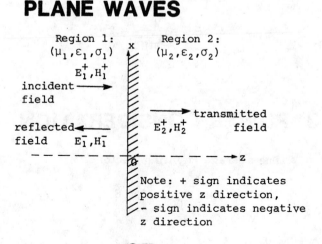

Defined: Reflection coefficient

$$\Gamma = \frac{E_{10}^-}{E_{10}^+} = \frac{\eta_2 - \eta_1}{\eta_2 + \eta_1}$$

A special case:

If region 1 is a perfect dielectric and region 2 is a perfect conductor then $\Gamma = -1$.

This means that all of the incident energy will be reflected by the perfect conductor. This phenomenon also agrees with the fact that no time-varying fields can exist in the perfect conductor.

Note: Γ may be a complex number.

Transmission coefficient $= \dfrac{E_{20}^+}{E_{10}^+} = \dfrac{2\eta_2}{\eta_2 + \eta_1}$

11.5 STANDING WAVES AND SWR

A standing wave will be produced when the incident waves in a perfect dielectric are combined with the total reflected waves which are reflected by the perfect conductor.

Standing wave in mathematical form:

$$\overline{E}_1 = 2E_{10}^+ \sin\beta_1 z \, \sin\omega t \, \overline{a}_x$$

Illustration of a standing wave.

Standing Wave Ratio:

The standing wave ratio is defined as the ratio of the maximum amplitude, $|E_{max}|$, to the minimum amplitude, $|E_{min}|$ occurring one quarter of a wave away.

$$SWR = \frac{E_{max}}{E_{min}}$$

$$= \frac{H_{max}}{E_{min}}$$

$$= \frac{1 + |\Gamma|}{1 - |\Gamma|}$$

Some special cases:

A) For $|\Gamma| = 1$ i.e., total reflection, SWR = ∞

B) For $|\Gamma| = 0$ i.e., no reflection, SWR = 1

C) For any other cases, SWR > 1 since $|\Gamma| \leq 1$

CHAPTER 12

TRANSMISSION LINES

12.1 TRANSMISSION-LINE EQUATIONS

Voltage and Current differential equations.

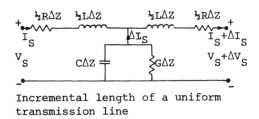

Incremental length of a uniform transmission line

Propagation is in $\bar{a}_z$ direction.

Voltage across the perimeter of the given circuit

$$V_s = (\tfrac{1}{2}R\Delta z + j\tfrac{1}{2}\omega L \Delta z)I_s + (\tfrac{1}{2}R\Delta z + j\tfrac{1}{2}\omega L \Delta z)(I_s + \Delta I_s)$$

$$+ V_s + \Delta V_s$$

$$\frac{\Delta V_s}{\Delta z} = -(R + j\omega L)I_s - \left[\frac{R}{2} + \frac{1}{2}j\omega L\right]\Delta I_s$$

$$\boxed{\frac{dV_s}{dz} = -(R + j\omega L)I_s}$$

C-78

letting $\Delta z \to 0$, and ΔI_s also approaches 0.

$$\boxed{\frac{dI_s}{dz} = -(G + j\omega C)V_s}$$

Maxwell's curl equations:

$$\nabla \times \overline{E}_s = -j\omega\mu \overline{H}_s$$

$$\frac{dE_{xs}}{dz} = -j\omega\mu H_{ys} \left[\begin{array}{l} \text{analogous to } \dfrac{dV_s}{dz} \\ \\ = -(R + j\omega L)I_s \end{array} \right]$$

Setting $\overline{E}_s = E_{xs}\, \overline{a}_x$, $\overline{H}_s = H_{ys}\, \overline{a}_y$

$$\nabla \times \overline{H}_s = (\sigma + j\omega\epsilon)\overline{E}_s$$

$$\frac{dH_{ys}}{dz} = -(\sigma + j\omega\epsilon)E_{xs}$$

$$\left[\text{analogous to } \frac{dI_s}{dz} = -(G + j\omega C)V_s \right]$$

Solution of two circuit equations:

$$V_s = V_0 \cdot e^{-\gamma z}$$

because $E_{xs} = E_{x0}\, e^{-\gamma z}$

The voltage wave propagates in the +z direction.

The propagation constant for the uniform plane wave

$$\gamma = [j\omega\mu(\sigma + j\omega\epsilon)]^{\frac{1}{2}} \text{ becomes}$$

$$\boxed{\gamma = \sqrt{(R + j\omega L)(G + j\omega C)}}$$

C-79

For a lossless line $R = G = 0$

$$\gamma = j\beta = j\omega \sqrt{LC}$$

Characteristics impedance Z_0:

$$H_{ys} = \frac{E_{x0}}{\eta} e^{-\gamma z} \quad \text{so} \quad I_s = \frac{V_0}{Z_0} e^{-\gamma z}$$

It is related to the positively travelling voltage wave by z_0 which is analogous to η.

$$\eta = \sqrt{\frac{j\omega\mu}{\sigma + j\omega\varepsilon}}$$

Hence,

$$\boxed{Z_0 = \sqrt{\frac{R + j\omega L}{G + j\omega C}}}$$

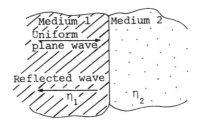

Reflection coefficient:

$$\Gamma = \frac{E_{x0}^-}{E_{x0}^+} = \frac{\eta_2 - \eta_1}{\eta_2 + \eta_1}$$

$$\boxed{\Gamma = \frac{V_0^-}{V_0^+} = \frac{Z_{02} - Z_{01}}{Z_{02} + Z_{01}}}$$

Standing wave ratio:

$$\boxed{SWR = \frac{1 + |\Gamma|}{1 - |\Gamma|}}$$

Input impedance:

$$\boxed{Z_{in} = Z_0 \frac{Z_L + jZ_0 \tan\beta\ell}{Z_0 + jZ_L \tan\beta\ell}}$$

12.2 TRANSMISSION-LINE PARAMETERS

Coaxial cable of inner radius a and outer radius b:

Capacitance $\boxed{C = 2\pi\varepsilon / \ln(b/a) \quad F/m}$

Shunt conductance $\boxed{= \frac{2\pi\sigma}{\ln(b/a)} \quad v/m}$

σ is the conductivity of the dielectric between the conductors at the operating frequency.

External inductance $\boxed{L_{ext} = \frac{\mu}{2\pi} \ln \frac{b}{a} \quad H/m}$

Any flux within either conductor was not taken into account.

Internal inductance $\boxed{L_{a,int} = \frac{\mu}{8\pi}}$

For very low frequencies where current distribution is uniform.

$$L_{LF} = \frac{\mu}{2\pi}\left[\ln\frac{b}{a} + \frac{1}{4} + \frac{1}{4(c^2-b^2)}\left(c^2 - 3b^2 + \frac{4b^4}{c^2-b^2}\ln\frac{c}{b}\right)\right]$$

Total inductance at low frequency

$$L_{HF} = \frac{\mu}{2\pi}\left[\ln\frac{b}{a} + \frac{\delta}{2}\left(\frac{1}{a} + \frac{1}{b}\right)\right]$$

Total high-frequency inductance. ($\delta \ll a$, $\delta \ll c-b$)

$$R = \frac{1}{2\pi\delta\sigma_c}\left(\frac{1}{a} + \frac{1}{b}\right) \quad \Omega/m$$

($\delta \ll a$, $\delta \ll c-b$)

Resistance/unit length.

Two-wire transmission line

$$C = \frac{\pi\varepsilon}{\cosh^{-1}(d/2a)} \quad F/m \quad \text{or} \quad C \simeq \frac{\pi\varepsilon}{\ln(d/a)} \quad (a \ll d)$$

$$L_{ext} = \frac{\mu}{\pi}\cosh^{-1}\frac{d}{2a} \quad H/m \quad \text{or} \quad L_{ext} \simeq \frac{\mu}{\pi}\ln\frac{d}{a}$$

$$L_{HF} = \frac{\mu}{\pi}\left(\frac{\delta}{2a} + \cosh^{-1}\frac{d}{2a}\right) \quad H/m \quad (\delta \ll a)$$

$$R = \frac{1}{\pi a \delta_c} \quad \Omega/m \quad (\delta \ll a)$$

$$G = \frac{\pi\sigma}{\cosh^{-1}(d/2a)} \quad v/m$$

conductor (σ_c)
dielectric (σ, ε, μ)

Planar transmission line

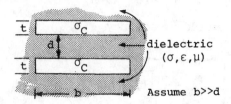

Assume b >> d.

$$C = \frac{\varepsilon b}{d} \quad F/m$$

$$L_{ext} = \mu \frac{d}{b} \quad H/m$$

$$L_{total} = \mu \frac{d}{b} + \frac{2}{\sigma_c \delta bw}$$

$$= \frac{\mu}{b}(d + \delta) \quad H/m \quad (\delta << t)$$

$$R = 2/\sigma_c \delta b \quad \Omega/m \quad \delta << t$$

$$G = \frac{\sigma b}{d} \quad v/m$$

12.3 TRANSMISSION-LINE EXAMPLE

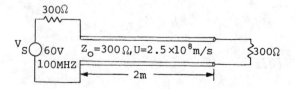

Load impedance is equal to the characteristic impedance, the line is matched, the reflection coefficient is zero, and the standing wave ratio is unity.

The wavelength on the line = 2.5m and the phase constant is 0.8π rad/m. Attenuation constant is zero.

Electrical length $\beta\ell = 2 \times 0.8\pi = 1.6\pi$ rad or

$$(= 288° \text{ or } 0.8 \text{ wavelength})$$

Input voltage $V_{in} = 30 \cos 2\pi \times 10^8 t$ Since there is no reflection, and no attenuation, the voltage at the load is 30V, but it is delayed in phase by 1.6π rad.

$V_L = 30 \cos(2\pi 10^8 t - 1.6\pi)$

Input current $I_{in} = V_{in}/3000$,

Load current $I_L = 0.1 \cos(2\pi 10^8 t - 1.6\pi)$.

Average power input by source = power delivered to load by line =

$$P_{in} = P_L = \frac{1}{2} \times 30 \times 0.1 = 1.5w$$

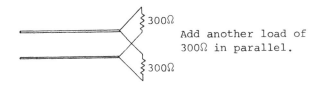

Add another load of 300Ω in parallel.

Now the load impedance is 150Ω,

$$\Gamma = \frac{150 - 300}{150 + 300} = -\frac{1}{3}$$

Standing wave ratio $= s = \frac{1 + (1/3)}{1 - (1/3)} = 2$.

$$Z_{in} = Z_0 \frac{Z_L + jZ_0 \tan\beta\ell}{Z_0 + jZ_L \tan\beta\ell} = 300 \times \frac{150 + j300 \tan 288°}{300 + j150 \tan 288°}$$

$= 510 \angle -23.8° = 466 - j206 \Omega$ (capacitive impedance)

$I_{s,in}$ (current flowing through the source)

$$= \frac{60}{766 - j206} = 0.075 \underline{/15°} \text{ A}$$

$$P_{in} = \frac{1}{2} \times (0.075)^2 \times 466 = 1.310 \text{w}.$$

Since the line is lossless, this will be delivered to the load. Thus each receiver receives only 0.655w. Input impedance of each receiver is 300Ω, the voltage across the receiver is

$$0.655 = \frac{1}{2} \frac{|V_{LS}|^2}{300} \quad \therefore \quad |V_{LS}| = 19.82\text{V}.$$

Relative phase of the input and load voltages:

The voltage at any point on the line:

$$\boxed{V_S = (e^{-j\beta z} + \Gamma e^{j\beta z}) V_0^+}$$

V_0^+ is the incident voltage amplitude.

$$\boxed{V_{S,in} = (e^{j\beta \ell} + \Gamma e^{-j\beta \ell}) V_0^+}$$

$$V_{LS} = V_{S,in} (1 + \Gamma) \cdot \frac{1}{e^{j\beta \ell} + \Gamma e^{-j\beta \ell}}$$

12.4 GRAPHICAL METHOD (SMITH CHART)

Graphical solutions using Smith Chart:

The Smith Chart helps in graphically finding the total field impedance Z from a given reflection coefficient Γ or vice versa.

A) Let Z be the normalized load impedance.

$$z = \frac{Z}{\eta} = \frac{1 + \Gamma}{1 - \Gamma}$$

B) Write Γ in terms of z.

$$\Gamma = \frac{z - 1}{z + 1}$$

C) Write Γ and Z in the rectangular complex forms

$$z = r + jx \qquad \Gamma = \Gamma_r + j\Gamma_i$$

D) Obtain r and x.

$$r = \frac{1 - \Gamma_r^2 - \Gamma_i^2}{(1 - \Gamma_r)^2 + \Gamma_i^2} \qquad x = \frac{2\Gamma_i}{(1 - \Gamma_r)^2 + \Gamma_i^2}$$

E) In calculations polar and rectangular forms are used alternatively.

The Smith chart contains the constant-r circles and constant-x circles, an auxiliary radial scale to determine $|\Gamma|$, and an angular scale for measuring ϕ.

An illustration of Smith Chart

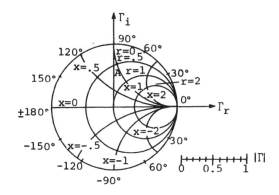

$Z = 25 + j50\,\Omega$ on a $50\,\Omega$ line and $z = 0.5 + j1$.
A is the intersection of $r = 0.5$ and $x = 1$ circles.
Reflection coefficient $\Gamma \cong .62$ at $\phi = 83°$.

CHAPTER 13

GUIDED WAVES

13.1 WAVES BETWEEN PARALLEL PLANES

TE, TM, and TEM Mode:

Assume +Z is the direction of propagation. Some definitions:

Transverse Magnetic (TM) Modes: Magnetic field perpendicular to propagation direction, $\bar{H}_z = 0$.

Transverse Electric (TE) Modes: Electric field perpendicular to propagation direction, $\bar{E}_z = 0$.

Transverse Electromagnetic (TEM) Modes: Both electric and magnetic field perpendicular to propagation direction, $\bar{E}_z = \bar{H}_z = 0$.

Important Points:

The behavior of waves between guides is predicted by Maxwell's equations, using the wave guide surfaces as boundaries.

The field configurations in the waveguides have characteristic quantities that involve integers m and n, which can take values from 0 to infinity. Only certain combinations of m and n are allowed, giving rise to limited modes of propagation.

For every mode, there is a lower cut-off frequency below which propagation cannot take place.

Maxwell's equations: (Space is enclosed by two conducting planes which are infinite in extent, and the waves are propagating in the z direction, varying sinusoidally with time.)

$$\overline{H}_x = \frac{-\gamma}{K^2} \frac{\partial \overline{H}_z}{\partial x} \qquad \overline{E}_x = \frac{-\gamma}{K^2} \frac{\partial \overline{H}_z}{\partial x}$$

$$\overline{H}_y = \frac{-j\omega\epsilon}{K^2} \frac{\partial \overline{E}_z}{\partial x} \qquad \overline{E}_y = \frac{j\omega\mu}{K^2} \frac{\partial \overline{H}_z}{\partial x}$$

$$K^2 = \gamma + \omega^2 \mu \epsilon$$

$$\gamma = \sqrt{j\omega\mu(\sigma + j\omega\epsilon)}$$

(The propagation constant)

13.2 TRANSVERSE ELECTRIC WAVES

$E_z = 0$, and considering the sinusoidal time variation, for the waves between perfectly conducting planes:

$$E_y = K_1 \sin\left(\frac{m\pi}{a}\right) x \cdot e^{j(\omega t - \gamma z)}$$

$$H_z = \frac{-m\pi}{j\omega\mu_0} K_1 \cos\left(\frac{m\pi}{a}\right) x \; e^{j(\omega t - \gamma z)}$$

$$H_x = \frac{-\gamma}{j\omega\mu_0} K_1 \sin\left(\frac{m\pi}{a}\right) x e^{j(\omega t - \gamma z)}$$

a is the separation between plates.

$$K = \frac{m\pi}{a} \quad m = 1, 2, 3, \ldots$$

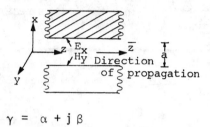

$$\gamma = \alpha + j\beta$$

For perfectly conducting planes γ is either real or purely imaginary.

13.3 TRANSVERSE MAGNETIC WAVES

$H_z = 0$, and considering sinusoidal time variation

$$\overline{H}_y = K_2 \cdot \cos\left(\frac{m\pi}{a}\right) x e^{j(\omega t - \gamma z)}$$

$$\overline{E}_x = \frac{\gamma}{j} K_2 \cdot \cos\left(\frac{m\pi}{a}\right) x e^{j(\omega t - \gamma z)}$$

$$E = \frac{jm\pi}{\varepsilon \omega a} K_2 \cdot \sin\left(\frac{m\pi}{a}\right) x e^{j(\omega t - \gamma z)}$$

$$m = 0, 1, 2, 3, \ldots$$

For both TE and TM waves:

$$\gamma = \left(\frac{m\pi}{a}\right)^2 - w^2 \mu \varepsilon$$

The cut-off frequency for the guide:

$$\boxed{f_{c_0} = \frac{m}{\sqrt{\mu\epsilon}\,2a} = \frac{mv}{2a}}$$

v is velocity of light in medium.

The wavelength in guide (λ_g)

$$\boxed{\lambda_g = \frac{2\pi}{\sqrt{w^2\mu\epsilon - (m\pi/a)^2}} = \frac{\lambda}{\sqrt{\epsilon_r - (\lambda/\lambda_{c_0})^2}} = \frac{2\pi}{\beta}}$$

where

$$\lambda_{c_0} = \frac{2a}{m}, \quad \beta = \sqrt{\omega^2\mu\epsilon - \left(\frac{m\pi}{a}\right)^2}$$

$$\frac{V}{f} = \frac{c}{f\sqrt{\epsilon_r}} = \frac{\lambda}{\sqrt{\epsilon_r}}$$

If $w^2\mu\epsilon \gg (m\pi/a)^2$, then phase velocity ($V_f$) approaches

$$V = 1/\sqrt{\mu\epsilon}$$

13.4 TRANSVERSE ELECTRO-MAGNETIC WAVES

Set $m = 0$ and $\gamma = \beta$ in $\overline{H}_y$, $\overline{E}_x$, and $\overline{E}$ expressions for the transverse magnetic waves.

$$\boxed{\begin{aligned}
\overline{H}_y &= K_2 \cdot e^{j(\omega t - \overline{\beta} z)} \\
\overline{E}_x &= \frac{\beta}{\omega\epsilon} K_2\, e^{j(\omega t - \overline{\beta} z)} \\
\overline{E}_z &= 0
\end{aligned}}$$

(Since E_z and H_z are zero, this wave is completely transverse.)

For m = 0, and an air dielectric

$$\lambda_g = \lambda = 2\pi/\omega\sqrt{\mu_0\varepsilon_0} = c/f$$

$$\gamma = j\omega\sqrt{\mu_0\varepsilon_0}$$

$$\beta = \omega\sqrt{\mu_0\varepsilon_0}$$

$$V_f = V = 1/\sqrt{\mu_0\varepsilon_0} = c = \text{velocity of light}$$

The relationship between λ and λ_g (wavelength in wave guide), V_f (phase velocity), and V_g (group velocity):

$$\lambda_g = \lambda / \sqrt{1 - (m\lambda/2a)^2}$$

$$v_f = c / \sqrt{1 - (m\lambda/2a)^2} \quad \text{the phase velocity}$$

$$V_g = c\sqrt{1 - (m\lambda/2a)^2} \quad \text{The group velocity.}$$

13.5 MISCELLANEOUS CONDITIONS IN TE, TM AND TEM MODES

Attenuation at a frequency below that of cut-off in a waveguide:

For TE waves

$$\alpha_m = \sqrt{\left(\frac{m\pi}{a}\right)^2 - \omega^2\mu\varepsilon_0}$$

if there are no losses.

$$\boxed{\alpha_m = 2\pi\sqrt{\left(\frac{m}{2a}\right)^2 - \left(\frac{f}{c}\right)^2}}$$

If the dielectric is air.

$$\boxed{= 2\pi\sqrt{\left(\frac{1}{\lambda_{c0}}\right)^2 - \left(\frac{1}{\lambda}\right)^2}}$$

λ_{c0} = The cut-off wavelength and

λ = The free-space wavelength.

If $\lambda_{c0} << \lambda$, then $\alpha_{c0} = 2\pi/\lambda_{c0}$.

Power transmitted between two parallel conducting planes in the TE_1 mode (m = 1).

For propagation, γ must be equal to $j\beta$, otherwise if γ is real, only attenuation occurs.

Expressions for E and H:

$$\overline{E}_y = K_1 \cos(\omega t - \beta z)\sin\left\{\frac{m\pi}{a}\right\} x$$

$$\overline{H}_x = \frac{\beta K_1}{\omega \mu} \cos(\omega t - \beta z)\sin\left\{\frac{m\pi}{a}\right\} x$$

For m = 1

$$\overline{P} = \text{The poynting vector} = \overline{E}_y \times \overline{H}_x = \frac{\beta K_1^2}{\omega \mu}\cos^2(\omega t - \beta z)\sin^2\frac{\pi x}{a}$$

$$\int_0^a \overline{P}dx = \int_0^a \frac{\beta K_1^2}{\omega \mu}\cos^2(\omega t - \beta z)\sin^2\frac{\pi x}{a}\, dx$$

$$P = \frac{\beta K_1^2 a}{2\omega\mu} \cos^2(\omega t - \beta z)$$

$$P_{av} = \frac{\beta K_1^2 a}{4\mu w} \quad \text{watts over one cycle.}$$

Attenuation of TE waves in waveguide:

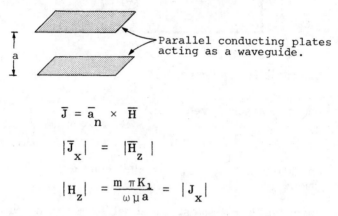

Parallel conducting plates acting as a waveguide.

$$\bar{J} = \bar{a}_n \times \bar{H}$$

$$|\bar{J}_x| = |\bar{H}_z|$$

$$|H_z| = \frac{m\pi K_1}{\omega\mu a} = |J_x|$$

The magnitude of current density is equal to the tangential component of the field intensity at the surface of both plates. i.e., H_z at $x = 0$ and at a.

Power loss per plate is $\frac{1}{2} J_x^2 R_w$ where R_w is the resistive compnent of the impedance Z_w, so

$$Z = \sqrt{\frac{j\omega\mu}{\gamma}}$$

$$R_w = \sqrt{\frac{\omega\mu}{2\gamma}}$$

$$\frac{1}{2} J_x^2 R_w = \frac{1}{2} \cdot \frac{m^2 \pi^2 K_1^2}{\omega^2 \mu^2 a^2} \sqrt{\frac{\omega\mu}{2\gamma}}$$

Power/unit area $= \frac{1}{2} R_e (\bar{E} \times \bar{H}^*) \cdot d\bar{s}$

where $\overline{H}^*$ is the complex conjugate

$$= \frac{-1}{2}(E_y H_x)dxdy$$

$$= \frac{K_1^2 \beta}{2\omega\mu} \sin^2\left(\frac{m\pi}{a}\right) x \, dxdy$$

Because

$$E_y = K_1 \cos(\omega t - \beta z)\sin\left(\frac{m\pi}{a}\right)x$$

$$H_x = -\frac{\beta K_1}{\omega\mu}\cos(\omega t - \beta z)\sin\left(\frac{m\pi}{a}\right)x$$

Total power transmitted for 1m of guide width

$$W_T = \int_0^a \frac{K_1^2 \beta}{2\omega\mu} \sin^2\left(\frac{m\pi}{a}\right) x \, dx$$

$$\boxed{W_T = K_1^2 \beta a / 4\omega\mu}$$

Attenuation

$$\boxed{\alpha_{TE} = \frac{\tfrac{1}{2}J_x^2 R_w}{W_T} = \frac{2m^2 \pi^2 \sqrt{\omega \mu_r / 2\gamma}}{\omega \cdot \mu \cdot \beta a^3}}$$

Expression for the specific wave impedance of a two plane conductor waveguide for TE and TM modes:

Specific wave impedance of a two plane conductor waveguide:

Specific impedance $Z_{wg} = \dfrac{E_y}{H_x} = \dfrac{j\omega\mu}{\gamma}$

$$Z_{wg} = \frac{\omega\mu}{\beta}$$

if there are no losses i.e., $\alpha = 0$ and $\gamma = j\beta$

$$= \frac{\omega \mu \lambda_g}{2\pi} = f \cdot \mu \cdot \lambda_g$$

$$\beta = 2\pi / \lambda_g$$

$$Z_{wg} = \frac{c}{\lambda} \mu \cdot \lambda_g$$

because $c = f \cdot \lambda$ = velocity of light

λ = wavelength outside of the guide.

$$c = \sqrt{\frac{1}{\mu_r \varepsilon_r}} \cdot \sqrt{\frac{1}{\mu_0 \varepsilon_0}}$$

$$Z_{wg} = \frac{\sqrt{\mu_0 \mu_0}}{\sqrt{\varepsilon_r \varepsilon_0}} \frac{\lambda_g}{\lambda}$$

$\sqrt{\frac{\mu_0}{\varepsilon_0}} \cong 377\,\Omega$ = impedance of free space.

$$Z_{wg} \text{ for TE modes} = \sqrt{\mu_r / \varepsilon_r} \; \frac{\lambda_g}{\lambda} \; 377\,\Omega$$

$$Z_{wg} \text{ for TM modes} = \sqrt{\frac{\mu_r}{\varepsilon_r}} \cdot \frac{\lambda}{\lambda_g} \; 377\,\Omega$$

The attenuation factor (for small attenuations):

The electric and magnetic vectors are attenuated at a rate:

$$\bar{E}_{out} = \bar{E}_{in} e^{-\alpha z} \quad \text{and} \quad \bar{H}_{out} = \bar{H}_{in} e^{-\alpha z}, \text{ respectively.}$$

$$P_{out} = \bar{E}_{in} \bar{H}_{in} e^{-2\alpha z} = P_{in} e^{-2\alpha z}$$

$$P_{lost} = P_{in} - P_{out} = P_{in}(1 - e^{-2\alpha})$$

so $\dfrac{P_{lost}}{P_{input}} = 1 - e^{-2\alpha} = 1 - \left[1 - 2\alpha + \dfrac{\alpha^2}{2!} + \dfrac{\alpha^3}{3!} + \cdots \right]$

$\dfrac{P_{lost}}{P_{input}} = 2\alpha$,

Thus,

$$\boxed{\alpha = \dfrac{P_{lost}}{2P_{input}}}$$

if α is very small.

Attenuation due to conductor losses between two planes: TM Mode.

$$\alpha = \dfrac{\text{Power absorbed}}{2(\text{Power transmitted})}$$

In the TM$_{10}$ mode:

$$P_{abs} = \tfrac{1}{2} J_y^2 R_w = \tfrac{1}{2} |\overline{H}_y|^2 R_{wg} = \tfrac{1}{2} K_2^2 \sqrt{\dfrac{\omega\mu_r}{2\sigma}}$$

per plate.

$$P_{tran} = \tfrac{1}{2} \text{Re}(\overline{E} \times \overline{H}) \cdot \overline{ds} \quad \text{per unit area.}$$

$$= \tfrac{1}{2} E_x H_y \, dxdy = \dfrac{K_2^2 \beta}{2\omega\varepsilon} \cos^2\left(\dfrac{m\pi}{a}\right) xdxdy$$

Total $P_{tran} = \dfrac{K_2^2 \beta}{2w\varepsilon} \displaystyle\int_0^b dy \int_0^a \cos^2 \dfrac{m\pi x}{a} dx$

$= K_2^2 \beta \, ab/4\omega\varepsilon$

$\beta = \sqrt{\omega^2\mu\varepsilon - K^2}, \quad K^2 = \omega_{c0}^2 \cdot \mu\varepsilon$

C-96

$$\boxed{\begin{aligned}\alpha &= \frac{\sqrt{\omega\,\mu_r/2}\,\gamma}{ab\,\beta/2\omega\epsilon} \\ &= \frac{1}{ab}\sqrt{\frac{2\epsilon}{\sigma}}\left[\frac{\omega^{3/2}}{\sqrt{\omega^2 - \omega_{c_0}^2}}\right]\end{aligned}}$$

where $\beta = \sqrt{\omega^2\mu\epsilon - K^2}$ and $K^2 = \omega_{c_0}^2 \cdot \mu\epsilon$

The frequency for minimum attenuation

$$= \omega = \sqrt{3}\,\omega_{c_0} \quad \text{or} \quad \boxed{f_{min} = f_{c_0}\sqrt{3}}$$

The field configurations of the TE_1 and TM_1 mode of propagation between plane-parallel conductors:

TE_1 mode:

$$E_y = K_1 \cdot \sin\frac{\pi}{a} \times \cos(\omega t - \beta z)$$

$$H_x = -\frac{\beta K_1}{\omega\mu}\sin\frac{\pi}{a} x\cos(\omega t - \beta z)$$

$$H_z = -\pi/\omega\mu a \cdot K_1 \cdot \cos\frac{\pi}{a} \cdot x\sin(\omega t - \beta z)$$

and $\quad \beta = \left[\omega^2\mu\epsilon - (\pi/a)^2\right]^{\frac{1}{2}}$

The $\bar{H}$ field, due to the displacement current caused by the time-varying E_y, will take the shape of small paths encircling the displacement current in the x-z plane. The slope of $\bar{H}$ at any point will be

$$\frac{H_z}{H_x} = \frac{dz}{dx} = \frac{\pi}{\beta a}\cot\left(\frac{\pi}{a}\right)x\tan(\omega t - \beta z)$$

i.e.,

$$\boxed{\sin \frac{\pi}{a} x \sin(\omega t - \beta z) = c_2}$$

after integrating both sides.

TM$_1$ mode: (m = 1)

$$H_y = K_2 \cos(\omega t - \beta z) \cos \pi x/a$$

$$E_x = \frac{K_2 \beta}{\omega \epsilon} \cos(\omega t - \beta z) \cos \pi x/a.$$

$$E_z = -\frac{K_2 \pi}{\omega \epsilon a} \sin(\omega t - \beta z) \sin \frac{\pi x}{a}$$

$$\beta = \left[\omega^2 \mu \epsilon - (\pi/a)^2 \right]^{\frac{1}{2}}$$

The slope of the loops at any point is found from:

$$E_x/E_z = dx/dz = \frac{-\beta a}{\pi} \cos(\omega t - \beta z) \cot \frac{\pi x}{a}$$

$$\boxed{\cos \frac{\pi x}{a} \sin(\omega t - \beta z) = C_4}$$

CHAPTER 14

WAVE GUIDES

14.1 PROPERTIES OF WAVE GUIDES

Assuming sinusoidally varying excitation, and using Maxwell's equations, for a rectangular guide with transmission in the z-direction:

$$\bar{H}_x = \frac{-\gamma}{K^2} \frac{\partial \bar{H}_z}{\partial x} + \frac{j\omega\varepsilon}{K^2} \frac{\partial \bar{E}_z}{\partial y}$$

$$\bar{H}_y = \frac{-\gamma}{K^2} \frac{\partial \bar{H}_z}{\partial y} - \frac{j\omega\varepsilon}{K^2} \frac{\partial \bar{E}_z}{\partial x}$$

$$\bar{E}_x = \frac{-\gamma}{K^2} \frac{\partial \bar{E}_z}{\partial x} - \frac{j\omega\mu}{K^2} \frac{\partial \bar{H}_z}{\partial y}$$

$$\bar{E}_y = \frac{-\gamma}{K^2} \frac{\partial \bar{E}_z}{\partial y} + \frac{j\omega\mu}{K^2} \frac{\partial \bar{H}_z}{\partial x}$$

$$K^2 = \gamma + \omega^2 \mu \varepsilon$$

and
$$\gamma = (K^2 + (j\omega\varepsilon)(j\omega\mu))^{\frac{1}{2}}$$

For the TE mode, the following equations are valid, assuming that for the perfectly conducting boundaries, E_y and E_z are zero at $x = 0$ and $x = a$ and E_x and E_z are zero at $y = 0$ and $y = a$:

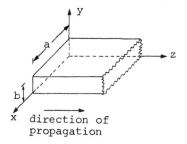

part of a rectangular wave guide

direction of propagation

$$\overline{H}_x = K_1 \frac{j\beta}{K^2} \frac{m\pi}{a} \sin\left(\frac{m\pi}{a}\right) x \cos\left(\frac{n\pi}{b}\right) y$$

$$\overline{H}_y = K_1 \frac{j\beta}{K^2} \frac{n\pi}{b} \cos\left(\frac{m\pi}{a}\right) x \sin\left(\frac{n\pi}{b}\right) y$$

$$\overline{H}_z = K_1 \cos\left(\frac{m\pi}{a}\right) x \cos\left(\frac{n\pi}{b}\right) y$$

$$\overline{E}_x = K_1 \left(\frac{j\omega\mu}{K^2}\right) \frac{n\pi}{b} \cos\left(\frac{m\pi}{a}\right) x \sin\left(\frac{n\pi}{b}\right) y$$

$$\overline{E}_y = K_1 \left(\frac{-j\omega\mu}{K^2}\right) \frac{m\pi}{a} \sin\left(\frac{m\pi}{a}\right) x \cos\left(\frac{n\pi}{b}\right) y$$

These are all propagation equations obtained by setting $\gamma = j\beta$.

and

$$K^2 = \gamma^2 + \omega^2\mu\epsilon \quad \text{Also} \quad K^2 = \left(\frac{m\pi}{a}\right)^2 + \left(\frac{n\pi}{b}\right)^2$$

$$\gamma^2 = \left(\frac{m\pi}{a}\right)^2 + \left(\frac{n\pi}{b}\right)^2 - \omega^2\mu\epsilon$$

The value of f (called f_{c_0}) or ω which makes $\beta = 0$ is a critical frequency above which propagation takes place. For $f > f_{c_0}$ becomes imaginary, so

$$\beta = \left[\omega^2\mu\epsilon - \left[\left(\frac{m\pi}{a}\right)^2 + \left(\frac{n\pi}{b}\right)^2\right]\right]^{\frac{1}{2}}$$

$$= \sqrt{\frac{\omega^2}{c^2} - \left(\frac{m\pi}{a}\right)^2 - \left(\frac{n\pi}{b}\right)^2} \quad \text{for air or vacuum.}$$

$$f_{c_0} = \frac{c}{2\pi\sqrt{\mu_r \varepsilon_r}} \cdot \sqrt{\left(\frac{m\pi}{a}\right)^2 + \left(\frac{n\pi}{b}\right)^2}$$

and λ_{c_0} (for air) $= \dfrac{c}{f_{c_0}} = \dfrac{2}{\sqrt{(m/a)^2 + (n/b)^2}}$

Wavelength in guide

$$\lambda_g = \frac{2\pi}{\sqrt{\omega^2\mu\varepsilon - (m\pi/a)^2 + (n\pi/b)^2}}$$
$$= \frac{\lambda}{\sqrt{\varepsilon_r - (\lambda/\lambda_{c_0})^2}}$$

ε_r is the dielectric constant filling the interior of the guide.

The phase velocity

$$V_f = \frac{2\pi f}{\beta} = \frac{\omega}{[\omega^2\mu\varepsilon - [(\frac{m\pi}{a})^2 + (\frac{n\pi}{b})^2]]^{\frac{1}{2}}}$$

The propagation equations: TM Modes

$$\overline{H}_x = K_2 j \left(\frac{\omega\varepsilon}{K^2}\right) \frac{n\pi}{b} \sin\left(\frac{m\pi}{a}\right) x \cos\left(\frac{n\pi}{b}\right) y$$

$$\overline{H}_y = K_2 \left(\frac{-j\omega\varepsilon}{K^2}\right) \frac{m\pi}{a} \cos\left(\frac{m\pi}{a}\right) x \sin\left(\frac{n\pi}{b}\right) y$$

$$\overline{E}_x = K_2 \left(\frac{-j\beta}{K^2}\right) \frac{m\pi}{a} \cos\left(\frac{m\pi}{a}\right) x \sin\left(\frac{n\pi}{b}\right) y$$

$$\overline{E}_y = K_2 \left(\frac{-j\beta}{K^2}\right) \frac{n\pi}{b} \sin\left(\frac{m\pi}{a}\right) x \cos\left(\frac{n\pi}{b}\right) y$$

$$E_y = K_2 \sin\frac{m\pi}{a} x \sin\left(\frac{n\pi}{b}\right) y$$

14.2 MISCELLANEOUS EXAMPLES

Nomenclature:

TM_{11} means m = 1 and n = 1

TE_{10} means m = 1 and n = 0

$\lambda_g = 2a \tan\theta$

where θ is the angle that the component wavefronts make with the wall of the guide.

$V_g = V \cos\theta$

V_g is the group velocity and

V is the velocity of component waves.

V_f = The phase velocity = $V/\cos\theta$.

$V_g \cdot V_f = V^2 = c^2$ for air or vacuum.

Expression for the propagation of power in the guide and the electric field intensity in TE_{10} mode:

The power in the waveguide:

Poynting vector $\overline{P} = \overline{E} \times \overline{H}$ watts/m².

$P_z = E_y H_x$

for dominant mode only E_y and H_z are considered.

$= \dfrac{E_y}{H_x} H_x^2$

$P_z = Z_{wg} H_m^2 \sin^2(\pi/a)x \sin^2(\omega t - \beta z)$

$Z_{wg} = \dfrac{E_y}{H_x}$ and $H_x = H_m \sin(\pi/a) \cdot x \cdot \sin(\omega t - \beta z)$

and

$$H_m = \frac{K_1 \beta}{K^2}$$

$$W_z = \frac{1}{2\pi} \int_0^{2\pi} P_z \, d(\omega t) = \frac{1}{2} Z_{wg} H_m^2 \sin^2\left(\frac{\pi}{a}\right) x$$

Total Power

$$W_T = \frac{1}{2} Z_{wg} H_m^2 \int_0^a \int_0^b \sin^2(\pi/a) x \, dx \, dy$$

$$\boxed{W_T = \frac{1}{4} Z_{wg} \cdot H_m^2 \, ab}$$

Peak electric field:

$$\boxed{E_y = Z_{wy} H_m = 2\sqrt{\frac{Z_{wg} W_T}{ab}}}$$

because

$$H_m = 2\sqrt{\frac{W_T}{Z_{wg} \, ab}}$$

Also

$$W_T = \frac{1}{4} \frac{E_y^2 \, ab}{Z_{wg}}$$

The characteristic impedance of a waveguide:

$$W_T = \frac{1}{4} Z_{wg} H_m^2 \, ab \qquad H_m = V/b \cdot Z_{wg}$$

$$= \frac{1}{4} V \left[\frac{a}{b} \cdot \frac{V}{Z_{wg}}\right]$$

$$V = b \cdot Z_{wg} \cdot H_m$$

$$Z_0 = V/I = b \cdot Z_{wg} \cdot H_m / \frac{a}{b}\left(\frac{b \cdot Z_{wg} \cdot H_m}{Z_{wg}}\right)$$

$$= \frac{b}{a} Z_{wg}$$

$$\boxed{Z_0 = 377 \frac{\lambda_g}{\lambda} \cdot \frac{b}{a}} \quad \text{for TE}_{10} \text{ mode}$$

Expression for the phase velocity in terms of ω_{c_0} and ω.

Above cut-off $\lambda = j\beta$, so

$$K^2 = -\beta^2 + \omega^2\mu\epsilon$$
$$\beta = (\omega^2\mu\epsilon - K^2)^{\frac{1}{2}}$$

$$\boxed{K^2 = \left(\frac{m\pi}{a}\right)^2 + \left(\frac{n\pi}{b}\right)^2 \text{ and } \omega_{c_0} = \frac{1}{\sqrt{\mu\epsilon}}\left[\left(\frac{m\pi}{a}\right)^2 + \left(\frac{n\pi}{b}\right)^2\right]^{\frac{1}{2}}}$$

$$\beta = \sqrt{\mu\epsilon}\ \omega\left[1 - \left(\frac{\omega_{c_0}}{\omega}\right)\right]^{\frac{1}{2}}$$

The phase velocity

$$V_f = \frac{\omega}{\beta} = \frac{V}{[1 - (\omega_{c_0}/\omega)^2]^{\frac{1}{2}}}$$

Derive group velocity,

$$V_g = v\sqrt{1 - (\omega_{c_0}/\omega)^2}$$

given that

$$V_g = d\omega/d\beta$$
$$\beta^2 = \frac{\omega^2}{v^2}\left[1 - \left(\frac{\omega_{c_0}}{\omega}\right)^2\right]$$

Differentiating, we get

$$2\beta d\beta = \frac{1}{v^2} 2\omega d\omega$$

$$\frac{d\omega}{d\beta} = \frac{\beta v^2}{\omega} = \frac{\omega\sqrt{\mu\epsilon}\sqrt{1-(\omega_{c_0}/\omega)^2}\, v^2}{\omega}$$

$$\therefore \frac{d\omega}{d\beta} = V_g = v\sqrt{1-(\omega_{c_0}/\omega)^2} \qquad v = \frac{1}{\sqrt{\mu\epsilon}}$$

To determine all the modes that can be transmitted in a rectangular waveguide:

The rectangular waveguide excited at 3×10^9 and 6×10^9 Hz

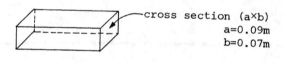
cross section (a×b)
a=0.09m
b=0.07m

$$f_{c_0} = \frac{c}{2\pi}\left[\left(\frac{m\pi}{a}\right)^2 + \left(\frac{n\pi}{b}\right)^2\right]^{\frac{1}{2}}$$

No modes can propagate where cut-off frequencies are below f_{c_0}. So $f_{c_0} = 3 \times 10^9$

$$3(10^9) > \frac{3(10^{10})}{2\pi}\left[\left(\frac{m\pi}{7}\right)^2 + \left(\frac{n\pi}{4}\right)^2\right]^{\frac{1}{2}}$$

$$\sqrt{\left(\frac{m\pi}{7}\right)^2 + \left(\frac{n\pi}{4}\right)^2} < \frac{(2\pi)}{10}$$

Squaring both sides and then simplifying we get

$$16m^2 + 49n^2 < 31.4$$

If $m = 1$ and $n = 0$ and propagation takes place therefore TE_{10} is the only possible mode.

If the waveguide is excited at 6×10^9 Hz, then

$16m^2 + 49n^2 < 125.5$.

By solving the inequality, we have the following modes

TE_{10}, TE_{01}, TE_{11}, TE_{20}, TM_{11}, TM_{20} AND TM_{21}

m	n
1	0
0	1
1	1
2	0
2	1

CHAPTER 15

RADIATION

15.1 NEAR AND FAR ZONE

The radiated power is determined by the radial component of the poynting vector P_r

$$P_r = E_\theta H_\phi$$

E_θ is the meridion component H_ϕ is the single magnetic component.

"Far-zone": A distance sufficiently far away from the oscillating current element.

For the far zone:

$$H_{\phi\,far} \cong \frac{Im \cdot h \cdot \sin\theta \; \omega}{4\pi v r} \cos\omega\left(t - \frac{r}{v}\right)$$

$$E_\theta \cong \frac{Im \cdot h}{4\pi\varepsilon r} \cdot \frac{\omega}{v^2} \cdot \sin\theta \cdot \cos\omega\left(t - \frac{r}{v}\right)$$

$$\cong \frac{Im \cdot h}{2r\lambda} \cdot \eta \cdot \sin\theta \cdot \cos\omega\left(t - \frac{r}{v}\right)$$

$$\frac{Im \cdot h \cdot \omega}{4\pi\varepsilon \, r \cdot v^2} = \frac{Im \cdot h}{4\pi r} \frac{2\pi\eta}{\lambda}$$

Radiated power per unit area of the spherical surface in the far zone:

$$P_r = \frac{Im^2 h^2 \omega^2}{(4\pi)^2} \frac{1}{\varepsilon r^2 v^3} \sin^2\theta \cdot \cos^2\omega\left(t - \frac{r}{v}\right)$$

$$P_{rav} = Im^2 \cdot h^2 \cdot \omega^2 \sin^2\theta / 32\pi^2 \cdot \varepsilon \cdot r^2 \cdot c^3$$

since average value of $(cosine)^2$ is $\frac{1}{2}$ and $v = c$ for radiation in free space.

Near zone:

For near zone:

$$\boxed{\begin{aligned} H_{(near)} &= \frac{Im \cdot h \cdot \sin\theta}{4\pi r^2} \sin\omega\left(t - \frac{r}{v}\right) \\ E_{(near)} &= Im \cdot h \cdot \sin\theta / 4\pi\varepsilon r^3 \omega \cdot \cos\omega\left(t - \frac{r}{v}\right) \end{aligned}}$$

From the above equations, we observed that at the near zone, the magnitude of the induction field is much greater than that of the radiated field and decreases very rapidly as we move away from the near zone to the far zone since its magnitude is proportional to $1/r^5$. As a result, the radiation field is dominant in the far zone.

15.2 TOTAL RADIATED POWER

$$dP_{rad} = P_{r(av)} 2\pi r^2 \cdot \sin\theta\, d\theta$$

where $2\pi r \sin\theta \cdot r d\theta$ is the differential element of surface.

$$= Im^2 h^2 \omega^2 \sin^3\theta\, d\theta / 16\pi\varepsilon c^3$$

$$\boxed{P_{rad} = \int_0^\pi dP_{rad}\, d\theta = \frac{Im^2 \cdot h^2 \cdot \omega^2}{12\pi\varepsilon c^3}}$$

15.3 RADIATION RESISTANCE

Total power radiated = $\begin{Bmatrix} \text{The effective} \\ \text{value of the} \\ \text{current} \end{Bmatrix}^2 \times$ Radiation resistance of the current element

$I_{RMS}^2 = \frac{1}{2} I_m^2$

R_{rad} = Radiation resistance
$= P_{rad} / \frac{1}{2} I_m^2$

$$\boxed{R_{rad} = 80 \pi^2 \cdot \frac{h^2}{\lambda^2} \; \Omega}$$

$\lambda = 2\pi c / \omega$
$\quad$ = wavelength

15.4 MISCELLANEOUS EXAMPLES

Expressions for E_r, E_θ and E_ϕ: (using Maxwell's equation)

$$\nabla \times \overline{H} = y\overline{E} + \varepsilon \frac{\partial \overline{E}}{\partial t} = \overline{J} + \varepsilon \frac{\partial \overline{E}}{\partial t}$$

$$= \varepsilon \frac{\partial \overline{E}}{\partial t}$$

Maxwell's equation for a perfect dielectric

$$\nabla (\nabla \cdot \overline{A}) = \nabla^2 \overline{A} + \nabla \times \nabla \times \overline{A}$$

$$\Delta V = \frac{1}{\varepsilon} \int \nabla^2 \overline{A} + \nabla \times \nabla \times \overline{A} \, dt$$

$$\Delta V = \frac{1}{\varepsilon} \int (\nabla \times \nabla \times \overline{A}) dt + \mu \frac{\partial \overline{A}}{\partial t}$$

$$\nabla \times H = \frac{1}{r^2 \sin\theta} \left[\frac{\partial}{\partial \theta} (r\sin\theta \, H_\phi) - \frac{\partial \, r H_\theta}{\partial \phi} \right] \overline{I}_r$$

$$+ \frac{1}{r\sin\theta} \left[\frac{\partial H_r}{\partial \phi} - \frac{\partial}{\partial r}(r\sin\theta \, H_\phi) \right] \bar{I}_\theta$$

$$+ \frac{1}{r} \left[\frac{\partial}{\partial r}(rH_\theta) - \frac{\partial}{\partial \theta} H_r \right] \bar{I}_\phi$$

$$\nabla \times H = \frac{I_m h \cos\theta}{2\pi r} \left[\frac{\omega}{rv} \cos\omega\left(t - \frac{r}{v}\right) + \frac{1}{r^2} \sin\omega\left(t - \frac{r}{v}\right) \right] \bar{I}_r$$

$$+ \frac{I_m H}{4\pi r} \sin\theta \left[\frac{-\omega^2}{v^2} \sin\omega\left(t - \frac{r}{v}\right) + \frac{\omega}{rv} \cos\omega\left(t - \frac{r}{v}\right) \right.$$

$$\left. + \frac{1}{r^2} \sin\omega\left(t - \frac{r}{v}\right) \right] \bar{I}_\theta$$

$$\nabla \times \bar{H} = \epsilon \frac{\partial \bar{E}}{\mu t} \quad \text{so} \quad \bar{E} = \frac{1}{\epsilon}(\nabla \times \bar{H})dt$$

$$\bar{E} = \text{(right-hand side of } \nabla \times \bar{H} \text{ equation)} \frac{dt}{\epsilon}$$

$$\boxed{E_r = \frac{I_m h \cos\theta}{2\pi\epsilon r} \left[\frac{1}{vr} \sin\omega\left(t - \frac{r}{v}\right) - \frac{1}{\omega r^2} \cos\omega\left(t - \frac{r}{v}\right) \right] \bar{I}_r}$$

$$\boxed{E_\theta = \frac{I_m h \sin\theta}{4\pi\epsilon r} \left[\frac{\omega}{v^2} \cos\omega\left(t - \frac{r}{v}\right) + \frac{1}{rv} \sin\omega\left(t - \frac{r}{v}\right) \right.}$$

After integrating $\nabla \times \bar{H}$ = equation w.r.t. and dividing by E

$$\boxed{\left. - \frac{1}{\omega r^2} \cos\omega\left(t - \frac{r}{v}\right) \right] \bar{I}_\theta \\ E_\phi = 0}$$

C-110

H_ϕ for the far field:

$$\overline{H} = \nabla \times \overline{A} = \begin{vmatrix} \overline{I}_r/r^2\sin\theta & \overline{I}_\theta/r\sin\theta & \overline{I}_\phi/r \\ \partial/\partial r & \partial/\partial\theta & \partial/\partial\phi \\ A_r & rA_\theta & r\sin\theta\, A_\phi \end{vmatrix}$$

$$H_\phi = \nabla_\phi \times \overline{A} = \frac{1}{r}\left[\frac{\partial}{\partial r}(rA_\theta) - \frac{\partial A_r}{\partial \theta}\right]$$

$$= \frac{1}{r}\left[\frac{\partial}{\partial r}(-r A_z \sin\theta) - \frac{\partial}{\partial \theta}(A_z \cos\theta)\right]$$

because $\quad A_r = A_z \cos\theta$

$\qquad A_\theta = -A_z \sin\theta$

$\qquad A_\phi = 0$

$$\boxed{H_\phi \simeq -\sin\theta\, \frac{\partial H_z}{\partial r}}$$

because A_z does not vary with θ. So

$$\frac{\partial A_z}{\partial \theta}\cos\theta = 0$$

Relationship between the vector potential and the magnetic and electric field intensities:

A current element of $I_m \cdot h \cdot e^{j\omega t}$ is assumed.

$$\overline{A} = \frac{1}{4\pi}\int_v \frac{J(t - r/v)dv}{r}$$

$$A_z = \frac{I_m \cdot h \cdot e^{j\omega(t-r/v)}}{4\pi r} = I_m \cdot h \cdot e^{u(\omega t - \beta r)/4\pi r}$$

$$\beta = \omega/v$$

$$\overline{H} = \nabla \times \overline{A} = \frac{1}{r}\left[\frac{\partial}{\partial r}(rA_\theta) - \frac{\partial A_r}{\partial \theta}\right]$$

$$H_\phi = \left(j\beta + \frac{1}{r}\right)I_m \cdot h \cdot e^{-j\beta r} \cdot \sin\theta/4\pi r$$

$$E_\theta = \eta I_m \cdot h \cdot \sin\theta \, e^{-j\beta r}\left(\beta + \frac{1}{jr} + \frac{1}{\beta r^2}\right)$$

$$E_r = \eta I_m h \cos\theta \cdot e^{-j\beta r}\left(\frac{2}{jr} + \frac{2}{\beta r^2}\right)$$

where $\beta = \frac{\omega}{v}$ and $\eta - \sqrt{\frac{\mu}{\varepsilon}} = \mu v$

Poynting vector due to an oscillating doublet:

$$P = P_\theta + P_r$$

$$P_\theta = -E_r H_\phi = \frac{I_m^2 \cdot h^2 \cdot \sin^2\theta}{16\pi^2 \varepsilon}\left\{-\frac{\cos 2\omega t'}{r4v} - \frac{\sin 2\omega t'}{2ar^5} + \frac{\omega \sin\omega t'}{2r^3 v^2}\right\}$$

where $t' = t - (r/v)$

P_θ is the pulsation of power in the θ direction, but there is no net current. $P_{\theta \, avg}$ is zero over a complete cycle.

$$P_r = E_\theta H_\phi = \frac{I_m^2 h^2 \sin^2\theta}{16\pi^2\varepsilon}\left[\frac{\sin 2\omega t'}{2\omega r^5} - \frac{\cos 2\omega t'}{r^2 v^3} + \frac{\omega \sin 2\omega t'}{r^3 v^2} + \frac{\omega}{2r^2 v^3} + \frac{\omega^2 \cos 2\omega t'}{2r^2 v^3}\right]$$

$$\boxed{P_{r(av)} = \frac{\omega^2 I_m^2 h^2 \sin^2\theta}{32\pi^2 r^2 v^2}\,\eta \text{ watts}/m^2}$$

where

$$\eta = \sqrt{\frac{\mu}{\varepsilon}}$$

Conclusion: The power output is proportional to the square of $\omega = 2\pi f$. Hence, the efficiency of transmission is improved as the frequency is raised.

Near field of an antenna:

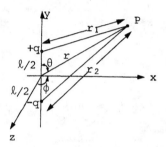

The oscillating dipole is considered as two equal and opposite oscillating charges separated by a distance ℓ.

The potential at point P due to two charges is given by:

$$V = \frac{q_1}{4\pi\varepsilon r_1} + \frac{q_2}{4\pi\varepsilon r_2}$$

where

$$r_1 \cong r - \frac{1}{2}\cos\theta$$

$$r_2 \cong r + \frac{1}{2}\cos\theta$$

$$V = \frac{1}{4\pi\varepsilon}\left[\frac{q}{r - \frac{1}{2}\cos\theta} + \frac{-q}{r + \frac{1}{2}\cos\theta}\right]$$

$i = I\sin\omega t$ The current in the dipole.

$$q = \int_0^t i\,dt = -\frac{I}{\omega}\cos\omega t$$

C-113

$$V = -\frac{I\cos\omega\left(t - \frac{r}{v}\right)(r_2 - r_1)}{4\pi\varepsilon\omega r_1 r_2}$$

$$= -I\ell\cos\theta\cos\omega(t - r/v)/4\pi\varepsilon\omega r^2.$$

since $r_2 - r_1 = I\cos\theta$ and $r_1 r_2 \cong r^2$.

$$E_r = -\frac{\partial V}{\partial r} = -\frac{I\ell}{4\pi\varepsilon r}\cos\theta\left[\frac{1}{vr}\sin\omega\left(t - \frac{r}{v}\right)\right.$$
$$\left. - \frac{1}{\omega r^2}\cos\omega\left(t - \frac{r}{v}\right)\right]$$

$$E_\theta = -\frac{1}{r}\frac{\partial V}{\partial\theta}$$

$$= \frac{I\ell\sin\theta}{4\pi\varepsilon\omega r^3}\cos\omega\left(t - \frac{r}{v}\right)$$

$$E_\phi = \frac{-1}{r\sin\theta}\cdot\frac{\partial V}{\partial\phi} = 0$$

because V is independent of ϕ.

CHAPTER 16

ANTENNAS

16.1 PROPERTIES OF ANTENNAS

An actual antenna, which is long enough to vary the current from one point to another, is subdivided into small elements, each one of these can be considered as an elementary oscillating dipole.

A flat top antenna consists of a horizontal wire with vertical lead-in as shown below:

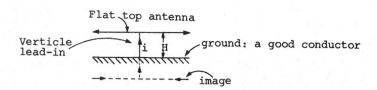

Analysis of a flat-top antenna:

If the ground can be considered as a good conductor, then it can be substituted by an image of the antenna, neglecting capacitance of the vertical wire.

Since the direction of current flow in the antenna and its image are opposite, the original horizontal wire antenna is a non-radiating member.

If $H \ll \lambda$ then the current along the vertical wire can be considered uniformly distributed.

So the antenna and its image forms a dipole of length ℓ = 2H and its radiation resistance is

$$(R_R)_H = \frac{R_R}{2} = \frac{80\pi^2}{2}\left(\frac{2H}{\lambda}\right)^2 \simeq 1580\left(\frac{H}{\lambda}\right)^2$$

In case of a center-fed, the field intensity is half of the above case, and therefore the radiated power will be one-fourth that of the above case (i.e., the current element).

$$R_R \text{ for the dipole} = \frac{1}{4} \cdot (R_R \text{ of the current element})$$
$$= 20\pi^2\left(\frac{h}{\lambda}\right)^2$$

valid for length $< \lambda/4$

Monopole:

It consists of a single short verticle wire of length L mounted on a metallic surface or ground which is considered a perfect reflector.

$$R_R(\text{monopole}) = \frac{1}{2} \cdot R_R \text{ of the dipole}$$
$$\cong 100\left(\frac{h}{\lambda}\right)^2 = 100\left(\frac{2H}{\lambda}\right)^2 \; \Omega$$

this is valid for $L < \frac{\lambda}{8}$

C-116

16.2 ELECTRIC FIELD INTENSITY AT A POINT P FAR FROM A PRACTICAL HALF-WAVE ANTENNA

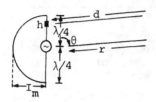

$$\overline{H}_{\phi \text{(far)}} \cong \frac{I_m \cdot h \cdot \omega \cdot \sin\theta}{4\pi v r} \cos\omega\left(t - \frac{r}{v}\right)$$

The effect of a current element at a distance far from the antenna.

$$i = I_m \cdot \cos\beta z$$

If the current has sinusoidal distribution.

$$\beta = 2\pi/\lambda$$

$$dH_\phi = \frac{\omega I_m \cdot h}{4\pi v r} \cos\beta z \sin\theta \cos\omega\left(t - \frac{r - z\cos\theta}{v}\right)$$

because $d \cong r - z\cos\theta$

$$H_\phi = \frac{I_m}{2\pi r} \frac{\cos\left(\frac{\pi}{2}\cos\theta\right)}{\sin\theta}$$

obtained after integrating dH_ϕ from $-\frac{\lambda}{4}$ to $\frac{\lambda}{4}$.

$$\boxed{E_\theta = \eta H_\phi = \frac{60 I_m}{r} \frac{\cos\left(\frac{\pi}{2}\cos\theta\right)}{\sin\theta}}$$

$$\eta = 120\pi$$

16.3 RADIATION FIELD OF AN ISOLATED, FULL WAVE ANTENNA AND RADIATION PATTERN

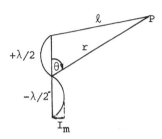

For a full-wave antenna, there is a node at the center.

$$E_\theta = \int_{-\frac{\lambda}{2}}^{\frac{\lambda}{2}} \eta \cdot \frac{I \cdot \sin\theta}{2d\lambda} \cos\omega\left(t - \frac{\ell}{v}\right) dx$$

assuming
$$I = I_m \cdot \sin\frac{2\pi x}{\lambda}$$

Since
$$\ell \cong r - x\cos\theta \quad \text{and} \quad \ell \cong r$$

Therefore
$$E_\theta = \frac{-\eta I_m \sin\omega(t-r/v)}{2\pi r}\left[\frac{\sin(\pi\cos\theta)}{\sin\theta}\right]$$

$$= \frac{-2\sin(\pi\cos\theta)}{(\omega/v)\sin^2\theta}\left[\sin\omega(t - r/v)\right]$$

Radiation pattern:

$$E_\theta = -E_{-\theta}$$

because
$$\frac{\sin(-\theta) = -\sin\theta}{\cos(-\theta) = \cos\theta}$$

C-118

$$E_{(\pi - \theta)} = -E_\theta$$

$$E_{(\pi + \theta)} = E_\theta$$

Conclusion: Pattern of quad III = quad I.

One quadrant is evaluated using the following table.

θ	cos θ	180°cos θ = y	sin y	siny/sinθ
0°	1	180°	0	0
15°	0.966	173.9°	0.106	0.410
30°	0.866	155.9°	0.408	0.816
45°	0.707	127.3°	0.795	1.124
60°	0.5	90°	1	1.155
75°	0.259	46.6°	0.727	0.753
90°	0	0°	0	0

Radiation pattern:

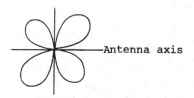
Antenna axis

16.4 EFFECTIVE LENGTH OF AN ANTENNA

It is defined as

$$L_{eH} = \frac{1}{I_{feed}} \int_{-L/2}^{L/2} I\,dL$$

Effective length of a $\frac{\lambda}{2}$ antenna:

$$L_{eff} = \int_{-\lambda/4}^{\lambda/4} \cos \frac{2\pi}{\lambda}\,dx$$

since $I_{(x)} = I_{feed} \cos \beta x$.

$$L_{eff} = \frac{\lambda}{\pi} \quad \text{for a } \frac{\lambda}{2} \text{ antenna.}$$

16.5 MISCELLANEOUS EXAMPLES

Radiation resistance, radiated power per square m at r and the total power radiated. For a short antenna

$$R_R = 80\pi^2 \left(\frac{L}{2\lambda}\right)^2$$

because for a short antenna $L_{eff} = \frac{L}{2}$.

Radiated power/m² at a distance r =

$$P_r = I_m^2 \cdot h^2 \cdot \omega^2 \cdot \sin^2\theta / 32\pi^2 \cdot \varepsilon \cdot r^2\, c^3$$

Total radiated power = $\boxed{P = \dfrac{I_m^2}{2} R_R}$.

$\left|\dfrac{E_t}{E_d}\right|$ for two antennas, acting as transmitter and receivers

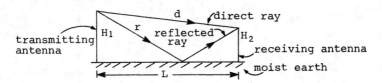

E_t = Total field of the receiving antenna

E_d = Field due to direct ray.

$$E_d = \dfrac{120\pi \cdot I_m \cdot h}{2d\,\lambda}\; \sin\theta \cos\omega\left(t - \dfrac{d}{v}\right)$$

Field intensity due to reflected ray is

$$E_r = -120\pi \cdot I_m \cdot h\big/2r\,\lambda\; \sin\theta \cos\omega\left(t - \dfrac{r}{v}\right)$$

negative sign due to reflection

$$E_t = E_d + E_r$$

$$= \dfrac{120\pi \cdot I_m \cdot h \cdot \sin\theta}{2r\,\lambda}\; 2\sin\left(\dfrac{2\pi}{\lambda L} H_1 H_2\right)$$

$$\boxed{\left|\dfrac{E_t}{E_d}\right| = 2\sin \dfrac{2\pi}{\lambda L} H_1 H_2}$$

Condition for receiving maximum signal at the receiving antenna:

$$\left|\dfrac{E_t}{E_d}\right| = \dfrac{2\sin 2\pi\, H_1 H_2}{\lambda\, L}$$

For the maximum signal

$$\frac{2\pi H_1 H_2}{\lambda L} = (2\eta - 1)\frac{\pi}{2}$$

The maximum value of $H_1 H_2$ will be found for $\eta = 1$.

$$\frac{2\pi H_1 H_2}{\lambda L} = \frac{\pi}{2}$$

$$\boxed{H_1 H_2 = \frac{\lambda L}{4}}$$

where H_1 = height of transmitting antenna.

H_2 = height of receiving antenna

Handbook of Electrical Engineering

SECTION D

Electronic Communications

CHAPTER 1

BASIC CIRCUIT PRINCIPLES

1.1 THE CAPACITOR

The unit of capacitance is the Farad, which is $\frac{\text{Coulomb}}{\text{Volt}}$. The voltage across the capacitor is given by $V = \frac{q}{C}$, \hfill (1.1)

where q is charge in coulombs and C is capacitance in farads.

Since $i = \frac{dq}{dt}$ and $q = \int i\, dt$ \hfill (1.2)

the current-voltage relationship is $V = \frac{1}{C}\int i\, dt$. \hfill (1.3)

Applying a definite integral to equation (1.2)
We obtain: $\quad q(t) - q(t_0) = \int_{t_0}^{t} i\, d\tau$

Substituting Eq. (1.1) $\quad$ At $t_0 = 0$

$$\boxed{V(t) = \frac{1}{C}\int_0^t i\, d\tau + \frac{q(0)}{C}}$$

where' $q(0)$ = initial charge.

Power and Energy in a Capacitor
From Eq.(1.3)
$$i = \frac{C dV}{dt}.$$

Since $P = VI$, then

$$P = \frac{CVdV}{dt} \quad (1.4)$$

To determine the total energy stored between t_0 and t, we integrate Eq. (1.4) to obtain

$$W(t) - W(t_0) = \frac{C}{2}\left[V^2(t) - V^2(t_0)\right] \quad (1.5)$$

If $V_c(t_0) = 0$

$$W(t) = \frac{C}{2}\left[V^2(t)\right] \quad (1.6)$$

If a sinusoidal voltage is applied, the net stored energy is zero. Since the capacitor stores energy and returns it to the circuit, the capacitor behaves as an inductor.

1.2 RC CIRCUITS

The important facts are:
- The volage across capacitor cannot change instantaneously (i.e. voltage at $t = 0^-$ equals voltage at $t = 0$).
- The capacitor behaves as an open circuit in dc steady state conditions.
- The sum of voltages around a closed loop is zero.

If the RC circuit has the form:

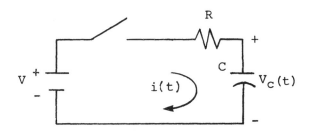

Use Kirchoff's Voltage Law (KVL) to obtain
$$V = Ri(t) + V_c(t), \text{ for } t > 0.$$

The homogeneous solution for the current in the circuit is given by

$i(t) = Ae^{(-1/RC)t}$ where A is a constant.

Using initial value conditions we find that $A = \dfrac{V}{R}$.

Thus
$$i(t) = \frac{V}{R} e^{-\left(\frac{1}{RC}\right)t}. \tag{1.7}$$

The time constant of the RC circuit is $\tau = RC$. This is the time it takes for the initial current to decay to 36.8 % of its initial value $I_0 = \left(\dfrac{V}{R}\right)$.

1.3 THE INDUCTOR

The unit of inductance is the henry, which is $\dfrac{1 \text{volt-sec}}{\text{ampere}}$

Voltage across an inductor is given by : $V = \dfrac{Ldi}{dt}$ (1.8)

From equation (1.8) we obtain $i = \dfrac{1}{L}\int V dt$

Using a definate integral for $t_0 \leq t$ on both sides we obtain :

$$i(t) = \frac{1}{L} \int_{t_0}^{t} V dt + i(t_0) \tag{1.9}$$

where $i(t_0)$ is the initial current.

Power and Energy in an Inductor
Since $P = VI$, for an inductor

$$P = Li \frac{di}{dt} \tag{1.10}$$

Since the integral of power is energy, integrating equation (1.10) for $t_0 \leq t$ leads to

$$\boxed{W(t) - W(t_0) = \frac{L}{2}\left[i^2(t) - i^2(t_0)\right]} \quad (1.11)$$

Assuming $t_0 = 0$ and that current through the inductor is zero for $t \leq 0$, we obtain

$$\boxed{W(t) = \frac{L}{2}\left[i^2(t)\right]} \quad (1.12)$$

The total energy supplied to the inductor is zero for alternating currrent. The inductor does not dissipate heat. Instead, energy is stored in its magnetic field.

1.4 RL CIRCUITS

The important facts are :
- The current through an inductor cannot change instantaneously (i.e. current is same for $t = 0^-$ as for $t = 0$).
- The inductor behaves as a short curcuit in D.C. steady state conditions.
- The sum of voltage drops around a closed loop is zero.
- The polarity of the voltage across an inductor is such that it opposes any change in current through it.

If the RL curcuit has the form :

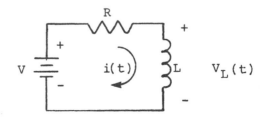

For such a circuit we obtain the following equation using KVL

$$V = i(t)R + L\frac{di}{dt}, \text{ for } t > 0. \quad (1.13)$$

The solution for the current in the circuit can be expressed as a sum of two terms:

$$i(t) = i_c(t) + I_p \quad (1.14)$$

$i_c(t)$ = homogeneous solution
I_p = particular solution.

We start by solving for $i_c(t)$. This is done by setting the forcing function to zero. The solution is of the form:

$$i_c(t) = Ae^{\alpha t}, \text{ for } t > 0 \quad (1.15)$$

where A and α are constants.
Substituting equation (1.15) in equation (1.13), we obtain
$0 = Ae^{\alpha t}(R + \alpha L)$.
But $(R + \alpha L)$ must be zero if $Ae^{\alpha t}$ is to be a solution.
Thus,

$$R + \alpha L = 0; \text{ and } \alpha = \frac{-R}{L}.$$

Finally obtaining the solution for the homogeneous solution:

$$\boxed{i_c(t) = Ae^{-\left(\frac{R}{L}\right)t}} \quad (1.16)$$

where A is a constant.
The particular solution is given by:

$$V = RI_p + Ae^{-\left(\frac{R}{L}\right)t}\left[\frac{-R}{L}L + R\right], \text{ for } t > 0 \quad (1.17)$$

Since the second term is zero, then

$$\boxed{I_p = \frac{V}{R}} \quad (1.18)$$

Thus the complete solution is

$$i(t) = \frac{V}{R} + Ae^{-\left(\frac{R}{L}\right)t} \qquad (1.19)$$

Assuming $i(0^-) = 0$ and since $i(0^+) = i(0^-)$, then $i(0^+) = 0$,

giving us $A = \frac{-V}{R}$ and the solution becomes

$$i(t) = \frac{V}{R}\left(1 - e^{-\left(\frac{R}{L}\right)t}\right) \qquad (1.20)$$

The time constant of the RL curcuit is $\tau = \frac{L}{R}$.

This is the time it takes for the current in the circuit (inductor) to reach 63% of its maximum value $I_m = \left(\frac{V}{R}\right)$.

Use of Nodal Analysis in Parallel RL Circuit :
If the circuit is of the form :

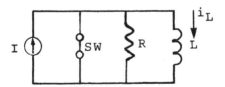

Using Kirchoff's Current Law (KCL) we arrive at

$$I = \frac{V(t)}{R} + \frac{1}{L}\int V(t)dt. \qquad (1.21)$$

Differentiating, we obtain a homogeneous equation with the solution of the form :

$V(t) = Ae^{\alpha t}$

where A and α are constants.

Substituting $i = \frac{1}{L}\int V(t)dt$ into Eq. (1.16) and manipulating we find that $A = IR$.
Thus,

$$\boxed{V(t) = I\,Re^{-\left(\frac{R}{L}\right)t}.}\qquad(1.22)$$

CHAPTER 2

FOURIER SERIES & FOURIER TRANFORMS

2.1 SIGNAL TERMINOLOGY

- Classification of signals :

    i) Periodic signals : A periodic signal is one that repeats itself in a predictable manner. The period T is such that
    $x(t+T) = x(t)$.

    ii) Nonperiodic signals : A nonperiodic signal has no finite period T.

    iii) Deteoministic signals : The instantaneous value of this signal can be given by a mathematical equation. The following is a deterministic signal

    e.g.
    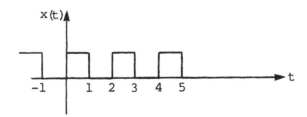

    iv) A random signal : Its instantaneous value can't be predicted at any given time.

(v) **Power and Energy in a Signal:** For an arbitrary signal $x(t)$, which may in general be complex its (normalized) power is defined by,

$$P = \lim_{T \to \infty} \frac{1}{2T} \int_{-T}^{T} |x(t)|^2 dt \qquad (2.1)$$

and its (normalized) energy is given by

$$E = \lim_{T \to \infty} \int_{-T}^{T} |x(t)|^2 dt = \int_{-\infty}^{\infty} |x(t)|^2 dt \qquad (2.2)$$

- $x(t)$ is a power signal, if and only if $0 < P < \infty$, thus implying that energy $E = \infty$.

- $x(t)$ is an energy signal, if and only if $0 < E < \infty$, so that the total power $P = 0$.

- Total energy in a power signal is infinite in the limit (i.e. over infinite time interval).

- All periodic signals are power signals but not all power signals are periodic.

- The total energy of a power signal is infinite but it has finite, nonzero average power $\overline{P}$. The finite $\overline{P}$ results due to division by the time-interval before it increases without limit.

- In an energy signal, the total energy contained in the signal is constant. The average power will be zero as the period approaches infinity.

e.g.

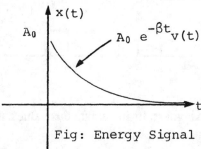

Fig: Energy Signal

Fig. : Energy Signal

E = The area under the curve = Net energy

$$= \int_0^t (x(t))^2 dt = \int_0^\infty A_0^2 e^{-2/3t} dt$$

$$\boxed{E = A_0^2/2\beta} \qquad (2.3)$$

This is an energy signal for $\beta > 0$.

However, if $\beta = 0$, then we find that

$$\boxed{P = \frac{1}{2}A_0^2} \qquad (2.4)$$

and the signal becomes a power signal.

2.2 FOURIER SERIES

Using Fourier series, any periodic signal can be represented by a series of sinusoidal components plus a dc term.

- The lowest frequency of the sinusoisal components is
  $f_0 = \dfrac{1}{T}$ (where T is the time period).

  = Fundamental frequency.
  Other frequencies are integer multiples of f_0.
- Mathematical forms of the Fourier series: -

  i) One-sided spectral representation, i.e. Sine-cosine form and the amplitude-phase form.
  ii) Two-sided spectral representation, i.e. complex exponential form
- Sine-cosine form :
  $$x(t) = \text{A periodic signal} = A_0 + \sum_{n=1}^{\infty} [A_n \cos n\omega_0 t + B_n \sin n\omega_0 t] \qquad (2.5)$$

  where $\quad \omega_0 = \dfrac{2\pi}{T} = 2\pi f_0$

$$A_0 = \frac{1}{T}\int_0^T x(t)\,dt$$

$$A_n = \frac{2}{T}\int_0^T x(t)\cdot \cos n\,\omega_0 t\,dt \quad \text{and}$$

$$B_n = \frac{2}{T}\int_0^T x(t)\cdot \sin(n\,\omega_0 t)\,dt$$

- Amplitude-phase form :

$$x(t) = A_0 + \sum_{n=1}^{\infty} C_n \cos(n\,\omega_0 t + \theta_n) \qquad (2.6)$$

where $C_n = \left[A_n^2 + B_n^2\right]^{1/2}$ and $\theta_n = \tan^{-1}\left(\frac{B_n}{A_n}\right)$.

- Complex exponential form :
  The general form of the complex exponential form of the Fourier series is

$$x(t) = \sum_{n=-\infty}^{\infty} C_n e^{jn\,\omega_0 t} \qquad (2.7)$$

where

$$C_n = \frac{1}{T}\int_0^T x(t) e^{-jn\,\omega_0 t}\,dt$$

- At a given frequency nf_o, $(n > 0)$, the spectral presentation is :

$$C_n e^{jn\,\omega_o t} + C_{-n} e^{-jn\,\omega_o t}$$

where $C_n = \dfrac{A_n - jB_n}{2}$ for $n \neq 0$

(where A_n and B_n are obtained from the sine-cosine form).
The dc component C_0 is A_0, and is the same in all Fourier forms.

Note :
The following are useful trigonometric integrals in determinimg the constants of the Fourier series.
n, k = 1, 2, 3, ...

T = one period

Since the average of one sinusoid is zero, the following holds:

$$\int_0^T \sin n\, \omega_o t\, dt = 0$$

and $\int_o^T \cos n\, \omega_0 t\, dt = 0$

Other useful integrals are:

$$\int_0^T \sin k\, \omega_0 t\, \cos n\, \omega_0 t\, dt = 0, \text{ for } k \neq n$$

$$\int_0^T \sin k\, \omega_0 t\, \sin n\, \omega_0 t\, dt = 0, \text{ for } k \neq n$$

$$\int_0^T \cos k\, \omega_0 t\, \cos n\, \omega_0 t\, dt = 0, \text{ for } k \neq n$$

Also:

$$\int_0^T \sin^2 n\, \omega_0 t\, dt = \frac{T}{2}$$

$$\int_0^T \cos^2 n\, \omega_0 t\, dt = \frac{T}{2}$$

2.3 CONVERGENCE OF FOURIER SERIES

Definition:
Any periodic waveform, i.e. one for which $f(t) = f(t+T)$, can be expressed by a Fourier series provided that:

(1) If it is discontinuous there are a finite number of discontinuities in the period T.
(2) It has a finite average value for the period T.

3) It has a finite number of positive and negative maxima.

2.4 SYMMETRY CONDITIONS

i) Even function : An even function is represented by

$$x(-t) = x(t). \qquad (2.8)$$

- Only cosine terms appear in the spectrum of the even function signal.

- Characteristics :

$$A_n = \frac{4}{T}\int_0^{T/2} x(t)\cos n\,\omega_0 t\, dt,$$
$$B_n = 0$$
$$C_n = \frac{2}{T}\int_0^{T/2} x(t)\cos n\,\omega_0 t\, dt \;\&\; \text{these are real.}$$

Example : Even function

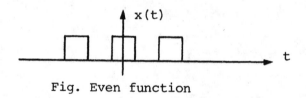

Fig. Even function

ii) Odd function :

- $x(-t) = x(t)$ \hfill (2.9)

- Only sine terms are present.
- Characteristics :

$$A_n = 0, B_n = \frac{4}{T}\int_0^{T/2} x(t)\sin n\,\omega_0 t\, dt$$
$$C_n = \frac{-2j}{T}\int_0^{T/2} x(t)\sin n\,\omega_0 t\, dt$$

Example :   Odd function

Fig. Odd function

iii) Half-wave symmetry :
Half-wave symmetry is characterized by

$$x\left(t + \frac{T}{2}\right) = -x(t) \qquad (2.10)$$

- This function will have only odd harmonics.

- Characteristics :

$$A_n = \frac{4}{T}\int_0^{T/2} x(t)\cos n\omega_0 t\, dt$$

$$B_n = \frac{4}{T}\int_0^{T/2} x(t)\sin n\omega_0 t\, dt$$

$$C_n = \frac{2}{T}\int_0^{T/2} x(t)\,\epsilon^{-jn\omega_0 t}\, dt$$

Example : Half-wave symmetry

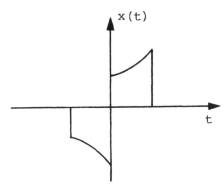

Fig: Half-wave symmetry

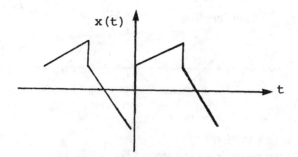

Fig: Half-wave symmetry function with dc component.

- If the function has a dc component, it is subtracted and then the symmetry condition is applied.

iv) **Full-wave symmetry** :

This function is given by
$$x\left(t + \frac{T}{2}\right) = x(t) \qquad (2.11)$$

- This function has only even harmonics including $n = 0$ harmonic (fundamental).

- Characteristics :

$$A_n = \frac{4}{T}\int_0^{T/2} x(t)\cos n\,\omega_0 t\, dt$$

$$B_n = \frac{4}{T}\int_0^{T/2} x(t)\sin n\,\omega_0 t\, dt$$

$$C_n = \frac{2}{T}\int_0^{T/2} x(t)e^{-jn\omega_0 t}\, dt$$

Example :   Full-wave symmetry

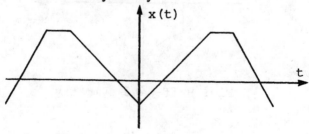

Fig. Full-wave symmetry

v) Important Points :
Any given function may be either even or odd (but not both) and the function may have either half-wave or full-wave symmetry (but not both).

2.5 FOURIER TRANSFORM

Definition : This is a mathematical function that gives the frequency spectrum of a non-periodic signal say $x(t)$.

Symbol : The symbolic representation of the Fourier transform operation is $F[x(t)] = X(f)$ (2.12)

The inverse Fourier transform operation is
$$F^{-1}[X(f)] = x(t). \qquad (2.13)$$

Mathematically,
1. The Fourier transform of signal $x(t)$ is expressed as
$$F[x(t)] = X(f) = \int_{-\infty}^{\infty} x(t) e^{-j 2\pi ft} dt \qquad (2.14)$$

2. The signal $x(t)$ is the inverse transform of $x(f)$ and can be expressed as
$$F^{-}[X(f)] = x(t) = \frac{1}{2\pi} \int_{-\infty}^{\infty} X(f) e^{j 2\pi ft} df \qquad (2.15)$$

$x(t)$ and $X(f)$ constitute a Fourier transform pair :

$X(f) = F[x(t)]$
$x(t) = F^{-1}[X(f)]$

2.6 PROPERTIES OF FOURIER TRANSFORMS

These properties are derived from the application of equations

(2.14) and (2.15).

1. Superposition (Linearity):

$$\boxed{ax_1(t) + bx_2(t) \leftrightarrow aX_1(f) + bX_2(f)} \qquad (2.16)$$

where a and b are constants and $x_1(t)$ and $x_2(t)$ are nonperiodic signals.

2. Time Scaling:

$$\boxed{x(at) \leftrightarrow \frac{1}{|a|}X\left(\frac{f}{a}\right)} \qquad (2.17)$$

where $x(at)$ represents $x(t)$ compressed by a factor 'a' and $x\left(\frac{f}{a}\right)$ represents x(f) expanded by 'a'.

3. Duality:

$$\boxed{X(t) \leftrightarrow x(-f)} \qquad (2.18)$$

This is obtained by interchanging f with t, after the inverse Fourier transform was performed on $x(-t) = \int_{-\infty}^{\infty} x(f) e^{-j2\pi ft} df$.

4. Time Shifting:

$$\boxed{x(t - t_0) \leftrightarrow e^{-j2\pi ft_0} X(f)} \qquad (2.19)$$

If the signal $x(t)$ is time shifted by an amount t_0, then its equivalent is the Fourier transform $X(f)$ multiplied by $e^{-j2\pi ft_0}$.

5. Frequency shifting:

$$\boxed{xe^{j2\pi f_c t} x(t) \leftrightarrow X(f - f_c)} \qquad (2.20)$$

Multiplying the signal $x(t)$ by a factor $e^{j2\pi f_c t}$ is equivalent to a shift in frequency of the Fourier transform by an amount f_c.

6. Area Under Signal $x(t)$:

If for $x(t) \leftrightarrow x(f)$ then

$$\int_{-\infty}^{\infty} x(t)dt = X(0). \tag{2.21}$$

The area under a function $x(t)$ = Fourier transform $x(f)$ at $f = 0$.

7. Area Under Transform $x(f)$:

If for $x(t) \leftrightarrow X(f)$ then

$$\int_{-\infty}^{\infty} X(f)dt = x(0) \tag{2.22}$$

At time $t = 0$, the area of $x(t)$ is the area under its Fourier transform.

8. Differentiation in the Time Domain:

For $x(t) \leftrightarrow X(f)$, and the assumption that the derivatives of $x(t)$ can be Fourier transformed, then

$$\frac{d^n x(t)}{dt^n} \leftrightarrow (j2\pi f)^n X(f) \tag{2.23}$$

where n is from 0 to ∞.

9. Integration in the Time Domain:

For $x(t) \leftrightarrow X(f)$, then

$$\int_{-\infty}^{t} x(\tau)d\tau \leftrightarrow \frac{1}{j2\pi f}X(f) + X(0)\delta(f) \tag{2.24}$$

This states that the integration of a time function $x(t)$ is equivalent to dividing its Fourier transform by a factor $j\,2\pi f$, plus $X(f)$ evaluated at $f = 0$, where δ represents the delta function.

10. Conjugate Functions :

For $x(t) \leftrightarrow X(f)$, then for a complex-valued function in the time domain,

$$\boxed{X^*(t) \leftrightarrow X^*(-f)} \qquad (2.25)$$

where the asterisk (*) inidcates complex conjugate.

11. Multiplication in the Time Domain :

For $x_1(t) \leftrightarrow X_1(f)$ and $x_2(t) \leftrightarrow X_2(f)$ then

$$\boxed{X_1(f)*X_2(f) \leftrightarrow \int_{-\infty}^{\infty} x_1(f')x_2(f-f')df'} \qquad (2.26)$$

Here, the integral is referred to as the convolution integral expressed in the frequency domain.

Therefore

$$\boxed{x_1(t) \cdot x_1(t) \leftrightarrow X_1(f)*X_2(f)} \qquad (2.27)$$

This states that when two signals are multiplied in the time domain, their product can be transformed into the convolution of their individual Fourier transforms.

12. Convolution in the Time Domain :

This is obtained by applying the property of duality to the property of multiplication in the Time Domain, and forming the Fourier transform pair.

Hence

$$\int_{-\infty}^{\infty} x_1(t') x_2(t - t') dt' \leftrightarrow X_1(f) X_2(f)$$

written as

$$\boxed{x_1(t) * x_2(t) \leftrightarrow X_1(f) X_2(f)} \qquad (2.28)$$

This states that multiplication of the Fourier transforms of two signals in the frequency domain is transformable to the convolution of these signals in the time domain.

2.7 LIST OF FOURIER TRANSFORM THEOREMS

NAME OF THEOREM	SIGNAL	FOURIER TRANSFORM
1. Superposition	$a_1 x_1(t) + a_2 x_2(t)$	$a_1 X_1(f) + a_2 X_2(f)$
2. Time delay	$x(t - t_0)$	$X(f) e^{-j\omega t_0}$
3a. Scale change	$x(at)$	$\|a\|^{-1} X(f/a)$
b. Time reversal	$x(-t)$	$X(-f) = X^*(f)$
4. Duality	$X(t)$	$x(-f)$
5a. Frequency translation	$x(t) e^{j\omega_0 t}$	$X(f - f_0)$
b. Modulation	$x(t) \cos \omega_0 t$	$\tfrac{1}{2} X(f - f_0) + \tfrac{1}{2} X(f + f_0)$
6. Differentiation	$\dfrac{d^n x(t)}{dt^n}$	$(j 2\pi f)^n X(f)$
7. Integration	$\int_{-\infty}^{t} x(t') dt'$	$(j 2\pi f)^{-1} X(f) + X(0) \delta(f)$
8. Convolution	$\int_{-\infty}^{\infty} x_1(t - t') x_2(t') dt'$ $= \int_{-\infty}^{\infty} x_1(t') x_2(t - t') dt'$	$X_1(f) X_2(f)$
9. Multiplication	$x_1(t) x_2(t)$	$\int_{-\infty}^{\infty} x_1(f - f') x_2(f') df'$ $= \int_{-\infty}^{\infty} x_1(f') x_2(f - f') df'$

2.8 FOURIER TRANSFORM PAIRS

SIGNAL		TRANSFORM

1. 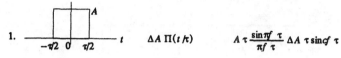 $\triangleq A\,\Pi(t/\tau)$ $A\tau\,\dfrac{\sin\pi f\,\tau}{\pi f\,\tau} \triangleq A\tau\,\mathrm{sinc}\,f\,\tau$

2. (triangle, peak B at 0, base $-\tau$ to τ) $\triangleq B\,\Lambda(t/\tau)$ $B\tau\,\dfrac{\sin^2 \pi f\,\tau}{(\pi f\,\tau)^2} \triangleq B\tau\,\mathrm{sinc}^2 f\,\tau$

3. $e^{-\alpha t}\,u(t)$ $\dfrac{1}{\alpha + j\,2\pi f}$

4. $\exp(-|t|/\tau)$ $\dfrac{2\tau}{1+(2\pi f\,\tau)^2}$

5. $\exp\left[-\pi(t/\tau)^2\right]$ $\tau\exp\left[-\pi(f\,\tau)^2\right]$

6. $\dfrac{\sin 2\pi Wt}{2\pi Wt} \triangleq \mathrm{sinc}\,2Wt$ (rect of height $1/2W$ from $-W$ to W) $\triangleq \dfrac{\Pi(f/2W)}{2W}$

7. $\exp[j(\omega_c t + \phi)]$ $\exp(j\phi)\,\delta(f-f_c)$

8. $\cos(\omega_c t + \phi)$ $\tfrac{1}{2}\delta(f-f_c)\exp(j\phi) + \tfrac{1}{2}\delta(f+f_c)\exp(-j\phi)$

9. $\delta(t-t_0)$ $\exp(-j\omega t_0)$

10. $\displaystyle\sum_{m=-\infty}^{\infty}\delta(t-mT_s)$ $\dfrac{1}{T_s}\displaystyle\sum_{n=-\infty}^{\infty}\delta\!\left(f-\dfrac{n}{T_s}\right)$

11. $\mathrm{sgn}\,t = \begin{cases} +1, & t>0 \\ -1, & t<0 \end{cases}$ $-\dfrac{j}{\pi f}$

12. $u(t) = \begin{cases} 1, & t>0 \\ 0, & t<0 \end{cases}$ $\tfrac{1}{2}\delta(f) + \dfrac{1}{j\,2\pi f}$

13. $\hat{x}(t)$ $-j\,\mathrm{sgn}(f)\,X(f)$

CHAPTER 3

LAPLACE TRANSFORMS

3.1 LAPLACE TRANSFORMS OF FUNCTIONS

The Laplace transform of a function $f(t)$, which is sectionally continuous on the interval $0 \leq t \leq A$ for any positive A, is denoted by $L[f(t)]$ or $F(s)$.

Thus: $$L[f(t)] = F(s) = \int_0^\infty e^{-st} f(t)\, dt \qquad (3.1)$$

where $f(t)$ is a function of a real variable t and is called the original function, $F(s)$ is called the image function and $s = r + jw$. If the attenuation, $r = 0$, then $s = jw$.

Properties of Laplace Transforms

$$L[f(t)] = \int_0^\infty e^{-st} f(t)\, dt = F(s)$$

$L[c f(t)] = cF(s)$ where C is a constant.

$$L[f(t) + g(t)] = L[f(t)] + L[g(t)]$$

$$= F(s) + G(s)$$

$$L[t\,f(t)] = -\frac{dF(s)}{dS}$$

$$L\left[\frac{1}{t}f(t)\right] = \int_s^\infty F(s)\,ds$$

$$L[f(t-a)] = e^{-as}F(s) \text{ where } a \text{ is a constant.}$$

$$L\left[\frac{df}{dt}\right] = sF(s) - f(0)$$

$$L\left[\frac{d^n f}{dt^n}\right] = -f^{(n-1)}(0) - sf^{(n-2)}(0)$$
$$- s^2 f^{(n-3)}(0) - \cdots$$
$$- s^{n-1} f(0) + s^n F(s)$$

$$L\left[\int f(\tau)d\tau\right] = \frac{1}{s}F(s) + \frac{1}{s}\int f(t)\,dt \bigg|_{t=0}$$

$$L[e^{at}f(t)] = F(s-a), \text{ where } a \text{ is a constant.}$$

The following is a table of often used Laplace transforms

$F(s)$	$f(t)$
0	0
$1/s$	1
$1/(s-a)$	e^{at}
$s/(s^2-a^2)$	$\cosh at$
$a/(s^2-a^2)$	$\sinh at$
$s/(s^2+\omega^2)$	$\cos \omega t$
$\omega/(s^2+\omega^2)$	$\sin \omega t$
$1/s^2$	t
$k!/s^{k+1}$	t^k
$\omega/((s-b)^2+\omega^2)$	$e^{bt}\sin \omega t$
$s/((s-b)^2+\omega^2)$	$e^{bt}\cos \omega t$
$k!/(s-b)^{k+1}$	$e^{bt}t^k$
$-a/s(s-a)$	$1-e^{at}$
$2s\omega/(s^2+\omega^2)^2$	$t\sin \omega t$
$(s^2-\omega^2)/(s^2+\omega^2)^2$	$t\cos \omega t$

3.2 INVERSE LAPLACE TRANSFORMS

The inverse Laplace transform of F(s), denoted by $L^{-1}[F(s)]$, may be found by using

$$L^{-1}[F(s)] = f(t) = \frac{1}{2\pi j}\int_{c-j\infty}^{c+j\infty} e^{ts}F(s)\,ds \quad (3.2)$$

However, very often other methods are used to find $L^{-1}[F(s)]$. One of these methods, called the method of partial-fraction expansion, is described as follows :

Step 1 : Decompose $F(s)$ into a sum of partial fractions by applying the technique of partial fraction expansions on $F(s)$ (if possible), i.e.,

$$F(s) = F_1(s) + F_2(s) + \ldots + F_n(s) \quad (3.3)$$

Step 2 : Use the table of Laplace transforms to find the inverse Laplace transform of each of the partial fractions, $F_i(s)$, that is

$$f(t) = L^{-1}[F(s)] = L^{-1}[F_1(s)]$$

$$+ L^{-1}[F_2(s)] + \ldots + L^{-1}[Fn(s)] \quad (3.4)$$

Other methods may also be used to decompose a given transform $F(s)$ into a sum of functions whose inverse transforms are already known or can be found directly from the table.

3.3 INITIAL AND FINAL VALUE THEOREMS AND CONVOLUTION

Initial Value Theorem

$$\boxed{\lim_{t \to 0} f(t) = \lim_{s \to \infty} sF(s)} \quad (3.5)$$

Final Value Theorem

$$\lim_{t \to \infty} f(t) = \lim_{s \to 0} sF(s) \qquad (3.6)$$

The convolution of the functions $f(t)$ and $g(t)$ is denoted by $f * g$ and defined as follows:

$$f(t) * g(t) = \int_0^t f(\tau) g(t - \tau) d\tau$$

where $f(t)$ and $g(t)$ are piecewise continuous on $0 \le t \le A$ for all possible A and are of exponential order. (i.e., if there exists a finite number σ_0 such that $|f(t)| < Ke^{\sigma_0}$ for $t \ge 0$, where k is a constant.)
Then

$$L[f(t) * g(t)] = L[f(t)]L[g(t)] \qquad (3.7)$$

The Convolution Formula

$$L^{-1}[F(s)G(s)] = \int_0^t f(\tau) g(t - \tau) d\tau$$
$$= f * g \qquad (3.8)$$

If we take the Laplace transforms of both sides of the equation above we obtain:

$$F(s)G(s) = L\left[\int_0^t f(\tau) g(t - \tau) d\tau\right]$$
$$= L[f * g] \qquad (3.9)$$

CHAPTER 4

SPECTRAL ANALYSIS

4.1 THE SAMPLING FUNCTION

The Sampling function is defined by,

$$S_a(x) = \frac{\sin x}{x} \qquad (4.1)$$

This function is plotted in the figure below. It is symmetrical about $x = 0$ and has the value $S_a(0) = 1$. The function is zero at $x = n\pi$, where n is an integer, but $n \neq 0$. Also, the function is minimax at $x = \left(n + \frac{1}{2}\right)\pi$, where n is an integer, but $n \neq 0$.

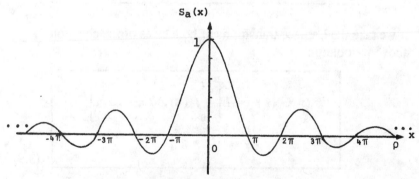

Fig. 4.1 : The function $S_a(x)$

4.2 RESPONSE OF A LINEAR SYSTEM

A block diagram of a linear time invariant system is shown in the diagram below. When the input is expressed as a Fourier series in exponential form, the output of the system is,

$$V_0(t) = \sum_{n=-\infty}^{\infty} H(\omega_n) V_n e^{j 2\pi n t / T_0} \quad (4.2)$$

Where, $H(\omega_n)$ = Transfer function of the system

V_n = Complex coefficients of the input waveform expressed in exponential Fourier series.

T_0 = Period of the input waveform

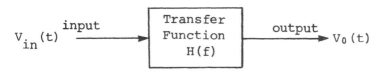

Fig. 4.2 Block diagram of a linear system

When Fourier transform representation is desired, the output is given by,

$$V_0(f) = H(f) V_i(f) \quad (4.3)$$

where $V_0(f)$ = Fourier transform of the output.

$V_i(f)$ = Fourier transform of the input.

$H(f)$ = Transfer function of the system.

4.3 NORMALIZED POWER

If $v(t)$ denotes a time variant waveform, then the quantity $\overline{v^2(t)}$ is referred to as the normalized power. In systems where two normalized powers S_1 and S_2 are encountered and if the ratio of powers is of interest, then it is expressed as,

$$K = 10 \log\left(\frac{S_2}{S_1}\right) dB \qquad (4.4)$$

If S_1 and S_2 are, respectively, the normalized power associated with sinusoidal signals of amplitude V_1 and V_2, then,

$$S_1 = \frac{V_1^2}{2} \text{ and } S_2 = \frac{V_2^2}{2}.$$

Therefore,

$$K = 10 \log \frac{V_2^2/2}{V_1^2/2} = 20 \log \frac{V_2}{V_1} dB \qquad (4.5)$$

Normalized Power in a Fourier Expansion

The normalized power associated with the entire Fourier series is

$$S = C_0^2 + \sum_{n=1}^{\infty} \frac{C_n^2}{2} \qquad (4.6)$$

where C_0 and C_n corresponds to the waveform $v(t)$ expressed as,

$$v(t) = C_0 + \sum_{n=1}^{\infty} C_n \cos\left(\frac{2\pi nt}{T_0} - \phi n\right) \quad (4.7)$$

Refer to Chapter 2, for a detailed discussion of the above coefficients C_0 and C_n.

The total normalized power in terms of exponential Fourier series coefficients is

$$S = \sum_{n=-\infty}^{n=+\infty} V_n V_n^* \quad (4.8)$$

where V_n^* is a conjugate.

4.4 POWER SPECTRAL DENSITY (PSD)

The power spectral density of a waveform is denoted by $G(f)$.

1. Periodic Waveform

$$G(f) = \sum_{n=-\infty}^{\infty} |V_n|^2 \delta(f - nf_0) \quad (4.9)$$

where G(f) denotes the power spectral density.
V_n is complex exponential Fourier series coefficients.
δ denotes the impulse function.

2. Non-Periodic Waveform

$$G_E(f) = |V(f)|^2 \quad (4.10)$$

where $V(f)$ is the Fourier transform of the input waveform. $G_E(f)$ denotes the energy spectral density since the input is non-periodic.

3. Effect of Transfer Function on PSD

$$\boxed{G_0(f) = G_i(f)|H(f)|^2} \qquad (4.11)$$

where $H(f)$ is the transfer function of the input.
$G_i(f)$ is the PSD of the input.
$G_0(f)$ is the PSD of the output.

4.5 RELATIONSHIP BETWEEN INPUT AND OUTPUT IN THE TIME DOMAIN

If $H(f)$ represents the transfer function of the system, then its inverse Fourier transform $h(t)$ represents the impulse response of the system. The input and output are related by the following equation :

$$\boxed{v_0(t) = \int_{-\infty}^{\infty} v_i(\tau)h(t-\tau)d\tau} \qquad (4.12)$$

where $v_i(t)$ is the input signal, $v_o(t)$ is the output signal and τ is a dummy variable.

4.6 PARSEVAL'S THEOREM

The equation that expresses Parseval's theorem is given by,

$$\boxed{E = \int_{-\infty}^{\infty} V(f) V^*(f) df = \int_{-\infty}^{\infty} |V(f)|^2 df = \int_{-\infty}^{\infty} [v(t)]^2 dt}$$
(4.13)

where E = Normalized energy.
$V(f)$ = Fourier transform of the waveform $v(t)$.

Note that Parseval's theorem is the extension of the nonperiodic case of Eqs. (4.8) and (4.10).

4.7 BANDLIMITING OF WAVEFORMS

In principle, periodic or non-periodic waveforms usually have spectral components extending to infinite frequency. When these waveforms are passed through networks discriminating among spectral components, the waveform will be distorted. A network which introduces no distortion is given by the transfer function,

$$H(f) = h_0 e^{-jwT_D}$$
(4.14)

where h_0 is a constant, and T_D is the time delay introduced by the network.

If a large part of the signal power or energy, say 99 percent, lies below a frequency f_M, then a network with high-frequency end f_M and low-frequency end f_L containing negligible fraction of the signal energy will produce negligible distortion in the waveform.

Bandlimiting at the Low Frequency End, f_L.

A low-pass RC circuit produces bandlimiting in the low frequency end. The transfer function of this network is given by,

$$\boxed{H(f) = \frac{1}{1 + jf/f_2}}$$
(4.15)

where $f_2 = \dfrac{1}{2\pi RC}$. At this frequency the response is down by a factor of $\frac{1}{\sqrt{2}}$ from its value at $f = 0$, this corresponds to -3 dB down.

Bandlimiting at the High Frequency End

A high-pass RC filter can be used to bandlimit waveforms at the high-frequency end. This filter has a transfer function given by,

$$H(f) = \frac{1}{1 - jf_1/f} \qquad (4.16)$$

where $f_1 = \frac{1}{2\pi RC}$

The frequency f_1 is the -3 dB frequency of the network and is generally described as the low frequency cut-off of the network. In order not to introduce any distortion it is required as a rule of thumb that the product of the low cut-off frequency, f_1, and the time constant, $\tau = RC$ should be $f_1 \tau = 0.02$.

4.8 CORRELATION BETWEEN WAVEFORMS

Correlation measures the similarity between two waveforms. The correlation between two waveforms $v_1(t)$ and $v_2(t)$ referred to as the average cross correlation is defined as,

$$R_{12}(\tau) = \lim_{T \to \infty} \frac{1}{T} \int_{-T/2}^{T/2} v_1(t) v_2(t + \tau) dt \qquad (4.17)$$

where τ is a dummy variable.

Correlation in Periodic Waveforms

If $v_1(t)$ and $v_2(t)$ are periodic with the same fundamental period T_0, then the average cross correlation is defined as,

$$R_{12}(\tau) = \frac{1}{T_0}\int_{-T_0/2}^{T_0/2} v_1(t)v_2(t+\tau)\,dt \quad (4.18)$$

Correlation of Nonperiodic Waveforms

If $v_1(t)$ and $v_2(t)$ are waveforms of finite energy, then the cross correlation is defined as,

$$R_{12}(\tau) = \int_{-\infty}^{\infty} v_1(t)v_2(t+\tau)\,dt \quad (4.19)$$

If $R_{12}(\tau) = 0$ for all τ, then the two waveforms are said to be uncorrelated.
In general $R_{12}(\tau) \neq R_{21}(\tau)$. Actually, it can be verified that $R_{21}(\tau) = R_{12}(-\tau)$.

Power Correlation :

If S_1 and S_2 represent the normalized power of the waveforms $v_1(t)$ and $v_2(t)$, then the normalized power S_{12} of $v_1(t) + v_2(t+\tau)$ is given by,

$$S_{12} = S_1 + S_2 + 2R_{12}(\tau) \quad (4.20)$$

4.9 AUTOCORRELATION

The correlation of a function with itself is called autocorrelation.

In this case $v_1(t) = v_2(t)$ and the cross correlation becomes,

$$R(\tau) = \lim_{T \to \infty} \frac{1}{T} \int_{-T/2}^{T/2} v(t)v(t+\tau) \, dt \quad (4.21)$$

Properties of Autocorrelation

1. $R(0) = \lim_{T \to \infty} \frac{1}{T} \int_{-T/2}^{T/2} [v(t)]^2 \, dt = S \quad (4.22)$

   where S represents the average power.

2. $R(0) \geq R(\tau)$

3. $R(\tau) = R(-\tau)$

Autocorrelation of Periodic Waveforms :

When the function is periodic, the autocorrelation function can be expressed in terms of its exponential Fourier series coefficient as :

$$R(\tau) = |V_0|^2 + 2 \sum_{n=1}^{\infty} |V_n|^2 \cos 2\pi n \frac{\tau}{\tau_0} \quad (4.23)$$

Relation between R(τ) and Power Spectral Density

If we apply the Fourier transformation to Eq. (4.23), we see that

$$F[R(\tau)] = \int_{-\infty}^{\infty} \left(\sum_{n=-\infty}^{\infty} |V_n|^2 e^{j 2\pi n \tau / \tau_0} \right) e^{-j 2\pi f \tau} \, d\tau \quad (4.24)$$

and for periodic waveforms Eq. (4.24) becomes the power spectral density of that waveform, i.e.

$$F[R(\tau)] = G(f) \quad (4.25)$$

Autocorrelation of Nonperiodic Waveforms

For nonperiodic waveforms,

$$\boxed{F[R(\tau)] = V(f)V^*(f) = |V(f)|^2} \quad (4.26)$$

where $V(f)$ = Fourier transform of the waveform.
$V^*(f)$ = Complex conjugate of the Fourier transform
$|V(f)|^2$ = Energy spectral density

CHAPTER 5

TRANSFER FUNCTION AND FILTERING

5.1 CONCEPT OF TRANSFER FUNCTION

- Let $x(t)$ be the input for a system and $y(t)$ be the output of the system. The input-output relationship can be given by

$$a_n \frac{d^n y}{dt^n} + a_{n-1} \frac{d^{n-1} y}{dt^{n-1}} \ldots = b_m \frac{d^m x}{dt^m} + b_{m-1} \frac{d^{m-1} x}{dt^{m-1}} \ldots \quad (5.1)$$

where n is the order of the system.

- Now $f\left[\dfrac{d^n x}{dt^n}\right] = (j\omega)^n X(f)$. This can be applied to both sides of the above equation resulting in

$$\overline{Y}(f)\left[a_n(j\omega)^n + a_{n-1}(j\omega)^{n-1} \ldots\right] = \overline{X}(f)\left[b_m(j\omega)^m \ldots\right]$$

$$\boxed{H(f) = \text{Fourier transfer function} = \frac{Y(f)}{X(f)} = \text{Steady state transfer function}}$$

- Procedure for determining $H(f)$ from the circuit :

    i) Convert each inductor L, each capacitor C and each resistor R to $j\omega L$, $-j/\omega c$ and R respectively.

ii) The input is represented by $X(f)$ and the desired output-variable by an arbitrary Fourier transform $Y(f)$.

iii) Using circuit analysis methods, the ouput $Y(f)$ is expressed in terms of $x(f)$ and the transfer function is determined by $H(f) = Y(f)/X(f)$.

5.2 IDEAL FILTER

- Mathematical Approach

$y(t) + n(t) \rightarrow \boxed{\text{ideal filter}} \rightarrow x(t)$ where $n(t)$ is the undesired signal.

The purpose of the filter is to eliminate the undesired signal from the input signal. The ideal output can be given as
$x(t) = K_0 y(t - \tau)$ where τ is the delay.

$X(f) = K_0 \cdot e^{-j\omega\tau} \cdot Y(f)$ ... after taking the Fourier transform.

$H(f) = K_0 / -\omega\tau = K_0 / -2\pi f \tau$

So, for the ideal filter considered above,

> - Amplitude response = K_0 and
> - Phase response = $-2\pi f \tau$

- Characteristics :
  i) An ideal filter is characterized by a constant-amplitude response and
  ii) A linear phase response over the frequency spectrum of the input signal.

- Frequency Bands :

  i) Passband : It is a frequency range in which all frequencies contained in the signal are allowed to transmit with the least amount of attenuation.

ii) **Stopband** : In this band, maximum attenuation is provided to the signal.

iii) **Transition band** : It is a frequency band between passband and the stop band.

- **Characteristics curve :**

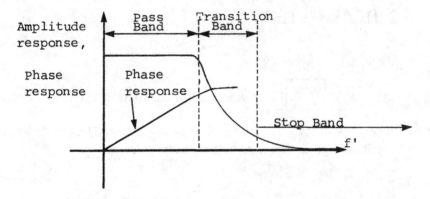

- **Phase and group delay :**

$$T_{ph}(f) = \frac{-\alpha(f)}{2\pi f} = \frac{\text{Phase response}}{\omega}$$

$$T_{gr}(f) = -\frac{d}{d\omega}\alpha(f)$$

Graphical representation

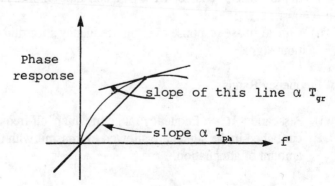

Important points :

 i) For an ideal filter, $T_{ph} = T_{gr} = \tau$ = exact delay of the signal.

 ii) Phase delay is the overall delay parameter and group delay represents the narrow-range delay.

5.3 RESPONSE OF AN IDEAL FILTER

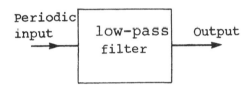

- Let the constant-amplitude and phase response be given by

 i) $A(f) = 1.0$ for $0 \le f < M\ f$.
    $= 0$ and

 ii) $\alpha(f) = -m\ 2\pi f_0 \cdot \tau_1$

- The periodic input signal and the output signal are given by

$$x(t) = \sum_{n=0}^{\infty} Cn \cdot \cos(n\omega_0 t + \phi_n) \quad (5.2)$$

and output $= \sum_{n=0}^{M} Cn \cdot \cos(n \cdot \omega_0 \cdot t + \phi_n - n \cdot 2\pi f_0 \cdot \tau_1)$ (5.3)

because the output is obtained by adding the input in the pass-band only.

- Filter response to common waveforms :

i) Square wave :

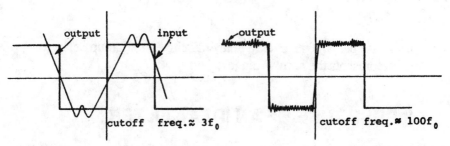

Fig: Response, considering up to third harmonic.

Fig: Components up to 99th harmonic

ii) Half-wave rectified sine wave :

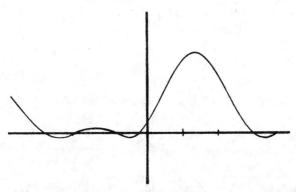

Fig: Fourier representation considering up to 3rd harmonics.

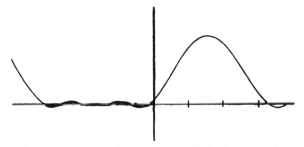

Fig: Considering up to 5th harmonic.

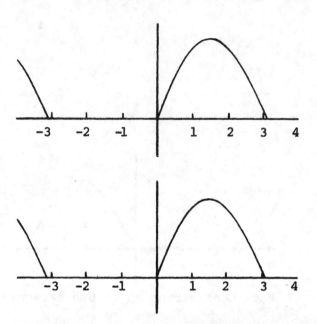

Fig: Fourier representation considering up to 30th & 100th harmonic respectively.

iii) Sawtooth pulse:

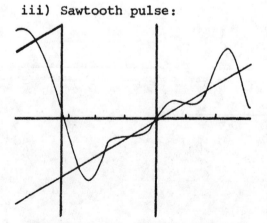

Fig: Fourier representation considering up to 3rd harmonic.

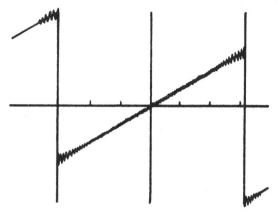

Fig: Fourier representation considering up to 100th harmonic.

5.4 FILTER APPROXIMATIONS

1) Butterworth response :

- This is also called "maximally flat response." The amplitude response of mth order is given by

$$A(f) = \frac{1}{\left[1 + \left(\frac{fc}{f}\right)^{-2m}\right]^{1/2}} \quad (5.4)$$

where f_c = cutoff frequency and corresponds to an attenuation of $M \cdot 3$ dB.

- Characteristics :
  i) As the order increases, the response tends to become " flatter " in the passband and there is a greater amount of attenuation in the stopband.

  ii) Beyond the cutoff, the amplitude response approaches an asymptote.

2) Chebyshev response :

The amplitude response is given by

$$A(f) = \frac{\gamma}{\left[\epsilon'^2 C_n^{\,2} \cdot \left(\frac{fc}{f}\right)^{-1} \right]^{\frac{1}{2}}} \quad (5.5)$$

where n represents the order of the corresponding transfer function.
fc_2 represents the cutoff frequency.
ϵ'^2 is a parameter which controls the ripples in the passband.

- Characteristics curves :

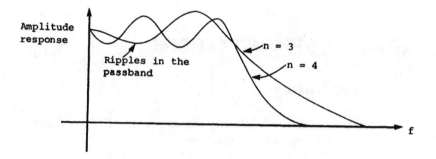

- Characteristics :

i) Total number of maxima and minima in the passband region is equal to the order of the filter.

ii) For a constant value of ϵ'_0, the attenuation increases in the stopband as the order of filter increases.

iii) For a given value of n, attenuation in the stopband increases as ripple increases.

3. Digital filter :

- A numerical algorithm that transforms an input data into an output data signal. These employ constant coefficient difference equations that relate the output stream to the input data stream.

5.5 TRANSMISSION OF ENERGY AND POWER SPECTRUM

- The concepts developed apply to both energy and power spectral density functions, and periodic and random power signals.

- For a system with transfer function H(f) :

$$Y(f) = H(f)X(f)$$

$$i.e. |Y(f)|^2 = |H(f)|^2 |X(f)|^2$$

The energy spectrum density function is $S(f')$ so

$S_x(f)$ = Input Energy Density Function = $X^2(f)$ and

$S_x(f') = X^2(f')$ so

$$\boxed{S_y(f) = (H(f))^2 S_x(f)} \qquad (5.6)$$

In this expression, $|H(f)|^2$ can be considered as a weighting function.

CHAPTER 6

RANDOM VARIABLES AND PROCESSES

6.1 DISCRETE RANDOM VARIABLES

Definition : A discrete random variable is a rule that maps any discrete point in its sample space on to a corresponding point on the real line. An example would be the rolling of fair dice.

6.2 SET THEORY

This is a useful tool in understanding Probability Theory.
Definition : A set is a collection of articles/things. Each article or thing is called an element, a member, or a point of the set.
For consideration, take a set comprised of all odd numbers between 0 and 12. This set is represented by the letter S where
$S = \{1, 3, 5, 7, 9, 11\}$ where the members are 1, 3, 5, 7, 9 and 11.
The rule of the above set example is : <u>all odd numbers between 0 and 12.</u>

<u>Universal Set</u> : Whenever a set contains all the components specified by the rule it is called the Universal Set, symbolized as S.

<u>Subset</u> : This is a set in which all of its elements are in the Universal Set.
e.g. $A = \{1, 3, 7\}$ $B = \{3, 7, 9\}$ $C = \{5, 7\}$ and $D = \{11\}$

Here A, B, C and D are all subsets of S above. The notation for

subset is the symbol ⊂ , read as "contained in." Therefore
$A \subset S$, $B \subset S$, $C \subset S$, and $D \subset S$.

Null Set : This is defined as a subset which does not contain any of the elements in the Universal Set. The symbol for a null set is ∅. Representation of the Universal Set, its subsets and elements can be graphically shown on a Venn Diagram. In this Figure reference is made to the set previously introduced.

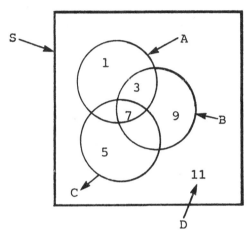

From the above Figure, Subsets B and C have over-lapping areas since element 7 is contained in both. On the other hand, subsets B and D have no common elements, and are referred to as mutually exclusive.

6.3 ALGEBRA OF SETS

Recall previously :

$S = \{1, 3, 5, 7, 9, 11\}$ $A = \{1, 3, 7\}$ $B = \{3, 7, 9\}$
$C = \{5, 7\}$ and $S = \{11\}$

Sum or Union : The sum or union of subsets B and C is given the notation $B + C$ or $B \cup C$, interpreted as "B or C". This results in a new set comprising all the elements of B and C,
i.e. $B + C = B \cup C = \{3, 5, 7, 9\}$ This relation is shown in the Venn diagrams on the next page.

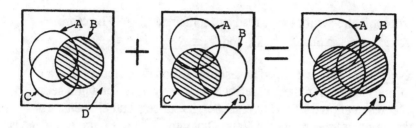

Intersection or Product : The intersection or product of subsets B and C is given the notation BC or $B \cdot C$ or $B \cap C$ and is read as "B and C". This is shown in the following diagrams :

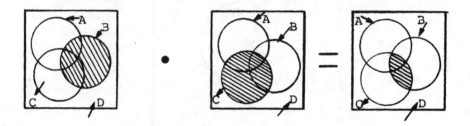

Difference : The difference of the subsets B and C is written as $B - C$ and is read as "B minus C". This relation is defined as a new set having the elements of B which are not in C, as shown :

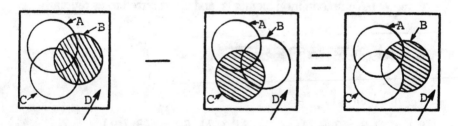

Complement : The complement of subset B, symbolized as $\overline{B}$, and read as "not B", is a new set comprising all elements of the Universal Set that are not in B.

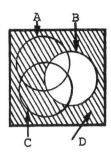

6.4 PROBABILITY THEORY

An assimilation of Set Theory and Probability Theory is given in the table below

Set Theory	Probability Theory
Universal Set S ↔	Sample Space
Subsets A, B, C & D ↔	Events A, B, C & D
Elements 1, 2, 3, 4 ↔	Outcomes $0_1 0_2 0_3 0_4$

<u>Marginal Probability</u> : This is defined as the ratio of the number of ways an event (say A) can occur to the number of ways the experiment can occur.

i.e. $\quad P(A) = \dfrac{n(A)}{N}$ where ideally $N \to \infty$ \quad (6.1)

<u>Joint Probability</u> : This is the probability of two events A and B (which are not mutually exclusive) occuring together.

i.e. $\quad P(AB) = \dfrac{n(AB)}{N}$ \quad (6.2)

The probability that either event A or B will occur is given by

$$P(A + B) = P(A) + P(B) - P(AB) \quad (6.3)$$

Conditional Probability : This is defined as the probability of an event A under the condition that an event B has occured. This is given by the relation

$$P\left(\frac{A}{B}\right) = P(AB)/P(B) \tag{6.4}$$

Bayes' Theorem : This is a special application of Conditional Probability. Let E = Universal Set, A, B, C = events each having outcomes $G, R,$ and W. If an outcome R occurred, it may be desirable to know if it was due to event A occurring. The probability that it was due to A occurring is :

$$P_{A/R} = \frac{P(A)P(R/A)}{P_{(A)}P(R/A) + P(B)P(R/B) + P(C)P(R/C)}$$

Note: A, B, and C are mutually exclusive and exhaustive.

This theorem can be applied to an experiment having any number of events which are mutually exclusive and exhaustive.

Statistical Independence : This is defined as the occurrence of an event A that is independent of the occurrence of another event B.

i.e. $\qquad P_{(AB)} = P(A)P(B)$

A and B are statistically independent.

General Formula :

$$P(A_i A_j A_k \ldots A_n) = P(A_i)P(A_j)P(A_k) \ldots P(A_n)$$
where $1 \leq i < j < k \ldots \leq n$

6.5 PROBABILITY DENSITY FUNCTION (PDF)

This is a plot of the probability that a random variable x will take on a value x_i, versus the values x_j, over a given sample space.

Binomial PDF
$$f_x(x) = \binom{n}{k} p^k (1-p)^{n-k} \delta(x-k)$$

Where n = number of events of the experiment, k = outcomes of the events and p = probability of the outcomes.

Poisson's PDF: This is obtained from the Binomial PDF under the conditions a) $n >> k$ b) $p << 1$ c) $np = \lambda$ where:

$$f_x(x) = \sum_{k=0}^{\infty} \frac{e^{-\lambda}\lambda^k}{k!} \delta(x-k)$$

Moments: These serve to give an overall behavior of a random variable but cannot predict exactly the next outcome of an experiment performed repeatedly.
Note: These are for discrete random variables.

1st Moment M_1: This is the mean value of the random variable.

$$M_1 = \sum_i x_i P(x_i)$$

where x_i is the outcome of the experiment and

$p(x_i)$ is the probability of the outcome.

2nd Moment M_2: This is the mean value of the square of the random variable (x)

$$M_2 = \sum_i x_i^2 P(x_i)$$

Second Central moment μ^2   This is the mean value of $(x - M_1)^2$

$$\mu_2 = \sum_i (x_i - M_1)^2 (px_i)$$

Characteristic Function $\phi_x(T)$: This is defined as the expected value of the function e^{jTx} where T is a dummy variable.

$$\phi_x(T) = E(e^{jTx}) = \sum_i e^{jTx_i} P(x_i)$$

where i covers all outcomes of the experiment.
where $j = \sqrt{-1}$ and $-\infty < t < \infty$

6.6 CONTINUOUS RANDOM VARIABLES AND PROCESSES

<u>Member Function</u> : Is a random variable which is a member of a set of random variables obtained by repeated experiments of a random process.

<u>Ensemble Averaging</u> : An arbitrary time (t_i) is chosen at which the values of all member functions $(f_1(t_i), f_2(t_i), f_3(t_i), \ldots f_n(t_i))$ are averaged.

<u>Time Averaging</u> : Any member function $(f_2(s))$ is chosen and its value is averaged for all time at $\Delta t_1, \Delta t_2, \Delta t_3 \ldots \Delta t_n$.

<u>Ergodicity</u> : The statistical averages of the random process are the same as the time averages.

6.7 ONE-DIMENSIONAL PROBABILITY DENSITY FUNCTION (PDF)

For Ensemble Averaging :

$$P\left[f_{b\,1} < f_\theta(s_1) < f_{b\,2}\right] = \frac{n_s}{N_B} = \int_{f_{b1}}^{f_{b2}} P(f)\,df \ ,$$

Here θ indicates the member function considered

$$= \frac{n_s}{N_B} \frac{\text{number of member functions in interval}}{\text{Total number of member functions}}$$

$P(f) = PDF$
$f_{b\,1}$ and $f_{b\,2}$ are points on the interval.

For Time Averaging : The same equation applies except that $\dfrac{n_s}{N_B}$ becomes $\dfrac{t_p}{T_W}$ See Figure 2

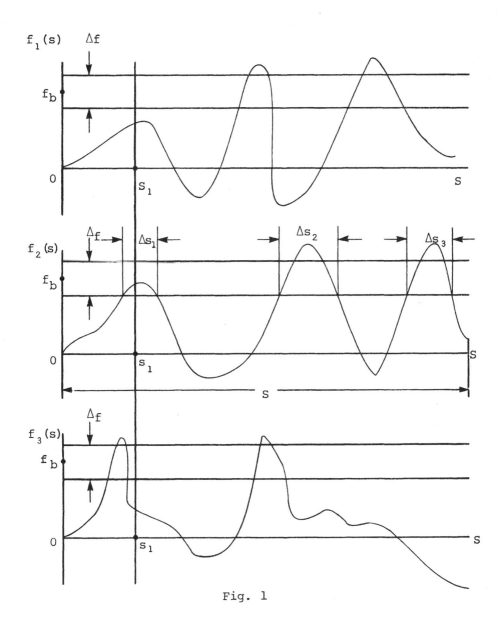

Fig. 1

In Fig. 1 : $f_1(s)$ $f_2(s)$ and $f_3(s)$ are member functions.
The member functions : $f_1(s)$, $f_2(s)$ and $f_3(s)$ are used to determine $f(s)$ by Ensemble Averaging at time S_1
The member function $f_2(s)$ is used to determine $f(s)$ by Time Averaging at times Δs_1, Δs_2, and Δs_3.

a)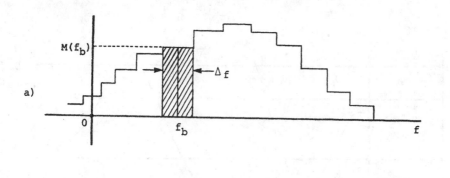

b)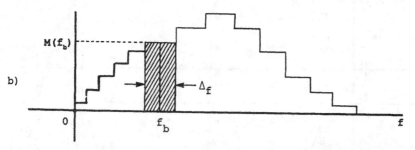

*The shaded area indicates:
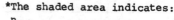
$$\frac{n_S}{N_B} = \frac{\text{\# of member functions in } \Delta_f}{\text{Total \# of member functions}}$$

c)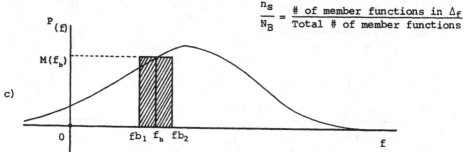

For equal intervals Δf along f, a number of different areas will result as shown in (a) above.

As $\Delta f \to 0$, the Probability Density Curve is shown in c)

* The same approach is used for Time Averaging where the shaded area indicates

$$\frac{\text{Time during which member function lies in} \Delta f}{\text{Total time } S} = \frac{t_p}{T_w}$$

6.8 MOMENTS FOR ONE-DIMENSIONAL RANDOM VARIABLES

First Moment M_{F1}: This is the average or expected value of the random variable (t).

$$M_{F1} = \int_{-\infty}^{\infty} t\, p(t)\, dt \qquad (6.6)$$

Note : $p(t)$ is the probability density function of T.

Second Moment Ms_2: This is the average value of the square of the random variable (t) or simply the variation of t at its origin.

$$Ms_2 = \int_{-\infty}^{\infty} t^2 p(t)\, dt \qquad (6.7)$$

Second Central Moment Mc_2 : This is the average value of $(t - M_{F1})^2$

$$M_{c2} = \int_{-\infty}^{\infty} (t - M_{F1})^2 p(t)\, dt$$

this is referred to also as the variance of t.

Moments obtained from : 1) Ensemble and 2) Time Averaging

1) a) $M_{F1} = \dfrac{1}{k}\sum\limits_{a=1}^{k} t_a(t_1)$ b) $M_{F2} = \dfrac{1}{k}\sum\limits_{a=1}^{k} t_a^2(t_1)$

c) $M_{c2} = \dfrac{1}{k}\sum\limits_{a=1}^{k} [t_a(t_1) - M_{F1}]^2$

Subscript a represents all member functions of variable t.

2) a) $M_{F1} = \langle t_a(t) \rangle = \dfrac{1}{2t_x}\displaystyle\int_{-t_x}^{t_x} t_a(t)\,dt$ where $t_x \to \infty$ for ideal conditions.

b) $M_{F2} = \langle t_a^{\,2}(t) \rangle = \dfrac{1}{2t_x}\displaystyle\int_{-t_x}^{t_x} t_a^{\,2}(t)\,dt$

c) $M_{c2} = \langle [t_a(t) - m_{F1}]^2 \rangle = \dfrac{1}{2t_x}\displaystyle\int_{-t_x}^{t_x} [t_a(t) - M_{F1}]^2\,dt$

6.9 STATISTICAL AVERAGES FOR ONE-DIMENSIONAL RANDOM VARIABLES

<u>Moments</u> : These give an overall behavioral pattern of the random variable.

<u>1st Moment M_{x1}</u>: This is defined as the average value of the random variable x.

$$M_{x1} = E[x] = \int_{-\infty}^{\infty} x f_x(x)\,dx$$

where x is the different values of the random variable X taken one at a time. $f_x(x)$ is the probability density function E reads as the "expectation operator."

<u>2nd Moment Mx_2</u> : This is defined as the mean-square value of x.

$$Mx_2 = E[x^2] = \int_{-\infty}^{\infty} x^2 f_x(x)\,dx$$

<u>Second Central Moment σ_x^2</u> : This is defined as the variance of X.

$$\sigma_x^2 = E\left[(x - Mx_1)^2\right] = \int_{-\infty}^{\infty} (x - Mx_1)^2 f_n(x)\,dx$$

Moments by a) Ensemble and b) Time Averaging

a) i) $Mx_1 = \dfrac{1}{k}\sum\limits_{a=1}^{k} x_a(t_1)$ ii) $Mx_2 = \dfrac{1}{k}\sum\limits_{a=1}^{k} x_a^2(t_1)$

iii) $\sigma_x^2 = \dfrac{1}{k}\sum\limits_{a=1}^{k} [x_a(t_1) - Mx_1]^2$

Subscript 'a' entails all member functions of the process, taken simultaneously at time t_1.

b) $Mx_1 = \dfrac{1}{2t_0}\int_{-t_0}^{t_0} x_a(t)\,dt \quad Mx_2 = \dfrac{1}{2t_0}\int_{-t_0}^{t_0} x_a^2(t)\,dt$

$\sigma_x^2 = \dfrac{1}{2t_0}\int_{-t_0}^{t_0} [x_a(t) - Mx_1]^2 dt$

Here time $t_0 \to \infty$ to make sure that $x_a(t)$ is sampled throughout.

6.10 CORRELATION FUNCTIONS

These are of two types :
a) <u>Autocorrelation</u> : For a single random process, this is defined as

$$R_{ff}(\tau) = \dfrac{1}{2t_a}\int_{-t_a}^{t_a} f(t)*(t - \tau)\,dt$$

$= E[f(t)f(t+\tau)]$ where $t \to \infty$
where $f(t)$ is the random variable.

b) <u>Cross correlation</u> : This is defined as

$$R_{xy}(t_1,t_2) = E[x(t_1)y(t_2)] = R^*_{yx}(t_2,t_1)$$

where $x(t)$ and $y(t)$ are the two processes observed at times t_1 & t_2.

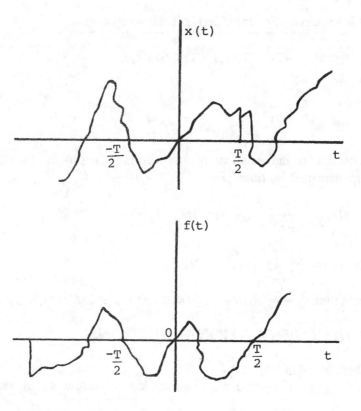

Wiener-Khintchine Theorem

This theorem states that there is an existence of a Fourier transform pair between the power spectral density $S_{xx}(\omega)$ and the autocorrelation function $R_{ff}(\tau)$ for the random variable $x(t)$.
Periodic function f(t) is formed from x(t) between $-\dfrac{T}{2} < t < \dfrac{T}{2}$

$$S_{xx}(\omega) = \int_{-\infty}^{\infty} R_{ff}(\tau) e^{-j\omega t} d\tau = F[R_{ff}(\tau)] \quad (6.8)$$

6.11 POWER SPECTRA DETERMINATION

This is obtained by analysing the behavior of a random variable to determine the spectral density $S_{xx}(\omega)$

Where $S_{xx}(\omega) = 4\pi M_{x1}^2 \delta(\omega) + b\sigma_x^2 \left[\dfrac{\sin b\,\omega/2}{b\dfrac{\omega}{2}}\right]^2$

and *b* is a series of equal intervals along the time axis.

Transmission of Power Spectra

This shows the relationship between the input and output spectral densities

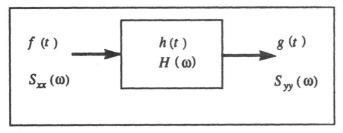

where h(t) is the unit impulse response and $H(\omega)$ the transfer function.

$$S_{yy}(\omega) = |H(j\omega)|^2 S_{xx}(w)$$

Transformation of APDF

This is the transformation of a random variable, where one is obtained from the other. Say *y* is obtained from *x* where *y* is considered to be a monotone-increasing differentiable function *g* of the random variable *x*.

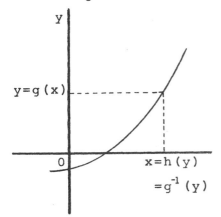

$y = g(x)$
The probability density function of the new variable y is:
$$f_y(y) = f_x[h(y)] \left|\dfrac{dh}{dy}\right|$$

6.12 THE GAUSSIAN PROBABILITY DENSITY

The gaussian or normal probability density function is defined as

$$F(x) = \frac{1}{\sqrt{2\pi\sigma^2}} e^{-(x-m)^2/2\sigma^2} \qquad (6.9)$$

where m = average value of the function.
σ^2 = variance of the function.
This density function is of spectral relevance to us, as many random phenomena in communication theory are exclusively characterized by this function.

The Central-Limit Theorem

This theorem indicates that the probability density of a sum of N independent random variables tends to approach a gaussian density as the number N increases. The mean and variance of this gaussian density are respectively the sum of the means and the sum of the variances of the N independent random variables.

6.13 SUM OF RANDOM VARIABLES

If x and y are two random variables with mean Mx, My and variance σ_x^2, σ_y^2, then $z = x + y$ is also a random variable with

$$\text{Mean} = M_z = M_x + M_y \qquad (6.10)$$

Also, if x and y are independent random variables, then the variance of z is

$$\sigma_z^2 = \sigma_x^2 + \sigma_y^2 \qquad (6.11)$$

The probability density function of z, provided x and y are independent is given by

$$f(z) = \int_{-\infty}^{\infty} f(x) f(z-x) dx \qquad (6.12)$$

Note that $f(z)$ is simply the convolution of $f(x)$ and $f(y)$.

CHAPTER 7

AMPLITUDE MODULATION

7.1 BASIC PRINCIPLES

In amplitude modulation, the baseband signal is translated to a higher frequency and then carried over the channel at that frequency. The translated signal is given by

$$v(t) = A_c[1 + m(t)]\cos\omega_c t \qquad (7.1)$$

where A_c = Carrier Amplitude
$m(t)$ = Baseband signal
ω_c = Carrier frequency

Note, that the waveform in Eq. (7.1) is modulated in amplitude.
Maximum Allowable Modulation

When the modulating signal is sinusoidal ; hence $m(t)$ in Eq. (7.1) is $M\cos\omega_m t$, then Eq. (7.1) becomes,

$$v(t) = A_c(1 + M\cos\omega_m t)\cos\omega_c t \qquad (7.2)$$

where M = a constant

In order to recover the signal at the receiver, it is required that $|M| \leq 1$. More generally we require in Eq.(7.1) that the maximum negative excursion of $m(t)$ be -1. In Fig. (7.1) a sinusoidally modulated carrier is shown for m > 1 and m < 1.

The extent to which a carrier has been amplitude-modulated is expressed in terms of a percentage modulation. If the modulation is symmetrical, the percentage modulation is defined as P, given by

$$\boxed{\frac{P}{100\%} = \frac{A_{c\,max} - A_c}{A_c} = \frac{A_c - A_{c\,min}}{A_c} = \frac{A_{c\,max} - A_{c\,min}}{2A_c}}$$
(7.3)

where
A_c = unmodulated carrier amplitude
$A_{c\,max}$ = maximum carrier aplitude
$A_{c\,min}$ = minimum carrier amplitude

In case of sinusoidal or tone modulation, $P = m \times 100$ percent.

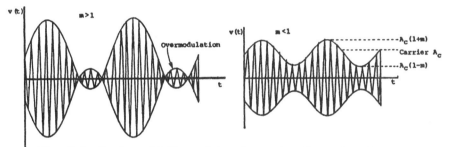

Fig. 7.1 A sinusoidally modulated carrier for m>1 and m<1.

Spectrum of an AM Signal

For the bandlimited nonperiodic baseband signal of finite energy, the spectrum is shown in Fig. 7.2 (a). If the baseband signal $m(t)$ is the super position of two sinusoidal components, i.e.
$m(t) = m_1 \cos 2\pi f_1 t + m_2 \cos 2\pi f_2$, then the one-sided spectrum appears in Fig. 7.2(b). The upper-sideband and lower-sideband are indicated in the figure.

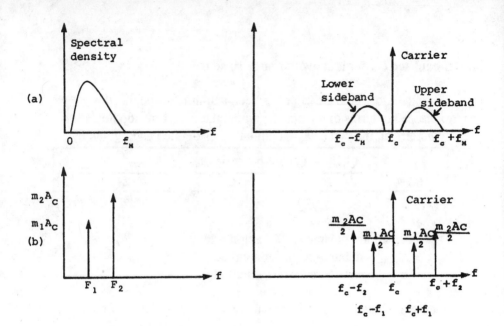

7.2 SQUARE-LAW DEMODULATOR

The baseband signal $m(t)$ which is amplitude modulated on a carrier, can be recovered by passing it through a nonlinear square-law device. The input-output relation for this device is,

$$y = kx^2 \tag{7.4}$$

where k is a constant.

If the applied signal to the demodulator is

$$x = A_o + A_c[1 + m(t)]\cos\omega_o t \tag{7.5}$$

then the output signal is,

$$y = \{A_o + A_c[1 + m(t)]\cos\omega_c t\}^2 \tag{7.6}$$

After dropping the ac terms and passing through a low-pass filter, the output of the device is,

$$S_0(t) = KA_c^2\left[m(t) + \frac{1}{2}m^2(t)\right] \tag{7.7}$$

Now, $m(t)$ can be recovered with some distortion.

7.3 BALANCED MODULATOR AND DOUBLE-SIDEBAND MODULATION

- A double-sideband signal is generated by multiplying the baseband modulating signal $y(t)$ by a very high-frequency signal $\cos \omega_c t$. This is done in a balanced modulator as shown below.

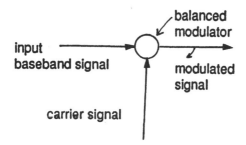

where $\omega_c \triangleq$ carrier frequency.

$$\text{Modulated signal } V_m(t) = m(t) \cdot \cos \omega_c t.$$

The magnitude of the carrier is unity.

$$V_m(t) = \frac{1}{2} m(t) e^{2\pi j \cdot f_c \cdot t} + \frac{1}{2} m(t) e^{-2\pi j \cdot f_c \cdot t} \quad (7.8)$$

$$\text{and } V_m(f) = \frac{1}{2} M(f - f_c) + \frac{1}{2} M(f + f_c) \quad (7.9)$$

- Generation of a DSB signal :

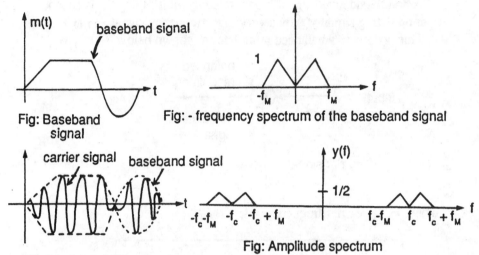

Fig: Baseband signal

Fig: - frequency spectrum of the baseband signal

Fig: Modulated Signal

Fig: Amplitude spectrum

Transmission bandwidth :

> The minumum transmission bandwidth for a DSB signal = 2w

i.e. The bandwidth for a DSB signal is twice the highest-modulating frequency.

The following figure illustrates a balanced modulator, its output $V_o(t)$ is a double-sideband suppressed-carrier signal.

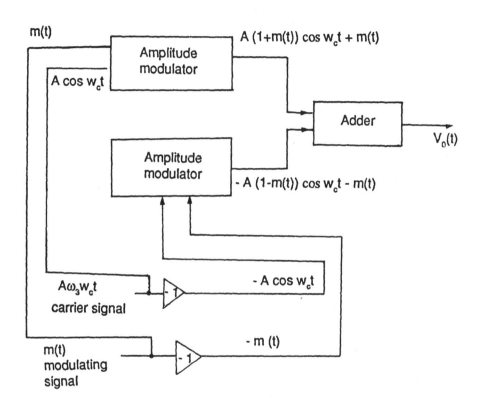

$$V_o(t) = 2m(t)A \cos \omega_c t \qquad (7.10)$$

7.4 SINGLE-SIDEBAND MODULATION

- Generation of a single-sideband signal :
  The single-sideband signal can be generated from a double-sideband signal by eliminating one sideband and transmitting the other sideband.

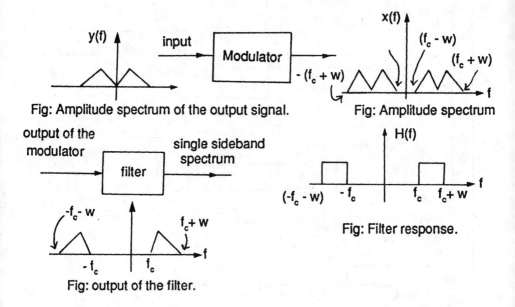

Fig: Amplitude spectrum of the output signal.
Fig: Amplitude spectrum
Fig: output of the filter.
Fig: Filter response.

Transmission bandwidth :

$$B_T = w$$

where w is the baseband bandwidth.

CHAPTER 8

FREQUENCY MODULATION

8.1 ANGLE MODULATION

If $\phi(t)$ is a function of the baseband signal, then the function defined as $v(t) = A \cos(\omega_c t + \phi(t))$ (where A and ω_c are constants) has a waveform that will oscillate between the limits of $+A$ and $-A$ as the frequency of the oscillations are dependent on the phase $\phi(t)$. Modulation of this type is called angle modulation.

8.2 PHASE AND FREQUENCY MODULATION

The two forms of angle modulation that we are interested here are phase modulation (PM) and frequency modulation (FM).
From the angle modulation we have

$$v(t) = A \cos(\omega_c t + \phi(t)) \qquad (8.1)$$

where $\phi(t)$ is a controlled function of a given baseband signal. The instantaneous total argument of $v(t)$ is $\omega_c t + \phi(t)$. Thus, $v(t)$ can be thought of as obtained by modulating the phase angle of $f(t) = A \cos \theta(t) = A \cos \omega_c t$ (namely $\theta(t) = \omega_c t$) by $\phi(t)$

$$\theta(t)_{modulated} = \theta(t) + \phi(t)$$
$$= \omega_c t + \phi(t),$$

and

$$v(t) = A \cos \theta(t)_{modulated}$$
$$= A \cos(\omega_c t + \phi(t))$$

Hence, $v(t)$ is a representation of a signal $f(t)$ which is modulated in phase. On the other hand, since the angular velocity associated with $v(t)$ is

$$\omega = \frac{d[\omega_c t + \phi(t)]}{dt} = \omega_c + \frac{d\phi(t)}{dt}$$

and the corresponding frequencies of ω_c and ω are, respectively,

$$f_c = \frac{\omega_c}{2\pi} \text{ and } f = \frac{\omega_c}{2\pi} + \frac{1}{2\pi}\frac{d\phi(t)}{dt} \quad (8.2)$$

v(t) is, thus modulated in frequency.
(i.e, v(t) is modulated by $\frac{1}{2\pi}\frac{d\phi(t)}{dt}$,

$$f \text{ modulated} = f = f_c + \frac{1}{2\pi}\frac{d\phi(t)}{dt}$$

$$= \frac{\omega_c}{2\pi} + \frac{1}{2\pi}\frac{d\phi(t)}{dt} \quad (8.3)$$

If we let the baseband (or modulating) signal be $m(t)$. Then by making $\phi(t)$ proportional to $m(t)$, i.e., $\phi(t) = \phi_p(t) = k_m(t)$ where k is a constant, we obtain the phase modulated (PM) waveform

$$v(t) = A\cos(\omega_c t + \phi_p(t))$$
$$= A\cos(\omega_c t + km(t))$$

whereas, by setting $\phi(t) = \phi_f(t) = k\int_{-\infty}^{t} m(\tau)d\tau$,

we obtain the frequency modulated (FM) waveform,

$$v(t) = A\cos(\omega_c t + \phi_f(t))$$
$$= A\cos\left(\omega_c t + k\int_{-\infty}^{t} m(\tau)d\tau\right)$$

Note, for FM, the frequency of $v(t)$ is

$$f = \frac{\omega}{2\pi} = \frac{d}{dt}\left[\frac{\omega_c t + \phi_f(t)}{[2\pi]}\right] \quad (8.4)$$

$$= \frac{\omega_c}{2\pi} + \frac{1}{2\pi}\frac{d\phi_f(t)}{dt}$$

$$= \frac{\omega_c}{2\pi} + \frac{1}{2\pi}\frac{d\left[k\int_{-\infty}^{t} m(\tau)d\tau\right]}{dt}$$

$$= \frac{\omega_c}{2\pi} + \frac{k}{2\pi}m(t)$$
$$= \frac{\omega_c}{2\pi} + k'm(t)$$
$$= f_c + k'm(t), \quad k' = \frac{k}{2\pi},$$

and $\phi_f(t)$ is made such that the deviation of the instantaneous frequency f from the carrier frequency f_c is proportional to the baseband signal $m(t)$.

That is, $f - f_c = \frac{k}{2\pi}m(t)$

Phase deviation is the maximum phase deviation of the total angle from the carrier angle $\omega_c t$, $(\phi_p(t))_{max}$.

Frequency deviation is the maximum departure of the instantaneous frequency from f_c, $(f - f_c)_{max}$.

For example, if $\phi_p(t) = \beta \sin \omega_m t$, $(\phi_p(t))_{max} = \beta$ and β is called the modulation index.

Further, if $v(t) = A \cos(\omega_c t + \beta \sin \omega_m t)$

The instantaneous frequency for $v(t)$ is

$$f = \frac{1}{2\pi} \frac{d[\omega_c t + \beta \sin \omega_m t]}{dt} \quad (8.5)$$
$$= \frac{\omega_c}{2\pi} + \frac{\beta \omega_m}{2\pi} \cos \omega_m t$$
$$= f_c + f_m \cos \omega_m t$$

The maximum frequency deviation is $(f - f_c)_{max} = \frac{\beta \omega_m}{2\pi}$

since $\frac{\omega_m}{2\pi} = f_m$, if we denote $(f - f_c)_{max}$ by Δf, we have

$\Delta f = \beta f_m$ or $\beta = \frac{\Delta f}{f_m}$.

The following diagram illustrates the relationship between phase and frequency modulation.

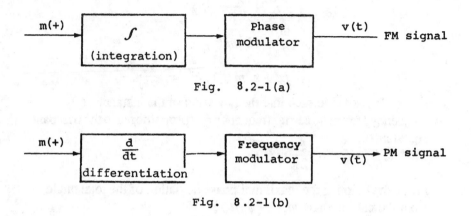

Fig. 8.2-1(a)

Fig. 8.2-1(b)

8.3 BANDWIDTH AND SPECTRUM OF AN FM SIGNAL

Consider a sinusoidally modulated FM signal
$v(t) = A \cos(\omega_c t + k \sin \omega_m t)$ where the modulating signal is $\sin \omega_m t$

For convenience we set A = 1, and we have

$$v(t) = \cos(\omega_c t + \beta \sin \omega_m t) \qquad (8.6)$$

$$= [\cos \omega_c t \cos(\beta \sin \omega_m t) - \sin \omega_c t \sin(\beta \sin \omega_m t)]$$

$$= \left[\cos \omega_c t \left[J_0(\beta) + \sum_{\substack{n=2 \\ n \text{ even}}}^{\infty} 2J_n(\beta) \cos n\omega_m t\right]\right.$$

$$\left. - \sin \omega_c t \left[\sum_{\substack{n=1 \\ n \text{ odd}}}^{\infty} 2J_n(\beta) \sin n\omega_m t\right]\right]$$

$$= \left[J_0(\beta)\cos \omega_c t + \cos \omega_c t \sum_{\substack{n=2 \\ n \text{ even}}}^{\infty} 2J_n(\beta) \cos n\omega_m t\right.$$

$$\left. - \sin \omega_c t \sum_{\substack{n=1 \\ n \text{ odd}}} 2J_n(\beta) \sin n\omega_m t\right]$$

$$= \left[J_o(\beta)\cos\omega_c t + \sum_{\substack{n=2 \\ n \text{ even}}}^{\infty} 2J_n(\beta)\cos\omega_c t \cos n\omega_m t \right.$$

$$\left. - \sum_{\substack{n=1 \\ n \text{ odd}}}^{\infty} 2J_n(\beta)\sin\omega_c t \sin n\omega_m t \right]$$

By using the identities

$$\cos\alpha \cos\gamma = \frac{1}{2}\cos(\alpha - \gamma) + \frac{1}{2}\cos(\alpha + \gamma)$$

$$\sin\alpha \sin\gamma = \frac{1}{2}\cos(\alpha - \gamma) - \frac{1}{2}\cos(\alpha + \gamma)$$

we obtain,

$$v(t) = J_o(\beta)\cos\omega_c t + \sum_{n=2}^{\infty} 2J_n(\beta)\left(\frac{1}{2}(\cos(\omega_c t - n\omega_m t) + \cos(\omega_c t + n\omega_m t))\right)$$

$$- \sum_{n=1}^{\infty} 2J_n(\beta)\left[\frac{1}{2}[\cos(\omega_c t - n\omega_m t) - \cos(\omega_c t + n\omega_m t)]\right]$$

$$= J_o(\beta)\cos\omega_c t + \sum_{\substack{n=2 \\ n \text{ even}}}^{\infty} J_n(\beta)[\cos(\omega_c - n\omega_m)t + \cos(\omega_c + n\omega_m)t]$$

$$- \sum_{\substack{n=1 \\ n \text{ odd}}}^{\infty} J_n(\beta)[\cos(\omega_c - n\omega_m)t - \cos(\omega_c + n\omega_m)t] \quad (8.7)$$

where $J_n(\beta)$ represents a Bessel function of the first kind, of order n and argument β. $J_n(\beta)$ has a unique real value for each given n and β value. By examining Eq. (8.7), we observe that the spectrum of this FM signal is composed of a carrier ($\cos\omega_c t$) with an amplitude $J_o(\beta)$ and a set of sidebands spaced symmetrically on either side of the carrier at frequency separations of ω_m, $2\omega_m$, $3\omega_m$, $4\omega_m$, $5\omega_m$ ...etc. Spectral frequencies greater than f_c represent the upper sideband components whereas the lower sideband components are represented by spectral frequencies less than f_c. Theoretically Eq. (1) suggests that there are an infinite number of sidebands in the spectrum of the FM signal, thus one would need an infinite bandwidth to transmit such a signal. Practically, however, due to the properties of the Bessel function $J_n(\beta)$, the various terms in Eq. (1) diminish rapidly beyond a certain point in the series for a given β so that the spectrum of Eq. (1) can be approximated quite accurately to a finite, practical bandwidth. Experiments have shown that the error resulting from bandlimiting for an FM signal would be tolerable as long as 98 % or more of the power is passed by the bandlimiting filter. This is the basis of one of the procedures for estimating transmission bandwidth for FM, which is referred to as Carson's rule. Carson's rule can be expressed as

$$B = 2(\Delta f + f_m) \text{ or} \quad (8.8)$$
$$B = 2(\beta + 1)f_m \quad (8.9)$$

where β is the bandwidth required to receive or transmit the FM signal with sinusoidal modulation, f_m is the modulating frequency, β is the modulation index and $\Delta f = \beta f_m$ is the maximum frequency deviation.

The bandwidth for narrow-band FM (NBFM) is $B \doteq 2f_m$. Similarly, for $B >> 1$ we obtain

$$B = 2(\beta + 1)f_m = 2\beta f_m$$

which is the bandwidth for wideband FM (WBFM). When β is not much greater (or much less) than 1 the bandwidth is approximated using

$$B = 2(1 + \beta)f_m \text{ or } B = 2(\Delta f + f_m)$$

Bandwidth Requirement For a Gaussian Modulated WBFM Signal :

When the carrier is modulated by a gaussian distribution, the resulting spectral density $G(f)$ will also be gaussian and is given by

$$G(f) = \frac{A^2}{4\sqrt{2\pi}\,\Delta f_{rms}}\left[e^{-(f-f_c)^2/2(\Delta f_{rms})^2} + e^{-(f+f_c)^2/2(\Delta f_{rms})^2}\right] \quad (8.10)$$

where f_c = mean value of distribution.

$(\Delta f_{rms})^2 \triangleq$ variance $\triangleq \sigma^2$

The required bandwidth B of the bandpass filter centered at f_c which will pass 98 % of the power of the waveform, is obtained from

$$.98 = \frac{2}{\sqrt{19}} \int_0^{B/2\sqrt{2}\Delta f_{rms}} e^{-x^2} dx$$

or $B = 2\sqrt{2}(1.645)\Delta f_{rms} = 4.6\Delta f_{rms}$

8.4 BESSEL FUNCTIONS

$$x^2 y'' + xy' + (x^2 - n^2)y = 0, n \geq 0 \qquad (8.11)$$

is called Bessel's differential equation whose solutions are called Bessel functions of order n.

$$J_n(x) = \sum_{k=0}^{\infty} \frac{(-1)^k \left(\frac{x}{2}\right)^{n+2k}}{k! \Gamma(n+k+1)}$$

is called Bessel functions of the first kind of order n.
We can show that

$$\cos(\beta \sin \omega_m t) = J_0(\beta) + \sum_{\substack{n=2 \\ \text{even}}}^{\infty} 2J_n(\beta)\cos n\, \omega_m t$$

$$= J_0(\beta) + 2J_2(\beta)\cos 2\omega_m t + 2J_4(\beta)\cos 4\omega_m t$$

$$+ \ldots + 2J_{2n}(\beta)\cos 2n\, \omega_m t + \ldots$$

and

$$\sin(\beta \sin \omega_m t) = \sum_{\substack{n=1 \\ \text{odd}}}^{\infty} 2J_n(\beta)\sin n\, \omega_m t$$

$$= 2J_1(\beta)\sin \omega_m t$$

$+ 2J_3(\beta)\sin 3\omega_m t + \ldots + 2J_{2n-1}(\beta)\sin(2n-1)\omega_m t + \ldots$

Furthermore,

$$J_o(\beta) = 1 - \frac{\beta^2}{2^2} + \frac{\beta^4}{2^2 \cdot 4^2} - \frac{\beta^6}{2^2 \cdot 4^2 \cdot 6^2} + \ldots$$

$$J_1(\beta) = \frac{\beta}{2} - \frac{\beta^3}{2^2 \cdot 4} + \frac{\beta^5}{2^2 \cdot 4^2 \cdot 6} - \frac{\beta^7}{2^2 \cdot 4^2 \cdot 6^2 \cdot 8} + \ldots$$

and for $\beta \ll 1$

$$J_o(\beta) \doteq 1 - \left(\frac{\beta}{2}\right)^2 \qquad (8.12)$$

$$J_n(\beta) \doteq \frac{1}{n!}\left(\frac{\beta}{2}\right)^n \qquad n \neq 0 \qquad (8.13)$$

The curves of $J_n(\beta)$ for $n = 0, 1, 2, 3, 4,$ and 5 are shown in the figure below.

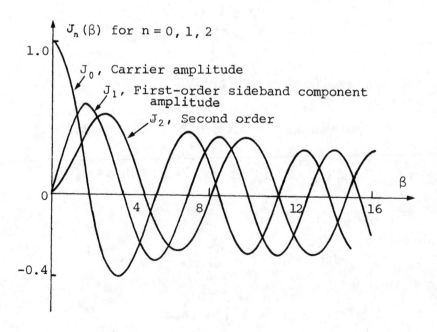

D-77

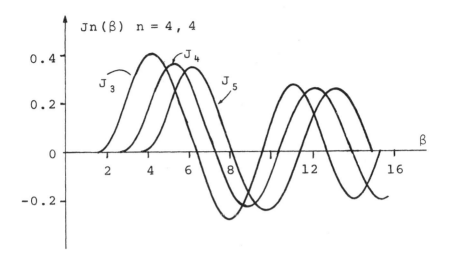

The values of $J_n(\beta)$ can also be found in texts of mathematical tables. The following table gives values of $J_n(\beta)$ for some n and β values.

n \ β	0.1	0.5	1	2	3	4	5	6	7	8	9	10
0	.9975	.9385	.7652	.2239	-.2601	-.3971	-.1776	.1506	.3001	.1717	-.09033	-.2459
1	.0499	.2423	.4401	.5767	.3391	-.06604	-.3276	-.2767	-.004683	.2346	.2453	.04347
2		.03125	.1149	.3528	.4861	.3641	.04657	-.2429	-.3014	-.1130	.1448	.2546
3			.01956	.1289	.3091	.4302	.3648	.1148	-.1676	-.2911	-.1809	.05838
4				.03400	.1320	.2811	.3912	.3576	.1578	-.1054	-.2655	-.2196
5					.04303	.1321	.2611	.3621	.3479	.1858	-.05504	-.2341
6					.01139	.04909	.1310	.2458	.3392	.3376	.2043	-.01446
7						.01518	.05338	.1296	.2336	.3206	.3275	.2167
8							.01841	.05653	.1280	.2235	.3051	.3179
9								.02117	.05892	.1263	.2149	.2919
10									.02354	.06077	.1247	.2075
11										.02560	.06222	.1231
12											.02739	.06337
13											.01083	.02897
14												.01196

Note: for a given β, only components greater than 1 percent of the peak carrier level are shown in this table.

8.5 SPECTRUM OF AN FM SIGNAL WITH SINUSOIDAL MODULATION

Consider a sinusoidally modulated FM signal.

$$v(t) = \cos(\omega_c t + B \sin \omega_m t) \quad (8.14)$$

We have found that Eq.(8.14) can be modified using Bessel functions as

$$v(t) = J_o(\beta)\cos\omega_c t + \sum_{\substack{n=2 \\ \text{even}}}^{\infty} J_n(\beta)[\cos(\omega_c - n\omega_m)t + \cos(\omega_c + n\omega_m)t]$$

$$- \sum_{\substack{n=1 \\ \text{odd}}}^{\infty} J_n(\beta)[\cos(\omega_c - n\omega_m)t - \cos(\omega_c + n\omega_m)t] \quad (8.15)$$

The spectrum of the above function indicates that, it is composed of a carrier with an amplitude $J_0(\beta)$ and a set of sidebands spaced symmetrically on either side of the carrier at frequency separation of ω_m, $2\omega_m$, $3\omega_m$, etc.

For $\beta << 1$,

$$J_o(\beta) \approx 1 - \left(\frac{\beta}{2}\right)^2$$

$$J_n(\beta) \approx \frac{1}{n!}\left(\frac{\beta}{2}\right)^n \quad \text{for} \quad n \neq 0$$

In relation to the spectrum of an FM signal, it means that for β very small, the spectrum is composed of a carrier and a single pair of sidebands with frequencies $\omega_c \pm \omega_m$. As β becomes somewhat larger, the amplitudes of successive sidebands become significant and they appear at frequencies $\omega_c \pm 2\omega_m$, $\omega_c \pm 3\omega_m$, etc.

Power in an FM Signal

Since the envelope of an FM signal has a constant amplitude, the power of such a signal is constant. If the carrier has a unit amplitude, the average power is given by $P_v = \frac{1}{2}$ and is independent of β.

Now when the carrier is modulated, the power in the sidebands may appear only at the expense of the power originally in the carrier. i.e.

$$J_o^2 + 2J_1^2 + 2J_2^2 + 2J_3^2 + \ldots = 1$$

and

$$P_v = \frac{1}{2}\left(J_o^2 + 2\sum_{n=1}^{\infty} J_n^2\right) = \frac{1}{2}$$

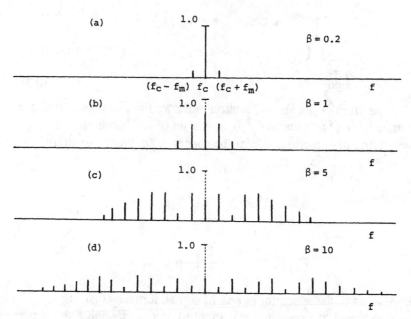

Fig. 8.5.1 The spectra of sinusoidally modulated FM signals for various values of β.

8.6 FREQUENCY MULTIPLICATION AND FM GENERATION

Any curcuit designed for generating an FM signal is called an FM modulator.

One of the methods for generating FM is to use a voltage-controlled oscillator (VCO) whose frequency is controlled by the modulating signal. An LC circuit shown below is an example of a VCO

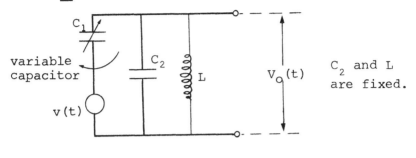

Fig.

In this circuit, $v(t)$ is the modulating signal and the value of capacitor C_1 depends on the value of $v(t)$. The instantaneous oscillator frequency f is given by

$$f = \frac{1}{2\pi\sqrt{(C_1+C_2)L}} \qquad (8.15)$$

As $v(t)$ changes, voltage across C_1 changes, which in turn causes a corresponding variation of the value of C_1 and thus a corresponding change in f.

Consider the equation

$$v(t) = A\cos(\omega_c t + \phi(t))$$

$v(t)$ is phase modulated if we let $\phi(t) = m(t)$ the modulating signal. If $m(t) << 1$ we have $\cos m(t) = 1$ and $\sin(m(t)) = m(t)$
Hence

$$\begin{aligned}v(t) &= A\cos(\omega_c t + \phi(t)) \\ &= A\cos(\omega_c t + m(t)) \\ &= A\cos\omega_c t \cos(m(t)) - A\sin\omega_c t \sin(m(t)) \\ &\doteq A\cos\omega_c t - A m(t)\sin\omega_c t\end{aligned}$$

A system which is based on the equation $v(t) \doteq A\cos w_c t - Am(t)\sin w_c t$ (for $m(t) << 1$) to generate PM signals is shown in Fig. 8.6-1. This system is often called the Armstrong system. It can easily be adapted for generating FM signals; all we need to do is to integrate $m(t)$ first (i.e., pass $m(t)$ through an integrator) before it is sent into the balanced modulator. In other words, we let

$$\phi(t) = k \int_{-\infty}^{t} m(\tau) d\tau \text{ instead of } \phi(t) = m(t)$$

Thus, we have $v(t) = A \cos(\omega_c t + \phi(t))$,

that is want to generate a narrowband FM,

$$v(t) = A \cos\omega_c t \; \phi(t) - A \sin\omega_c t \, \sin\phi(t)$$
$$\doteq A \cos\omega_c t - A \phi(t) \sin\omega_c t$$

where $\phi(t) = \left(k \int_{-\infty}^{t} m(\tau) d\tau \right)$ and $m(t) << 1$.

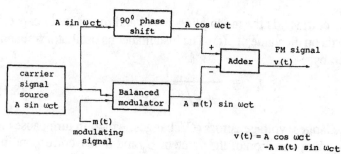

Fig. 8.6-1 Generation of a PM signal using Armstrong method.

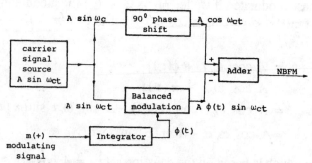

Fig. 8.6-2 Generation of a NBFM signal using Armstrong method.

In general, it is possible to use an FM generator to produce PM and vice versa.

D-83

Frequency Multiplication

The process in which the spectrum of a given signal is shifted to a different frequency range, without changing the amplitude of the spectrum, is called frequency translation.

Frequency translation is often achieved through a balanced modulator (when designed especially for frequency translation purposes, it is often referred to as a mixer.) followed by a filter to remove the unwanted sidebands.

Frequency multiplication is a process in which the center frequency f of a given signal is multiplied by an integer number n so that a new signal with center frequency nf is generated. In frequency multiplication, unlike in frequency translation, the modulated intelligence may be altered.

A frequency multiplier usually consists of a nonlinear element and a bandpass filter. For a input given signal, the nonlinear element will output a signal with same fundamental frequency as the input signal but is rich in higher-frequency harmonics.

The bandpass filter will filter out all other components of the output of the nonlinear element of the frequency multiplier except the component at frequency nf.

8.7 FM DEMODULATION

The following fig. illustrates the principle of FM demodulation.

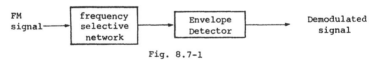

Fig. 8.7-1

The frequency selective network can be a simple RC circuit with a frequency dependent transfer function

$$|H(jw)| = \left|\frac{V_o}{V_{in}}\right|$$

which has the properties shown by the plot of its magnitude in fig. 8.7-2.

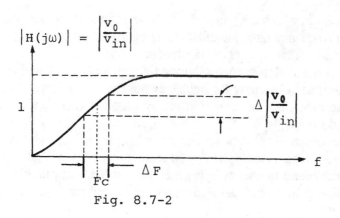

Fig. 8.7-2

FM detector circuits are often referred to as FM discriminators. A differentiator is an approximate example of a FM discriminator. The idea of a differentiator is that it has a transfer function of the form $|H(j\omega)| = k\omega$ where k is a constant. We often apply the signal to an amplitude limiter before passing it through a frequency discriminator. The function of the limiter is to remove the amplitude variations in the signal since the frequency discriminators often responds not only to a change in the instantaneous frequency of the input signal which is desired, but also to its amplitude variations which is not desirable at all. The input-output characteristic of an ideal limiter is shown in Fig. 8.7-3.

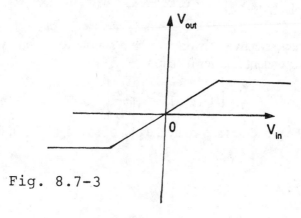

Fig. 8.7-3

CHAPTER 9

PULSE MODULATION SYSTEMS

9.1 SAMPLING THEOREM

(a) Low Pass Signals
If a band limited signal $m(t)$ of frequency f_m is periodically sampled every T_s seconds, where $T_s < \dfrac{1}{2f_m}$, then these samples can be used to reconstruct the original signal without any distortion.

T_s = Sampling Time
f_s = Sampling Frequency $\Big\}$ Nyquist Rate
f_m = Highest Frequency Spectral Component of $m(t)$

The sampling procedure is explained by Fig. 9.1-1, and the resulting spectrum is drawn in Fig. 9.1-2 and Fig. 9.1-3. For different values of f_s, note that, when $f_s < 2f_m$, there is a overlap of spectrum of $m(t)$ and the exact recovery of $m(t)$ is not possible.

(b) Bandpass Signals
For a signal $m(t)$ whose highest-frequency component is f_m and lowest frequency component is f_c, the sampling frequency is simply given by

$$f_s = 2(f_m - f_c) \qquad (9.1)$$

However it is necessary that either f_m or f_l be a harmonic of f_s, so that the above equation is valid for bandpass signals. Note that for low-pass signals $f_l = 0$.

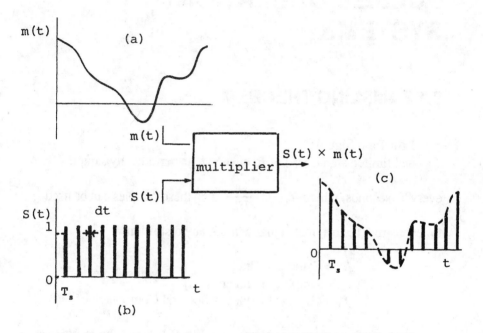

Fig. 9.1.1 (a) Signal m(t) to be sampled.
(b) Sampling function S(t).
(c) Sampled function S(t) m(t).

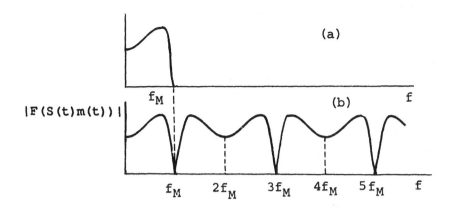

Fig. 9.1.2 (a) Spectral density of M(t)
(b) Spectral density of the sampled function M(t)·S(t)

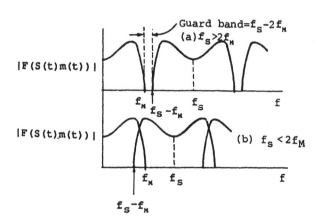

Fig. 9.1.3

(a) Plot of spectral density when $f_s > 2f_M$

(b) Plot of spectral density when $f_S < 2f_M$ overlapping occurs.

9.2 PULSE AMPLITUDE MODULATION

A sampled signal consists of a train of pulses, where each pulse corresponds to the amplitude of the signal $m(t)$ at the corresponding sampling time. Thus the signal is modulated in amplitude and hence the scheme is called PULSE AMPLITUDE MODULATION (PAM). Several PAM signals can be multiplexed, as long as they are kept distinct and are recoverable at the receiving end separately as they were sampled at different times. This system is an example of time division multiplexing (TDM), where the sampling theorem is invoked.

Fig 9.2-1 explains the above method, where at the transmitting end several bandlimited signals are connected to the contact point of a rotary switch (commutator). The rotary switch at the receiving end (Decommutator) is synchronous with the switch at the transmitting end. The switch can be mechanical or electronic, depending upon the sampling interval. Fig 9.2-2 explains how 2 baseband signals are interlaced. If f_m is the highest frequency present in any base band signals, then the switch must revolve at 2 f_m per second.

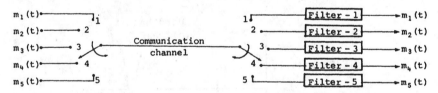

Fig. 9.2.1. Several band emitted baseband signals are transmitted over a single channel, using the sampling principle.

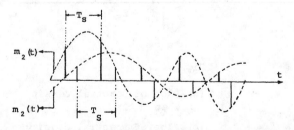

Fig. 9.2.2 The interlacing of 2 baseband signals $m_1(t)$, $m_2(t)$. Sampling period T_S is the same for both signals.

Channel Bandwidth and Crosstalk

If N signals are multiplexed and all are bandlimited to a frequency say f_m. Then the channel bandwidth is given by,

$$\text{BW Channel} \triangleq f_c = Nf_m$$

$$T_s \triangleq \text{ sampling period} \leq 1/2f_m$$

The interval of separation between successive samples of different baseband signals is given by $1/N2f_m$. If the bandwidth of the channel is restricted, then the channel response to each sample or a electrical impulse will be a waveform which may stay with amplitude long after the selection of samples. In such a case, at the receiving end the previous sample of other signals contribute significant amplitude to a signal at the given sampling time. Such combination of baseband signals at the output end is called <u>crosstalk</u> and is not desired at all.

However, it can be proven that when the channel bandwidth is exactly Nf_m, crosstalk can be avoided. Hence the bandwidth requirement for PAM system is exactly the same as in single side band transmission which uses the frequency division multiplexing scheme.

9.3 NATURAL SAMPLING AND FLAT-TOP SAMPLING

(a) Natural Sampling:

The samples considered in last section were instantaneous. However these instantaneous samples have the property that at the transmitting end of the channel, may have infinitesimal energy and when it propogates through a band limited channel, it gives rise to signals having infinitesimal small peak values which are easily lost in background noise.

In this section we consider samples which have a pulse duration τ and sampling period T_s. This sampling procedure is called natural sampling, where the sampled signals consist of a train of pulses of varying amplitude, which follow the waveform or the signal $m(t)$.

The sampling frequency is given by,
$$f_s \geq 2f_m$$
The baseband signal $m(t)$ and the sampled signal $m(t)\,s(t)$ are shown in fig 9.3-1. The Fourier series representation of sampling waveform $s(t)$ is given by,

$$s(t) = \frac{\tau}{T_s} + \frac{2\tau}{T_s}\left(A_1 \cos 2\pi \frac{t}{T_s} + A_2 \cos 4\pi \frac{t}{T_s} + \ldots\right) \quad (1)$$

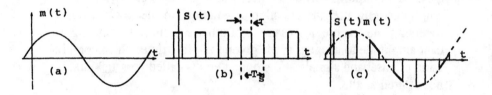

Fig.9.3.1 (a) Baseband signal m(t) (b) Sampling signal S(t) with pulse-time duration τ (c) Sampled signal S(t)m(t)

The coefficient A_n given by

$$A_n = \frac{\sin\left(n\pi \frac{\tau}{T_s}\right)}{n\pi \frac{\tau}{T_s}} \quad (9.2)$$

Hence the sampled signal $s(t)\,m(t)$ is given by,

$$s(t)m(t) = \frac{\tau}{T_s}m(t) + \frac{2\tau}{T_s}[m(t)A_1\cos 2\pi(2f_m)t$$
$$+ m(t)A_2\cos 2\pi(4f_m)t + ...]$$

when low pass filtered with a cutoff frequency of f_m, the signal is recovered in the form,

$$S_r(t) = \frac{\tau}{T_s}m(t) \qquad (9.3)$$

Note that, in eq. (3) the amplitude of the received signal increases as pulsetime duration τ increases. However, it is not possible to supress the crosstalk generated in a channel, when the samples have finite duration. If N samples are multiplexed, it is required to have sample duration τ much less than $\frac{T_s}{N}$ to avoid severe crosstalk.

(b) Flat-Top Sampling

Pulses shown in Fig. 9.3-1 (b), which follows the waveform of the signal, required a complicated and expensive circuitry. Instead the flat-topped pulses are very frequently employed as shown in Fig. 9.3-(2)a In sampling of this time, the baseband signals $m(t)$ can be recovered only with some distortion. However the circuitry to perform this simplifies the design procedure. The transform of the flat-topped sampled signal can be considered as the instantaneous sampled signal passing through a system which broadens the duration of pulse τ as shown in Fig. 9.3-2 (b).

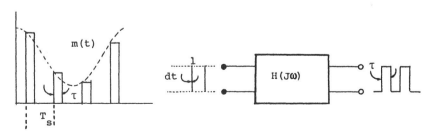

Fig. 9.3.2 (a)
Flat-topped sampling

Fig. 9.3.2 (b)
System which broadens the pulse duration of the sample.

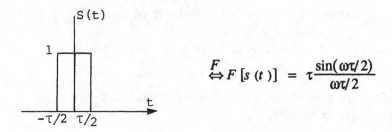

Therefore the transfer function of network shown in Fig 9.3-2 (b) is,

$$H(j\omega) = \frac{\text{output}}{\text{input}} = \frac{\tau}{dt}\frac{\sin(\omega\tau/2)}{\omega\tau/2} \quad (9.4)$$

Assuming the baseband signal bandlimited to f_m area and sampled with a sampling period $T_s \geq \frac{1}{2f_m}$, the transform of the flat-topped sampled signal is given by,

F[flat-topped sampled $m(t)$] = $H(j\omega)$ m $(j\omega)$

$$= \frac{\tau}{T_s}\frac{\sin(\omega\tau/2)}{\omega\tau/2}m(j\omega) \quad 0 \leq f \leq f_m \quad (9.5)$$

The spectrum of the flat-top sampled signal is given in fig. 9.3-3 (d) for the $m(j\omega)$ shown in fig. 9.3-3 (a). Observe that the frequency range of interest, i.e. 0 to f_m, the spectrum of the flat-top sampled signal differs with a small amount of distortion from the original signal. This distortion results from the fact that the spectrum is multiplied by the spectrum of the sampling function which has a $\frac{\sin x}{x}$ envelope. The distortion can be minimized by choosing $T << 1/f_m$. If N signals are multiplexed, then $T \leq 1/2f_m N$.

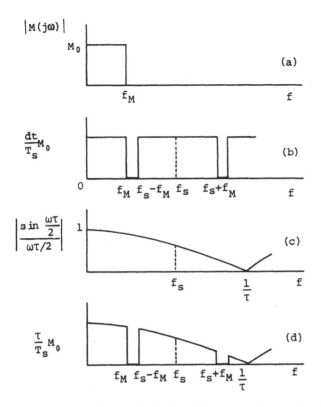

Fig. 9.3.3 (a) An ideal baseband spectrum (b) Spectrum or signal when sampled instantaneously (c) The $(\sin x)/x$ form introduced by flat-top sampling which creates distortion in the recovery or base band signal (d) Spectrum of the sampled signal.

9.4 SIGNAL RECOVERY THROUGH HOLDING AND CROSSTALK

In the case of natural and flat-topped sampling, we observed that as we make τ large, the amplitude of the output signal increases — which is highly desirable, but at the same time distortion increases. Also since $\tau/\tau_s \leq 1/N$, as N increases, the amplitude of the output signal also increases. This section will give an alternative method, which increases the output amplitude of the signal, but with limited distortion.

Fig. 9.4-1 illustrates this method, where at the receiving end, the sample value of each baseband signal is left until the next sample arrives. Hence, the output waveform consists of an up and down staircase waveform as shown in the figure 9.4-1. The spectral density of the recovered waveform is given by

$$F[m(t), \text{ sampled and held }] \frac{\sin(\omega T_s/2)}{\omega T_s/2} \; m(j\omega) \quad 0 \leq f \leq f_m$$

Note that in the above equation there is no τ/T_s term as was present in flat-top sampling. Hence the amplitude is increased by a factor of T_s/τ

(a) Crosstalk Due to High Frequency Cut-Off of the Channel

In section 9.2, the idea of crosstalk was introduced. This section will analyze the topic in detail. We have noticed before that the crosstalk in channel can be completely eliminated by instantaneous sampling. Since the instantaneous sampling is not practical, finite duration sampling is often employed which in turn gives rise to cross-talk as the channel has a limited bandwidth. For the purpose of analysis we introduce a term TIME-SLOT, which is the entire time interval consisting of the individual pulse and a guard time which separates one pulse from the other as shown in Fig. 9.4-1(b).

For calculation purposes we approximate the high-frequency channel behavior as an RC low pass filter shown in Fig. 9.4-1 (a). This RC circuit has a 3-db cutoff frequency at,

$$F_c = \frac{1}{2\pi RC} \text{ and time constant } \tau_c = RC$$

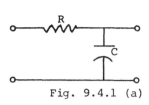

Fig. 9.4.1 (a)

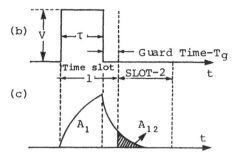

Fig. 9.4.1 (b) Time Slot
(c) Response of channel in 2 adjacent channels

Note that in Fig. 9.4-1 (c) the channel time constant $\tau_c = RC$ must be very small in comparison with the guard time interval τ_g to keep crosstalk to a minimum. Since the sample duration τ is usually greater than τ_g, we also have $\tau >> \tau_c$.

(b) Crosstalk Factor

Due to crosstalk, the message in channel 1 appears in channel 2. If channel 2 is empty, the effect of crosstalk is more severe than if a strong signal is present in channel 2. The crosstalk factor k is defined as the ratio of the signal leaked from channel 1 to the message signal present in channel 2 and is given by

$$\text{CROSSTALK FACTOR} = K = \frac{\tau_c}{\tau} e^{-\tau_g/\tau_c} \quad (9.6)$$

for a pariticular value of τ_c, the Channel Bandwidth is given by,

$$\text{BANDWIDTH}_{\text{CHANNEL}} = f_c = \frac{1}{2\pi RC} = \frac{1}{2\pi \tau_c}$$

(c) Crosstalk Due to Low Frequency Cutoff of the Channel

The low frequency behavior of the channel may be represented by a first order RC high pass filter and the resulting overlap, i.e. crosstalk, in adjacent channels is shown in Fig. 9.4-2. Here the sample pulse in time slot 1 develops a tilt in time slot 2, which is given by

$$\text{Accumulation - Tilt} = \Delta = v\left(1 - e^{-\tau/\tau_c}\right) \qquad (9.7)$$

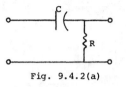

Fig. 9.4.2(a)
An RC high pass filter represents the low frequency characteristic of the communication channel.

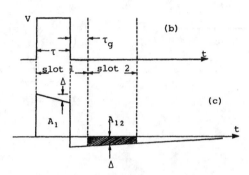

Fig. 9.4.2 (b) Two time slots
(c) Slot 1 develops a tilt in slots to follow which causes crosstalk.

In order to reduce crosstalk i.e. to make Δ small, we require $\tau << \tau_c$. Hence the tilt is very nearly linear and decays very slowly and crosstalk extends to many time slots.
The crosstalk factor is given by

$$K = \tau/\tau_c \quad \text{where} \quad \Delta << V \qquad (9.8)$$

The low frequency cutoff of the channel is given by,

$$f_c = \frac{1}{2\pi RC} = \frac{1}{2\pi\tau_c} \qquad (9.9)$$

9.5 PULSE TIME MODULATION

In pulse amplitude modulation, the baseband signal $m(t)$ modulates the amplitude of a pulse. In pulse time modulation (PTM), the timing of the pulse is varied rather than the amplitude. At each sampling of the baseband signal, a constant amplitude pulse is generated whose duration is modulated by $m(t)$. Modulation of this type is called pulse duration modulation, where an unmodulated pulse has a width τ_0, and the departure of pulse width from τ_0 is proportional to $m(t)$. This type of modulation is shown in Fig. 9.5-1 (b). It is also possible to modulate the leading or trailing edge of the pulse. In Fig. 9.5-1 (c) the baseband signal $m(t)$ modulates the position within the time slot of a constant duration and amplitude pulse. Modulation of this type is called pulse position modulation (PPM).

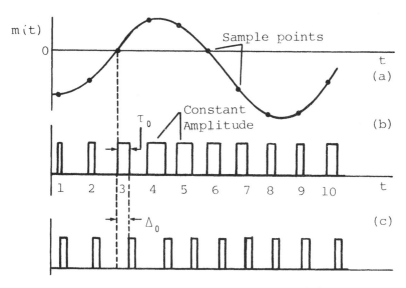

Fig. 9.5.1 (a) Base band signal m(t)
(b) Pulse duration modulated (PDM) signal
(c) Pulse position modulated signal (PPM)

(a) Method of Generating Pulse-Time Modulation Signal

First, a baseband signal $m(t)$ is flat-top sampled as shown in Fig. 9.5-2(a). In Fig 9.5-2 (b) a linear ramp wave from $R(t)$ is generated synchronously with the PAM samples. In Fig.9.5-2 (c) the PAM samples and the synchronized ramps are added and then passed through a comparator with the reference level as shown. Note that, the comparator's output has two voltage level characteristics under the following conditions :
1. When the input signal is less than the reference level.
2. When the input signal is larger than reference level.

In fig. 9.5-2 (d), a pulse-duration modulated signal is generated. Notice that the leading edge of a pulse output of the comparator is the first crossing of the reference level by the waveform $S_A(t) + R(t)$ and the trailing edge is the second crossing of the reference level. The procedure is shown in the block diagram in Fig.9.5-1 (e). A PPM signal can be obtained from PDM by triggering on the negative-going pulse edges.

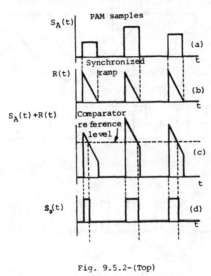

Fig. 9.5.2-(Top)

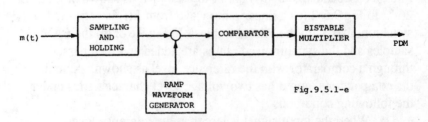

Fig. 9.5.2 Method of generating a PDM signal.

CHAPTER 10

PULSE CODE MODULATION

10.1 QUANTIZATION OF SIGNALS

The quantized signal $m(t)$ is an approxiamation of $m(t)$. However in most instances, $m_q(t)$ is easily separable from additive noise. As shown in Fig. 10.1-1, signal $m(t)$ is divided into M equal intervals. If $m(t)$ swings from a maximum value of V_H to a minimum V_L then the step size is defined as,

$$S = \frac{V_H - V_L}{M} \qquad (10.1)$$

The center of each step size is called the quantization level $M_0, M_1, \ldots M_3$, whenever $m(t)$ is in the range of Δ_1 the quantized signal $m_q(t)$ stays at the constant level M_1. Note that,

Distance between each level = S
Distance or extremes V_H and V_L from the nearest quantization level
= $\frac{S}{2}$.

The quality of approximation is improved by increasing the number of levels, i.e. reducing the step size.

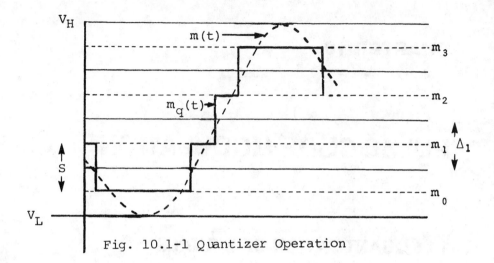

Fig. 10.1-1 Quantizer Operation

Quantization Error :

We define quantization error $\triangleq e$

INSTANTANEOUS $|e| \leq \dfrac{S}{2}$

The difference $m(t) - m_q(t)$ is regarded as quantization noise. This noise along with the additive noise during transmission, causes the error in the transmitted baseband signal.

$$\boxed{\begin{array}{c}\text{Mean Square Quantization Error}\\[4pt] \overline{e^2} = \sum_{i=1}^{M} \int_{m_i - S/2}^{m_i + S/2} F(m)(m - m_i)^2\, dm\end{array}}\quad (10.2)$$

where $f(m)\,dm \triangleq$ Probability density function

and $\int_{-\infty}^{\infty} f(m)\,dm = 1$

When the number of quantization levels is very large, so that the step size is very small, the mean square error is approximated as,

$$\overline{e^2} = \dfrac{S^2}{12} \qquad (10.3)$$

10.2 PULSE CODE MODULATION

As we have noted earlier, prior to transmission the baseband signal $m(t)$ is sampled and then to reduce the effect of noise, it is quantized. Instead of transmitting these quantized levels directly, very often a unique code number for each quantized level is transmitted. Usually these codes are based on the binary arithmetic system and they are transmitted as pulses. This type of transmission is known as Pulse Code Modulation or simply PCM. Fig 10.2-1 explains the procedure of this modulation technique. Here each code number is represented by its binary equivalent and then corresponding electrical pulses are generated. Note that the code number assignment can be arbitrary, but the receiver must be aware of the assignment.

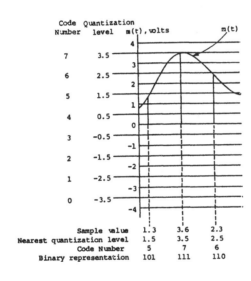

Fig. 10.2-1 (b)

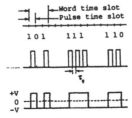

Fig. 10.2-1 (a)

(a) For each code number, its binary equivalent and the corresponding pulse representation.

The PCM System

In Fig. 10.2-2 a PCM communication system is shown. Here the transmitter consists of a sampler, quantizer and an encoder. The encoder generates a unique binary pulse to each quantized level. At the receiving end, the encoded signal is requantized to get rid of the additive noise introduced by the channel. This receiver quantizer decides whether a positive or a negative pulse is received and then transmits the result to the decoder. The decoder, also regarded as a digital-to-analog converter, performs the inverse operation of an encoder. The filter then separates the baseband signal from other high frequency components. The final output is not an exact replica of the analog signal $m(t)$, due to the presence of quantization noise and the error in the decision making process of the receiver quantizer.

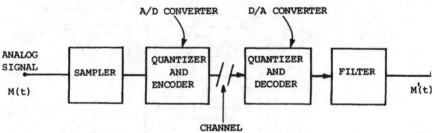

Fig. 10.2-2 The PCM Communication System.

10.3 COMPANDING

If there are M quantization levels, then the code number requires N bits, where $M = 2^N$. The ratio of output signal power to the quantization noise is given by

$$\frac{S_o}{N_q} = \left(2^N\right)^2 = 2^{2N} \tag{10.4}$$

The above equation is valid with the assumption that the output signal power $S_o \approx S_i$, which is the input signal power. Equation (10.4) can be expressed in terms of decibels as,

$$\boxed{\left[\frac{S_o}{N_q}\right]_{db} = 10 \log_{10}\left[\frac{S_o}{N_q}\right] = 6N \ db} \tag{10.5}$$

The above equation can be interpreted as, when a signal $m(t)$ swings through all quantization levels, without overshooting, the SNR is $6N db$, where N is the number of bits used in the quantization code. In summary, if the input signal power is reduced such that when the number of bits needed to represent the code is decreased, the output SNR decreases by $6db$/bit. Therefore the number of quantization levels should be increased when the input signal has a low amplitude and decreased when it has high amplitude. This procedure is called companding.

For voice transmission it is required that S_o/N_q be no less than 30dB. So if an N = 8 bit system is used, the dynamic range is $6N - 30 = 18db$. But if it is companded using a μ-law compandor, the dynamic range is increased from 18 db to 48 db with the tradeoff that the maximum SNR is only 38db, instead of 48 db.

10.4 DELTA MODULATION

Delta modulation (DM) is another PCM scheme which requires much simpler circuitry than other PCM techniques. A DM communication system is shown in Fig 10.4-1. Here, note that the modulator output is not the coded signal itself, but the difference between the waveform $m(t)$ and its approximation $\tilde{m}(t)$. The modulator output is the input pulse train multiplied by the polarity of difference signal $\Delta(t)$. $\tilde{m}(t)$ is an approxiamition of $m(t)$ as the modulator output is integrated before being fed back to the difference amplfier. At the receiving end the quantizer makes the decision whether a received pulse is negative or positive. The low pass filter output recovers the baseband signal $m(t)$. The recovered signal is corrupted due to the receiver quantizer noise and due to the stepwise approximation of Delta Modulation.

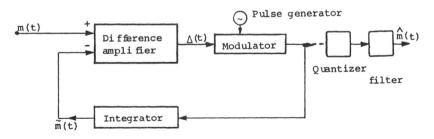

Fig. 10.4-1 A Delta-Modulation System.

Limitations of DM

Delta modulation exhibits an additional overloading besides the obvious large excursion of amplitude of the baseband signal This additional overloading, which is unique in DM, appears when the modulating signal changes between sampling by an amount greater than the stepsize, i.e. $\tilde{m}(t)$ deviates from original waveform $m(t)$ in the region of increased slope. Hence, the DM system is said to be slope-overloaded. To avoid slope-overloading the following condition is to be satisfied.

$$A \leq \frac{Sf_s}{2\pi f}$$

where A = Peak amplitude of $m(t)$
S = Step Size
f = Frequency of $m(t)$

10.5 BINARY PHASE SHIFT KEYING (BPSK)

A binary waveform can be transmitted using AM, FM, or PM techniques. BPSK is a Phase Modulation (PM) technique. In BPSK, when the data is a at one level, e.g. logic level 1, the transmitted signal is a sinusoid of fixed amplitude. Now, when the data is at the other level, i.e. logic level 0, the transmitted sinusoid is phase shifted by 180.° Hence, the transmitted signal has the form given by,

$$m_{BPSK}(t) = A \cos[\omega_o t + \phi(t)] \quad (10.6)$$

where A = constant amplitude
$\phi = 0$ for $m(t) = +1V$
$\quad = \pi$ for $m(t) = -1V$ }(Assumption)

Hence eq. (1) can be modified to

$$m_{BPSK}(t) = b(t)A\cos\omega_0 t$$

Now, $B_{BPSK} \triangleq$ BANDWIDTH = $2f_b$, where f_b = Bit rate

Power Spectrum of BPSK

If T_b is the pulse duration associated with each pulse of the binary waveform $m(t)$, the power spectral density is given by,

$$G_{BPSK}(f) = \frac{P_s T_b}{2}\left[\frac{\sin\pi(f-f_o)T_b}{\pi(f-f_o)T_b}\right]^2 + \left[\frac{\sin\pi(f+f_o)T_b}{\pi(f+f_o)T_b}\right] \quad (10.7)$$

where $P_s = \frac{1}{2}A^2$ = power associated with the carrier.

Note that power spectral density has a $\frac{\sin x}{x}$ form and occupies the whole frequency spectrum. However, more than 90 % of the total power is contained within the range $2f_b$ about the carrier frequency. Hence the higher frequency component can be filtered to utilize the spectrum effecively.

10.6 DIFFERENTIAL PHASE SHIFT KEYING

Differential phase shift keying (DPSK) has the advantage over BPSK that it avoids the need of a synchronous carrier at the receiving end, and it also eliminates the ambiguity of whether the received data is inverted or not. The DPSK transmitter and the receiver are shown in Fig.10.6-1 and 10.6-2 respectively.

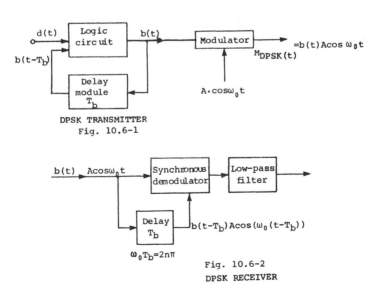

DPSK TRANSMITTER
Fig. 10.6-1

Fig. 10.6-2
DPSK RECEIVER

The transmitted signal is,
$$m_{DPSK}(t) b(t) A \cos \omega_o t$$
$$= I A \cos \omega_o t \qquad (10.8)$$

Hence, when $d(t) = 0$ the carrier phase does not change at the beginning of the bit interval, while the phase is changed by 180° when $d(t) = 1$.

The synchronous demodulator output at the receiving end is filtered to obtain the product $b(t) b(t - T_b)$.

$b(t) b(t - T_b) = 1$ if $d(t) = 0$
$b(t) b(t - T_b) = -1$ if $d(t) = 1$.

In DPSK, the received bit is determined from two successive bit intervals. Hence, noise in one bit interval will produce error for two bit determinations.

$$\boxed{BW_{DPSK} = 2f_b} \text{ where } f_b = \frac{1}{T_b} = \text{Bit rate}$$

10.7 FREQUENCY SHIFT KEYING

As the name suggests, here the frequency of carrier is modulated instead of the phase. Here the carrier signal is given by,

$$m_{BFSK}(t) = A \cos[\omega_o t + d(t) \Omega t]$$

Depending on the value of data waveform $d(t)$, the above equation can be modified to,

$$m_{BFSK}(t) = A \cos[\omega_o t + \Omega t] \text{ for } d(t) = +1V \text{(i.e. Logic 1)}$$
$$= A \cos[\omega_o t + \Omega t] \text{ for } d(t) = -1V \text{ (i.e. logic 0)}$$

Hence the carrier has an angular frequency offset by a constant $+ \Omega$ or $- \Omega$ about the nominal carrier frequency ω_0, depending on the data waveform $d(t)$.

The bandwidth required by binary frequency shift keying is $BW = 4f_b$, which is twice the bandwidth of BPSK. However, there is an advantage of a lower probability of error than for BPSK.

CHAPTER 11

MATHEMATICAL REPRESENTATION OF NOISE

11.1 FREQUENCY DOMAIN REPRESENTATION AND SPECTRAL COMPONENTS OF NOISE

The frequency domain characteristics of noise can be analyzed by first considering a sample function of noise in an interval of length T. A periodic function is generated by repeating the sample function every T seconds. This periodic function is next expanded in a Fourier Series. Now, the true characteristics of the sample function from $t = -T/2$ to $T/2$ can be obtained in the limit as T goes to infinity. Hence, the noise $n(t)$ can be represented as

$$n(t) = \lim_{\Delta f \to 0} \sum_{k=1}^{\infty} (a_k \cos 2\pi k \, \Delta f t + b_k \sin 2\pi k \, \Delta f t) \quad (11.1)$$

$$\text{where } \Delta f = \frac{1}{T}$$

or in exponential form

$$n(t) = \lim_{\Delta f \to 0} \sum_{k=1}^{\infty} C_k \cos(2\pi k \, \Delta f t + \theta_k) \quad (11.2)$$

$$\text{where } C_k^2 = a_k^2 + b_k^2 \text{ and } \theta_k = -\tan^{-1} \frac{b_k}{a_k}$$

Note : a_k, b_k, c_k and θ_k are all random variables, not fixed numbers. Also as $\Delta f \to 0$, the discrete spectral lines in the Fourier series representation get closer and finally form a continuous spectrum. The power spectral density of noise $n(t)$, can now be written as

$$G_n(f) = \lim_{\Delta f \to 0} \frac{\overline{C_k^2}}{4\Delta f} = \frac{\overline{a_k^2} + \overline{b_k^2}}{4\Delta f} \qquad (11.3)$$

where C_k^2's are replaced by $\overline{C_k^2}$, i.e. the expected value or ensemble average of the square of the random variable C_k.

Now the power in the frequency range f_1 to f_2 can be obtained by integrating the power spectral density function.

$$P_{f_1 \to f_2} = 2\int_{f_1}^{f_2} G_n(f) df \qquad (11.4)$$

Total Power $\quad P_T = 2\int_0^\infty G_n(f) df \qquad (11.5)$

Since $G_n(f) = G_n(-f)$

Spectral Component of Noise

As mentioned before, the components of $n(t)$, i.e. a_k, b_k, c_k and θ_k are random variables not fixed numbers. The coefficients associated with the kth frequency interval as $\Delta f \to 0$, can be written as

$$n_k(t) = a_k \cos 2\pi k \Delta f t + b_k \sin 2\pi k \Delta f t \qquad (11.6)$$

or in exponential form as

$$n_k(t) = C_k \cos(2\pi k \Delta f t + \theta_k) \qquad (11.7)$$

Since the components a_k and b_k are random variables, variance of $n_k(t)$ is determined by taking the average over the ensemble of $[n_k(t)]^2$. Since $n_k(t)$ is a stationary process, we have the result

$$\overline{[n_k(t)]^2} = Pk \underline{\underline{\Delta}} \text{ NORMALIZED POWER} = \overline{a_k^2} = \overline{b_k^2}$$

In order to be consistent with the above result, we set

$$\overline{a_k b_k} = 0$$

Thus, a_k and b_k are uncorrelated; we may also establish that a_k and b_k are both gaussian. Since a_k and b_k have no DC components, then mean value of a_k, $\overline{a_k} = 0$

similarly $b_k = 0$.

Since $c_k^2 = a_k^2 + b_k^2$, and a_k and b_k are both gaussian, then the c_k's have a Rayleigh probability density given by,

$$f(c_k) = \frac{c_k}{P_k} e^{-c_k^2/2P_k} \quad c_k \geq 0, \; P_k = \text{Normalized Power in } \Delta f .$$

and θ_k has a uniform probability density given by,

$$f(\theta_k) = \frac{1}{2\pi} \quad -\pi \leq \theta_k \leq \pi$$

11.2 FILTERING

I. Narrow band Filtering :

As noted earlier, noise can be represented as a superposition of discrete, two sided spectral components. Hence when the noise is passed through a narrow band filter, the output will be somewhat sinusoidal. This is due to the fact that the Fourier transform of a sinusoid is two discrete components at the sinusoid's frequency, one at f_c and the other at $-f_c$. In Fig.10.2-1 the frequency response of the narrow band filter is shown. Note that the output is not a pure sinusoid, i.e. its amplitude is a random variable. The envelope of the filter output occupies a spectral range from $-B/2$ to $+B/2$, where B is the filter bandwidth. It is important to note that as B becomes smaller, the output of the filter becomes more and more sinusoidal.

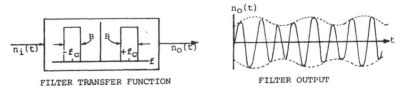

FILTER TRANSFER FUNCTION   FILTER OUTPUT

Fig. 11.2-1: Response of a narrow band filter to noise

II. Response of RC Filter to White Noise :

White noise is the noise whose power spectral density is uniform over the entire frequency range. Assume that the power spectral density over the entire frequency range is given by,

$$G_n(f) = \frac{\eta}{2} \qquad (11.8)$$

where η is a constant.
Now, the transfer function of an RC filter, with the cutoff frequency at f_c is given by,

$$H(f) = \frac{1}{1+jf/f_c} \qquad (11.9)$$

Using the relation

$$\boxed{G_{n_o}(f) = G_{n_i}(f)|H(f)|^2} \qquad (11.10)$$

where $G_{n_o}(f)$ = Output noise spectral density
$G_{n_i}(f)$ = Input noise spectral density

Substituting (11.8) and (11.9) in (11.10) we obtain for the RC-low pass filter

$$\boxed{G_{n_o}(f) = \frac{\eta}{2}\frac{1}{1+(f/f_c)^2}} \qquad (11.11)$$

Now the filter output noise power N_o is obtained by integrating the $G_{n_o}(f)$ given in Eq.(11.11).

$$\boxed{N_o = \int_{-\infty}^{\infty} G_{n_o}(f)df = \frac{\pi}{2}\eta f_c} \qquad (11.12)$$

III. Response to an Ideal Low-Pass Filter :

An ideal (rectangular) low- pass filter has a transfer function given by

$$H(f) = \begin{cases} 1 & |f| \le B \\ 0 & \text{elsewhere} \end{cases}$$

If the input noise is white, the output-power spectral density is given by,

$$G_{n_o}(f) = \begin{cases} \dfrac{\eta}{2} & |f| \le B \\ 0 & \text{elsewhere} \end{cases}$$

And the output noise power is

$$N_o = \int_{-\infty}^{\infty} G_{n_o}(f)\,df = \eta B \tag{11.13}$$

IV. Response to a Rectangular Bandpass Filter :

A rectangular bandpass filter has the following transfer function,

$$H(f) = \begin{cases} 1 & -f_2 < f < -f_1 \text{ and } f_1 < f < f_2 \\ 0 & \text{elsewhere} \end{cases}$$

So the output noise power spectral density is given by

$$G_{n_o}(f) = \begin{cases} \dfrac{\eta}{2} & -f_2 < f < -f_1 \text{ and } f_1 < f < f_2 \\ 0 & \text{elsewhere} \end{cases}$$

Therefore output noise power is,

$$N_o = \eta(f_2 - f_1) \tag{11.14}$$

V. Response to a Differentiator :

A differentiator has the property that its output waveform is the time derivative of the input waveform. Its transfer function is given by,

$$H(f) = j\,2\pi\tau f \tag{11.15}$$

where τ is a constant of proportionality.

If white noise is the input to the filter, the output noise power spectral density is given by,

$$G_{n_o}(f) = |H(f)|^2 G_{n_i}(f) = 4\pi^2 \tau^2 f^2 \frac{\eta}{2} \tag{11.16}$$

If the differentiator is followed by a rectangular low-pass filter having a bandwidth B, then the output noise power is given by,

$$N_o = \int_{-B}^{B} 4\pi^2 \tau^2 f^2 \frac{\eta}{2}\,df = \frac{4\pi^2}{3}\eta\tau^2 B^3 \tag{11.17}$$

VI. Response to an Integrator :

A network which integrates a function over an interval T, has a transfer function given by,

$$H(f) = \frac{1 - e^{-j\omega T}}{j\omega \tau}$$

where τ is a constant.

If the noise is white with a constant power spectral density $\frac{\eta}{2}$, the output noise power is given by,

$$\boxed{N_o = \frac{\eta T}{2\tau^2}} \qquad (11.18)$$

11.3 NOISE BANDWIDTH

We will denote B_N as the noise bandwidth. The noise bandwidth is the bandwidth of an ideal rectangular filter which passes the same noise power as does a real filter. Fig. 11.4-1 illustrates this concept.

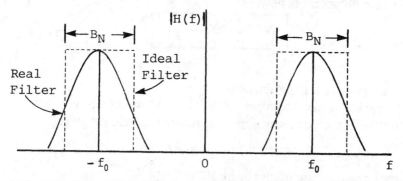

Fig.11.3-1 Illustration of the noise bandwidth of a filter.

For white noise, a rectangular filter i.e. an ideal low-pass filter with $H(f) = 1$ over its passband B_N, would yield an output noise power

$$N_o(\text{Rectangular}) = \frac{\eta}{2} \cdot 2B_N = \eta B_N \qquad (11.19)$$

Now we will evaluate the noise bandwidth of filters discussed in section 11.3.

I. Noise Bandwidth of RC Filter

By definition of noise Bandwidth,
N_o (RC) = N_o (Rectangular)
i.e. $\frac{\pi}{2}\eta f_c = \eta B_N$

$$B_N = \frac{\pi}{2} f_c \qquad (11.20)$$

II. Noise Bandwidth Of An Ideal Bandpass Filter.

$$N_o(\text{Bandpass}) = N_o(\text{Rectangular})$$

i.e. $\eta(f_2 - f_1) = \eta B_N$

$$\boxed{B_N = f_2 - f_1} \qquad (11.21)$$

11.4 SHOT NOISE

The term shot noise comes from considering the nature of current flow in a circuit containing semiconductors. This noise consists of identically shaped pulses with random amplitudes and random time occurences. To calculate the noise power spectral density we make the assumption that the random process is stationary. Fig. 11.4-1 shows a pulse train which represents shot noise.

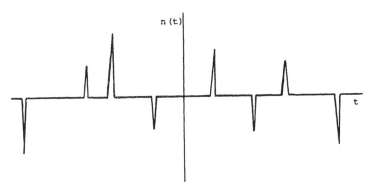

Fig. 11.4-1 Shot noise. Pulses of random amplitude and time of occurrence.

The spectral density of shot noise $n(t)$ is given by,

$$G_n(f) = \frac{1}{T_s}\overline{|P(f)|^2} \quad -\infty < f < \infty \quad (11.22)$$

where $P(f)$ is the Fourier Transform of a single pulse.
If $P(t)$ is an impulse of strength I, then $|P(f)| = I$ and the power spectral density is

$$G_n(f) = \frac{I^2}{T_s} \quad -\infty < f < \infty \quad (11.23)$$

from the above equation we conclude that, when the shot noise consists of impulses that occur randomly, the noise power spectral density is white.

CHAPTER 12

NOISE IN COMMUNICATION SYSTEMS

12.1 NOISE IN AM SYSTEMS

I.) The Amplitude Modulation Receiver
The input signal to an AM receiver might be obtained by a receiving antenna which receives its signal from a transmitting antenna. The carrier of the received signal is called an RF (Radio Frequency) carrier. The signal then is amplified by an RF amplifier and then goes through a mixer. The mixer generates sum and difference frequency components about the local oscillator frequency. They are given by $f_{osc} + f_{rf}$ = sum frequency
$$f_{osc} - f_{rf} = \text{difference frequency}$$
where f_{rf} = input radio frequency carried by the RF carrier.
   f_{osc} = sinusoidal waveform generated by a local oscillator.
f_{osc} is always chosen to be much greater than f_{rf}, hence this system is often referred to as a superhetrodyne system. The sum frequency is rejected by a filter and the difference frequency carrier, usually referred as the IF (Intermediate Frequency) carrier is applied to an IF amplifier. This process of replacing a modulated RF carrier by a modulated IF carrier is called conversion. The IF amplifier output is then passed through an IF carrier filter to the demodulator. Finally, the baseband signal is recovered by passing the demodulator output through a baseband filter. The entire process is shown in Fig. 12.1.1.

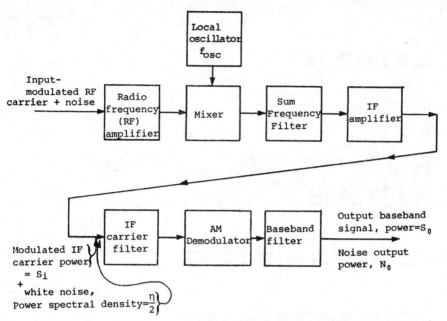

Fig. 12.1.1 A receiving system for an amplitude-modulated signal.

There are two important advantages to mixing the RF carrier with a local oscillator.

a) The generated IF carrier frequency is substantially lower than the input RF carrier, thus the gain and flat-topped filters required by the receiver can be more easily synthesized at low frequency than at high frequency.

b) Only a fractional change in the f_{osc} could accomodate a large range of RF frequencies.

II) Noise in Single-Sideband (SSB) Modulation

The SSB receiver is shown in Fig.12.1.2. The input signal $S_i(t)$ is given by,

$$S_i(t) = A \cos\left[2\pi(f_c + f_m)t\right] \qquad (12.1)$$

where A = amplitude of the carrier
f_c = carrier frequency
f_m = frequency of a sinusoidal baseband signal

D-117

Hence the input signal power

$$S_I = \frac{1}{2}(A)^2 = \frac{A^2}{2} \qquad (12.2)$$

The baseband filter output can derived to be,

$$S_o(t) = \frac{A}{2}\cos 2\pi f_m(t) \qquad (12.3)$$

Hence the output signal power $S_o = \frac{1}{2}\left(\frac{A}{2}\right)^2 = \frac{A^2}{8} = \frac{S_I}{4}$

$$(12.4)$$

Note that for analysis purposes we assumed the baseband signal to be a sinusoid. But the result obtained is entirely general, as one can easily apply the theory of Fourier transform which says any signal can be decomposed into a combination of sines and cosines.
In a similar manner the noise power can be calculated. If we assume that the noise is white with a constant power spectral density $\frac{\eta}{2}$ then the output noise power N_o is given by,

$$N_o = \text{output noise power} = \frac{\eta f_m}{4} \qquad (12.5)$$

where $\frac{\eta}{2}$ = input white noise power
$2f_m$ = bandwidth of the baseband filter.
Thus the signal-to-noise ratio (SNR) can be obtained by,

$$\frac{S_o}{N_o} = \frac{S_I/4}{\eta f_m/4} = \frac{S_I}{\eta f_m} \qquad (12.6)$$

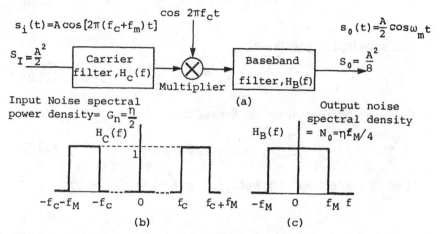

Fig.12.1.2(a) A synchronous demodulator operating on a single-sideband single-tone signal.(b)The bandpass range of the carrier filter. (c) The passband of the lowpass baseband filter.

III) Noise in Double Sideband Suppressed Carrier (DSB SC) Modulation

We will first calculate the signal power, assuming that the baseband signal is a sinusoid. However, the result obtained will be entirely general, as discussed in the last section. In order to make the received power the same as in the SSB case, the input signal in DSB-SC can be written as,

$$S_I(t) = \frac{A}{\sqrt{2}} \cos[2\pi(f_c + f_m)t] + \frac{A}{\sqrt{2}} \cos[2\pi(f_c - f_m)t] \quad (12.7)$$

where A = amplitude of the carrier
f_c = carrier frequency
f_m = frequency of a sinusoidal baseband signal

Hence

$$S_I = \text{input signal power} = \frac{1}{2}\left(\frac{A}{\sqrt{2}}\right)^2 + \frac{1}{2}\left(\frac{A}{\sqrt{2}}\right)^2 = \frac{A^2}{2} \quad (12.8)$$

The output of the baseband filter which has two passbands will give a signal,

$$S_o(t) = \frac{A}{\sqrt{2}} \cos 2\pi f_m t$$

D-119

Thus

$$S_o = \text{output signal power} = \frac{1}{2}\left(\frac{A}{\sqrt{2}}\right)^2 = \frac{A^2}{4} = \frac{S_I}{2} \quad (12.9)$$

$$N_o = \text{output noise power} = \frac{\eta f_m}{2}$$

where $\frac{\eta}{2}$ = power spectral density if input white noise

f_m = Bandwidth of any one passband of the baseband filter.

Thus

$$\boxed{\text{SNR} = \frac{S_o}{N_o} = \frac{S_I/2}{\eta f_m/2} = \frac{S_I}{\eta f_m}} \quad (12.10)$$

which is exactly the SNR obtained in SSB modulation.

IV) Noise in Double Sideband with Carrier

The carrier is used as a transmitted reference to obtain the reference signal $\cos 2\pi f_c t$, at the receiving end. As noted before, the inclusion of the carrier increases the total input signal power, but makes no contribution to the output signal power. Here the received signal is,

$$\begin{aligned} S_i(t) &= A[1 + m(t)]\cos 2\pi f_c t \\ &= A\cos 2\pi f_c t + A\, m(t)\cos 2\pi f_c t \end{aligned} \quad (12.11)$$

where A = amplitude of the carrier

f_c = carrier frequency

$m(t)$ = baseband signal being modulated.

Hence the input power is,

$$S_I = \frac{A^2}{2} + \frac{A^2}{2}\overline{m^2(t)} \quad (12.12)$$

where $\overline{m^2(t)}$ = time average of the square of the baseband signal. The output signal power is the input signal power associated with the sideband only.

Thus

$$S_o = \frac{\overline{m^2(t)}}{1 + \overline{m^2(t)}} \cdot S_I \quad (12.13)$$

Hence SNR in DSB with carrier is

$$\frac{S_o}{N_o} = \frac{\overline{m^2(t)}}{1 + \overline{m^2(t)}} \frac{S_I}{\eta f_m} \qquad (12.14)$$

If the modulating baseband signal is a sinusoid, then equation can be modified to,

$$\frac{S_o}{N_o} = \frac{M^2}{2 + M^2} \frac{S_I}{\eta f_m}$$

where M = constant = Modulation Index
Note that SNR is considerably less here than the previous methods, due to the presence of the carrier.

12.2 NOISE IN FREQUENCY MODULATION SYSTEMS

A) The FM Receiver

The block diagram of an FM receiver is the same as that of an AM receiver, except that the AM demodulator is now replaced by an FM demodulator. The FM demodulator consists of a limiter and a FM discriminator as shown in Fig. 12.2-1.

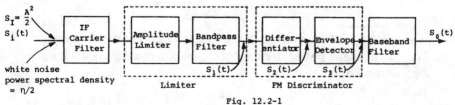

Fig. 12.2-1

a) The Limiter

Since in the FM system the message is carried by the carrier frequency, not the amplitude, then any fluctuation in the amplitude of the carrier will be due to noise. The limiter is exclusively used to suppress such amplitude fluctuations. A typical limiter input-output characteristic is shown in Fig. 12.2-2.

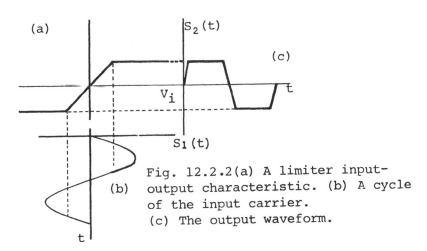

Fig. 12.2.2(a) A limiter input-output characteristic. (b) A cycle of the input carrier. (c) The output waveform.

Note that the output follows the input only over a limited range and as a result the output is more or less a square wave. The frequency of the square wave is entirely dependent on the carrier frequency. The bandpass filter following the limiter selects the fundamental frequency and as a result the filter output is more or less a sinusoid. Thus if

$$S_i(t) = A \cos[\omega_c t + \phi(t)] \quad (12.15)$$

Then the limiter output $S_1(t)$ will be given by,

$$S_1(t) = A_L \cos[\omega_c t + \phi(t)] \quad (12.16)$$

where $S_i(t)$ = Input signal to the IF carrier filter

$$\phi(t) = k \int_{-\infty}^{t} m(\lambda) d\lambda$$

where $m(t)$ = frequency-modulating baseband waveform.
A_L = limited amplitude of the carrier, which is independent of the input amplitude.

b) The Discriminator

Like the limiter, the FM-discriminator consists of two components. First, there is a network which behaves as a differentiator or a frequency-to-amplitude modulation converter over a range of frequencies. When a constant amplitude FM signal passes through this linear network, the output is amplitude modulated with time corresponding to the instantaneous frequency of the carrier. Hence the output of the differenitator, whose input is $S_i(t)$, is given by,

$$S_2(t) = \sigma \frac{d}{dt} S_1(t) = -\sigma A_L \left[\omega_c + \frac{d}{dt}\phi(t) \right] \sin\left[\omega_c t + \phi(t) \right] \quad (12.17)$$

where σ = a constant and corresponds to the slope of the differentiator.

The output of the differentiator is passed through an envelope detector and then a baseband filter recovers the baseband signal $m(t)$. The output of the envelope detector can be written as,

$$S_3(t) = \sigma A_L \left[\omega_c + \frac{d}{dt}\phi(t) \right] = \alpha \omega_c + \alpha \frac{d}{dt}\phi(t) \quad (12.18)$$

where $\alpha = \sigma A_L$ = constant;

Note that, the envelope detector output also referred to as the FM-discriminator output is proportional to the input frequency.

B) Calculation of Signal-To-Noise Ratio (SNR)

Recalling that,

$$\phi(t) = k \int_{-\infty}^{t} m(\lambda) d\lambda \quad (12.19)$$

and

$$S_3(t) = \alpha \omega_c + \alpha \frac{d}{dt}\phi(t) = \alpha \omega_c + \alpha k m(t) \quad (12.20)$$

the output signal power is given by,

$$S_o = \alpha^2 k^2 \overline{m^2(t)} \quad (12.21)$$

where $\alpha = \sigma A_L$ = a constant
A_L = amplitude of the carrier imposed by the limiter
k = a constant in $\phi(t)$ term

$\overline{m^2(t)}$ = time average of the square of the baseband modulating waveform.

If we assume that noise is white with power spectral density $\frac{\eta}{2}$, the output power spectral density can be derived as,

$$G_{n_3}(f) = \frac{\alpha^2 \omega^2 \eta}{A^2} \qquad -B/2 \leq f \leq B/2 \qquad (12.22)$$

where $\frac{\eta}{2}$ = a constant power spectral density of white noise
A = amplitude of the input carrier.
B = bandwidth of the IF carrier filter
    = $2(\Delta f + f_m)$
where Δf = maximum frequency deviation
    f_m = bandwidth of the baseband signal

The power spectral density $G_{n_3}(f)$ is plotted in Fig.12.2-3.

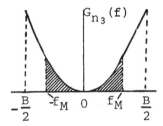

Fig. 12.2.3 The variation with frequency of the power spectral density at the output of an FM demodulator.

Now the output noise power N_o can be calculated by integrating the power spectral density $G_{n_3}(f)$ over the desired range of frequency, i.e.

$$N_o = \int_{-f_m}^{f_m} G_{n_3}(f) df = \frac{8\pi^2}{3} \cdot \frac{\alpha^2 \eta}{A^2} f_m^3 \qquad (12.23)$$

Note that the noise power increases parabolically with the frequency as seen in Fig. 12.2-3. Hence the output SNR from Eq. (12.21) and Eq. (12.23) is,

$$SNR = \frac{S_o}{N_o} = \frac{\alpha^2 k^2 \overline{m^2(t)}}{8\pi^2/3 (\alpha^2 \eta/A^2) f_m^3} = \frac{3}{4\pi^2} \cdot \frac{k^2 \overline{m^2(t)}}{f_m^2} \cdot \frac{A^2/2}{\eta f_m} \qquad (12.24)$$

D-124

If the modulating signal $m(t)$ is sinusoidal and produces a frequency deviation of Δf, then the output SNR is,

$$\text{SNR} = \frac{S_o}{N_o} = \frac{3}{2}\beta^2 \frac{S_I}{N_M} \quad (\text{for } m(t) = \text{a sinusoid})$$

where

$$\beta = \frac{\Delta f}{f_m} = \text{modulation index}$$
$$S_I = \text{input signal power} = \frac{A^2}{2}$$
$$N_M = \eta f_m = \text{noise power at the input in the baseband bandwidth } f_m$$

C) Comparison of AM and FM SNR

This section will compare the SNR of a FM system with a DSB modulation with carrier AM. If we assume 100 % modulation, when the baseband signal is a sinusoid the ratio of the figure of merit of FM to AM can be written as,

$$\frac{\upsilon_{FM}}{\upsilon_{AM}} = \frac{9}{2}\beta^2 \quad \text{where } \upsilon \triangleq \text{ figure of merit.}$$

As β is increased, the SNR is greatly improved. However as β increases, the bandwidth requirement also increases. This is due to the fact that $\beta = \Delta F / F_M$. A simple calculation will show the fact that, each increase in bandwidth by a factor of 2 increases $\upsilon_{FM}/\upsilon_{AM}$ by a factor of 4 (6dB).

12.3 NOISE IN PCM SYSTEMS

A) PCM Transmission System

Fig. 12.3-1 shows a binary PCM system. The baseband waveform $m(t)$ is first quantized. The quantized signal $m_q(t)$ can be expressed as,

$$m_q(t) = m(t) + e(t) \quad (12.25)$$

where $m_q(t)$ = quantized baseband signal
$e(t)$ = error signal resulted by quantization

Next the quantized signal is sampled by an impulse train $s(t)$ represented by,

$$s(t) = I \sum_{k=-\infty}^{\infty} \delta(t - kT_s) \quad (12.26)$$

$$T_s = \frac{1}{2f_m} = \text{sampling interval}$$

where
I = impulse strength
Therefore the sampled quantized baseband signal is,

$$m_{qs}(t) = m_s(t) + e_s(t)$$

$$= m(t)I \sum_{k=-\infty}^{\infty} \delta(t - kT_s) + e(t)I \sum_{k=-\infty}^{\infty} \delta(t - kT_s) \quad (12.27)$$

Then the quantized strength-modulated impulse train is transmitted over a communication channel, after being passed through an A/D converter. Transmission may be direct or by using PSK, FSK as discussed in earlier chapters. In any event, the received signal is detected by the use of a matched filter and thereafter passed to a D/A converter. We denote the output of the D/A converter as $\tilde{m}_s(t)$ and the output of baseband lowpass filter as $m(t)$. This output signal $m_o(t)$ has exactly the waveform of the original baseband signal with a possible difference in amplitude. The signal it also contains a noise waveform $n_q(t)$ due to quantization and $n_o(t)$ which is due to thermal noise.

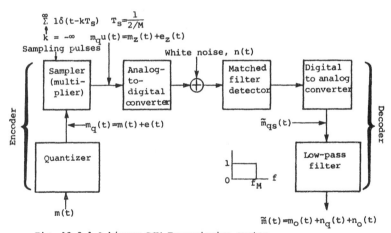

Fig. 12.3-1 A binary PCM Transmission system.

B) Calculation of Quantization And Thermal Noise
I. Quantization Noise :
From equation (12.27), we note that the sampled quantization error waveform is given by,

$$e_s(t) = e(t)I \sum_{k=-\infty}^{\infty} \delta(t - kT_s) \qquad (12.28)$$

where I = impulse strength of the sampling function
T_S = sampling period

The power spectral density of the sampled quantization error $e_s(t)$ can be derived as,

$$G_{e_s}(f) = \frac{I^2}{T_s}\overline{e^2(kT_s)} \qquad (12.29)$$

In Chapter 11, we found that if the quantization levels are separated by amount s, then the quantization error is given by,

$$\overline{e^2(t)} = \frac{s^2}{12}$$

Since the density function of $e(t)$ is stationary, the variance of $e(t)$ is equal to the variance of $e(kT_s)$

Hence, from (12.29) $G_{R_s}(f) = \frac{I^2}{T_s} \frac{s^2}{12} \qquad (12.30)$

Finally, the quantization noise power N_q can be obtained by integrating (12.30),

$$N_q = \int_{-f_m}^{f_m} G_{e_s}(f)\,df = \frac{I^2}{T_s}\frac{s^2}{12}2f_m = \frac{I^2}{T_s^2}\frac{s^2}{12} \qquad (12.31)$$

where $2f_m = \frac{1}{T_s}$

II..Thermal Noise :

The effect of additive thermal noise causes the matched filter detector to make an occasional error in determining whether the transmitted bit was a 1 or a 0. The probability of such an error depends on the relation E_s/η, where E_s is the transmitted signal energy during a bit and $\frac{\eta}{2}$ is power spectral density of the noise. The error probability also depends on the type of modulation used i.e. PSK, FSK etc.

The power spectral density of thermal noise error impulse train is,

$$G_{tn}(f) = \frac{I^2\overline{(\Delta m_s)^2}}{T} = \frac{NP_e I^2\overline{(\Delta m_s)^2}}{T_s} = \frac{2^{2N}s^2 P_e I^2}{3T_s} \qquad (12.32)$$

D-127

where I = impulse strength
P$_e$ = probability of a bit error
N = number of digits per word
T$_s$ = sampling interval separating each word
S = amount of incorrect determination in the quantized value of the sampled signal, which causes an error in the least significant bit.

$$(\Delta m_s)^2 = \frac{1}{N}\left[s^2 + (2s)^2 + (4s)^2 + \ldots + (2^{N-1}s)^2\right]$$
$$= \frac{2^{2N}-1}{3N}s^2 \cong \frac{2^{2N}s^2}{3N} \qquad (12.33)$$

where $2s$ = amount corresponding to an error in 2nd least significant bit.
$4s$ = amount corresponding to an error in the next higher bit and so on.

$$T = \frac{T_s}{NP_e} = \text{mean time between words which are in error.}$$

Finally, the output thermal noise power can be obtained by integrating equation (12.32),

$$N_{t_n} = \int_{-f_m}^{f_m} G_{tn}(f)\,df = \frac{2^{2N}s^2 P_e I^2}{3T_s^2} \qquad (12.34)$$

$$\text{where } T_s = \frac{1}{2f_m}$$

C) Calculation of Output Signal Power

From Eq. (12.27), the sampled signal which appears at the top of the baseband filter in Fig.12.3-1, is given by,

$$m_s(t) = m(t)I \sum_{k=-\infty}^{\infty} \delta(t - kT_s) \qquad (12.35)$$

Hence, the signal at the output of the filter, can be written as,

$$m_o(t) = \frac{I}{T_s}m(t) \qquad (12.36)$$

Therefore the normalized signal power from (12.36) is

$$\overline{M_o^2} = \frac{I^2}{T_s^2}\overline{m^2(t)} \qquad (12.37)$$

where $\overline{m^2(t)} = \frac{M^2 s^2}{12}$,

M = number of quantization levels
S = stepsize
Hence the output signal power is,

$$S_o = \overline{m_o^2(t)} = \frac{I^2}{T_s^2}\frac{M^2 s^2}{12} \qquad (12.38)$$

D) Output Signal to Noise Ratio :
Summarizing the results, we have,

$$S_o = \frac{I^2}{T_s^2}\frac{M^2 s^2}{12}$$

$$N_q = \frac{I^2}{T_s^2}\frac{s^2}{12}$$

$$N_{in} = \frac{2^{2N} s^2 P_e I^2}{3T_s^2}$$

Therefore the signal-to-quantization ratio, $\dfrac{S_o}{N_q} = M^2 = (2^N)^2 = 2^{2N}$

Signal-to-thermal noise ratio, $\dfrac{S_o}{N_{in}} = \dfrac{1}{4P_e}$

Signal-to-noise ratio, $\dfrac{S_o}{N_o} = \dfrac{S_o}{N_q + N_{in}} = \dfrac{2^{2N}}{1 + 4P_e 2^{2N}}$

E) SNR in the Case of PSK and FSK Modulations

$$\left(\frac{S_o}{N_o}\right)_{PSK} = \frac{2^{2N}}{1 + 2^{2N+1} \operatorname{erf}_c \sqrt{(1/2N)}(s_i/\eta f_m)} \qquad (12.39)$$

$$\left(\frac{S_o}{N_o}\right)_{FSK} = \frac{2^{2N}}{1 + 2^{2N+1} \operatorname{erf}_c \sqrt{(\cdot 3/N)}(s_i/\eta f_m)} \qquad (12.40)$$

where $\quad erf_c(u) = \dfrac{2}{\sqrt{\pi}} \displaystyle\int_u^\infty e^{-u^2} du \quad$ and is discussed in Chapter 6.

S_i = received signal power
N= number of bits corresponding to each code word.

12.4 NOISE IN DELTA MODULATION

A delta modulation system, including a thermal noise source is shown in Fig. 12.4-1. In the absence of thermal noise, the output of both integrators is identical. The modulator output is a sequence of pulses $P_o(t)$ whose polarity depends on the polarity of difference signal $\Delta(t) = m(t) - \tilde{m}(t)$, where $\tilde{m}(t)$ is the integrator output. The output noise power in the baseband frequency range is,

$$N_q = \frac{s^2}{3}\frac{f_m}{f_c} = \frac{s^2 \tau f_m}{3}$$

where $f_c = \dfrac{1}{\tau}$ baseband filter cutoff frequency

$\dfrac{s^2}{3}$ = quantization noise power in a frequency range f_c.

When thermal noise is present, the matched filter in the receiver of Fig. 12.4-1 will occasionally make an error in determining the polarity of the transmitted waveform. The thermal noise output is,

$$N_{tn} = \frac{2s^2 P_e}{\pi^2 T}\left(\frac{1}{f_i} - \frac{1}{f_m}\right) \cong \frac{2s^2 P_e}{\pi^2 f_i \tau} (\text{if } f_i \leq f_m) \quad (12.41)$$

where $\omega = 2\pi f$
$\quad$ s = step size
$\quad f_1$ = low frequency cutoffs of baseband filter
The maximum signal output power is,

$$S_o = \frac{s^2}{2\omega_m^2 \tau^2}$$

where $\omega_m = 2\pi f_m$

f_m = upper limit of the baseband frequency range
τ = interval between steps

Therefore the output signal-to-quanitzation ratio is,

$$\frac{S_o}{N_q} = \frac{3/8\pi^2}{(f_m \tau)^3} \approx \frac{3/80}{(f_m \tau)^3} \quad (12.42)$$

For a fixed channel bandwidth, we require that

$$\tau = \frac{1}{2f_m^N}$$

From (12.42) $\frac{S_o}{N_q} = 0.3 N^3$ (N = number of bits in a code word)

Hence, the output signal-to-noise ratio

$$\frac{S_o}{N_o} = \frac{S_o}{N_q + N_{tn}} = \frac{s^2/2\omega_m^2 \tau^2}{(s^2 \tau f_m /3) + (2s^2 P_e/\pi^2 f_1 \tau)}$$

$$= \frac{3\pi/(\omega_m \tau)^3}{1 + \{12/[(\omega_m \tau)(\omega_1 \tau)]\} erf_c \sqrt{s_i \tau/\eta}} \quad (12.43)$$

where $S_i \tau$ = received signal power = E_s

$P_e = \frac{1}{2} erf_c \sqrt{\frac{E_s}{\eta}}$ for PSK or direct transmission

In Fig. 12.4-2, a plot of SNR of PCM and DM is given. Note that in this example i.e. for N = 8, at high input, the output SNR for PCM is 26dB greater than for DM. Hence PCM always yields a superior performance as compared with delta modulation (DM), but at the expense of a complex circuitry.

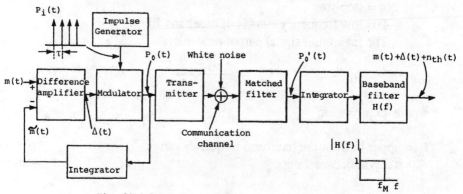

Fig. 12.4.1 A Delta-modulation system.

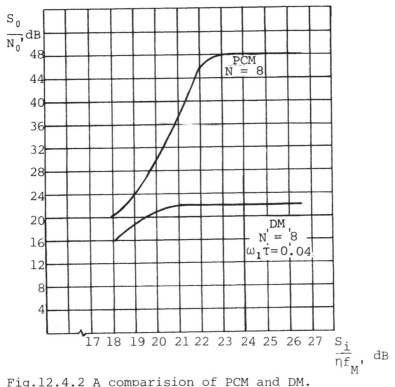

Fig.12.4.2 A comparision of PCM and DM.

CHAPTER 13

NOISE CALCULATIONS

13.1 RESISTOR NOISE

Resistor noise results from the random motion of electrons, which are free to move within the resistor.
It has been determined experimentally that the noise voltage $v(t)$ is gaussian and has a mean square value of

$$\overline{v_n^2} = 4kTR\, df \qquad (13.1)$$

where T = temperature in degrees kelvin
k = Boltzmann constant = 1.37×10^{-23} Joule /°K
df = narrow frequency band under consideration
R = resistance in ohms ──

Experiments also indicate that v_n^2 in Eq.(13.1) is independent of the center frequency of the filter of bandwidth df ranging from 0Hz to 10 GHz. Thus, excluding optical communication systems, for all communication systems, the normalized power supplied by $v_n(t)$ is also independent of this center frequency and hence the power spectral density of the noise source is white and given by,

$$G_v(f) = \frac{\overline{v_n^2}}{2df} = 2kTR \qquad (13.2)$$

where, $G_v(f)$ = power spectral density of the noise source.
For calculation of the noise power generated by a combination of resistos, it's worth remembering that the sum of two or more independent gaussian processes is itself a gaussian random process and further,

the variance of the sum is equal to the sum of the variances of individual processes. For example, the mean square voltage measured across the series combination of two resistors R_1 and R_2 is given by,

$$\overline{v_n^2} = \overline{v_{n_1}^2} + \overline{v_{n_2}^2} = 4kT(R_1+R_2)df \qquad (13.3)$$

13.2 NOISE IN A NETWORK WITH REACTIVE ELEMENTS

A physical resistor of resistance R at a temperature T has a mean square voltage given in Eq. (13.1), can be interpreted as a noiseless resistor in series with a voltage source $\overline{v_n^2}$, as shown in Fig. 13.2-1 and 13.2-2.

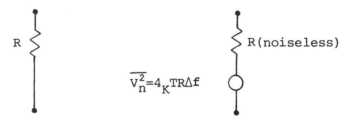

Figure 13.2-1
A physical resistor

Figure 13.2-2
Thevenin equivalent consisting of a noise- less resistor and a noise source.

similarly a network composed of resistors, inductors and capacitors in Fig. 12.2-3 can be represented by an equivalent network in Fig. 13.2-4. The impedance $Z(f)$ looking back into the network has real and imaginary part and is a function of frequency. That is,

$$Z(f) = R(f) + jX(f) \qquad (13.4)$$

where $R(f)$ = Real part
$X(f)$ = Imaginary part

The capictor and inductor have reactance given by,

$$X_C = \frac{1}{2\pi f C} \text{ and } X_L = 2\pi f L$$

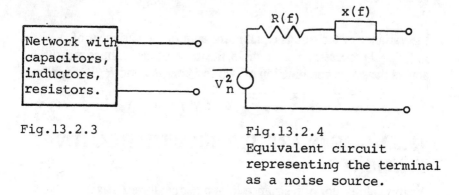

Fig.13.2.3

Fig.13.2.4
Equivalent circuit representing the terminal as a noise source.

The mean square voltage of Fig.13.2-4 is given by,

$$\overline{v_n^2} = 4kTR(f)df \quad (13.5)$$

Correspondingly, the two sided power spectral density of the open circuit voltage in Fig.13.2-4 is,

$$G_v(f) = 2kTR(f) \quad (13.6)$$

13.3 AVAILABLE POWER

The available power is defined as the maximum power that can be obtained in an external load, corresponding to a source and an internal resistance. The available power depends only on the resistive component of the source impedance. For the network drawn in Fig.13.3-1 the available power is,

$$P_a = \frac{v_s^2}{4R_s} \quad (13.7)$$

using Eq. (13.7), the available thermal noise power of a resistor R in the frequency range df is,

$$P_a = \frac{4kTRdf}{4R} = kTdf \quad (13.8)$$

Hence, the two sided thermal-noise PSD is,

$$G_a = \frac{P_a}{2df} = \frac{kT}{2} \quad (13.9)$$

It is important to note that, G_a in Eq. (13.9) does not depend on the resistance of the resistor, but only on the physical constant k and on the temperature. Similarly for a network with reactive element, the two-sided noise PSD is $\frac{kT}{2}$, which is independent of circuit components and the circuit configuration. Fig. 13.3-2 shows forms of one-sided and two-sided power spectral density.

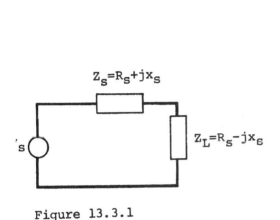

Figure 13.3.1

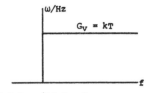

(a) One-sided noise power spectral density.

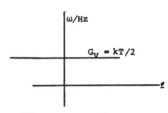

(b) Two-sided noise power spectral density.

Figure 13.3.2

13.4 NOISE TEMPERATURE

From Eq. (13.9), solving for T, we obtain,

$$T = \frac{P_a}{kdf} \quad (13.10)$$

For a frequency band B, if the available noise power is P_a, the noise temperature is,

$$T = \frac{P_a}{kB} \quad (13.11)$$

Effective Noise Temerature:
For two part networks, i.e. networks having input and output terminals, there will be noise sources inside the network. Hence the SNR at the output will be lower than at the input. In order to describe the extent to which the signal is corrupted in passing through a two-port network, there are two commonly employed methods. In one method, it is characterized in terms of effective input noise temperature. The other

method is the noise figure method, which will be discussed in the next section. We define the available gain $g_a(f)$ for two-port networks as,

$$g_a(f) = \frac{\text{Available PSD at the two-port output}}{\text{Available PSD at the source input}}$$

The transfer function $H(f)$ of a circuit has a similar meaning to that of $g_a(f)$, except that $g_a(f)$ relates to the ratio of power spectral densities. When N two-port networks are cascaded, the overall available gain is the product,

$$g_a = g_{a_1} \cdot g_{a_2} \cdot g_{a_3} \cdots g_{a_N} \qquad (13.12)$$

where $g_{a_1}, g_{a_2} \cdots g_{a_N}$ are the available gains for each two-port network. If the available gain of the two-port network is g_a, and if the two-port network is noise free, then the available two sided PSD at two-port output would be,

$$G'_{a_o} = g_a(f)\frac{kT}{2} \qquad (13.13)$$

However, because the network will add noise, the true PSD, G_{a_o} at the output will be greater than G'_{a_o}. This can be interpreted as if the two-port network is noise free and the increased noise is due to a temperature rise at the source i.e.

$$G_{a_o} = g_a(f)\frac{k(T+T_e)}{2} \qquad (13.14)$$

where T_e = effective input-noise temperature.
This effective temperature depends not only on the source, but also on the network itself.

13.5 NOISE FIGURE

The two-port noise figure is defined as,

$$F(f) = \frac{G_{a_o}}{G'_{a_o}} = \frac{G_{a_o}}{g_a(f)(KT_o/2)} \qquad (13.15)$$

where $F(f)$ = noise figure at a particular spot in the frequency spectrum
G_{a_o} = noise PSD at the output of a two-port network
G'_{a_o} = noise PSD at the output, if the network is noiseless
T_o = room temprature, usually 290 $\overset{\circ}{K}$.
Actual noise PSD, G_{a_o}, is always greater than G'_{a_o}. Hence, F is bounded by $1 < F < \infty$. An ideal noisefree circuit would result in $F = 1$. The larger the value of F, the more the SNR is degraded by the

circuit. Thus, the circuit with the lowest noise figure would be the most desirable if all other factors were equal. The noise figure in decibels is given as,

$$F_{dB} = 10 \log_{10} F \tag{13.16}$$

In the decibel form the ideal two-port network would have $F_{dB} = 0 dB$.

Noise Figure and Effective Temperature
From Eq. (13.14) with $T = T_o$, and Eq. (13.15), the noise figure and effective temprature T_e are related by,

$$T_e = T_o (F - 1) \tag{13.17}$$

or, inversely

$$F = 1 + \frac{T_e}{T_o} = \frac{T_o + T_e}{T_o} \tag{13.18}$$

The average noise figure over a frequency range from F_1 to F_2 is related to the spot noise figure in Eq. (13.15) as,

$$\overline{F} = \frac{\int_{F_1}^{F_2} g_a(f) F(f) df}{\int_{F_1}^{F_2} g_a(f) df} \tag{13.19}$$

If the power spectral densities of signal to noise are uniform over a frequency range from F_1 to F_2, then the average noise figure in Eq. (13.19) implies that,

$$\overline{F} = \frac{S_i / N_i}{S_o / N_o} \tag{13.20}$$

where S_i = Total input available signal power
N_i = Total input available noise power
S_o = Output available signal power
N_o = Output available noise power

If the available gain g_a is constant over the frequency range of interest, so that $\overline{F} = F$, then $S_o = g_a S_i$ and the following useful formulas can be deduced from Eq. (13.20).

$$F = \frac{1}{g_a}\frac{N_o}{N_i} \qquad (13.21)$$

$$N_o = g_a N_i + N_{tp} \qquad (13.22)$$

$$F = 1 + \frac{N_{tp}}{g_a N_i} \qquad (13.23)$$

$$N_{tp} = g_a(F-1)N_i \qquad (13.24)$$

where $g_a N_i$ = output noise due to the noise present at the input
N_{tp} = Additional noise due to the two port itself.
For a cascade of k two-port networks, the noise figure is

$$F = F_1 + \frac{F_2 - 1}{g_{a_1}} + \frac{F_3 - 1}{g_{a_1} g_{a_2}} + \ldots + \frac{F_k - 1}{g_{a_1} g_{a_2} \cdots g_{a_{(k-1)}}} \qquad (13.25)$$

and equivalent temprature T_e of k cascade stages is,

$$T_e = T_{e_1} + \frac{T_{e_2}}{g_{a_1}} + \frac{T_{e_3}}{g_{a_1} g_{a_2}} + \ldots + \frac{T_{e_k}}{g_{a_1} g_{a_2} \cdots g_{a_{(k-1)}}} \qquad (13.26)$$

13.6 NOISE BANDWIDTH

The noise bandwidth of a filter or amplifier is obtained from the formula,

$$B_N = \frac{1}{2 g_{a_o}} \int_{-\infty}^{\infty} g_a(f)\, df \qquad (13.27)$$

where g_{a_o} is the constant value of $g(f)$ over the passband of the filter.

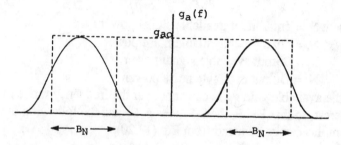

Fig. 13.6.1 Concept of noise bandwidth in a two-sided PSD representation.

Fig. 13.6-1 illustrates the concept of noise bandwidth.
The noise bandwidth, is in general, a different value than the usual bandwidth parameters, such as the 3dB bandwidth. The noise bandwidth may be very close to twice the usual 3dB bandwidth, for filters having sharp cutoffs. However, for a relatively slow cutoff, it is best to determine the noise bandwidth from Eq. (13.27). As the number of poles in the filter increases, the ratio of B_N to the 3dB bandwidth definitions becomes closer to unity for filters having relatively flat passband characteristics. An abbreviated set of noise bandwidth values for low-pass Butterworth filters up to 10 poles is provided in Table 13.6.1. In each case F_1 is the 3dB bandwidth and $g_a(f)$ has a maximum at DC and equals 1 for a unity gain filter.

TABLE 13.6-1
Equivalent Noise Bandwidths of Butterworth Filters

POLES	1	2	3	4	5	6	7	8	9	10
B_N/F_1	1.571	1.111	1.047	1.026	1.017	1.012	1.008	1.006	1.005	1.004

CHAPTER 14

DATA TRANSMISSION

14.1 AN INTEGRATE AND DUMP PCM RECEIVER

This receiver is shown in Fig. 14.1. A binary encoded PCM baseband signal consists of a time sequence of voltage levels, +V, 0, -V volts. When the noise is mixed with the input signal, the sampled value is generally different from 0 or ±V. Since there is an equal likelihood of transmitting +V, 0, -V volts, their probabilities are equal i.e.

$$P[+V] = P[0] = P[-V] = \frac{1}{3} \qquad (14.1)$$

The probability of error can be found by processing the received signal plus noise, in such a manner that the sample voltage at the output of receiver due to the signal is emphasized relative to the sample voltage due to the noise.

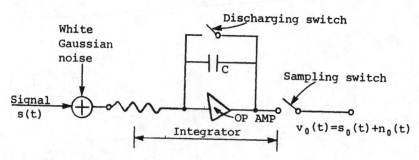

Fig. 14.1 An integrate and dump PCM receiver.

D-141

The output of the receiver is

$$V_0(T_0) = s_0(T_0) + n_0(T_0) \tag{14.2}$$

where T_0 is the output sampling time.

If we assume that the noise is gaussian, then the following results for the probability of error are obtained. The probability of error will be independent of the signal if and only if,

$$P_e[-V] = P_e[0] = P_e[+V] \tag{14.3}$$

and the probability of error is,

$$P_e = \frac{1}{2}erf_c\left(\frac{V^2 T}{\eta}\right)^{1/2} = \frac{1}{2}erf_c\left(\frac{E_s}{\eta}\right)^{1/2} \tag{14.4}$$

where T = bit interval
V = peak value of the input signal voltage
$\frac{\eta}{2}$ = PSD of gaussian or white noise

$E_s = V^2 T$ = signal evergy of a bit

Several values of $erf_c(x)$ are tabulated at the end of book.

14.2 THE MATCHED FILTER

When the input noise is white, the matched filter has a transfer function,

$$H(f) = \frac{KP^*(f)}{\eta/2} e^{-j 2\pi f T} \tag{14.5}$$

where
$$P(f) = F[p(t)] = F[s_1(t) - s_2(t)]$$

and $s_1(t), s_2(t)$ are the transmitted binary signals.
K = a constant
$\frac{\eta}{2}$ = power spectral density of white noise.

Now, the impulse response of the filter in Eq. (14.5) is,

$$h(t) = \frac{2k}{\eta}[s_1(T-t) - s_2(T-t)] \tag{14.6}$$

where T = one bit interval.

Error Probability

The probability of error for the matched filter is given by,

$$P_e = \frac{1}{2}\text{erf}_c\left[\frac{P_o(T)}{8\sigma_o^2}\right]^{1/2} \quad (14.7)$$

and

$$(P_e)_{minimum} = \frac{1}{2}\text{erf}_c\left[\frac{E_s}{\eta}\right]^{1/2} \quad (14.8)$$

where $P_o(T) = S_{o1}(T) - S_{o2}(T)$
E_s = signal energy of a bit
σ_o^2 = noise variance
S_{o1}, S_{o2} = Received signals corresponding to the transmitted signals.

From Eq. (14.8), note that the probability of error depends only on the signal's energy and not on the signal's waveshape.

14.3 COHERENT RECEPTION

The coherent reception system is an alternate to the integrate and dump system, discussed in section 14.1. As shown in Fig.14.2, the input to the system is a binary data waveform $s_1(t)$ or $s_2(t)$ corrupted by noise $n(t)$. The received signal plus noise $V_i(t)$ is multiplied by a locally generated waveform $s_i(t) - s_2(t)$. The output of the multiplier is passed through an integrator whose output is sampled at $t = T$, where T is the bit length. Immediately after each sampling, at the beginning of each new bit interval, the energy stored in the integrator is discharged. This type of receiver is also called a correlator, since the received signal and noise is correlated with the waveform $s_1(t) - s_2(t)$. The output signal and noise of the correlator are,

$$S_o(T) = \frac{1}{T}\int_0^T S_i[s_1(t) - s_2(t)]dt \quad (14.9)$$

$$n_o(T) = \frac{1}{\tau}\int_0^T n(t)[s_1(t) - s_2(t)]dt \quad (14.10)$$

Note that $S_o(T)$ and $n_o(T)$ in Eq. (14.9) and Eq. (14.10) are the same as in the case of the matched filter. Hence the performance of the two systems are identical. The correlators and matched filter are two independent techniques which happen to yield the same result. In the following sections, several applications of coherent reception systems are discussed.

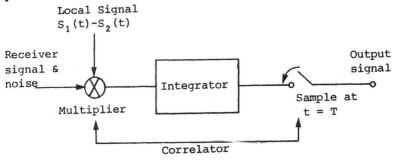

Fig. 14.2 A coherent system of signal reception.

14.4 PHASE-SHIFT KEYING

In phase-shift keying (PSK), the input signals are

$$s_1(t) = A \cos \omega_o t$$
$$s_2(t) = -A \cos \omega_o t$$

Error Probability

$$Pe = \tfrac{1}{2} \text{erfc} \sqrt{\frac{E_s}{\eta}} = \tfrac{1}{2} \text{erfc} \sqrt{\frac{A^2 T}{2\eta}} \quad (14.11)$$

where E_s = signal energy = $\dfrac{A^2 T}{2}$

T = one bit interval

(a) Imperfect Phase Synchronization

If the local signal $s_1(t) - s_2(t) = 2A \cos \omega_o t$ has a fixed phase offset i.e., the local signal becomes $2A \cos(\omega_o t + \phi)$ then the probability of error is,

$$P_e = \frac{1}{2}\mathrm{erf}_c\sqrt{\frac{E_s}{\eta}\cos^2\phi}$$

$$= \frac{1}{2}\mathrm{erf}_c\sqrt{\frac{A^2T\cos^2\phi}{2\eta}}$$
(14.12)

Thus, the phase shift ϕ increases the error probability.

(b) Imperfect Bit Synchronization

The bit synchronizer is used in the integrator to assure that the integration starts at $t = 0$ and ends at $t = T$. If, for some reason the integration ends at $t = T + \tau$ instead of at $t = T$, two different bits may be overlapped resulting in a increase in error probability, which is given by,

$$P_e = \frac{1}{2}\mathrm{erf}_c\sqrt{\left(\frac{E_s}{\eta}\right)\left(1 - \frac{2|\tau|}{T}\right)}$$
(14.13)

where τ = amount of overlap

Note that the output noise voltage will not be affected by this overlap since the integration time T remains unchanged. If both phase error and timing error are present, then the error probability is,

$$P_e = \frac{1}{2}\mathrm{erf}_c\left[\left(\frac{E_s}{\eta}\right)(\cos^2\phi)\left(1 - \frac{2|\tau|}{T}\right)^2\right]^{1/2}$$
(14.14)

14.5 FREQUENCY-SHIFT KEYING

In frequency-shift keying the received signal is either,
$$s_1(t) = A\cos(\omega_o + \Omega)t$$
or
$$s_2(t) = A\cos(\omega_o - \Omega)t$$

Hence, the local waveform used in the coherent reception system is,

$$P(t) = s_1(t) - s_2(t) = A\cos(\omega_o + \Omega)t - A\cos(\omega_o - \Omega)t$$
(14.15)

Error Probability

$$P_e \approx \frac{1}{2}\text{erfc}\left(0.6\frac{E_s}{\eta}\right)^{1/2} \quad (14.16)$$

where E_s = signal energy = $\frac{A^2 T}{2}$

14.6 FOUR-PHASE PSK (QPSK)

In QPSK, one of four possible waveforms is transmitted during each bit interval T. These waveforms are

$$S_i(t) = A\cos\left(\omega_o t + [2m-1]\frac{\pi}{4}\right) \quad m = 1, 2, 3, 4 \quad (14.17)$$

The receiver system is shown in Fig. 14.3. Note that two correlators are needed and hence the transmitted signal may be recognized from a determination of the output of both correlators. In the presence of noise there will be error due to one or both correlators. The probability that correlator 1 or correlator 2 will make an error is

$$P'_{e_1} = P'_{e_2} = \frac{1}{2}\text{erfc}\sqrt{\frac{A^2 T_s}{4\eta}} \quad (14.18)$$

where $T_s = 2T$
and T = duration of a single bit.

Error Probability

$$P_e(QPSK) = 1 - P_c \approx 2P'_{e_1} = \text{erfc}\sqrt{\frac{A^2 T}{4\eta}} \quad (14.19)$$

where P_c = probability that the QPSK system will correctly identify the transmitted signal $\approx 1 - 2P'_{e_1}$

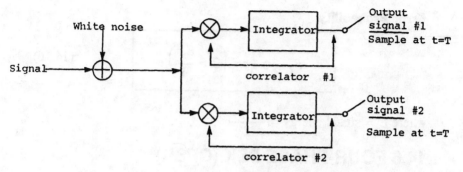

Fig. 14.3 A correlation receiver for QPSK

CHAPTER 15

INFORMATION THEORY AND CODING

15.1 CONCEPT OF INFORMATION

If $m_1, m_2 \ldots$ are allowable messages with probability of occurence $P_1, P_2, \ldots$ respectively in a communication system, then by definition the system has conveyed an amount of information I_K given by,

$$I_K \equiv \log_2 \frac{1}{P_K} \quad (15.1)$$

and $\quad P_1 + P_2 + \ldots = 1 \quad (15.2)$

15.2 ENTROPY AND INFORMATION RATE

Entropy
Entropy or average information per message interval, represented by symbol H, is given by

$$H = \frac{I_{total}}{L} = \sum_{K=1}^{M} P_K \log_2 \frac{1}{P_K} \quad (15.3)$$

where M represents number of different and independent messages with probability of occurence $P_1, P_2, \ldots$ and L represents the number of message generated during a time period and I_{total} is given by

$$I_{total} = P_1 L \log_2 \frac{1}{P_1} + P_2 L \log_2 \frac{1}{P_2} + \ldots \quad (15.4)$$

It is important to note that the entropy associated with an extremely unlikely message, as well as an extremely likely message, is zero, since

$$\lim_{P \to 0} P \log \frac{1}{P} = 0 \qquad (15.5)$$

Entropy becomes a maximum when all the messages are equally likely, i.e.

$$H_{max} = \sum_{k=1}^{M} \frac{1}{M} \log_2 M \qquad (15.6)$$

Information Rate

If the message source generates a message at the rate r messages per second, then the information rate is defined as,

$R = rH$ = average number of bits of information /second

15.3 SHANNON'S THEOREM

Theorem : Given a source of M equally likely messages, with $M >> 1$, which is generating information at R, and a channel capacity C, then if $R \leq C$ there exists which is generating information at a R. Given a channel capacity C. Then, if $R \leq C$ there exists a coding technique, such that the output of the source may be transmitted over the channel with a probability of error in the received message which may be made arbitrarily small.

Gaussian Channel

The Shannon-Hartley theorem, which is complementary to Shannon's theorem in Sec. 15. 3-1, is applicable to the channel in which the noise is gaussian. The theorem says that the channel capacity of a white, bandlimited gaussian channel is

$$\boxed{C = B \log_2\left(1 + \frac{S}{N}\right) \; (bits\,/\,sec)} \qquad (15.7)$$

where B = channel bandwidth
       S = signal power
       N = total noise within the channel bandwidth = ηB
       where $\frac{\eta}{2}$ = two sided PSD of white noise.

15.4 BANDWIDTH AND SNR TRADEOFF

Eq. (15.7) can be used to calculate the bandwidth SNR tradeoff. With fixed signal power and in the presence of white gaussian noise, the channel capacity approaches an upper limit with increasing bandwidth. From Eq. (15.7), this limit is known as Shannon's limit and can be calculated as

$$C_\infty = \lim_{B \to \infty} C = 1.44 \, s/n_o \qquad (15.8)$$

where S = signal power
$\frac{\eta}{2}$ = two sided PSD of white noise.

There is no lower limit on the bandwidth for SNR and bandwidth tradeoff.

15.5 CODING AND ERROR DETECTION

Coding

Generally, coding allows us in principle (up to the Shannon limit), to design a communication system in which both information bit rate and error rate are independently and arbitrarily specified but subject to a constraint on bandwidth. Coding accomplishes its purpose through the deliberate introduction of redundancy in to messages. The following sections describe the important features of commonly employed codes.

Repetition Code and the Binary Symmetric Channel

In Binary Symmetric Channel (BSC), shown in Fig. 15-1, the probability of transmission of any bit is,

$$P[0] = P[1] = \frac{1}{2} \qquad (15.9)$$

If a bit 1 or a 0 is replaced by a sequence of M 1's or M 0's respectively, then the probability of error of the transmission is,

$$P_e(M) = \sum_{i=(M+1)/2}^{M} \binom{M}{i} p^i (1-p)^{M-i} \qquad (15.10)$$

where p = probability of error in transmission

and $\binom{M}{i}$ represents the number of combination of M things taken i at a time.

The error probability in Eq. (15.10) decreases as M increases. However as M increases, the information rate is reduced by a factor $\frac{1}{M}$.

In order to keep the information rate a constant, the bit duration must be reduced by a factor of $\frac{1}{M}$ which in turn increases the bandwidth by the factor M.

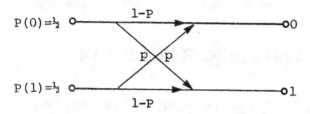

Fig. 15.1 The BSC (binary symmetric channel)

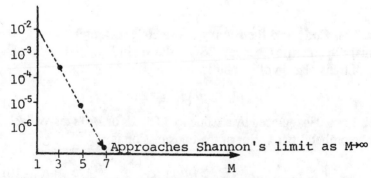

Fig. 15.2 Plot of error probability of BSC as a function of number of repetition M.

15.6 BLOCK CODES

Definition : Codes formed by taking a block of k information bits and adding r redundant bits to form a codeword of length $n = r + k$, are called block codes.

These kinds of block codes are also said to be systematic. A nonsystematic code has n bits in the codeword and k information bits which are not explicitly presented in the codeword.

Rate of the Code
By definition it is represented by

$$\boxed{R_C = \frac{k}{n}} \qquad (15.11)$$

if T_b and T_c represents the bit duration in the uncoded and coded word, then

$$\frac{f_c}{f_b} = \frac{T_b}{T_c} = \frac{n}{k} = \frac{1}{R_c} \qquad (15.12)$$

Hamming Distance and Weight
The distance between a pair of codewords is the number of places in which they differ from one another. The minimum value of this distance is called the hamming distance denoted by d_{min}. The hamming distance d_{min} established an upper limit to effectiveness of a code.

The two important properties of hamming distance are,

1) A code can detect D errors, if and only if

$$D \leq d_{min} - 1 \qquad (15.13)$$

2) A code can correct errors, if and only if

$$d_{min} \geq 2t + 1 \qquad (15.14)$$

The weight of a code is defined by,

$$W(c_i) = \sum_{j=1}^{n} d_{ij} \neq 0$$

in words, it is simply the number of non-zero places in the codeword. The hamming weight, is the minimum weight in the entire collection of code words.

Error Probability in Decoding
When soft-decision decoding is employed, i.e. a decision whether the received word is a codeword or not after all the bits are received, the error probability is given by,

$$P_{e_{soft}} \leq \frac{2^k}{2} \mathrm{erfc}\left[\left(\frac{E_b}{\eta}\right)\left(\frac{k}{n}\right) d_{min}\right]^{1/2} \quad (15.15)$$

If hard decision is employed, i.e. a decision is made about each received bit, then the error probability is,

$$P_{e_{hard}} \leq \frac{2^k}{2} \mathrm{erfc}\left[\left(\frac{E_b}{\eta}\right)\left(\frac{k}{n}\right) \frac{d_{min}}{2}\right]^{1/2} \quad (15.16)$$

where $E_b = P_s T_b$ = normalized bit energy in a bit

P_s = normalized signal power

$\frac{\eta}{2}$ = PSD of white noise

k = length of uncoded word

n = length of coded word.

A soft decision decoding has 2 to 3 dB improvement over hard decision decoding at the expense of extremely complex hardware.

Coding of Block Codes
If there are k information bits and r parity bits, then the codeword has $n = k + r$ bits.

$M = 2^K$ = number of valid codewords.

Define the parity check matrix H, whose elements are

$$H = \left| \begin{array}{cccc|ccccc} h_{11} & h_{12} & \cdots & h_{1k} & 1 & 0 & 0 & \cdots & 0 \\ h_{21} & h_{22} & \cdots & h_{2k} & 0 & 1 & 0 & \cdots & 0 \\ \vdots & \vdots & & \vdots & \vdots & & & & \vdots \\ \vdots & \vdots & & & & & & & \\ h_{r1} & h_{r2} & \cdots & h_{rk} & 0 & 0 & 0 & \cdots & 1 \end{array} \right|$$

$\qquad\qquad\qquad$ rxk $\qquad\qquad$ rxr Identity matrix

The h's in the above matrix are constant logical variables with value 0 or 1. The generator matrix G can be obtained from the H matrix i.e.,

$$G = \begin{vmatrix} 1 & 0 & 0 & \cdots & 0 & h_{11} & h_{11} & \cdots & h_{11} \\ 0 & 1 & 0 & \cdots & 0 & h_{12} & h_{12} & \cdots & h_{12} \\ \vdots & & & & & \vdots & & & \\ 0 & 0 & 0 & \cdots & & h_{1k} & h_{1k} & & h_{1k} \end{vmatrix}$$

$\underbrace{\phantom{k \times k \text{ identity matrix}}}_{k \times k \text{ identity matrix}} \quad \underbrace{}_{k \times r}$

The relationship between the generator matrix and parity check matrix is such that,

$$GH^T = 0 \qquad (15.17)$$

Decoding the Received Codeword

Let T = transmitted codeword
 R = received word = $T + E$
where E = error word
If the syndrome = S = HR^T = 0, then one of the following is true,

 1) There are no errors

or,

 2) There are so many errors, that a transmitted codeword has been changed to a different codeword.

If the syndrome $S = HR^T \neq 0$, then errors have been made in one of more places of the transmitted bit. Now

$$S = HR^T = H(T^T + E^T) = HT^T + HE^T = HE^T$$

This $S = HE^T$ provides information about the bit position in which errors have been made. If S corresponds to any particular column of the parity check matrix, then an error has been made in the corresponding bit position of the transmitted word.

CHAPTER 16

ANTENNAS

16.1 INTRODUCTION

The antenna uses voltage and current from a transmission line or the electric and magnetic field from a waveguide to "launch" an electromagnetic waveform into free space or into the local environment. The antenna acts as a transducer to match the transmission line or waveguide to the medium surrounding the antenna. The launching process is known as radiation, and the launching antenna is the transmitting antenna. If a wavefront is intercepted by an antenna, some power is absorbed from the wavefront, and this antenna acts as a receiving antenna.

In free space, the wavelength λ of a propagating waveform is determined from the equation

$$\lambda = \frac{c}{f} \tag{16.1}$$

where c = speed of light = $3 \times 10^8 (m/\sec)$
f = frequency in Hertz

Power Density

At any point on the imaginary sphere enclosing the antenna at the center, the power density is the power per unit area radiating through an infinitesimal unit of area. The radiated waveform spreads as the inverse square of distance from the antenna. The power density concept is illustrated in Fig. 16-1.

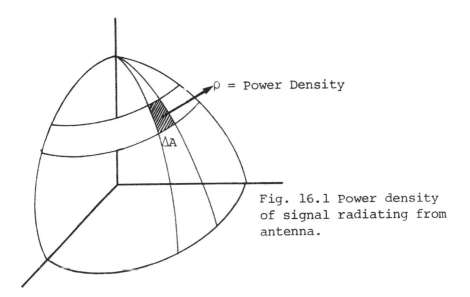

Fig. 16.1 Power density of signal radiating from antenna.

Radiation Pattern

The radiation pattern is a plot of the relative strength or intensity of the antenna radiation as a function of the orientation in a given plane. Two examples of radiation patterns are shown in Fig. 16.2.

Omnidirectional : In this case the radiation is constant at a given distance from the antenna and is independent of the angular orientation with respect to the antenna.

Directional: Here the radiation pattern is more directive towards any one side. This antenna is characterized by a major lobe and some minor lobes. There is almost no radiation in the direction opposite to the peak of the major lobe. There will be a small radiation in the direction of minor lobes.

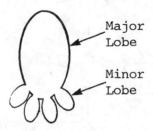

a) Omnidirectional   b) Directional

Fig. 16.2

16.2 CONCEPT OF GAIN AND BEAMWIDTH

Antenna gain is a measure of the concentration of power density in the radiated wavefront in a given direction. Gain is usually compared to an isotropic antenna, one which radiates equally in all direction and thus has a gain of unity. The antenna gain is defined as

$$G = \frac{\dfrac{\text{maximum power radiated}}{\text{unit solid angle}}}{\dfrac{\text{total power delivered to antenna}}{4\pi}}$$

Directivity is defined as,

$$G_D = \frac{\dfrac{\text{maximum power radiated}}{\text{unit solid angle}}}{\dfrac{\text{total power radiated}}{4\pi}}$$

The gain and directivity are the same for an ideal antenna, one for which there is no loss.

Ideal Antenna :

$$G = G_D = \frac{4\pi}{\Omega_B}$$

where Ω_B = beam solid angle (in steradians) measured between 3 dB points.

Beamwidth

The beamwidth is related to beam solid angle by the formula:

$$\Omega_B = \left(\frac{\pi}{4}\right)\theta_B^{\,2} \text{ (steradians)}$$

where θ_B = beamwidth

The concept of beam solid angle for an ideal antenna is illustrated in Fig.16-3. The beam area is defined as

$$\text{Beam Area} = r^2\Omega_B = r^2\left(\frac{\pi}{4}\right)\theta_B^2$$

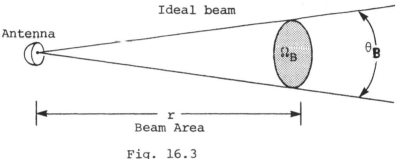

Fig. 16.3

16.3 ANTENNA CHARACTERISTICS

Capture Area or Aperture

The capture area or aperture A_e of an antenna is an effecitve "area" parameter for the antenna. It may or may not correspond to the physical area. The aperture is defined as,

$$A_e = \frac{\text{received power } (W)}{\text{maximum power density of antenna } (W/m^2)}$$

The aperture and the gain of an antenna are related by the following formula.

$$A_e = \frac{\lambda^2}{4\pi}G$$

Antenna Impedance

A given antenna at a given frequency will exhibit a certain input impedance having a real part and an imaginary part. In the proper operating range, the imaginary part should ideally be zero, and the real part should ideally be equal to the characteristic resistance of the transmission line if reflection are to be avoided.

Radiation Resistance

The radiation resistance R_{rad} is the real part of the complex input impedance to the antenna and is the value that accounts for the actual power accepted from the input signal. Thus, the input power P_{in} to an antenna is

$$P_{in} = I^2 R_{rad}$$

where I = RMS value of the input current.

16.4 POWER RELATIONS

Power is related to the voltage and impedance by

$$P = \frac{V^2}{Z}$$

where V = voltage
Z = impedance of the system

The power received or transmitted by an antenna can be expressed as

$$S_R = \frac{S_T}{4\pi d^2} G_T G_R \frac{\lambda^2}{4\pi} = \frac{S_T G_T G_R}{\left(4\pi \frac{d}{\lambda}\right)^2}$$

where S_R = received power at the antenna
S_T = transmitted power at the antenna
G_T = transmitter antenna gain
G_R = receiver antenna gain
λ = radiated wavelength
d = distance between sending and receiving points

The ratio of S_R to S_T is related to the effective aperture by the formula

$$\frac{S_R}{S_T} = \frac{A_{Tem} A_{Rem}}{(d\lambda)^2}$$

where A_{Tem} = transmitter maximum effective aperture
A_{Rem} = receiver maximum effective aperture
λ = wavelength
d = distance between two antennas

Transmission Path Loss

The basic transmission path loss, L is defined as the reciprocal of the ratio S_R to S_T, expressed in decibels,

$$L = 10 \log_{10} \left(\frac{4\pi d}{\lambda}\right)^2 \quad (db)$$

where L = transmission path loss.

The Radar Equation

The simplest form of radar system is one in which the power is transmitted in short bursts, and the time delay is measured for the return energy to be detected back at the source location. From a knowledge of time delay, T_d the distance d to the object may be computed as

$$d = \frac{cT_d}{2}$$

where c = speed of propogation of the waves
T_d = time delay

The model for developing the radar equation is shown in Fig. 16-4. In this analysis, we will assume that the same antenna is used for both transmitting and receiving. The power density P of the transmitted signal is

$$P = \frac{gP_t}{4\pi d^2}$$

where g = antenna gain
P_t = power transmitted from isotropic antenna
d = distance between the antenna and the object

The backscatter cross-section parameter denoted by σ, is defined as

$$\sigma = \frac{\text{power backscattered in direction of source } (W)}{\text{incident } (W/m^2)}$$

Note that σ has a dimension of square meters, so it is an area parameter. The backscatter power P_{bs} is determined as

$$P_{bs} = \sigma P = \frac{\sigma P_t g}{4\pi d^2}$$

Back at the receiver, the power density P_l resulting from the backscattered energy is

$$P_l = \frac{P_{bs}}{4\pi d^2} = \frac{\sigma P_t g}{(4\pi)^2 d^2}$$

Due to the aperture effect of antennas, the power captured P_r is expressed as

$$\boxed{P_r = P_l A_e = \frac{\sigma P_t g A_e}{(4\pi)^2 d^2} = \frac{\sigma P_t g^2 \lambda^2}{(4\pi)^3 d^4}}$$

where A_e = effective aperture = $\dfrac{\lambda^2 g}{4\pi}$

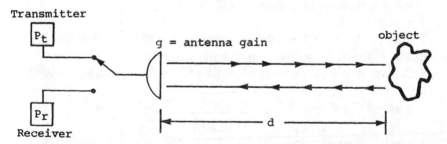

Fig. 16.4 Model used in developing the radar range equation

CHAPTER 17

TRANSMISSION LINES

17.1 EQUATIONS FOR LINE VOLTAGE AND CURRENT

The parameters of a transmission line are its resistance, inductance, conductance, and capacitance per unit length of the line pair. These parameters are typically given as

$$R \quad (\Omega m^{-1})$$
$$L \quad (Hm^{-1})$$
$$G \quad (\mho m^{-1})$$
$$C \quad (Fm^{-1})$$

As indicated in Fig. 17-1, uniform transmission lines may assume a variety of different cross-section geometries.

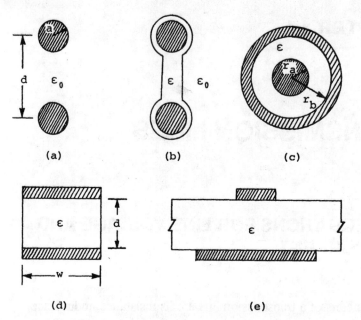

Figure 17.1 Cross sections of various types of transmission lines: (a) parallel lines; (b) parallel wires imbedded in dielectric material; (c) coaxial conductors; (d) planar; and (e) strip line.

The uniform transmission line is modeled schematically in Fig. 17-2, but note that the lumped elements R, L, G, and C are actually distributed parameters.

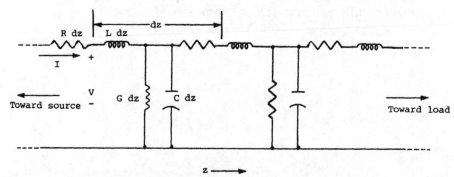

Fig. 17.2 Lumped constant representation of a transmission line.

Characteristic Impedance

The ratio $\sqrt{Z/Y}$ is called the characteristic impedance of the line, designated Z_o.

$$Z_o = \sqrt{\frac{Z}{Y}} = \sqrt{\frac{R + j\omega L}{G + j\omega C}} \quad (17.1)$$

Line Voltage and Current Expressions

At a distance l from the load, $l = -z$, the equations for voltage and current may be written independently of any coordinate axis as

$$v(l) = v^+ e^{\gamma l} + v^- e^{-\gamma l}$$
$$= v^+(e^{\gamma l} + Pe^{-\gamma l}) \quad (17.2)$$

and

$$I(l) = \frac{v^+}{Z_o}(e^{\gamma l} - Pe^{-\gamma l}) \quad (17.3)$$

where $P = \dfrac{v^-}{v^+}$

γ = propagation constant = $\sqrt{ZY}$
 = $\sqrt{(R + j\omega l)(G + j\omega c)} \ (m^{-1})$

where Z = series impedance per unit line length
and Y = shunt admittance per unit line length.

Input Impedance

It is also possible to express the load voltage and current in terms of the input voltage and current. Referring to Fig. 17-2, the input impedance to the line at a distance l from the load is

$$Z_{in} = Z(l) = \frac{v(l)}{I(l)} = Z_o \frac{Z_L + Z_o \tanh \gamma l}{Z_o + Z_L \tanh \gamma l} \ (\Omega) \quad (17.4)$$

where

$$Z_L = \frac{V_L}{I_L} = Z_o \frac{1+P}{1-P} \quad (17.5)$$

17.2 DISTORTIONLESS LINE

A propogation constant with a real part shows that there is attenuation as well as phase shift as the voltage and current waves propagate along the line. For reference, the propagation constant γ is given by

$$\gamma = \sqrt{ZY} = \sqrt{(R + jwL)(G + jwC)} = (\alpha + j\beta) \quad (17.6)$$

where α is referred as the attenuation constant.

In order to have a distortionless transmission, the attenuation constant α must not be a function of frequency, and β must be directly proportional to frequency. When the characteristic impedance Z_o is real, i.e.,
if

$$Z_o = \frac{R + jwL}{G + jwC} \text{ is real}$$

then

$$\frac{L}{R} = \frac{C}{G}$$

and

$$Z_o = \sqrt{\frac{R}{G}} = \sqrt{\frac{L}{C}} \quad (\Omega) \quad (17.7)$$

and also

$$\gamma = \sqrt{RG} + jw\sqrt{LC} = (\alpha + j\beta) \quad (m^{-1}) \quad (17.8)$$

17.3 LOSSLESS LINES

A lossless line is a transmission line in which the series resistance and shunt coductance are both zero. The lossless line is also a distortionless line according to the criteria established in the previous section. The voltage and current for a lossless line are given by

$$V = v^+\left(e^{jw\sqrt{LC}\,l} - Pe^{-jw\sqrt{LC}\,l}\right) \quad (V) \qquad (17.9)$$

$$I = \frac{v^+}{Z_o}\left(e^{jw\sqrt{LC}\,l} - Pe^{-jw\sqrt{LC}\,l}\right) \quad (A) \qquad (17.10)$$

The characteristic impedance Z_o is given by

$$Z_o = \sqrt{\frac{Z}{Y}} = \sqrt{\frac{0 + jwL}{0 + jwC}} = \sqrt{\frac{L}{C}} \quad (\Omega) \qquad (17.11)$$

and propagation constant

$$\gamma = \sqrt{ZY} = jw\sqrt{LG} = j\beta \quad (m^{-1}) \qquad (17.12)$$

17.4 INPUT IMPEDANCE AND STANDING WAVES (FOR LOSSLESS LINES)

The input impedance at a distance from the load for a lossless line is given by

$$Z_{in} = Z_o \frac{Z_L + jZ_o \tan\beta l}{Z_o + jZ_L \tan\beta l} \quad (\Omega) \qquad (17.13)$$

The maxima and minima of Z_{in} occurs at the maxima and minima of the voltage and current standing waves.

$$(Z_{in})_{max} = \frac{V_{max}}{I_{min}} \quad (\Omega) \qquad (17.14)$$

$$(Z_{in})_{min} = \frac{V_{min}}{I_{max}} \quad (\Omega) \qquad (17.15)$$

The standing wave ratio is designated as

$$S = \frac{V_{max}}{V_{min}} = \frac{I_{max}}{I_{min}} = \frac{1 + |P|}{1 - |P|} \qquad (17.16)$$

where

$$P = \frac{Z_L - Z_o}{Z_L + Z_o} \qquad (17.17)$$

The maxima and minima of Z_{in} can now be expressed in terms of the standing wave ratio.

$$(Z_{in})_{max} = SZ_o \quad (\Omega) \qquad (17.18)$$

$$(Z_{in})_{min} = \frac{Z_o}{S} \quad (\Omega) \qquad (17.19)$$

Power Flow

If V( ) and I( ) are the total voltage and current at a distance from the load, the average input power at this point is

$$P_{av,in} = \left[\left(\frac{1}{2}\right)\frac{|v^+|^2}{Z_o} - \left(\frac{1}{2}\right)\frac{|v^+|^2|P|^2}{Z_o}\right] \quad (\omega) \qquad (17.20)$$

The first term in the above equation is the incident power, and second term is the reflected power. The incident power is the power in the forward traveling wave, and it is the power that would be absorbed by the load if the reflection coefficient were zero, i.e., if $Z_L = Z_0$. Thus for lossless lines,

$$\boxed{P_{in} = P_{net} = P_{load} \quad (P_{incident} - P_{reflected})}$$

17.5 SKIN EFFECT, AND HIGH AND LOW LOSS APPROXIMATION

In the equations developed in an earlier section, it was assumed that R and G are independent of frequency. Realistically, because of skin effect, the effective cross-sectional area of the conductor is reduced when the frequency is increased; thus R must increase with frequency. An exact analysis of skin effect in conductors of circular cross-section involves the use of Bessel functions, but if the skin depth δ is small compared to radius of the conductor, the approximate determination of R is illustrated in Fig. 17-3. For a coaxial line, where δ is calculated as for a plane surface

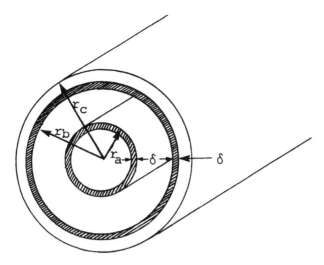

Figure 17.3 Illustration of the reduction in current-carrying cross-sectional area of coaxial conductors due to skin effect.

The total series resistance R is, approximately,

$$R \approx \frac{1}{2} \frac{\sqrt{f_M}}{\pi \sigma} \left(\frac{1}{r_a} + \frac{1}{r_b} \right) \quad (\Omega m^{-1}) \quad (17.21)$$

where σ = conductivity (υ /m)
M = permeability (H/M)

If the transmission line losses are small, i.e. $RG << w^2 LC$, then,

$$\alpha_{low-loss} = \frac{1}{2} \left(R \sqrt{\frac{C}{L}} + G \sqrt{\frac{L}{C}} \right) \quad (m^{-1}) \quad (17.22)$$

and

$$\beta_{low-loss} \approx w \sqrt{LC} \quad (m^{-1})$$

If the series resistance and shunt conductance are appreciable such that $RG >> w^2 LC$, then the high loss approximations are,

and
$$\alpha_{high-loss} \approx \sqrt{RG} \quad (m^{-1})$$

$$B_{high-loss} \approx \frac{1}{2} w \left(C \sqrt{\frac{R}{G}} + L \sqrt{\frac{G}{R}} \right) \quad (m^{-1}) \quad (17.23)$$

Handbook of Electrical Engineering

SECTION E
Laplace Transforms

CHAPTER 1

THE LAPLACE TRANSFORM

1.1 INTEGRAL TRANSFORMS

A class of transformations, which are called **integral transforms**, are defined by

$$T\{f(t)\} = \int_{-\infty}^{\infty} K(s,t)\, f(t)\, dt = F(s) \qquad (1)$$

Given a function $K(s, t)$, called the **kernel** of the transformation, equation (1) associates with each function $f(t)$, of the class of functions for which the above exists, a function $F(s)$ defined by (1).

Various particular choices of the kernel function $K(s, t)$ in (1) have led to a special transformation, each with its own properties to make it useful in specific circumstances. The transform defined choosing the kernel

$$K(s,t) = \begin{cases} 0 & \text{for } t < 0 \\ e^{-st} & \text{for } t \geq 0. \end{cases} \qquad (2)$$

is called **Laplace transform**, which is the one to which this book is devoted.

1.2 DEFINITION OF LAPLACE TRANSFORM

Let $f(t)$ be a function for $t > 0$. The Laplace transform of $f(t)$, denoted by $L\{f(t)\}$, is defined by

$$L\{f(t)\} = \int_0^\infty e^{-st} f(t)\, dt \tag{3}$$

1.3 NOTATION

The integral in (3) is a function of the parameter s that is called $F(s)$. It is customary to denote the functions of t by the lower case letters f, g, h, k, y, etc., and their Laplace transforms by the corresponding capital letters. Also some texts use capital letter L and show the Laplace transform of $f(t)$ by $L\{f(t)\}$. Therefore we may write

$$\mathcal{L}\{f(t)\} = L\{f(t)\} = \int_0^\infty e^{-st} f(t)\, dt = F(s) \tag{4}$$

If $f(x)$ is a function of $x \geq 0$, then its Laplace transform is denoted by $L\{f(x)\}$.

Example

Find $L\{\sin at\}$.

By definition

$$L\{\sin at\} = \int_0^\infty e^{-st} \sin at\, dt.$$

By employing integration by parts

$$e^{bx} \sin mx\, dx = \frac{e^{bx}(b \sin mx - m \cos mx)}{b^2 + m^2} + c.$$

Therefore

$$L\{\sin at\} = \left[\frac{e^{-st}(-s \sin at - a \cos at)}{(-s)^2 + a^2} \right]_{t=0}^{t=\infty}$$

E-2

Since for positive s, $e^{-st} \to 0$ at $t \to \infty$, and $\sin at$ and $\cos at$ are bounded functions, therefore the above yields

$$\mathcal{L}\{\sin at\} = 0 - \frac{-a}{s^2 + a^2} = \frac{a}{s^2 + a^2}$$

1.4 LAPLACE TRANSFORMATION OF ELEMENTARY FUNCTIONS

The following table shows Laplace transforms of some elementary functions.

Note: Factorial n. For every integer $n > 0$,

$$\text{factorial } n = n! = 1 \cdot 2 \cdot \ldots \cdot n$$

and by definition $0! = 1$.

Table 1

$f(t)$	$\mathcal{L}\{f(t)\} = F(s)$	
1	$\frac{1}{s}$	$s > 0$
t	$\frac{1}{s^2}$	$s > 0$
t^n for $n = 0, 1, 2, \ldots$	$\frac{n!}{s^{n+1}}$	$s > 0$
e^{at}	$\frac{1}{s-a}$	$s > a$
$\sin at$	$\frac{a}{s^2 + a^2}$	$s > 0$
$\cos at$	$\frac{s}{s^2 + a^2}$	$s > 0$
$\sinh at$	$\frac{a}{s^2 - a^2}$	$s > |a|$
$\cosh at$	$\frac{s}{s^2 - a^2}$	$s > |a|$
$\sqrt{t}$	$\frac{1}{2s}\sqrt{\frac{\pi}{s}}$	$s > 0$

1.5 SECTIONALLY OR PIECEWISE CONTINUOUS FUNCTIONS

The function $f(t)$ is said to be **piecewise continuous** or **sectionally continuous** over an interval $a < t < b$ if that interval can be divided into a finite number of intervals $c < t < d$ such that

1. $f(t)$ is continuous in the open interval $c < t < d$,

2. $f(t)$ approaches a finite limit as t approaches each end point within the interval $c < t < d$; that is, the limits

$$\lim_{\varepsilon \to 0} f(d-\varepsilon) \text{ and } \lim_{\varepsilon \to 0} f(c+\varepsilon)$$

exist and are finite, where $\varepsilon > 0$.

An example of a sectionally continuous function is shown in the following Figure 1.1.

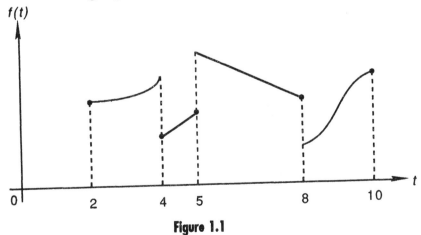

Figure 1.1

1.6 FUNCTIONS OF EXPONENTIAL ORDER

The function $f(t)$ is said to be of **exponential order** as $t \to \infty$ if there exists two constants $M > 0$ and b and a fixed t_o

such that

$$|f(t)| < M e^{bt} \quad \text{for all } t \geq t_0 \qquad (5)$$

We also say $f(t)$ is of **exponential order** b as $t \to \infty$, and we write

$$f(t) = 0(e^{bt}), \quad t \to \infty. \qquad (6)$$

Example 1

Most elementary functions such as all power functions

$$f(t) = t^n \text{ for every } n \geq 0,$$

$$\sin at, \quad \cos at, \quad e^{at}, \quad \sinh at, \quad \text{and} \quad \cosh at$$

are all of exponential order. Because for example $t^n \leq e^{nt}$, for all $t > 0$, therefore $f(t)$ is of exponential order n (one could show that the exponential order of the function t^n is as low as any positive number).

Example 2

The functions

$$e^{t^2} \text{ and } e^{t^3}$$

are some examples of functions that are not of exponential order. Because for any given number $b > 1$ we have

$$e^{t^3} > e^{t^2} > e^{bt}$$

for all $t > b$.

1.6.1 SUFFICIENT CONDITION FOR EXPONENTIAL ORDER FUNCTIONS

Let $f(t)$ be a function on $t \geq 0$, if a constant b exists such that the limit

$$\lim_{t\to\infty}[e^{-bt}f(t)]$$

exists, then the function $f(t)$ is of exponential order.

Example

The function $f(t) = t^5 + e^{2t}$ is of exponential order 2, since

$$\left|e^{-2t}f(t)\right| = \frac{t^5 + e^{2t}}{e^{2t}} \to 1 \quad \text{as} \quad t \to \infty.$$

1.7 EXISTENCE OF LAPLACE TRANSFORM

Theorem 1.1

If for all finite $N > 0$ the integral

$$\int_0^N e^{-st} f(t)\, dt$$

exists and $f(t)$ is of exponential order b as $t \to \infty$, then the Laplace transform

$$\mathcal{L}\{f(t)\} = \int_0^\infty e^{-st} f(t)\, dt = F(s)$$

exists for $s > b$.

1.7.1 BEHAVIOR OF LAPLACE TRANSFORMS AT INFINITY

Theorem 1.2

If $f(t)$ is sectionally continuous over every finite interval in the range $t \geq 0$ and if $f(t)$ is of exponential order,

$$f(t) = 0(e^{bt}) \quad \text{as} \quad t \to \infty,$$

then the Laplace transform $\mathcal{L}\{f(t)\}$ exists for $s > b$.

Theorem 1.3

If $f(t)$ is sectionally continuous over every finite interval

in the range $t \geq 0$, and is of exponential order as $t \to \infty$, and if
$$\mathcal{L}\{f(t)\} = F(s)$$
then

$$\lim_{s \to \infty} F(s) = 0$$

Remark: It is important to realize that the conditions stated in Theorem 1.2 are sufficient to guarantee the existence of the Laplace transform. If the conditions are not satisfied, however, the Laplace transform may or may not exist. For example the function $t^{-1/2}$ is not sectionally continuous in every interval in the range $t \geq 0$. But

$$\mathcal{L}\{t^{-1/2}\} = (\pi/s)^{1/2}, \quad s > 0$$

exists.

1.8 SOME IMPORTANT PROPERTIES OF LAPLACE TRANSFORMS

In the following theorems we assume that all functions satisfy the sufficient conditions in Theorem 1.2 so that their Laplace transforms exist.

1.8.1 LINEARITY PROPERTY

Theorem 1.4

Let $f_1(t)$ and $f_2(t)$ be two functions with Laplace transforms $F_1(s)$ and $F_2(s)$ correspondingly. Then for all constants c_1 and c_2 we have

$$\mathcal{L}\{c_1 f_1(t) + c_2 f_2(t)\} = c_1 \mathcal{L}\{f_1(t)\} + c_2 \mathcal{L}\{f_2(t)\} \qquad (7)$$
$$= c_1 F_1(s) + c_2 F_2(s)$$

Example

$$\mathcal{L}\{\sin 5t - 3e^{-2t} + 6t^5\} = \mathcal{L}\{\sin 5t\} - 3\mathcal{L}\{e^{-2t}\} + 6\mathcal{L}\{t^5\}$$

$$= \left(\frac{5}{s^2 + 25}\right) - 3\left(\frac{1}{s+2}\right) + 6\left(\frac{5!}{s^6}\right)$$

$$= \frac{5}{s^2 + 25} - \frac{3}{s+2} + \frac{720}{s^6}$$

1.8.2 CHANGES OF SCALE PROPERTY

Theorem 1.5

Let $f(t)$ be a function with the Laplace transform $F(s)$. Then for every $a > 0$, we have

$$\mathcal{L}\{f(at)\} = \frac{1}{a} F\left(\frac{s}{a}\right) \qquad (8)$$

Example

Since $\mathcal{L}\{\cos t\} = \dfrac{s}{s^2 + 1}$, then

$$\mathcal{L}\{\cos 5t\} = \frac{1}{5} \frac{(s/5)}{(s/5)^2 + 1} = \frac{s}{s^2 + 25}$$

1.8.3 FIRST SHIFT PROPERTY

Theorem 1.6

Let $\mathcal{L}\{f(t)\} = F(s)$, then

$$\mathcal{L}\{e^{at} f(t)\} = F(s - a) \qquad (9)$$

for all real numbers a, Figure 1.2, following.

Example

Find $\mathcal{L}\{t^3 e^{5t}\}$. Apply the first shift property for the function

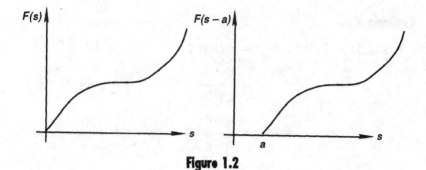

Figure 1.2

$f(t) = t^3$ and $a = 5$. Since

$$\mathcal{L}\{t^3\} = \frac{3!}{s^{3+1}} = \frac{6}{s^4} \quad \text{for} \quad s > 0$$

then

$$\mathcal{L}\{t^3 e^{5t}\} = \frac{6}{(s-5)^4} \quad \text{for} \quad s > 5.$$

1.8.4 SECOND SHIFT PROPERTY

Theorem 1.7

If

$$\mathcal{L}\{f(t)\} = F(s) \quad \text{and} \quad g(t) = \begin{cases} f(t-a) & \text{if } t \geq a \\ 0 & \text{if } t < a, \end{cases}$$

then

$$\mathcal{L}\{g(t)\} = e^{-as} F(s) \qquad (10)$$

Figure 1.3, following.

Example

Find the Laplace transform of the function

$$g(t) = \begin{cases} \sin(t-5) & t \geq 5 \\ 0 & t < 5 \end{cases}$$

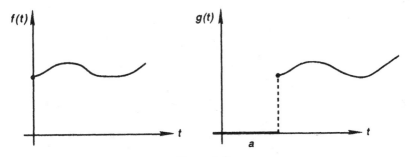

Figure 1.3

Since $\mathcal{L}\{\sin t\} = \dfrac{1}{s^2 + 1}$, then

$$\mathcal{L}\{\sin t - 5\} = e^{-5s}\mathcal{L}\{\sin t\}$$

$$= \dfrac{1}{e^{5s}(s^2 + 1)}.$$

1.8.5 LAPLACE TRANSFORM OF DERIVATIVE

Theorem 1.8

If $f(t)$ is continuous for $t \geq 0$ and of exponential order as $t \to \infty$, and if $f'(t)$ is sectionally continuous, then

$$\mathcal{L}\{f(t)\} = F(s)$$

implies that

$$\mathcal{L}\{f'(t)\} = sF(s) - f(0) \qquad (11)$$

Example

Since $\mathcal{L}\{\sqrt{t}\} = \dfrac{1}{2s}\sqrt{\dfrac{\pi}{s}}$ (Table 1) and $(\sqrt{t})' = \dfrac{1}{2\sqrt{t}}$, then we have

$$\mathcal{L}\left\{\dfrac{1}{\sqrt{t}}\right\} = 2\mathcal{L}\left\{\dfrac{1}{2\sqrt{t}}\right\},$$

by linearity

$$= 2(sL\{\sqrt{t}\} - 0)$$

$$= \sqrt{\frac{\pi}{s}}.$$

Theorem 1.9

If in Theorem 1.8 $f(t)$ fails to be continuous at a point $a > 0$, then

$$L\{f'(t)\} = sF(s) - f(0) = e^{-as}\{f(a+) - f(a-)\} \quad (12)$$

Figure 1.4.

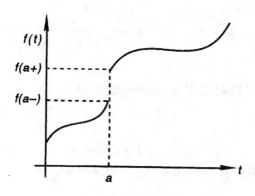

Figure 1.4

Theorem 1.10

If in Theorem 1.9 the function is discontinuous at $t = 0$, then

$$L\{f'(t)\} = sF(s) - f(0+) \quad (13)$$

Theorem 1.11

If $f(t), f'(t), f''(t), \ldots, f^{(n-1)}(t)$ are continuous and of exponential order as $t \to \infty$, and $f^{(n)}(t)$ is sectionally continuous, then

$$\mathcal{L}\{f(t)\} = F(s)$$

implies that

$$\mathcal{L}\{f^{(n)}(t)\} = s^n F(s) - s^{n-1} f(0) - s^{n-2} f^{(1)}(0) - s^{n-3} f^{(2)}(0) - \ldots - f^{(n-1)}(0) \tag{14}$$

Example

Specially

$$\mathcal{L}\{f''(t)\} = s^2 F(s) - s f(0) - f'(0),$$

and

$$\mathcal{L}\{f^{(3)}(t)\} = s^3 F(s) - s^2 f(0) - s f'(0) - f''(0).$$

1.8.6 DERIVATIVE OF LAPLACE TRANSFORMS

Theorem 1.12

If $f(t)$ is sectionally continuous and is of exponential order as $t \to \infty$, then

$$\mathcal{L}\{f(t)\} = F(s)$$

implies that

$$F'(s) = \mathcal{L}\{-t f(t)\} \tag{15}$$

and more general

$$F^{(n)}(s) = \mathcal{L}\{(-t)^n f(t)\} \tag{16}$$

Example

Since
$$\mathcal{L}\{e^{5t}\} = \frac{1}{s-5} \text{ and } \left(\frac{1}{s-5}\right)' = \frac{-1}{(s-5)^2},$$

we have
$$\mathcal{L}\{te^{5t}\} = -\mathcal{L}\{-te^{5t}\}$$
$$= -\left(\frac{1}{s-5}\right)'$$
$$= \frac{1}{(s-5)^2}.$$

1.8.7 PERIODIC FUNCTIONS

The function $F(t)$ is said to be **periodic with period** T if we have

$$\boxed{f(t+T) = F(t)}$$

The functions

$$\sin at, \quad \cos at, \quad \sec at, \quad \text{and} \quad \csc at$$

are periodic with period $T = 2\pi/a$, and the functions

$$\tan at \quad \text{and} \quad \cot at$$

are periodic with period $T = \pi/a$.

Theorem 1.13

If $f(t)$ has Laplace transform and if $f(t) = f(t+w)$, then

$$\mathcal{L}\{f(t)\} = \frac{\int_0^w e^{-st} f(t)\, dt}{1 - e^{-sw}} \qquad (17)$$

CHAPTER 2

INVERSE LAPLACE TRANSFORM

2.1 DEFINITION OF INVERSE LAPLACE TRANSFORM

If

$$\mathcal{L}\{f(t)\} = F(s),$$

we say that $f(t)$ is an **Inverse Laplace transform**, or **Inverse transform**, of $F(s)$ and we write

$$f(t) = \mathcal{L}^{-1}\{F(s)\} \qquad (1)$$

Example

Since

$$\mathcal{L}\{\sin 5t\} = \frac{5}{s^2 + 25},$$

we can write

$$\mathcal{L}^{-1}\left\{\frac{5}{s^2 + 25}\right\} = \sin 5t.$$

2.2 UNIQUENESS OF INVERSE LAPLACE TRANSFORM

2.2.1 NULL FUNCTIONS

A function $n(t)$ is called a **null function** if for every positive number a we have

$$\int_0^a n(t)\, dt = 0 \qquad (2)$$

In general, a function that is zero for all but a countable set of real numbers is a null function. Notice that a set is called countable if one can put its elements in a one to one correspondence with the set of positive integers.

Example

The function

$$n(t) = \begin{cases} 1 & \text{if } t \text{ is integer} \\ 0 & \text{otherwise} \end{cases}$$

is a null function.

Theorem 2.1 — Lerch's Theorem

If

$$\mathcal{L}\{f_1(t)\} = \mathcal{L}\{f_2(t)\},$$

then for some null function $n(t)$ we have

$$f_1(t) - f_2(t) = n(t). \qquad (3)$$

That is, an inverse Laplace transform is unique except for the addition of an arbitrary null function.

Example

Let

$$f_1(t) = e^t \text{ and } f_2(t) = \begin{cases} e^t & \text{if } t \neq 1 \\ 5 & \text{if } t = 1 \end{cases}$$

The difference of these two functions is a null function:

$$N(t) = f_1(t) - f_2(t) = \begin{cases} 0 & \text{if } t \neq 1 \\ -5 & \text{if } t = 1 \end{cases}$$

while their Laplace transforms are the same

$$L\{f_1(t)\} = L\{f_2(t)\} = \frac{1}{t-1}.$$

2.3 SOME INVERSE LAPLACE TRANSFORMS

The following table shows some inverse Laplace transforms.

Table 2

$F(s)$	$L^{-1}\{F(s)\} = f(t)$
$\dfrac{1}{s}$	1
$\dfrac{1}{s^2}$	t
$\dfrac{1}{s^{n+1}} \quad n = 0, 1, 2, \ldots$	$\dfrac{t^n}{n!}$
$\dfrac{1}{s-a}$	e^{at}
$\dfrac{1}{s^2 + a^2}$	$\dfrac{\sin at}{a}$
$\dfrac{s}{s^2 + a^2}$	$\cos at$
$\dfrac{1}{s^2 - a^2}$	$\dfrac{\sinh at}{a}$
$\dfrac{s}{s^2 - a^2}$	$\cosh at$

2.4 SOME PROPERTIES OF INVERSE LAPLACE TRANSFORMS

2.4.1 LINEARITY PROPERTY

Theorem 2.2

If c_1 and c_2 are constant numbers then

$$\mathcal{L}^{-1}\{c_1 F_1(s) + c_2 F_2(s)\} = c_1 \mathcal{L}^{-1}\{F_1(s)\} + c_2 \mathcal{L}^{-1}\{F_2(s)\} \quad (4)$$

Example

Using linearity property and Table 2, we can write

$$\mathcal{L}^{-1}\left\{\frac{5}{s^2} + \frac{6s}{s^2 - 25} - \frac{12}{s - 6}\right\}$$

$$= 5\mathcal{L}^{-1}\left\{\frac{1}{s^2}\right\} + 6\mathcal{L}^{-1}\left\{\frac{1}{s^2 - 25}\right\} - 12\mathcal{L}^{-1}\left\{\frac{1}{s - 6}\right\}$$

$$= \frac{5}{t} + 6\cosh 5t - 12e^{6t}.$$

2.4.2 FIRST TRANSLATION OR SHIFTING PROPERTY

Theorem 2.3

$$\mathcal{L}^{-1}\{F(s)\} = e^{-at}\, \mathcal{L}^{-1}\{F(s - a)\} \quad (5)$$

Or

$$\mathcal{L}^{-1}\{F(s - a)\} = e^{at}\, \mathcal{L}^{-1}\{F(s)\} \quad (6)$$

Example

Since we have

$$\mathcal{L}^{-1}\left\{\frac{1}{s^2 + 1}\right\} = \sin t,$$

we have

$$\mathcal{L}^{-1}\left\{\frac{1}{s^2 - 6s + 10}\right\} = \mathcal{L}^{-1}\left\{\frac{1}{(s-3)^2 - 9 + 10}\right\}$$

$$= \mathcal{L}^{-1}\left\{\frac{1}{(s-3)^2 + 1}\right\}$$

$$= e^{3t}\mathcal{L}^{-1}\left\{\frac{1}{s^2 + 1}\right\}$$

$$= e^{3t}\sin t.$$

2.4.3 SECOND TRANSLATION OR SHIFTING PROPERTY
Theorem 2.4

If

$$\mathcal{L}^{-1}\{F(s)\} = f(t),$$

then

$$\mathcal{L}^{-1}\{e^{-as}F(s)\} = \begin{cases} f(t-a) & \text{for } t \geq a \\ 0 & \text{for } t < a. \end{cases} \quad (7)$$

Example

Since

$$\mathcal{L}^{-1}\left\{\frac{1}{s^7}\right\} = \frac{t^6}{6!},$$

we have

$$\mathcal{L}^{-1}\left\{\frac{e^{-5s}}{s^7}\right\} = \begin{cases} \dfrac{(t-5)^6}{6!} & \text{for } t \geq 5 \\ 0 & \text{for } t < 5. \end{cases}$$

2.4.4 CHANGE OF SCALE PROPERTY
Theorem 2.5

If
$$\mathcal{L}^{-1}\{F(s)\} = f(t),$$

then for every constant number $k > 0$ we have

$$\mathcal{L}^{-1}\{F(ks)\} = \frac{1}{k} f\left(\frac{t}{k}\right) \tag{8}$$

Example

Since
$$\mathcal{L}^{-1}\left\{\frac{1}{s^2 - 1}\right\} = \sinh t,$$

we have
$$\mathcal{L}^{-1}\left\{\frac{1}{s^2 - 25}\right\} = \mathcal{L}^{-1}\left\{\frac{1}{25\left(\frac{s}{5}\right)^2 - 25}\right\}$$

$$= \mathcal{L}^{-1}\left\{\frac{1}{25} \frac{1}{\left(\frac{s}{5}\right)^2 - 1}\right\}$$

$$= \frac{1}{25} \mathcal{L}^{-1}\left\{\frac{1}{\left(\frac{s}{5}\right)^2 - 1}\right\}$$

$$= \frac{1}{5} \sinh 5t.$$

2.4.5 INVERSE LAPLACE TRANSFORM OF DERIVATIVES
Theorem 2.6

If
$$\mathcal{L}^{-1}\{F(s)\} = f(t)$$

then

$$\mathcal{L}^{-1}\{F'(s)\} = -t f(t) \qquad (9)$$

and more general

$$\mathcal{L}^{-1}\{F^{(n)}(s)\} = (-t)^n f(t) \qquad (10)$$

Example

Since

$$\mathcal{L}^{-1}\left\{\frac{s}{s^2+1}\right\} = \cos t \quad \text{and} \quad \left(\frac{s}{s^2+1}\right)' = \frac{1-s^2}{(s^2+1)^2},$$

we have

$$\mathcal{L}^{-1}\left\{\frac{1-s^2}{(s^2+1)^2}\right\} = (-t)\cos t = -t\cos t.$$

2.4.6 INVERSE LAPLACE TRANSFORM OF INTEGRALS

Theorem 2.7

If

$$\mathcal{L}^{-1}\{F(s)\} = f(t),$$

then

$$\mathcal{L}^{-1}\left\{\int_s^\infty F(x)\,dx\right\} = \frac{f(t)}{t} \qquad (11)$$

Example

Since

$$\mathcal{L}^{-1}\left\{\frac{1}{s(s-1)}\right\} = \mathcal{L}^{-1}\left\{-\frac{1}{s} + \frac{1}{s-1}\right\}$$

$$= -\mathcal{L}^{-1}\left\{\tfrac{1}{s}\right\} + \mathcal{L}^{-1}\left\{\tfrac{1}{s-1}\right\}$$

$$= -1 + e^t,$$

and

$$\int_s^\infty \frac{1}{x(x-1)} dx = \ln\left(\frac{s}{s-1}\right),$$

we have

$$\mathcal{L}^{-1}\left\{\ln\left(\frac{s}{s-1}\right)\right\} = \frac{-1+e^t}{t}.$$

2.5 THE CONVOLUTION PROPERTY

2.5.1 DEFINITION

We call the integral

$$\int_0^t f(u) g(t-u)\, du = f*g(t) \qquad (12)$$

the **convolution integral**, or **convolution**, of f and g.

Example

Find the convolution $f*g(t)$ if

$$f(t) = t \quad \text{and} \quad g(t) = t^2.$$

Since $f(u) = u$ and $g(t-u) = (t-u)^2 = t^2 - 2tu + u^2$, we have

$$f*g(t) = \int_0^t u(t^2 - 2tu + u^2)\, du$$

$$= \int_0^t (t^2 u = 2tu^2 + u^3)\, du$$

$$= \left[\tfrac{1}{2} t^2 u^2 - \tfrac{2}{3} tu^3 + \tfrac{1}{4} u^4\right]_{u=0}^{u=t}$$

$$= \frac{1}{2}t^4 - \frac{2}{3}t^4 + \frac{1}{4}t^4 - 0$$

$$= \frac{1}{12}t^4$$

2.5.2 PROPERTIES OF CONVOLUTION

Theorem 2.8

Convolution of the functions f and g obeys the following laws.

(a) Commutative law

$$f*g = g*f \qquad (13)$$

(b) Associative law

$$f*\{g*h\} = \{f*g\}*h \qquad (14)$$

(c) Distributive law

$$f*\{g + h\} = f*g + f*h \qquad (15)$$

(d) Derivative property

$$(f*g)' = f'*g = f*g' \qquad (16)$$

2.5.3 INVERSE LAPLACE TRANSFORM OF THE CONVOLUTION

Theorem 2.9

If

$$L^{-1}\{F(s)\} = f(t) \quad \text{and} \quad L^{-1}\{G(s)\} = g(t),$$

then

$$\mathcal{L}^{-1}\{F(s)G(s)\} = \int_0^t f(u)\,g(t-u)\,du = f*g(t) \quad (17)$$

Example

To find
$$\mathcal{L}^{-1}\left\{\frac{1}{s^2(s+1)^2}\right\},$$
one can use convolution. We have
$$\mathcal{L}^{-1}\left\{\frac{1}{s^2}\right\} = t \quad \text{and} \quad \mathcal{L}^{-1}\left\{\frac{1}{(s+1)^2}\right\} = te^{-t}.$$
Then using integration by part we have

$$\mathcal{L}^{-1}\left\{\frac{1}{s^2(s+1)^2}\right\} = \int_0^t (ue^{-u})(t-u)\,du$$

$$= \left[(ut - u^2) - (t - 2u) + (-2)(-e^{-u})\right]_{u=0}^{u=t}$$

$$= te^{-t} + 2e^{-t} + t - 2.$$

CHAPTER 3

SOME SPECIAL FUNCTIONS

3.1 THE GAMMA FUNCTION

We use the Gamma function to calculate the Laplace transform of non-integral powers of t.

If $r > 0$, we define the **Gamma function** by

$$\Gamma(r) = \int_0^\infty t^{r-1} e^{-t}\, dt. \tag{1}$$

Example

Find $\Gamma(1)$.

$$\begin{aligned}
\Gamma(1) &= \int_0^\infty t^{1-1} e^{-t}\, dt] \\
&= \int_0^\infty e^{-t}\, dt \\
&= [-e^{-t}]_{t=0}^{t=\infty} \\
&= 1.
\end{aligned}$$

Therefore

$$\Gamma(1) = 1 \tag{2}$$

3.1.1 PROPERTIES OF THE GAMMA FUNCTION

Theorem 3.1

For $r > 0$, we have

$$\Gamma(r+1) = r\,\Gamma(r). \qquad (3)$$

Example

$$\Gamma\!\left(\frac{3}{2}\right) = \Gamma\!\left(\frac{1}{2}+1\right)$$
$$= \frac{1}{2}\Gamma\!\left(\frac{1}{2}\right).$$

$\Gamma(1/2)$ is calculated in example of Theorem 3.3.

Theorem 3.2

For any positive integer n we have

$$\Gamma(n+1) = n! \qquad (4)$$

Example

$$\Gamma(6) = 5! = 120.$$

Theorem 3.3

For every number $r > -1$ we have

$$\mathcal{L}\{t^r\} = \frac{\Gamma(r+1)}{s^{r+1}}\,;\quad s > 0. \qquad (5)$$

Example

Find $\Gamma(1/2)$. Since

$$\mathcal{L}\!\left\{\sqrt{\frac{1}{t}}\right\} = \sqrt{\frac{\pi}{s}},$$

if in (4) we put $r = -1/2$, we have

$$\mathcal{L}\left\{\sqrt{\frac{1}{t}}\right\} = = \frac{\Gamma\left(\frac{1}{2}\right)}{s^{(1/2)}}.$$

Or

$$\Gamma\left(\frac{1}{2}\right) = s^{1/2} \mathcal{L}\left\{\sqrt{\frac{1}{t}}\right\}$$

$$= s^{1/2} \sqrt{\frac{\pi}{2}}$$

$$= \sqrt{\pi}.$$

Hence

$$\Gamma\left(\frac{1}{2}\right) = \sqrt{\pi} \qquad (6)$$

3.2 BESSEL FUNCTIONS

The **Bessel function of order** n is defined by

$$J_n(t) = \sum_{k=0}^{\infty} \frac{(-1)^k \left(\frac{1}{2}t\right)^{2k}}{k!\,\Gamma(k+n+1)} \qquad (7)$$

This function is also called the **Bessel function of first kind** and of **index** n. We will use $J_n(t)$ in an application of the differential equations method of chapter 5.

Example

For $n = 0$

$$J_0(t) = \sum_{k=0}^{\infty} \frac{(-1)^k \left(\frac{1}{2}t\right)^{2k}}{k!\,\Gamma(k+1)}$$

$$= \sum_{k=0}^{\infty} \frac{(-1)^k \left(\frac{1}{2}t\right)^{2k}}{k!\,k!}.$$

And

$$J_0(2) = \sum_{k=0}^{\infty} \frac{(-1)^k}{(k!)^2}.$$

3.2.1 MODIFIED BESSEL FUNCTION

For every number n the **modified Bessel function** of order n is defined by

$$I_n(t) = (-i)^{-n} J_n(it) \qquad (8)$$

where i is the imaginary unit, i.e.

$$i = \sqrt{-1}.$$

3.2.2 SOME PROPERTIES OF BESSEL FUNCTIONS

Theorem 3.4

Bessel functions satisfy the following properties

(i) For all positive n

$$J_{-n}(t) = (-1)^n J_n(t) \qquad (9)$$

(ii) For all n

$$J_{n+1}(t) = {}^{2n}/_t\, J_n(t) - J_{n-1}(t) \qquad (10)$$

(iii) For all $n \neq 0$

$$\frac{d}{dt}[t^n J_n(t)] = t^n J_{n-1}(t) \qquad (11)$$

(iv)
$$J'_0(t) = -J_1(t) \qquad (12)$$

(v) The following formula is called the **Generating function** for the Bessel function.

$$e^{(t/2)(u-1/u)} = \sum_{n=-\infty}^{\infty} J_n(t) u^n \qquad (13)$$

(vi) $J_n(t)$ satisfies the **Bessel's differential equation**

$$t^n y'' + t y' + (t^2 - n^2) y = 0 \qquad (14)$$

3.3 THE ERROR FUNCTION

The **Error function**, abbreviated **erf**, is defined by

$$\text{erf}(t) = \frac{2}{\sqrt{\pi}} \int_0^t e^{-u^2} du \qquad (15)$$

We use erf(t) to find the inverse Laplace transform of some certain simple functions.

Example

Since

$$\mathcal{L}^{-1}\left\{\sqrt{\frac{1}{s}}\right\} = \frac{1}{\sqrt{\pi t}},$$

and therefore by the first shifting property, formula (5) of chapter 2, we have

$$\mathcal{L}^{-1}\left\{\frac{1}{\sqrt{s+1}}\right\} = \frac{e^{-1}}{\sqrt{\pi t}}.$$

Then the convolution theorem yields

$$\mathcal{L}^{-1}\left\{\frac{1}{s}\frac{1}{s+1}\right\} = \int_0^t 1 \cdot \frac{e^{-u}}{\sqrt{\pi u}}\, du$$

$$= \frac{2}{\sqrt{\pi}} \int_0^{\sqrt{t}} e^{-z^2}\, dz$$

$$= \operatorname{erf}(\sqrt{t}).$$

3.3.1 SOME PROPERTIES OF erf(t)

Theorem 3.5

erf(t) satisfies the following properties

(i)
$$\frac{d}{dt}\operatorname{erf}(t) = \frac{2}{\sqrt{\pi}} e^{-t^2} \quad (16)$$

(ii)
$$\operatorname{erf}(t) = \frac{2}{\sqrt{\pi}} \sum_{n=0}^{\infty} \frac{(-1)^n t^{2n+1}}{(2n+1)\, n!} \quad (17)$$

(iii)
$$\lim_{t \to \infty} \operatorname{erf}(t) = 1 \quad (18)$$

3.4 THE COMPLEMENTARY ERROR FUNCTION

The **Complementary Error function**, abbreviated **erfc**, is defined by

$$\operatorname{erfc}(t) = 1 - \operatorname{erf}(t) = 1 - \frac{2}{\sqrt{\pi}} \int_0^t e^{-u^2}\, du$$
$$= \frac{2}{\sqrt{\pi}} \int_t^{\infty} e^{-u^2}\, du \quad (19)$$

3.4.1 A PROPERTY OF erfc(t)

Theorem 3.6

For every constant m

$$\lim_{t \to \infty} t^m \, \text{erfc}(t) = 0 \qquad (20)$$

3.5 THE SINE AND COSINE INTEGRALS

The **Sine Integral function** is defined by

$$si(t) = \int_0^t \frac{\sin u}{u} du = \int_0^1 \frac{\sin(tv)}{v} dv. \qquad (21)$$

And the **Cosine Integral function** is defined by

$$ci(t) = \int_t^\infty \frac{\cos u}{u} du = \int_1^\infty \frac{\cos(tv)}{v} dv. \qquad (22)$$

3.6 THE EXPONENTIAL INTEGRAL

The **Exponential Integral function** is defined as

$$ei(t) = \int_t^\infty \frac{e^{-u}}{u} du = \int_1^\infty \frac{e^{-tv}}{u} dv. \qquad (23)$$

3.7 THE UNIT STEP FUNCTION OR THE HEAVISIDE FUNCTION

The **unit step function** or the **Heaviside function** is defined as

$$h_1(t - t) = \begin{cases} 0 & \text{if} \quad t < t_0 \\ 1 & \text{if} \quad t \geq t_0 \end{cases} \qquad (24)$$

where we assume that $t_0 \geq 0$. A generalization for the unit step function is

$$h_c(t-t) = \begin{cases} 0 & \text{if} \quad t < t_0 \\ c & \text{if} \quad t \geq t_0 \end{cases} \qquad (25)$$

Example

One can write the **square-wave function,** Figure 3.1, in terms of Heaviside functions.

$$f(t) = h_c(t) - h_c(t-1) + h_c(t-2) \ldots$$
$$= \sum_{k=0}^{\infty} (-1)^k h_c(t-k). \qquad (26)$$

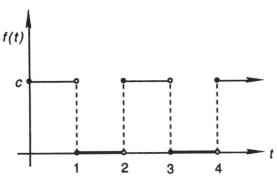

Figure 3.1

3.7.1 SOME PROPERTIES OF THE UNIT STEP FUNCTION

Theorem 3.7

The unit step function satisfies the following properties

(i)
$$h_c(t-t_0) = c\, h_1(t-t_0) \qquad (27)$$

E-32

(ii) The graph of $h_1(t-t)$ is given in Figure 3.2.

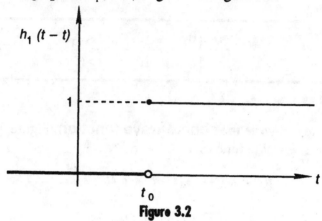

Figure 3.2

(iii)

$$\mathcal{L}\{h_1(t-t_0)\} = \frac{e^{-st_0}}{s} \qquad (28)$$

(iv)

$$\mathcal{L}\{h_1(t)\} = \frac{1}{s} \qquad (29)$$

(v)

$$\{h_c(t-t_0)\} = \frac{c\,e^{-st_0}}{s} \qquad (30)$$

(vi) If $\mathcal{L}[f(t)] = F(s)$, then

$$\{f(t-t_0)h_1(t-t_0)\} = e^{-st_0}\,{}^\circ F(s) \qquad (31)$$

Example

Express the function

$$f(t) = \begin{cases} 6 & \text{if} \quad 0 \le t < 4 \\ 2t+1 & \text{if} \quad t \ge 4 \end{cases}$$

E-33

in terms of the Heaviside function, h, and find $\mathcal{L}\{f(t)\}$.

Since $h_1(t) = 0$ for $t < 0$ and $h_1(t) = 1$ for $t \geq 0$, we build $f(t)$ in the following way. We write

$$f_1(t) = 6\, h_1(t)$$

that works for $0 < t < 4$, to knock off 6 for $t > 4$ we write

$$f_2(t) = 6\, h_1(t) - 6\, h_1(t-4).$$

Then we add the term $(2t + 1)\, h_1(t - 4)$ and finally have

$$f(t) = 6\, h_1(t) - 6\, h_1(t-4) + (2t+1)\, h_1(t-4)$$

or

$$f(t) = h_6(t) - h_5(t-4) + th_2(t-4).$$

Now we calculate the Laplace transform

$$\mathcal{L}\{f(t)\} = \mathcal{L}\{h_6(t) - h_5(t-4) + th_2(t-4)\}$$

$$= \mathcal{L}\{h_6(t)\} - \mathcal{L}\{h_5(t-4)\} + \mathcal{L}\{th_2(t-4)\}.$$

Hence since

$$\mathcal{L}\{h_6(t)\} = \frac{6}{s},\ \mathcal{L}\{h_5(t-4)\} = \frac{5e^{-4s}}{s},$$

and

$$\mathcal{L}\{th_2(t-4)\} = 2e^{-4s}\, \mathcal{L}\{t\} = 2e^{-4s}\, \frac{1}{s^2},$$

we have

$$\mathcal{L}\{f(t)\} = \frac{6}{s} - \frac{5e^{-4s}}{s} + \frac{2e^{-4s}}{s^2}$$

$$= \frac{6s - 5s\, e^{-4s} + 2e^{-4s}}{s^2}.$$

3.8 THE UNIT IMPULSE FUNCTION

Consider the function

$$f_\varepsilon(t) = \begin{cases} \dfrac{1}{\varepsilon} & \text{if} \quad 0 \leq t \leq \varepsilon \\ 0 & \text{if} \quad t > \varepsilon \end{cases}, \qquad (32)$$

that is shown in the figure 3.3.

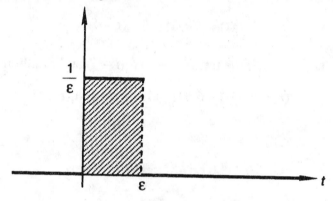

Figure 3.3

It is true that for all value of $\varepsilon > 0$, the area of the shaded region is always equal to 1, i.e.

$$\int_0^\infty f_\varepsilon(t)\, dt = 1. \qquad (33)$$

The **unit impulse function** or **Dirac delta function**, denoted by $\delta(t)$ is defined as the following limiting function

$$\delta(t) = \lim f_\varepsilon(t) \text{ as } \delta \to 0. \qquad (34)$$

3.8.1 SOME PROPERTIES OF THE UNIT IMPULSE FUNCTION

Theorem 3.8

$\delta(t)$ satisfies the following properties.

(i)
$$\int_0^\infty \delta(t)\,dt = 1 \qquad (35)$$

(ii) For any continuous function g(t)
$$\int_0^\infty \delta(t)\,g(t)\,dt = g(0) \qquad (36)$$

(iii) For any continuous function g(t)
$$\int_0^\infty \delta(t - t_0)\,g(t)\,dt = g(t_0) \qquad (37)$$

(iv) $\delta(t)$ is the identity element for the convolution operator, i.e. for any function g(t) we have
$$\delta * g = g \qquad (38)$$

(v)
$$\mathcal{L}\{\delta(t)\} = 1 \qquad (39)$$

(vi)
$$\mathcal{L}\{\delta(t - t_0)\} = e^{-st_0} \qquad (40)$$

3.9 THE BETA FUNCTION

If $m > 0$ and $n > 0$, the **Beta function** is defined by
$$B(m,n) = \int_0^1 u^{m-1}(1-u)^{n-1}\,du. \qquad (41)$$

Theorem 3.9

The Beta function satisfies the following properties.

(i)
$$B(m,n) = \frac{\Gamma(m)\,\Gamma(n)}{\Gamma(m+n)} \qquad (42)$$

(ii)
$$B(m,n) = 2\int_0^{\pi/2} \sin^{2m-1} u \cos^{2n-1} u\, du \qquad (43)$$

Example

$$\int_0^{\pi/2} \sin^7 u \cos^5 u\, du = \frac{1}{2} B(4,3)$$

$$= \frac{1}{2} \frac{\Gamma(4)\,\Gamma(3)}{\Gamma(4+3)}$$

$$= \frac{1}{2} \frac{(3!)(2!)}{6!} = \frac{1}{120}.$$

3.10 MORE PROPERTIES OF INVERSE LAPLACE TRANSFORMS

3.10.1 MULTIPLICATION BY s^a

Theorem 3.10

In Laplace transform, multiplication by s effects differentiation on $f(t)$. If $\mathcal{L}^{-1}\{F(s)\} = f(t)$, then

$$\mathcal{L}^{-1}\{s\,F(s)\} = f'(t) \quad \text{when} \quad f(0)=0, \qquad (44)$$

and

$$\mathcal{L}^{-1}\{s\,F(s)\} - f(0)\} = f'(t) \quad \text{when} \quad f(0) \neq 0, \quad (45)$$

or

$$\mathcal{L}^{-1}\{s\,F(s)\} = f'(t) + f(0)\,\delta(t) \quad \text{when} \quad f(0) \neq 0. \quad (46)$$

Example

Since
$$\mathcal{L}^{-1}\left\{\frac{1}{s^2+1}\right\} = \sin t \quad \text{and} \quad \sin 0 = 0,$$

we have
$$\mathcal{L}^{-1}\left\{\frac{1}{s^2+1}\right\} = (\sin t)' = \cos t.$$

Example

Since
$$\mathcal{L}^{-1}\left\{\frac{1}{s^2+1}\right\} = \cos t \quad \text{and} \quad \cos 0 = 1,$$

we have
$$\mathcal{L}^{-1}\left\{\frac{s^2}{s^2+1}\right\} = (\cos t)' + (\cos 0)\,\delta(t)$$
$$= -\sin t + \delta(t).$$

3.10.2 DIVISION BY S

Theorem 3.11

In Laplace transform, division by s effects integration on $f(t)$. If $\mathcal{L}^{-1}\{F(s)\} = f(t)$, then

$$\mathcal{L}^{-1}\left\{\frac{F(s)}{s}\right\} = \int_0^t f(u)\,du. \quad (47)$$

Or

$$\mathcal{L}^{-1}\left\{\frac{F(s)}{s}\right\} = \int_0^t \{F(s)\}(u)\,du. \qquad (48)$$

From (48) we obtain the following more general formula

$$\mathcal{L}^{-1}\left\{\frac{F(s)}{s}\right\} = \int_0^t \mathcal{L}^{-1}\left\{\frac{F(s)}{s^{n-1}}\right\}(u)\,du. \qquad (49)$$

for $n = 1, 2, 3, \ldots$

Example 1

Using (47) since

$$\mathcal{L}^{-1}\left\{\frac{1}{s-1}\right\} = e^t,$$

we have

$$\mathcal{L}^{-1}\left\{\frac{1}{s(s-1)}\right\} = \int_0^t e^u\,du$$

$$= [e^u]_{u=0}^{u=t}$$

$$= e^t - 1.$$

Example 2

Using (49) three times and example 1, we have

$$\mathcal{L}^{-1}\left\{\frac{1}{s^3(s-1)}\right\} = \int_0^t \mathcal{L}^{-1}\left\{\frac{1}{s^2(s-1)}\right\}(u)\,du$$

$$= \int_0^t \left[\int_0^u \mathcal{L}^{-1}\left\{\frac{1}{s(s-1)}\right\}(v)\,dv\right] du$$

$$= \int_0^t \left[\int_0^u \left(\int_0^v \mathcal{L}^{-1}\left\{\frac{1}{(s-1)}\right\}(w)\,dw\right) dv\right] du$$

$$= \int_0^t \int_0^u \int_0^v (e^w - 1) \, dw \, dv \, du$$

$$= e^t - \frac{t^3}{6} - \frac{t^2}{2} - t - 1$$

Theorem 3.12

$$\mathcal{L}^{-1}\left\{\frac{F(s)}{s^n}\right\} = \int_0^t \mathcal{L}^{-1}\left\{\frac{F(s)}{s^{n-1}}\right\}(u_1) \, du_1$$

$$= \int_0^t \int_0^{u_1} \mathcal{L}^{-1}\left\{\frac{F(s)}{s^{n-2}}\right\}(u_2) \, du_2 \, du_1$$

or

$$= \ldots$$

$$\boxed{\mathcal{L}^{-1}\left\{\frac{F(s)}{s^n}\right\} = \int_0^t \int_0^{u_1} \int_0^{u_2} \ldots \int_0^{u_{n-1}} f(u_n) \, du_n \, du_{n-1} \ldots du_1} \tag{50}$$

Example

Since

$$\mathcal{L}^{-1}\left\{\frac{1}{s^2 - 1}\right\} = \sinh t,$$

we have

$$\mathcal{L}^{-1}\left\{\frac{1}{s^3(s^2 - 1)}\right\}$$

$$= \int_0^t \mathcal{L}^{-1}\left\{\frac{1}{s^2(s^2 - 1)}\right\}(u) \, du$$

$$= \int_0^t \int_0^u \mathcal{L}^{-1}\left\{\frac{1}{s(s^2 - 1)}\right\}(v) \, dv \, du$$

$$= \int_0^t \int_0^u \int_0^v \mathcal{L}^{-1}\left\{\frac{1}{(s^2 - 1)}\right\}(w) \, dw \, dv \, du$$

$$= \int_0^t \int_0^u \int_0^v \sinh w \, dw \, dv \, du$$

$$= \int_0^t \int_0^u (\cosh v - 1) \, dv \, du$$

$$= \int_0^t (\sinh u - u) \, du$$

$$= \cosh t - \frac{1}{2} t^2 - 1.$$

CHAPTER 4

APPLICATION TO ORDINARY LINEAR DIFFERENTIAL EQUATIONS

4.1 ORDINARY DIFFERENTIAL EQUATION WITH CONSTANT COEFFICIENTS

The Laplace transformation transforms a linear differential equation with constant coefficients into an algebraic equation.

To solve an **initial value problem (IVP)**:

$$a_0 \frac{d^n y}{dt^n} + a_1 \frac{d^{n-1} y}{dt^{n-1}} + \ldots + a_{n-1} \frac{dy}{dt} + a_n y = f(t) \quad (1)$$

$$y(0) = c_0, y'(0) = c_1, \ldots, y^{(n-1)}(0) = c_{n-1} \quad (2)$$

by means of the Laplace transform, we perform the following procedure.

a. Let $Y(s)$ be the Laplace transform of the unknown function $y(t)$, and $F(s)$ be the Laplace transform of the known function $f(t)$, i.e.

$$Y(s) = \mathcal{L}\{y(t)\} \quad \text{and} \quad F(s) = \mathcal{L}\{f(t)\}.$$

b. Take the Laplace transform of both sides of (1).

c. Use the linearity property, Theorem 1.4, to separate the terms and use the derivative property, Theorem 1.11,

$$\mathcal{L}\{y^{(n)}\} = s^n \mathcal{L}\{y(t)\} - s^{n-1}y(0) - s^{n-2}y'(0) - \ldots - y^{(n-1)}(0)$$

$$= s^n Y(s) - c_0 s^{n-1} - c_1 s^{n-2} - \ldots - c_{n-1}.$$

d. Form the algebraic equation

$$[a_0 s^n + a_1 s^{n-1} + \ldots + a_{n-1} s + a_n] Y(s)$$

$$- c_0 [a_0 s^{n-1} + a_1 s^{n-2} + \ldots + a_{n-1}]$$

$$- c_1 [a_0 s^{n-2} + a_1 s^{n-3} + \ldots + a_{n-2}]$$

$$\vdots$$

$$- c_{n-2}[a_0 s + a_1] - c_{n-1} a_0 = F(s).$$

e. Solve the resulting equation of step d to determine $Y(s)$.

f. Use the Laplace inverse transform to determine the solution

$$y(t) = \mathcal{L}^{-1}\{Y(s)\}.$$

Example 1

Solve the initial value problem

$$y'' - 2y' + y = 0, \qquad (3)$$

$$y(0) = 1, \; y'(0) = 0. \qquad (4)$$

Taking the Laplace transform of (3), we have

$$\mathcal{L}\{y'' - 2y' + y\} = \mathcal{L}\{0\}$$

or

$$\mathcal{L}\{y''\} - 2\mathcal{L}\{y'\} + \mathcal{L}\{y\} = 0.$$

Now we substitute the initial values (4), to obtain

$$s^2 Y(s) - s\, y(0)\, y'(0) - 2(s\, Y(s) - 1) + Y(s) = 0$$

or

$$(s^2 - 2s + 1)\, Y(s) - s + 2 = 0.$$

Therefore

$$Y(s) = \frac{s-2}{s^2 - 2s + 1} = \frac{s-2}{(s-1)^2}.$$

Hence

$$y(t) = \mathcal{L}^{-1}\left\{\frac{s-2}{(s-1)^2}\right\}$$

$$= \mathcal{L}^{-1}\left\{\frac{1}{s-1}\right\} - \mathcal{L}^{-1}\left\{\frac{1}{(s-1)^2}\right\}$$

$$= e^t - te^t.$$

Example 2

Solve the IVP

$$y'' - 6y' + 8y = t\sin t + \cos t \tag{5}$$

$$y(0) = 0,\ y'(0) = 1. \tag{6}$$

Taking the Laplace transform of the equation (5), we have

$$\mathcal{L}\{y'' - 6y' + 8y\} = \mathcal{L}\{t\sin t + \cos t\}.$$

Separating the terms, we obtain

$$\mathcal{L}\{y''\} - 6\mathcal{L}\{y'\} + 8\mathcal{L}\{y\} = \mathcal{L}\{t\sin t\} + \mathcal{L}\{\cos t\}.$$

Using the derivative property and the table of transforms, we find

$$s^2 Y(s) - s\, y(0) - y'(0) - 6\,[s\, Y(s) - y(0)] + 8\, Y(s)$$
$$= \frac{2s}{(s^2+1)^2} + \frac{s}{s^2+1}.$$

Use the initial values (6) and simplify to find

$$(s^2 - 6s + 8)Y = \frac{2s}{(s^2+1)^2} + \frac{s}{s^2+1} + 1.$$

Therefore

$$Y = \frac{2s}{(s^2+1)^2(s-2)(s-4)} + \frac{s}{(s^2)(s-2)(s-4)}$$
$$+ \frac{1}{(s-2)(s-4)}.$$

Now to prepare $Y(s)$ for the inverse Laplace transform, we use the partial fractions method of chapter 5 to write

$$Y(s) = \frac{85963}{173400}\frac{1}{s^2+1} - \frac{8712}{14450}\frac{1}{(s^2+1)^2} + \frac{365}{578}\frac{1}{s-4}$$
$$- \frac{39}{50}\frac{1}{s-2}.$$

Therefore

$$y(t) = \mathcal{L}^{-1}\{Y(s)\}$$
$$= \frac{85963}{173400}\mathcal{L}^{-1}\left\{\frac{1}{s^2+1}\right\} - \frac{8712}{14450}\mathcal{L}^{-1}\left\{\frac{1}{(s^2+1)^2}\right\}$$
$$+ \frac{365}{578}\mathcal{L}^{-1}\left\{\frac{1}{s-4}\right\} - \frac{39}{50}\mathcal{L}^{-1}\left\{\frac{1}{s-2}\right\}.$$

Thus, using the general table of Laplace transforms, we have

$$y(t) = \frac{85963}{173400} \sin t - \frac{8712}{14450} \frac{\sin t - t \cos t}{2} + \frac{365}{578} e^{4t}$$
$$- \frac{39}{50} e^{2t}.$$

Or

$$y(t) = \frac{9736699}{50112600} \sin t + \frac{4356}{14450} t \cos t + \frac{365}{578} e^{4t}$$
$$- \frac{39}{50} e^{2t}.$$

4.2 ORDINARY DIFFERENTIAL EQUATIONS WITH VARIABLE COEFFICIENTS

The method of Laplace transform also can be used to solve some differential equations with variable coefficients. By Theorem 1.12, we have

$$\mathcal{L}\{t^n y(t)\} = (-1)^n Y^{(n)}(s), \tag{7}$$

i.e., multiplying the function by powers of t effects the transform in derivatives, the method of Laplace transform is more effective for linear differential equations with terms of the following form:

$$t^m \frac{d^n}{dt^n} y(t). \tag{8}$$

Example

We use the Laplace transform to solve the **Laguerre equation**

$$t \frac{d^2 y}{dt^2} + (1-t) \frac{dy}{dt} + 2y = 0. \tag{9}$$

We take the Laplace transform of the equation (9)

$$L\left\{t\frac{d^2y}{dt^2}+(1-t)\frac{dy}{dt}+2y\right\}=0.$$

We separate the terms to obtain

$$L\left\{t\frac{d^2y}{dt^2}\right\}+L\left\{\frac{dy}{dt}\right\}-L\left\{t\frac{dy}{dt}\right\}+2L\{y\}=0.$$

Using the derivative property of Laplace transforms, formula (15) of chapter 1, the above equation will be written as

$$-\frac{d}{ds}L\left\{\frac{d^2y}{dt^2}\right\}+L\left\{\frac{dy}{dt}\right\}+\frac{d}{ds}L\left\{\frac{dy}{dt}\right\}+2L\{y\}=0.$$

Using formula (14) of chapter 1, we will have

$$-\frac{d}{ds}\left[s^2Y(s)-s\,y(0)-y'(0)\right]+\left[sY(s)-y(0)\right]$$

$$+\frac{d}{ds}\left[sY(s)-y(0)\right]+2Y(s)=0.$$

This equation is simplified in the following form

$$s(1-s)\frac{d}{ds}Y(s)+(3-s)Y(s)=0.$$

Or

$$\frac{\frac{d}{ds}Y(s)}{Y(s)}=\frac{3-s}{s(s-1)}=-\frac{3}{s}+\frac{2}{s-1}$$

Solving this differential equation, we obtain the general solution

$$Y(s)=\frac{(s-1)^2}{s^3}=C\left[\frac{1}{s}-\frac{2}{s^2}+\frac{1}{s^3}\right].$$

Therefore

$$y(t) = C\left[\mathcal{L}^{-1}\left\{\frac{1}{s}\right\} - 2\mathcal{L}^{-1}\left\{\frac{1}{s^2}\right\} + \mathcal{L}^{-1}\left\{\frac{1}{s^3}\right\}\right].$$

Thus by the Laplace inverse formulas we have

$$y(t) = C\left[1 - 2t + \frac{1}{2}t^2\right]$$

where C is a constant number that is determined by using an initial value.

4.3 SYSTEMS OF LINEAR DIFFERENTIAL EQUATIONS

A system of linear differential equations involves two or more linear differential equations of two or more unknown functions of a single independent variable. A simple example of a system of linear differential equations is

$$\begin{cases} x' = x + 5y &, x(0) = 0 \\ y' = -x - y &, y(0) = 1 \end{cases} \quad (10)$$

where x and y are unknown functions of the variable t. Another example of a system of linear differential equations is

$$\begin{cases} 2x' + y - y = t &, x(0) = 1 \\ x' + y' = t^2 &, y(0) = 0. \end{cases} \quad (11)$$

When the initial conditions are specified, the Laplace transform will reduce a system of linear differential equations with constant coefficients to a set of simultaneous algebraic equations, in the transformed functions.

Example 1

Solve the system of linear equations (10) by means of the Laplace transforms.

We denote

$$X(s) = \mathcal{L}\{x(t)\} \quad \text{and} \quad Y(s) = \mathcal{L}\{y(t)\}.$$

Taking Laplace transform of each equation of the system (10), we obtain

$$\begin{cases} sX(s) - x(0) = X(s) + 5Y(s) \\ sY(s) - y(0) = -X(s) - Y(s). \end{cases}$$

Or

$$\begin{cases} (s-1)X - 5Y = 0 \\ X + (s+1)Y = 1. \end{cases} \quad (12)$$

There are several ways to solve an algebraic system of equations. Using the substitution method and solving the first equation of (12) for Y in terms of X, we obtain

$$Y = \frac{s-1}{5} X \quad (13)$$

Substituting (13) into the second equation of (12), we obtain an equation for X

$$X + \frac{(s+1)(s-1)}{5} X = 1. \quad (14)$$

Simplifying (14) and solving for X, we get

$$X(s) = \frac{5}{s^2 + 4}. \quad (15)$$

Therefore

$$x(t) = \mathcal{L}^{-1}\{X(s)\}$$

$$= \mathcal{L}^{-1}\left\{\frac{5}{s^2 + 4}\right\}$$

$$= \frac{5}{2} \mathcal{L}^{-1}\left\{\frac{2}{s^2 + 2^2}\right\} = \frac{5}{2} \sin 2t.$$

To find y(t), one may substitute X(s) in one of the equations of (12) to find Y(s), then using Y(s) we can find y(t) by the inverse Laplace transform. However, it is easier to substitute x(t) in the first equation of (10), then by solving the resulting equation

$$(5/_2 \sin 2t)' = 5/_2 \sin 2t + 5 y(t),$$

Or

$$5 \cos 2t = 5/_2 \sin 2t + 5 y(t),$$

we find

$$y(t) = \cos 2t - 1/_2 \sin 2t.$$

Example 2

To solve the system of linear equations (11), we take the Laplace transform of its equations to get

$$\begin{cases} 2[sX(s) - x(0)] + [sY(s) - y(0)] - Y(s) = \dfrac{1}{s^2} \\ [sX(s) - x(0)] + [sY(s) - y(0)] = \dfrac{2}{s^3} \end{cases} \quad (16)$$

where $X(s) = L\{x(t)\}$ and $Y(s) = L\{y(t)\}$. Use the initial conditions $x(0) = 1$ and $y(0) = 0$ to find

$$\begin{cases} 2sX(s) + (s - 1)Y(s) = 2 + \dfrac{1}{s^2} \\ sX(s) + sY(s) = 1 + \dfrac{2}{s^3} \end{cases} \quad (17)$$

Solving the algebraic system (17) for X and Y, we obtain

$$Y(s) = \frac{4 - s}{s^3(s + 1)}.$$

To prepare Y(s) for the Laplace inverse transform we use the partial fractions method of chapter 5 to write

$$Y(s) = \frac{5}{s} - \frac{5}{s^2} + \frac{4}{s^3} - \frac{5}{s+1}$$

Thus

$$y(t) = 5\mathcal{L}^{-1}\left\{\frac{1}{s}\right\} - 5\mathcal{L}^{-1}\left\{\frac{1}{s^2}\right\} + 2\mathcal{L}^{-1}\left\{\frac{2!}{s^3}\right\}$$

$$- 5\mathcal{L}^{-1}\left\{\frac{1}{s+1}\right\}$$

or

$$y(t) = 5 - 5t + 2t^2 - 5e^{-t}.$$

By substituting $Y(s)$ in the second equation of (17), we have

$$X(s) = -Y(s) + \frac{1}{s} + \frac{2}{s^4},$$

therefore

$$x(t) = -\mathcal{L}^{-1}\{Y(s)\} + \mathcal{L}^{-1}\left\{\frac{1}{s}\right\} + \frac{1}{3}\mathcal{L}^{-1}\left\{\frac{3!}{s^4}\right\}$$

$$= -y(t) + t + \frac{1}{3}t^3.$$

or

$$x(t) = -4 + 5t - 2t^2 + 1/3\, t^3 + 5e^{-t},$$

$$y(t) = 5 - 5t + 2t^2 - 5e^{-t}.$$

4.4 THE VIBRATION OF SPRING

Suppose a mass m is attached to the lower end of a light coil spring, suspended from a point on a support (figure 4-1(a)), and brought to the point of equilibrium, E, where it remains at rest (Figure 4-1(b)). Once the mass m is moved from the point of equilibrium E (figure 4-1(c)), the motion of the mass m will be determined by a differential equation with some initial conditions.

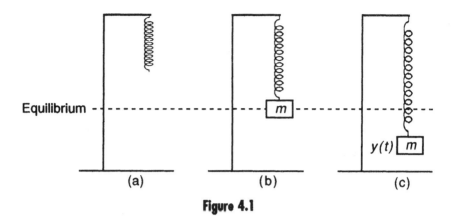

Figure 4.1

If $y(t)$ denotes the instantaneous displacement of m at time t from the equilibrium or rest position, then by the Hooke's law there will be a **restoring force acting on m**

$$F = -k\, y(t) \qquad (18)$$

where k is a constant depending on the spring that is called the **spring constant**. On the other hand by the Newton's second law of motion **the net force acting on m is equal to the mass times acceleration**, i.e.

$$F = m\, \frac{d^2 y(t)}{dt^2} \qquad (19)$$

Therefore by equating (18) and (19) **the differential equation of the motion of the mass** is

$$m\, y'' = -k\, y \qquad (20)$$

Or since usually the time depending derivatives are shown by $\dot{y}$ we have

$$\ddot{y} + \frac{k}{m}\, y = 0 \qquad (21)$$

since the mass was set to motion by first displacing it to an initial position y_0, and then giving it an initial velocity v_0. Thus, together with the differential equation (21), we have two initial conditions

$$y(0) = y_0, \quad \dot{y}(0) = v_0. \qquad (22)$$

4.4.1 DAMPED VIBRATIONS

In practice, a vibrating spring is most often subjected to frictional and some other forces, such as air resistance, which act to retard (dampen) the motion to come to rest. Experiments have shown that

the magnitude of the damping force is approximately proportional to the velocity of the mass,

provided that the velocity of the mass is small. Since the damping force acts in opposite direction of the motion, we can express the **damping force** as

$$-b\frac{dy}{dt} \qquad (23)$$

where b is called the **damping constant**. Thus the **differential equation** of the **damped vibration motion** is

$$m\ddot{y} + b\dot{y} + ky = 0 \qquad (24)$$

Furthermore, when some time varying **external force** also acts on m, the equation of the **damped vibration motion with external force** is

$$m\ddot{y} + b\dot{y} + ky = f(t) \qquad (25)$$

Thus, in general, a mathematical model for the **linear mass-spring system** formulated in an initial value problem of the form

$$m y'' + b y' + k y = f(t) \ (t > 0) \qquad (26)$$
$$y(0) = y_0, \ y'(0) = y'_0$$

where $m > 0$ is the mass, $b \geq 0$ is the damped constant, $k > 0$ is the spring constant, and $f(t)$ is a time-varying external force. One can solve the IVP (26) by the means of Laplace transformation.

Damped Vibration is the linear mass system (26) when $b > 0$.

4.4.2 UNDAMPED VIBRATION

Undamped Vibration is the linear mass system (26) when $b = 0$.

4.4.3 FREE VIBRATION

If in a linear spring system there is no external force, i.e. in equation (26) $f(t) = 0$, then the system is called a **free vibration system.**

Solution of the free vibration system.

$$my'' + by' + ky = 0. \qquad (27)$$
$$y(0) = 0, \ y'(0) = y'_0$$

Case 1 (undamped)

If in (27) $b = 0$.

$$my'' + ky = 0. \qquad (28)$$
$$y(0) = 0, \ y'(0) = y'_0$$

Taking the Laplace transform of (28) and using the initial conditions, we will have

$$y(t) = c\cos(\omega_0 t - \delta)$$
$$\omega_0 = \sqrt{\frac{k}{m}}, \quad c = \sqrt{y_0^2 + (y'_0/\omega_0)^2} \quad (29)$$
$$\cos\delta = y_0/c, \quad \sin\delta = y'_0/(\omega_0 c)$$

It represents **simple harmonic motion** with amplitude c, frequency ω_0, and period

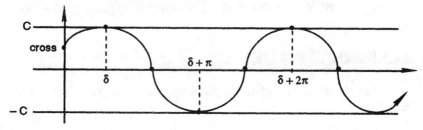

Figure 4.2

Case 2 (damped)

If in (27) $b > 0$. The general solution of the **damped free vibration model**:

$$my'' + by' + ky = 0$$

depends on the sign of the quantity $b^2 - 4km$, that is given in three cases.

(a) If $b^2 - 4km > 0$ (overdamping) (Figure 4-3(a)):

$$y(t) = c_1 e^{z_1 t} + c_2 e^{z_2 t}, \quad z_1 < 0, \quad z_2 < 0$$
$$z_1 = \frac{-b + \sqrt{b^2 - 4km}}{2m}, \quad z_2 = \frac{-b - \sqrt{b^2 - 4km}}{2m} \quad (30)$$

(b) If $b^2 - 4km = 0$ (critical damping) (Figure 4-3(b)):

$$y(t) = c_1 e^{z_1 t} + c_2 t e^{z_1 t}, \quad z_1 = -\frac{b}{2m} < 0. \tag{31}$$

(c) If $b^2 - 4km < 0$ (underdamping) (Figure 4-3(c)):

$$y(t) = c\, e^{-bt/2m} \cos(\lambda_0 t - \delta)$$

$$\lambda_0 = \sqrt{\frac{4km - b^2}{4m^2}}, \quad c^2 = c_1^{\,2} + c_2^{\,2} \tag{32}$$

$$\cos \delta = c_1/c,\ \sin \delta = c_2/c$$

In this case of damping each solution is an **oscillation** with **amplitude** $ce^{-bt/2m}$ and period $2\pi / \lambda_0$; it is decreasing to zero.

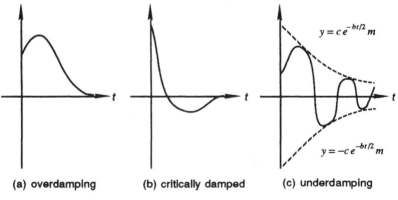

(a) overdamping (b) critically damped (c) underdamping

Figure 4.3

4.5 RESONANCE

An important example in the problem of undamped vibration of a spring is

$$my''(t) + ky(t) = F_0 \cos \omega t$$
$$y(0) = y_0,\ y'(0) = \dot{y}_0 \tag{33}$$

If we let

$$w_0 = \sqrt{\frac{k}{m}} \qquad (34)$$

then the problem will turn into the following standard form

$$y''(t) + \omega_0^2 y(t) = \frac{F_0}{m}\cos\omega t \qquad (35)$$
$$y(0) = y_0,\ y'(0) = \dot{y}_0$$

where ω_0^2 is called the **natural frequency** and ω is called the **applied frequency** of the undamped mass-spring system.

By putting

$$Y(s) = \mathcal{L}\{y(t)\},$$

taking the Laplace transform of the equation of (35), and using the initial values of (35), we will obtain

$$Y(s) = \frac{sy_0 + \dot{y}_0}{s^2 + \omega_0^2} + \frac{F_0}{m}\cdot\frac{\omega}{(s^2 + \omega_0^2)(s^2 + \omega^2)}. \qquad (36)$$

To find the position function $y(t)$, we have to find the Laplace inverse transform of (36), that differs according to whether $\omega = \omega_0$ or $\omega \neq \omega_0$.

Case 1

If $\omega \neq \omega_0 = \sqrt{k/m}$, the solution of the undamped vibration model (33) is

$$y(t) = y_0 \cos\omega_0 t + \frac{\dot{y}_0}{\omega_0}\sin\omega_0 t$$
$$= \frac{F_0}{m(\omega_0^2 - \omega^2)}[\cos\omega t = \cos\omega_0 t] \qquad (37)$$

Case 2

If $\omega = \omega_0 = \sqrt{k/m}$, the solution of (33) is

$$y(t) = y_0 \cos \omega_0 t + \frac{\dot{y}_0}{\omega_0} \sin \omega_0 t + \frac{F_0}{2m\omega_0} t \sin \omega_0 t. \quad (38)$$

In this case the solution $y(t)$ obtained by (38) may be regarded as an oscillation with frequency ω_0 and amplitude

$$\frac{F_0 t}{(2m\omega_0)}$$

which increases with t (Figure 4.4). This phenomenon is called **resonance**. It occurs when the applied frequency is equal to the natural frequency of an undamped spring system.

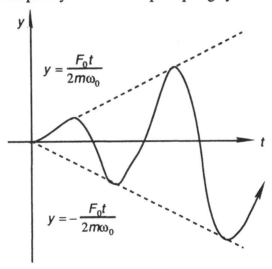

Figure 4.4

4.6 THE SIMPLE PENDULUM

A pendulum is made of a rod of length L feet that is suspended by one end so that it can swing freely in a vertical plane. Let a weight of m pounds be attached to the free end of the rod, and let the weight of the rod, compared to the weight m, be negligible.

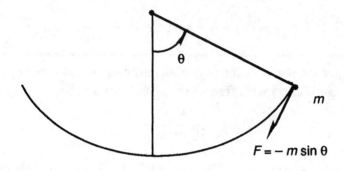

Figure 4.5

Let θ (radians) be the angular displacement from the vertical position (Figure 4.5), at time t (second). The tangential component of the force m is $m \sin \theta$ tending to decrease θ. Therefore, along the circular trajectory curve, the mass m is under the force

$$F = -m \sin \theta \qquad (39)$$

On the other hand by the Newton's second law of motion

$$F = \frac{m}{g} \frac{d^2 s}{dt^2} \qquad (40)$$

where $s = L\theta$ is the arc length from the vertical position. Thus equating (39) and (40) we conclude that

$$\frac{m}{g} \frac{d^2 s}{dt^2} = -m \sin \theta.$$

Since $s = L\theta$, from this equation it follows that

$$\frac{d^2 \theta}{dt^2} + \frac{g}{L} \sin \theta = 0 \qquad (41)$$

The second order nonlinear differential equation (41) governs the motion of the pendulum; its solution involves an elliptic integral which is not easy. If θ is small, however, sin θ and θ are approximately equal and (41) closely approximated by the second order linear differential equation

$$\frac{d^2\theta}{dt^2} + \frac{g}{L}\theta = 0. \qquad (42)$$

Requiring $-.3 < \theta < .3$ (radian), the solution of (42) with pertinent initial values is easily obtained by the means of Laplace transformation.

4.7 ELECTRIC CIRCUITS

The notations used in simple electrical circuits for circuit elements are:

1. t (second) for time.

2. q (coulombs) quantity of electricity; could be charge on a capacitor.

3. i (amperes) current, time rate of flow of electricity.

4. E (volts) electromotive force, could be supplied by a generator or battery.

5. R (ohms) resistance by a resistor.

6. L (henrys) inductance by an inductor.

7. (farads) capacitance by a capacitor.

8. K for key or switch.

A circuit is treated as a network containing only one closed path.

Example

An "RLC" circuit is shown in Figure 4-6.

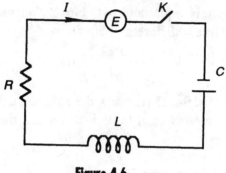

Figure 4.6

When the switch K is closed a charge q will flow to the capacitor plates, then the following relations between the elements will be set:

(a) By definition current is

$$i = \frac{dq}{dt} \qquad (43)$$

(b) Voltage drop across a resistor is

$$R\,i = R\,\frac{dq}{dt} \qquad (44)$$

(c) Voltage drop across an inductor is

$$L\,\frac{di}{dt} = L\,\frac{d^2q}{dt^2} \qquad (45)$$

(d) Voltage drop across a capacitor is

$$\frac{q}{C} \qquad (46)$$

E-61

(e) Voltage drop across a generator is

$$- \text{voltage rise} = -E \qquad (47)$$

Problem

An important problem in an electrical circuit is to find the charge on the capacitor and the currents as functions of time. For this purpose we find a differential equation by using the Kirchhoff's laws:

Kirchhoff's Laws:

(i) In an electrical circuit the algebraic sum of the currents flowing toward any junction point is equal to zero.

(ii) The algebraic sum of the potential drops, or voltage drops, around any closed loop is equal to zero.

Example

From the Kirchoff's laws for the RLS circuit of Figure 4.6, we get the following equation

$$L\,i(t) + R\,i(t) + \tfrac{1}{c}\,q(t) = E(t) \quad (t \geq 0). \qquad (48)$$

And if $i(0) = 0$, then

$$i'(0) = \frac{1}{L[E(0) - 1/C\,q(0)]}.$$

Differentiating this equation and using $q'(t) = i(t)$ we obtain the IVP

$$L\,i''(t) + R\,i'(t) + \tfrac{1}{c}\,i(t) = E'(t) \qquad (49)$$
$$i(0) = 0,\ i'(0) = \tfrac{1}{L}(E(0) - \tfrac{1}{c}\,q(0)).$$

Notice that if L, R, and C are positive constants, the IVP (44) is the same as the IVP (27) for the linear damped-spring system.

If in (9) we use $i(t) = q'(t)$ and $i'(t) = q''(t)$, we obtain the IVP

$$L q''(t) + R q'(t) + \tfrac{1}{c} q(t) = E(t),$$
$$q(0) = q_0, \; q'(0) = i(0) = 0.$$
(50)

4.8 BEAMS

Consider a beam of length l, (Figure 4.5). Denote distance from one end of the beam by x, and the deflection of the beam by y. If the beam is subjected to a vertical load $w(x)$, the deflection y must satisfy the equation

$$E I \frac{d^4 y}{dx^4} = w(x) \quad \text{for} \quad 0 < x < 2C.$$
(51)

where the quantity $(E\,I)$ is called the **flexural rigidity** of the beam.

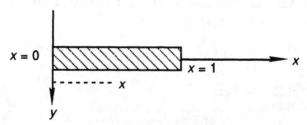

Figure 4.7

The boundary conditions associated with the differential equation (46) depend on the way that beam is supported. The common boundary conditions are of the following types:

(i) Beam imbedded in a support:

$$y = 0, \; y' = 0 \text{ at the point.}$$

(ii) Beam simply supported:

$$y = 0, y'' = 0 \text{ at the point.}$$

(iii) Beam free:

$$y'' = 0, y''' = 0 \text{ at the point.}$$

CHAPTER 5

METHODS OF FINDING LAPLACE TRANSFORMS AND INVERSE TRANSFORMS

5.1 INITIAL VALUE THEOREM

Theorem 5.1

If the limits exist, then

$$\lim_{t \to 0+} f(t) = \lim_{s \to \infty} s F(s) \tag{1}$$

5.2 FINAL VALUE THEOREM

Theorem 5.2

If limits exist, then

$$\lim_{t \to \infty} f(t) = \lim_{s \to 0} s F(s) \tag{2}$$

5.3 METHODS OF FINDING LAPLACE TRANSFORMS

In the following we list some common methods for determining Laplace transforms.

5.3.1 DIRECT METHOD

This involves direct use of the definition of the Laplace transforms, formula (3) of chapter 1:

$$\mathcal{L}\{f(t)\} = \int_0^\infty e^{-st} f(t)\, dt. \qquad (3)$$

Example.

Find $\mathcal{L}\{f(t)\}$ if

$$f(t) = \begin{cases} 7 & \text{if} \quad 0 < t < 10 \\ 0 & \text{if} \quad t \geq 10 \end{cases}$$

Using (3), we have

$$\mathcal{L}\{f(t)0\} = \int_0^\infty e^{-st} f(t)\, dt$$

$$= \int_0^{10} e^{-st} \cdot 7\, dt + \int_{10}^\infty e^{-st} \cdot 0\, dt$$

$$= 7 \int_0^{10} e^{-st}\, dt$$

$$= 7 \frac{e^{-st}}{-s} \bigg|_0^{10} = 7\, \frac{(1 - e^{-10s})}{s}.$$

5.3.2 POWER SERIES METHOD

If $f(t)$ has a power series expansion (Taylor series) given by

$$f(t) = a_0 + a_1 t + a_2 t^2 + \ldots = \sum_{n=0}^\infty a_n t^n \qquad (4)$$

where

$$a_n = \frac{f^{(n)}(0)}{n!} \qquad (5)$$

then

$$\mathcal{L}\{f(t)\} = \frac{a_0}{s} + \frac{a_1}{s^2} + \frac{2!\, a_2}{s^3} + \ldots = \sum_{n=0}^{\infty} \frac{n!\, a_n}{s^{n+1}} \qquad (6)$$

provided that (6) is convergent for $s > b$, for some $b > 0$, or

$$\mathcal{L}\{f(t)\} = \sum_{n=0}^{\infty} \frac{f^{(n)}(0)}{s^{n+1}}. \qquad (7)$$

Example 1

Find $\mathcal{L}\{e^{at}\}$.

We use the power series method. The Taylor series for e^t is

$$e^t = 1 + \frac{t}{1!} + \frac{t^2}{2!} + \frac{t^3}{3!} + \ldots = \sum_{n=0}^{\infty} \frac{t^n}{n!}.$$

Therefore by substituting (at) for t in this, we will have

$$e^{at} = 1 + at + a^2 \frac{t^2}{2!} + a^3 \frac{t^3}{3!} + \ldots = \sum_{n=0}^{\infty} a^n \frac{t^n}{n!}.$$

Thus

$$\mathcal{L}\{e^{at}\} = \frac{1}{s} + a \frac{1}{s^2} + a^2 \frac{2!}{2!\, s^3} + a^3 \frac{3!}{3!\, s^4} + \ldots$$

$$= \frac{1}{s} \sum_{n=0}^{\infty} \left(\frac{a}{s}\right)^n = \frac{1}{s} \frac{1}{1 - \frac{a}{s}}$$

$$= \frac{1}{s - a}.$$

Example 2

Find $\mathcal{L}\{\sin \sqrt{t}\}$.

Since

$$\sin t = t - \frac{t^3}{3!} + \frac{t^5}{5!} - \frac{t^7}{7!} + \ldots = \sum_{n=0}^{\infty} (-1)^n \frac{t^{2n+1}}{(2n+1)!}$$

Then by replacing t by $\sqrt{t}$ in this, we will have

$$\sin\sqrt{t} = \sqrt{t} - \frac{\sqrt{t}^3}{3!} + \frac{\sqrt{t}^5}{5!} - \ldots = \sum_{n=0}^{\infty}(-1)^n \frac{t^{(2n+1)/2}}{n!}.$$

Since $\mathcal{L}\{t^p\} = \dfrac{\Gamma(p+1)}{s^{p+1}}$, the Laplace transform is

$$\mathcal{L}\{\sin\sqrt{t}\} = \sum_{n=0}^{\infty}(-1)^n \frac{\Gamma((2n+1)/2+1)}{n!\, s^{(2n+1)/2+1}} \qquad (8)$$

Now using (3) of chapter 3, we have

$$\Gamma\left(\frac{2n+3}{2}\right) = \frac{2n+1}{2}\Gamma\left(\frac{2n+1}{2}\right)$$
$$= \frac{2n+1}{2}\frac{2n-1}{2}\Gamma\left(\frac{2n-1}{2}\right) = \ldots$$
$$= \frac{(2n+1)!}{2^{2n+1}n!}\Gamma\left(\frac{1}{2}\right) = \frac{(2n+1)!}{2\cdot 4^n n!}\sqrt{\pi}$$

using this in (8), we will obtain

$$\mathcal{L}\{\sin\sqrt{t}\} = \frac{\sqrt{\pi}}{2\, s^{3/2}}\sum_{n=0}^{\infty}(-1)^n\frac{1}{n!}\left(\frac{1}{4s}\right)^n.$$

This is the Taylor series for

$$\mathcal{L}\{\sin\sqrt{t}\} = \frac{\sqrt{\pi}}{2s^{3/2}} = \frac{\sqrt{\pi}}{2s^{3/2}}e^{=1/4s}.$$

5.3.3 METHOD OF DIFFERENTIAL EQUATIONS

This method involves finding a suitable differential equation satisfying the given function $f(t)$, and using the properties of Laplace transforms, such as derivative and linearity properties of chapter 1.

Example

To find $\mathcal{L}\{J_0(t)\}$ where $J_0(t)$ is the Bessel function of order zero, we use the method of differential equations. Using the properties of the Bessel function given in chapter 4, the function

$$y = J_0(t)$$

satisfies the initial value problem

$$t y''(t) + y(t) + t y(t) = 0, \qquad (9)$$
$$y(0) = 1, \; y'(0) = 0.$$

Let

$$Y(s) = \mathcal{L}\{y(t)\}.$$

Taking the Laplace transform of the equation of (9), we have

$$-\frac{d}{ds}[s^2 Y - s y(0) - y'(0)] + [sY - y(0)] - \frac{dY}{ds} = 0$$

which, using the initial conditions of (7), reduces to

$$-\frac{d}{ds}[s^2 Y - s] + [sY - 1] - \frac{dY}{ds} = 0$$

or

$$\frac{dY}{Y} = -\frac{s \, ds}{s^2 + 1}$$

Integrating this equation, we obtain

$$Y = \frac{c}{\sqrt{s^2 + 1}}$$

To evaluate c, we use the following facts

$$\lim_{s \to \infty} sY(s) = \frac{cs}{\sqrt{s^2 + 1}} = c \; \text{ and } \; \lim_{t \to 0} J_0(t) = 1$$

E-69

and the initial value theorem, Theorem 5.1 formula (1),

$$\lim_{s \to \infty} sY(s) = \lim_{t \to 0} J_0(t),$$

to find $c = 1$. Therefore

$$\mathcal{L}\{J_0(t)\} = \frac{1}{\sqrt{s^2 + 1}}.$$

5.3.4 METHOD OF DIFFERENTIATION WITH RESPECT TO A PARAMETER

Another method of finding the Laplace transform of a function is to differentiate with respect to a parameter. The idea is through the following fact. If

$$\mathcal{L}\{f(at)\} = \int_0^\infty e^{-st} f(at) \, dt$$

Then

$$\frac{d}{da} \mathcal{L}\{f(at)\} = \frac{d}{da} \int_0^\infty e^{-st} f(at) \, dt$$

$$= \int_0^\infty e^{-st} t \, f'(at) \, dt$$

$$= \mathcal{L}\{t f'(at)\}.$$

Therefore

$$\mathcal{L}\{t f'(at)\} = \frac{d}{da} \mathcal{L}\{f(at)\} \qquad (10)$$

Example

Since

$$\mathcal{L}\{\sin at\} = \frac{a}{s^2 + a^2} \quad \text{and} \quad \frac{d}{da} \sin at = t \cos at,$$

therefore we have

$$\mathcal{L}\{t\cos at\} = \frac{d}{da}\mathcal{L}\{\sin at\}$$

$$= \frac{d}{da}\frac{a}{s^2+a^2}$$

$$= \frac{s^2+a^2-2a^2}{(s^2+a^2)^2}$$

$$= \frac{s^2-a^2}{(s^2+a^2)^2}.$$

5.3.5 MISCELLANEOUS METHODS – MULTIPLICATION BY T^n PROPERTY AND DIVISION BY T PROPERTY

This method uses the properties like convolution (formula (17) of Chapter 1) and **multiplication by t^n property**, formula (16) of chapter 1, that is:

$$\mathcal{L}\{t^n f(t)\} = (-1)^n F^{(n)}(s) \qquad (11)$$

Also, it uses the **division by t property**, formula (11) of chapter 1, that is:

$$\mathcal{L}\left\{\frac{f(t)}{t}\right\} = \int_s^\infty F(x)\,dx \qquad (12)$$

Example 1

To find $\mathcal{L}\{t\cos at\}$, we may use the formula (11) and

$$\mathcal{L}\{\cos at\} = \frac{s}{s^2+a^2}$$

to find

$$\mathcal{L}\{t\cos at\} = (-1)\frac{d}{ds}\mathcal{L}\{\cos at\}$$

$$= -\frac{d}{ds}\frac{s}{s^2+a^2}$$

$$= -\frac{s^2 + a^2 - 2s^2}{(s^2 + a^2)^2}$$

$$= \frac{s^2 - a^2}{(s^2 + a^2)^2}.$$

Example 2

To find $\mathcal{L}\{e^{-t^2}\}$ we can use the division by t property, formula (12), to write

$$\mathcal{L}\{e^{-t^2}\} = \mathcal{L}\left\{\frac{te^{-t^2}}{t}\right\} = \int_s^\infty u e^{-u^2} du$$

$$= \frac{1}{2} e^{-u^2} \Big|_{u=s}^{u=\infty} = \frac{1}{2} e^{-s^2}.$$

5.4 METHODS OF FINDING INVERSE TRANSFORMS

5.4.1 PARTIAL FRACTION METHOD

Rational functions. A quotient or ratio of two polynomial functions is called a **rational function**. Therefore a rational function $R(s)$ looks like

$$R(s) = \frac{P(s)}{Q(s)} = \frac{a_n s^n + a_{n-1} s^{n-1} + \ldots + a_0}{b_m s^m + b_{m-1} s^{m-1} + \ldots + b_0} \quad (13)$$

where a_i and b_j for all $i = 0, 1, \ldots, n$ and $j = 0, 1, \ldots, m$, are real numbers.

Proper rational function. We say that the rational function (13) is **proper** if the degree of its numerator is lower than the degree of its denominator, i.e. $n < m$.

The idea of partial fraction method is based on a procedure and some rules such as the following:

Partial fractions procedure.

(i) By eliminating the common factors of numerator and denominator, reduce the fraction to its lowest form.

(ii) If the result is not a proper fraction, then we divide $Q(s)$ into $P(s)$ so as to obtain a quotient plus a remainder which is a proper fraction.

(iii) Factor the denominator of the remainder fraction, and write the fraction as a sum of fractions of the forms:

$$\frac{A}{(s-a)^k} \quad \text{and} \quad \frac{Bs+C}{(s^2+bs+c)^l} \qquad (14)$$

Partial fractions rules.

1. If a factor $(s-a)$ appears k times in the factorization of $Q(s)$, then assume that the partial fractions of $\frac{P(s)}{Q(s)}$ contains the terms

$$\frac{A_1}{(s-a)} + \frac{A_2}{(s-a)^2} + \ldots + \frac{A_k}{(s-a)^k}. \qquad (15)$$

2. If an irreducible factor $s^2 + bs + c$ appears l time in factorization of $Q(s)$, then assume the partial fraction of $\frac{P(s)}{Q(s)}$ contains the terms

$$\frac{B_1 s+c_1}{(s^2+bs+c)} + \frac{B_2 s+c_2}{(s^2+bs+c)^2} + \ldots + \frac{B_l s+c_l}{(s^2+bs+c)^l}. \qquad (16)$$

Example 1

Find

$$\mathcal{L}^{-1}\left\{\frac{2s^2-4}{(s+1)(s-2)(s-3)}\right\}.$$

We write

$$\frac{2s^2-4}{(s+1)(s-2)(s-3)} = \frac{A}{s+1} + \frac{B}{s-2} + \frac{C}{s-3}. \quad (17)$$

To determine the coefficients A, B, and C we may use one of the following two methods:

Method 1

Multiply (17) by $(s+1)(s-2)(s-3)$, to obtain

$$2s^2 - 4 = A(s-2)(s-3) + B(s+1)(s-3) \\ + C(s+1)(s-2). \quad (18)$$

Letting $s = -1$, this will give

$$2 - 4 = A(-3)(-4) + 0 + 0,$$

i.e., $A = -1/6$.

Similarly, by letting $s = 2$ and $s = 3$, (18) will give $B = -4/3$ and $C = 7/2$ respectively.

Method 2

Multiplying both sides of (17) by $(s+1)(s-2)(s-3)$, we obtain (18). Then simplifying (18) we have

$$2s^2 - 4 = (B+C)s^2 + (A - 2B - C)s + (-2A - 3B - 2C).$$

Equating the coefficients, we will have the system of linear equations

$$\begin{cases} B + C = 2 \\ A - 2B - C = 0 \\ -2A - 3B - 2C = -4 \end{cases}$$

Solving this system, we obtain the coefficients A, B, and C.

Thus, using the values of A, B, and C from the method 1 in the formula (17), we have

$$\mathcal{L}^{-1}\left\{\frac{2s^2-4}{(s+1)(s-2)(s-3)}\right\}$$

$$=\mathcal{L}^{-1}\left\{\frac{-1/6}{s+1}+\frac{-4/3}{s-2}+\frac{7/2}{s-3}\right\}$$

$$=-\frac{1}{3}e^{-t}-\frac{7}{2}t^2e^{2t}+4te^{2t}+\frac{1}{3}e^{2t}.$$

Example 2

Find

$$\mathcal{L}^{-1}\left\{\frac{3s+1}{(s-1)(s^2+1)}\right\}.$$

We write

$$\frac{3s+1}{(s-1)(s^2+1)}=\frac{A}{s-1}+\frac{Bs+c}{s^2+1}.$$

Multiplying this by $s-1$ and letting $s=1$, then we will have $A=2$. Now we have

$$\frac{3s+1}{(s-1)(s^2+1)}=\frac{2}{s-1}+\frac{Bs+c}{s^2+1}$$

To determine B and C, we substitute for s any number other than 1, for example $s=0$ and 2; then

$$-1=-2+C \quad \text{and} \quad \frac{7}{5}=2+\frac{2B+C}{5}$$

from which we obtain $C=1$ and $B=-2$. Thus

$$\mathcal{L}^{-1}\left\{\frac{3s+1}{(s-1)(s^2+1)}\right\}$$

$$=\mathcal{L}^{-1}\left\{\frac{2}{s-1}+\frac{-2s+1}{s^2+1}\right\}$$

$$= 2\mathcal{L}^{-1}\left\{\frac{1}{s-1}\right\} - 2\mathcal{L}^{-1}\left\{\frac{2}{s^2+1}\right\} + \mathcal{L}^{-1}\left\{\frac{1}{s^2+1}\right\}$$

$$= 2e^t - 2\cos t + \sin t.$$

5.4.2 THE HEAVISIDE EXPANSION FORMULA

Let $R(s)$ be a proper rational function

$$R(s) = \frac{P(s)}{Q(s)}$$

where $Q(s)$ has m distinct real zeros a_k, $k = 1, 2, \ldots, m$, i.e.

$$Q(s) = (s - a_1)(s - a_2) \ldots (s - s_m).$$

Then we have the **Heaviside expansion formula**:

$$L^{-1}\left\{\frac{P(s)}{Q(s)}\right\} = \sum_{k=1}^{m} \frac{P(a_k)}{Q'(a_k)} e^{a_k t}. \tag{19}$$

Example

Since

$$s^3 - 6s^2 + 11s - 6 = (s-1)(s-2)(s-3), \text{ and}$$
$$(s^3 - 6s^2 + 11s\ 6)' = 3s^2 - 12s + 11,$$

We have

$$\mathcal{L}^{-1}\left\{\frac{s^2+1}{s^3 - 6s^2 + 11s - 6}\right\} = \frac{2}{2}e^t + \frac{5}{-1}e^{2t} + \frac{10}{2}e^{3t}$$

$$= e^t - 5e^{2t} + 5e^{3t}.$$

CHAPTER 6

FOURIER TRANSFORMS

6.1 FOURIER SERIES

Let $f(x)$, $-L < x < L$, be a real-valued function. The **Fourier series** of f is the **trigonometric series**

$$a_0 + \sum_{n=1}^{\infty} \left(a_n \cos \frac{n\pi x}{L} + b_n \sin \frac{n\pi x}{L} \right) \quad (1)$$

where (a_n, b_n) are called the **Fourier coefficients** and are defined by

$$a_0 = \frac{1}{2L} \int_{-L}^{L} f(x)\, dx$$
$$a_n = \frac{1}{L} \int_{-L}^{L} f(x) \cos \frac{n\pi x}{L}\, dx \quad (2)$$
$$b_n = \frac{1}{L} \int_{-L}^{L} f(x) \cos \frac{n\pi x}{L}\, dx$$

for $n = 1, 2, \ldots$

Theorem 6.1 EVEN FUNCTION

If f is an **even function**, i.e.

$$f(x) = f(-x),$$

then
$$b_n = 0 \text{ for } n = 1, 2, \ldots \tag{3}$$

ODD FUNCTION

If f is an **odd function**, i.e.
$$f(-x) = -f(x)$$
then
$$A_n = 0 \text{ for } n = 0, 1, 2, \ldots \tag{4}$$

Example 1

Let $f(x) = x$, $-L < x < L$. Then $f(x)$ is odd and we have
$$A_n = 0 \text{ for all } n$$
$$B_n = \frac{1}{L} \int_{-L}^{L} x \sin \frac{n\pi x}{L} \, dx$$
$$= \frac{2}{L} \int_{0}^{L} x \sin \frac{n\pi x}{L} \, dx$$

Using integration by parts, it follows that
$$B_n = \left(\frac{2L}{n\pi}\right)(-1)^{n+1},$$
therefore the Fourier series of $f(x) = x$, $-L < x < L$, is
$$f(x) = \frac{2L}{\pi} \sum_{n=1}^{\infty} \frac{(-1)^{n+1}}{n} \sin \frac{n\pi x}{L}. \tag{5}$$

PARTIAL SUM OF ORDER N

The **partial sum of order N** of the trigonometric series (1) is the function

$$\boxed{f_N = A_0 + \sum_{n=1}^{N} \left(A_n \cos \frac{n\pi x}{L} + B_n \sin \frac{n\pi x}{L} \right)} \tag{6}$$

Example 2

The partial sums f_N for $N = 1, 2, 3$ of the Fourier series of $f(x) = x$, $-\pi < x < \pi$, are shown in the Figure 6.1. These are found from Example 1 using $L = \pi$.

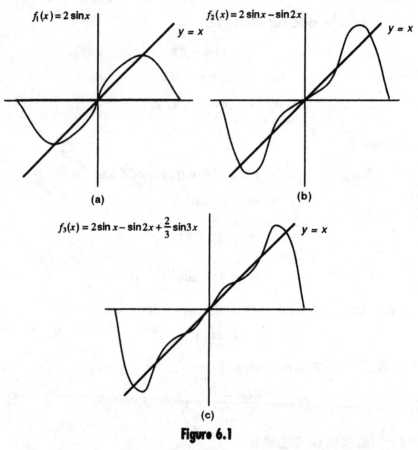

Figure 6.1

6.2 FOURIER SINE AND COSINE SERIES

Suppose a function $f(x)$ is defined for $0 < x < L$ and we want a Fourier series representation of it. To get this, we can extend f to the interval $(-L, L)$ in two ways which follow.

6.2.1 ODD EXTENSION

Odd extension is denoted by $f_o(x)$ and is defined by

$$f_o(x) = \begin{cases} f(x) & \text{for } 0 < x < L \\ -f(-x) & \text{for } -L < x < 0 \\ 0 & \text{for } x = 0 \end{cases} \quad (7)$$

which is an odd function, thus we have a **Fourier sine series** representation for $f(x)$:

$$\sum_{n=1}^{\infty} B_n \sin\frac{n\pi x}{L}; \quad 0 \leq x \leq L$$

$$B_n = \frac{2}{L}\int_0^L f(x)\sin\frac{n\pi x}{L}dx. \quad (8)$$

6.2.2 EVEN EXTENSION

Even extension of f from the interval $(0, L)$ to the interval $(-L, L)$ is denoted by f_E and defined by

$$f_E(x) = \begin{cases} f(x) & \text{for } 0 < x < L \\ -f(-x) & \text{for } -L < x < 0 \\ 0 & \text{for } x = 0 \end{cases} \quad (9)$$

which is an even function, thus we have the **Fourier cosine series** representation for $f(x)$:

$$a_0 + \sum_{n=1}^{\infty} a_n \cos\frac{n\pi x}{L}; \quad 0 \leq x \leq L$$

$$a_0 = \frac{1}{L}\int_0^L f(x)dx, \quad (10)$$

$$a_n = \frac{2}{L}\int_0^L f(x)\cos\frac{n\pi x}{L}dx \text{ for } n = 1, 2, \ldots$$

6.3 PIECEWISE SMOOTH FUNCTIONS

A function $f(x)$, $0 , x < b$, is said to be **piecewise smooth** if f and all its derivatives are piecewise continuous.

Example

The function

$$f(x) = |x|; -\pi < x < \pi$$

is continuous on the entire interval, while its derivative

$$f'(x) = \begin{cases} 1 & \text{if } x > 0 \\ -1 & \text{if } x < 0 \end{cases}$$

is piecewise continuous, with

$$f'(0+0) = 1 \text{ and } f'(0-0) = -1.$$

All higher derivatives, for $n = 2, 3, \ldots$, are

$$f^{(n)}(x) = 0 \text{ for } x \neq 0$$

which are piecewise continuous on the interval. Thus f is piecewise smooth on $(-\pi, \pi)$.

6.4 PERIODIC EXTENSIONS

The **periodic extension** of a function $f(x)$, defined on $a < x < b$, Figure 6.2(a), is denoted by $\bar{f}$ and is defined by setting

$$\boxed{\bar{f}[x + n(b-a)] = f(x) \qquad (11)}$$

for $a < x < b$ and $n = \ldots, -2, -1, 0, 1, 2, \ldots$, Figure 6.2(b).

6.5 THEOREM 2 (CONVERGENCE THEOREM)

Let f be a piecewise smooth function on $(-L, L)$, then the

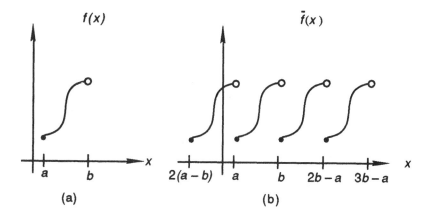

Figure 6.2

Fourier series of f converges for all x to the average value

$$\tfrac{1}{2}[\overline{f}(x+0) + \overline{f}(x-0)] \tag{12}$$

where $\overline{f}$ is the periodic extension of f and

$$\overline{f}(x+0) = \lim_{\varepsilon \to 0+} f(x+\varepsilon) \quad \text{and} \quad \overline{f}(x-0) = \lim_{\varepsilon \to 0-} f(x-\varepsilon) \tag{13}$$

Thus

$$\tfrac{1}{2}[\overline{f}(x+0) + \overline{f}(x-0)] \tag{14}$$
$$= a_0 + \sum_{n=1}^{\infty}\left(a_n \cos\frac{n\pi x}{L} + b_n \sin\frac{n\pi x}{L}\right)$$

6.5.1 FIRST CRITERION FOR UNIFORM CONVERGENCE

Let $f(x)$, $-L < x < L$, be a piecewise smooth function. If the Fourier coefficients $\{a_n\}$, $\{b_n\}$ satisfy

$$\sum_{n=1}^{\infty}(|a_n| + |b_n|) < \infty, \tag{15}$$

then the Fourier series converges uniformly.

Example

The Fourier sine series

$$\sum_{n=1}^{\infty} \frac{\sin nx}{n^2}$$

converges uniformly.

6.5.2 SECOND CRITERION FOR UNIFORM CONVERGENCE

Assume that a function $f(x)$, $-L < x < L$, satisfies the **Direchlet's conditions:**

> (i) f is piecewise smooth on $(-L, L)$,
>
> (ii) f is continuous and $(-L, L)$, and (16)
>
> (iii) $f(-L+0) = f(L-0)$

Then the Fourier series converges uniformly.

6.6 GENERAL CRITERION FOR DIFFERENTIATION

Theorem 6.3

If $f(x)$ satisfies the Direchlet's conditions (16), then

$$\frac{1}{2}[f'(x+0)+f'(x-0)] \qquad (17)$$

$$= \sum_{n=1}^{\infty} \frac{n\pi}{L}\left(b_n \cos \frac{n\pi x}{L} - a_n \sin \frac{n\pi x}{L}\right)$$

Example

We know that the Fourier series of the even function $f(x) = x^2$, $-\pi < x < \pi$, is of the following form

$$a_0 + \sum_{n=1}^{\infty} a_n \cos nx .$$

To determine the coefficients $\{a_n\}$, we use (17) to find

$$2x = -\sum_{n=1}^{\infty} n a_n \sin nx.$$

Now we use the Fourier series of $f(x) = x$, formula (5) for $L = \pi$, to obtain

$$2x = 4\sum_{n=1}^{\infty} (-1)^{n+1} \frac{\sin nx}{n}$$

Therefore $a_n = 4(-1)^n/n^2$ for $n = 1, 2, \ldots$. To compute a_0 we must return to the definition

$$a_0 = (1/2\pi)\int_{-\pi}^{\pi} x^2 dx = \pi^2/3.$$

Therefore we have

$$x^2 = \frac{\pi^2}{3} + \sum_{n=1}^{\infty} \frac{(-1)^n}{n^2} \cos nx \quad \text{for } -\pi < x < \pi.$$

6.7 PARSEVAL'S THEOREM AND MEAN SQUARE ERROR

Let $f(x)$, $-L < x < L$, be a piecewise smooth function with Fourier series

$$a_0 + \sum_{n=1}^{\infty} \left(a_n \cos \frac{n\pi x}{L} + b_n \sin \frac{n\pi x}{L} \right).$$

Then:

Parseval's theorem states that

$$\frac{1}{2L}\int_{-L}^{L} [f(x)]^2 dx = a_0^2 + \frac{1}{2}\sum_{n=1}^{\infty} (a_n^2 + b_n^2) \quad (18)$$

Mean square error σ_N^2 is defined by

$$\sigma_N^2 = \frac{1}{2L}\int_{-L}^{L} [f(x) - f_N(x)]^2 dx \quad (19)$$

where $f_N(x)$ is the N-th partial sum, defined by (6). We have

$$\sigma_N^2 = \frac{1}{2}\sum_{n=N+1}^{\infty} (a_n + b_n) \quad (20)$$

Example

Let $f(x) = x$, then we have

$$a_n = 0, \ b_n (-1)^{n-1}(2/n)$$

therefore the mean square error is

$$\sigma_N^2 = \frac{1}{2}\sum_{n=N+1}^{\infty}\frac{4}{m^2} = 2\sum_{n=N+1}^{\infty}\frac{1}{m^2}.$$

6.8 COMPLEX FORM OF FOURIER SERIES

With the convention of the **De Moivre's formula**

$$e^{i\theta} = \cos\theta + i\sin\theta \ ; \ i = \sqrt{-1} \qquad (21)$$

the Fourier series of $f(x)$, $-L < x < L$, in complex notation, assumes the form

$$f(x) = \sum_{-\infty}^{\infty} \alpha_n e^{in\pi x/L} \qquad (22)$$

where the coefficients $\{\alpha_n\}$ for $n = 0, +1, -1, +2, -2, \ldots$, are defined by

$$\alpha_n = \frac{1}{2L}\int_{-L}^{L} f(x)e^{-(in\pi x/L)}dx \qquad (23)$$

6.9 PARSEVAL'S IDENTITY

Parseval's Identity will take the form:

$$\frac{1}{2L}\int_{-L}^{L}|f(x)|^2 dx = \sum_{-\infty}^{\infty}|\alpha_n|^2. \qquad (24)$$

Example

Compute the complex Fourier series of

$$f(x) = e^{2x}, \quad -\pi < x < \pi$$

$$\alpha_n = \frac{1}{2\pi}\int_{-\pi}^{\pi} e^{2x}e^{-inx}dx = \frac{1}{2\pi}\int_{-\infty}^{\infty} e^{(2-in)x}dx$$

$$= \frac{(-1)^n}{2\pi(2-in)}(e^{2\pi} - e^{-2\pi}) = \frac{1}{\pi}\sinh 2\pi \frac{(-1)^n(2+in)}{4+n^2}.$$

Therefore the complex Fourier series of $f(x) = e^{2x}, -\pi < x < \pi$, is

$$f(x) = \frac{1}{\pi}\sinh 2\pi \sum_{-\infty}^{\infty} \frac{(-1)^n(2+in)}{4+n^2}e^{inx}$$

6.10 FOURIER INTEGRAL TRANSFORMS

We assume that $f(x)$ is a real-valued piecewise smooth function on every bounded interval (a,b) and that all the relevant integrals are absolutely convergent. Then we define the **Fourier integral transform**, or simply **Fourier transform**, of f by

$$\mathcal{F}(\mu) = \frac{1}{2\pi}\int_{-\infty}^{\infty} f(x)e^{-i\mu x}dx \qquad (25)$$

then the function $f(x)$ is recovered by the **Fourier Inversion formula**

$$f(x)\int_{-\infty}^{\infty} \mathcal{F}(\mu)e^{i\mu x}d\mu \qquad (26)$$

The functions $f(x)$ and $\mathcal{F}(\mu)$, (25) and (26) respectively, are called **Fourier transform pairs**.

6.10.1 PARSEVAL'S THEOREM

Parseval's theorem for Fourier transforms is

$$\int_{-\infty}^{\infty} |f(x)|^2 dx = 2\pi \int_{-\infty}^{\infty} |\mathcal{F}(\mu)|^2 d\mu \qquad (27)$$

6.10.2 INVERSION THEOREM FOR FOURIER TRANSFORMS

Also called **convergence theorem for Fourier transforms**.

Let $f(x)$, $-\infty < x < \infty$, be a piecewise smooth function on each interval with

$$\int_{-\infty}^{\infty} |f(x)|\, dx$$

finite. Define the Fourier transform $\mathcal{F}(\mu)$ by (25). Then for each x we have

$$\boxed{\lim_{L \to \infty} \int_{-L}^{L} \mathcal{F}(\mu) e^{i\mu x}\, dx = \frac{1}{2}[f(x+0) + f(x-0)]} \quad (28)$$

Example

(Square wave function). We find the Fourier transform of the square wave function

$$f(x) = \begin{cases} 0 & \text{for} \quad x < a \\ 1 & \text{for} \quad a \leq x \leq b \\ 0 & \text{for} \quad x > b \end{cases} \quad (29)$$

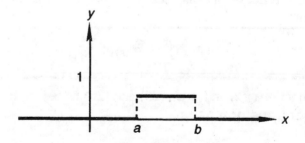

Figure 6.3

We have

$$\mathcal{F}(\mu) = \frac{1}{2\pi}\int_{-\infty}^{\infty} f(x) e^{-i\mu x}\, dx = \frac{1}{2\pi}\int_{a}^{b} e^{-i\mu x}\, dx$$

$$= \begin{cases} \dfrac{e^{-i\mu b} - e^{-i\mu a}}{-2\pi i\mu} & \text{for } \mu \neq 0 \\ \dfrac{b-a}{2} & \text{for } \mu = 0 \end{cases}$$

By the Fourier inversion theorem we have

$$\lim_{L\to\infty}\int_{-L}^{L} \mathcal{F}(\mu)e^{i\mu x}dx = \lim_{L\to\infty}\int_{-L}^{L} \frac{\left(e^{-i\mu b}-e^{-i\mu a}\right)e^{-i\mu x}}{-2\pi i\mu}d\mu$$

$$= \begin{cases} 0 & \text{for} & x < a \\ 1/2 & \text{for} & x = a \\ 1 & \text{for} & a < x < b \\ 1/2 & \text{for} & x = b \\ 0 & \text{for} & x > b \end{cases}$$

And the Parseval's theorem states that

$$\frac{1}{2\pi}\int_{-\infty}^{\infty} \frac{\left|e^{-i\mu b} - e^{-i\mu a}\right|^2}{\mu^2} d\mu = b - a.$$

6.11 FOURIER COSINE FORMULAS

Let $f(x)$ be defined for all $x > 0$. We extend $f(x)$ to an even function (even extension) by

$$f(x) = f(-x) \text{ for } x < 0,$$

then we have the **Fourier cosine formulas**

$$\boxed{f(x) = \int_0^{\infty} \mathcal{F}_c(\mu)\cos\mu x\, dx, \ \mathcal{F}_c(\mu) = \frac{2}{\pi}\int_0^{\infty} f(x)\cos\mu x\, dx} \quad (30)$$

6.12 FOURIER SINE FORMULAS

Let $f(x)$ be defined for all $x > 0$. We extend f to a an odd function (odd extension) by

$$f(x) = -f(-x) \text{ for all } x < 0,$$

then we have the **Fourier sine formulas**

$$f(x) = \int_0^\infty \mathcal{F}_s(\mu) \sin \mu x \, dx, \quad \mathcal{F}_s(\mu) = \frac{2}{\pi} \int_0^\infty f(x) \sin \mu x \, dx \quad (31)$$

6.13 THE CONVOLUTION THEOREM

If $h(x)$ is the convolution of $f(x)$ and $g(x)$, i.e.

$$h(x) = (f*g)(x) = \int_{-\infty}^\infty f(u) \, g(x-u) \, du \quad (32)$$

Then

$$\int_{-\infty}^\infty h(x) e^{-i\mu x} dx = \left\{ \int_{-\infty}^\infty f(x) e^{-i\mu x} dx \right\} \left\{ \int_{-\infty}^\infty g(x) e^{-i\mu x} dx \right\} \quad (33)$$

or

$$H(\mu) = \mathcal{F}(\mu) \, G(\mu) \quad (34)$$

where H, $\mathcal{F}$, and G are the Fourier transforms of h, f, and g respectively.

6.14 RELATIONSHIP OF FOURIER AND LAPLACE TRANSFORMS

Let the function $f(t)$ be defined for all $t > 0$. We consider the function

$$g(t) = \begin{cases} e^{-xt} f(t) & \text{for } t > 0 \\ 0 & \text{for } t < 0 \end{cases} \quad (35)$$

then from (25), with μ replace by y, we see that the Fourier transform of $g(t)$ is

$$\mathcal{F}\{g(t)\} = \int_0^\infty e^{-(x+iy)t} f(t)\, dt = \int_0^\infty e^{-st} f(t)\, dt \quad (36)$$

where

$$s = x + iy \quad (37)$$

The right side of (36) is the Laplace transform of $f(t)$ in complex variable. This result gives a relationship between Fourier and Laplace transforms

$$\mathcal{F}\{e^{-xt} f(t)\} = \mathcal{L}\{f(t)\} \quad (38)$$

or

$$\mathcal{F}\{f(t)\} = \mathcal{L}\{e^{xt} f(t)\} \quad (39)$$

CHAPTER 7

APPLICATIONS OF LAPLACE TRANSFORMS TO INTEGRAL AND DIFFERENCE EQUATIONS

7.1 INTEGRAL EQUATIONS

An **integral equation**, is an equation of the form

$$y(t) = f(t) + \int_a^b k(x,t)\, y(x)\, dx \qquad (1)$$

where $f(t)$ and $k(x,t)$ are known functions, and a and b are either constant numbers or some known functions of t. The function $y(t)$ is an unknown which has to be determined.

Kernel

The function $k(x,t)$ in (1) is called the kernel of the integral equation.

7.1.1 FREDHOLM INTEGRAL EQUATION

This equation is an integral equation of type (1) where a and b are constant numbers.

7.1.2 VOLTERRA INTEGRAL EQUATION

This equation is an integral equation of type (1) where a

is a constant while $b = t$, i.e., it is of the form

$$y(t) = f(t) + \int_0^t k(x,t)\, y(x)\, dx \qquad (2)$$

7.1.3 INTEGRAL EQUATION OF CONVOLUTION TYPE

If in the integral equation (1) we put

$$a = 0,\ b = t \text{ and } k(x,t) = k(t - x)$$

then we have an **Integral equation of convolution type**, i.e. it is of the form

$$y(t) = f(t) + \int_0^t k(t - x)\, y(x)\, dx \qquad (3)$$

This integral equation of convolution type is also written as

$$y(t) = f(t) + k * y(t) \qquad (4)$$

SOLUTION OF AN INTEGRAL EQUATION OF CONVOLUTION TYPE

To solve the integral equation (3) or (4), we assume

$$Y(s) = L\{y(t)\},\ F(s) = L\{f(t)\},\ \text{and}\ K(s) = L\{k(t)\}$$

then we take the Laplace transform of both sides of (4) to obtain

$$Y(s) = F(s) + K(s)\, Y(s) \qquad (5)$$

solving (5) for $Y(s)$, we find

$$Y(s) = \frac{F(s)}{1 - K(s)} \qquad (6)$$

Thus by Laplace inversion

$$y(t) = \mathcal{L}^{-1}\left\{\frac{F(s)}{1-K(s)}\right\} \qquad (7)$$

Example

To solve

$$y(t) = t^2 + \int_0^t \sin(t-x)\, y(x)\, dx,$$

we write

$$y(t) = t^2 + (y * \sin)(t)$$

Thus

$$\mathcal{L}\{y(t)\} = \mathcal{L}\{t^2 + (y * \sin)(t)\}$$

or

$$Y(s) = \frac{2}{s^3} + \frac{Y(s)}{s^2 + 1}$$

or

$$Y(s) = \frac{2}{s^3} + \frac{Y(s)}{s^5}.$$

Hence

$$y(t) = \mathcal{L}^{-1}\left\{\frac{2}{s^3}\right\} + \mathcal{L}^{-1}\left\{\frac{2}{s^5}\right\}$$

$$= t^2 + \frac{1}{12}t^4.$$

7.2 INTEGRO-DIFFERENTIAL EQUATIONS

An equation involving an integral and some derivatives of an unknown function $y(t)$ is called an **integro-differential equation**.

Example

The following is an integro-differential equation

$$y'(t) - y(t) = t + \int_0^t \sin(t-x)\, y(x)\, dx \qquad (8)$$

The solution of an integro-differential equation like (8) is easily obtained by taking the Laplace transform of both sides of the equation. It is then solvable subject to some given initial conditions.

Example

To solve (8), we write it in the following convolution form

$$y'(t) + y(t) = t + (\sin * y)(t)$$

with initial condition

$$y(0) = 0.$$

We take the Laplace transform of the equation, to obtain

$$\mathcal{L}\{y'(t) + y(t)\} = \mathcal{L}\{t + (\sin * y)(t)\}$$

or

$$sY(s) - y(0) + Y(s) = \frac{1}{s^2} + \frac{Y(s)}{s^2+1}.$$

Therefore using the initial condition $y(0) = 0$, we have

$$(s+1)Y - \frac{Y}{s^2+1} = \frac{1}{s^2}$$

Solving this equation, we will obtain

$$Y = \frac{s^2+1}{s^3(s^2+s+1)}$$

$$= \frac{1}{s^3} - \frac{1}{s^2} + \frac{1}{s} - \frac{s}{s^2+s+1}.$$

Thus

$$y(t) = \mathcal{L}^{-1}\left\{\frac{1}{s^3} - \frac{1}{s^2} + \frac{1}{s} - \frac{s}{s^2+s+1}\right\}$$

$$= \mathcal{L}^{-1}\left\{\frac{1}{s^3}\right\} - \mathcal{L}^{-1}\left\{\frac{1}{s^2}\right\} + \mathcal{L}^{-1}\left\{\frac{1}{s}\right\} - \mathcal{L}^{-1}\left\{\frac{s}{s^2+s+1}\right\}$$

$$= \frac{t^2}{2} - t + e^{-1/2t}\left(\cos\frac{\sqrt{3}}{t}t - \frac{\sqrt{3}}{3}\sin\frac{\sqrt{3}}{2}t\right).$$

7.3 DIFFERENCE EQUATIONS

An equation involving an unknown function $y(t)$ with one or more shifted forms of it like $y(t-a)$, where a is a constant, is called a difference equation.

Example

Solve the difference equation

$$y(t) - y(t-1) = t.$$

We take the Laplace transform of the equation, to have

$$\mathcal{L}\{y(t)\} - \mathcal{L}\{y(t-1)\} = \mathcal{L}\{t\}$$

or

$$Y(s) - \mathcal{L}\{y(t-1)\} = \frac{1}{s^2}.$$

By the second shifting theorem, property 4 of chapter 1 formula (10), we have

$$\mathcal{L}\{y(t-1)\} = e^{-s}Y(s).$$

Therefore

$$Y(s) - e^{-s}Y(s) = \frac{1}{s^2}.$$

Thus

$$Y(s) = \frac{1}{s^2(1 - e^{-s})}.$$

Hence we have

$$y(t) = \mathcal{L}^{-1}\left\{\frac{1}{s^2(1-e^{-s})}\right\}$$

$$= \mathcal{L}^{-1}\left\{\frac{1}{s^2} + e^{-s}\frac{1}{s^2} + e^{-2s}\frac{1}{s^2} + \ldots\right\}.$$

Now we use the second translation property, property 3 formula (7) of chapter 2, to obtain

$$y(t) = \begin{cases} t & \text{for} \quad t < 1 \\ t + (t-1) & \text{for} \quad 1 \leq t < 2 \\ t + (t-1) + (t-2) & \text{for} \quad 2 \leq t < 3 \\ \vdots & \end{cases}$$

$$= \begin{cases} t & \text{for} \quad t < 1 \\ 2t - 1 & \text{for} \quad 1 \leq t < 2 \\ 3t - 3 & \text{for} \quad 2 \leq t < 3 \\ 4t - 7 & \text{for} \quad 3 \leq t < 4 \\ \vdots & \\ nt - (n-1)n/2 & \text{for} \quad n \leq t < (n+1) \end{cases}$$

7.4 DIFFERENTIAL-DIFFERENCE EQUATIONS

A difference equation involving some derivatives of the unknown function $y(t)$ is called a **differential-difference equation**.

Example

$$y''(t) = y(t-2) + 5 \sin t$$

is a differential-difference equation where by taking Laplace transform of its both sides, using formula (10) of chapter 1, and using derivative property formula (14) of chapter 1, we will obtain

$$s^2Y(s) - sy(0) - y'(0) = e^{-2s}Y(s) + \frac{5}{s^2+1},$$

where by assuming some initial values, e.g.

$$y(0) = 0 \text{ and } y'(0) = 1,$$

it will reduce to

$$s^2Y(s) - 1 = e^{-2s}Y(s) + \frac{5}{s^2+1}$$

or

$$Y(s) = \frac{1}{s^2 - e^{-2s}} + \frac{5}{(s^2+1)(s^2 - e^{-2s})}.$$

Therefore

$$y(t) = \mathcal{L}^{-1}\left\{\frac{1}{s^2 - e^{-2s}}\right\} + \mathcal{L}^{-1}\left\{\frac{5}{(s^2+1)(s^2 - e^{-2s})}\right\}.$$

CHAPTER 8

APPLICATIONS TO BOUNDARY-VALUE PROBLEMS

8.1 FUNCTIONS OF TWO VARIABLES

A function of two variables is denoted by

$$u = u(x,y)$$

and its **partial derivatives** by

$$u_x = \frac{\partial u}{\partial x},\ u_y = \frac{\partial u}{\partial y},\ u_{xx} = \frac{\partial^2 u}{\partial x^2},\ u_{xy} = \frac{\partial^2 u}{\partial x \partial y},\ u_{yy} = \frac{\partial^2 u}{\partial y^2}$$

Example

If $u(x,y) = \sin x + \cos y + x y$, then

$$u_x = \cos x + y,\ u_y = -\sin y + x,$$

$$u_{xx} = -\sin x,\ u_{xy} = u_{yx} = 1,\ \text{and}\ y_{yy} = -\cos y.$$

8.2 PARTIAL DIFFERENTIAL EQUATION

A **linear second order partial differential equation** is an equation of the form

$$A u_{xx} + 2 B u_{xy} + C u_x + D u_y + E u = G \qquad (1)$$

where A, B, C, D, E, and G are all functions of x and y.

If in (1) $G = 0$, then the equation is called **homogeneous**.

8.3 SOME IMPORTANT PARTIAL DIFFERENTIAL EQUATIONS

Laplace's equation.

$$u_{xx} + u_{yy} = 0 \qquad (2)$$

The wave equation.

$$u_{tt} - C^2 u_{xx} = 0 \qquad (3)$$

The heat equation.

$$u_t - K u_{xx} = 0 \qquad (4)$$

The telegraph equation.

$$u_{tt} - C^2 u_{xx} + 2\beta u_t + \propto u = 0 \qquad (5)$$

Poisson's equation.

$$u_{xx} + u_{yy} = G \qquad (6)$$

8.4 CLASSIFICATION OF SECOND-ORDER PARTIAL DIFFERENTIAL EQUATIONS

Depending on the second-order coefficients A, B, and C, the equation (1) is classified in the following three types:

$$\begin{aligned} &\text{Elliptic if } AC - B^2 > 0 \\ &\text{Hyperbolic if } AC - B^2 < 0 \quad\quad (7) \\ &\text{Parabolic if } AC - B^2 = 0 \end{aligned}$$

Example

Poisson's and Laplace's equations are elliptic. Wave and telegraph equations are hyperbolic. The heat equation is parabolic.

8.5 BOUNDARY CONDITIONS

Unlike ordinary differential equations, it is hard to formulate a general method of solving for all second-order partial differential equations. Instead, we specify a solution in terms of some certain boundary conditions. Natural types of boundary conditions are:

8.5.1 DIRICHLET PROBLEM

It is to specify the values of $u(x,y)$ on the boundary of a bounded plane region with smooth boundary. It is used for elliptic equations.

Example

In a physical problem of determining the electrostatic potential function u in a cylinder where the charge density is specified in the interior and the boundary is required to be an equipotential surface, the Dirichlet's problem is

$$\begin{cases} u_{xx} + u_{yy} = -\rho & \text{for } x^2 + y^2 < R^2, \\ u(x,y) = C & \text{for } x^2 + y^2 = R^2. \end{cases}$$

8.5.2 CAUCHY PROBLEM

It is naturally used for a hyperbolic equation. It is to specify the solution $u(x,t)$ and its time derivative $u_t(x,t)$ on the line $t = 0$. Also, we may need to specify boundary conditions at the end of a finite interval or, in case of infinite intervals, to restrict the growth of $u(x,t)$ when x becomes very large.

Example

The following Cauchy problem is treated for the **vibrating string** with fixed ends.

$$u_{tt} - C^2 u_{xx} = 0 \quad \text{for} \quad t > 0,\ 0 < x < L$$

$$u(x,0) = f_1(x) \quad \text{for} \quad 0 < x < L$$

$$u_t(x,0) = f_2(x) \quad \text{for} \quad 0 < x < L$$

$$u(0,t) = 0 \quad \text{for} \quad t > 0$$

$$u(L,t) = 0 \quad \text{for} \quad t > 0.$$

Where the functions f_1 and f_2 correspond respectively to the initial position and velocity of the vibrating string.

8.6 SOLUTION OF BOUNDARY-VALUE PROBLEMS BY LAPLACE TRANSFORMS

8.6.1 ONE-DIMENSIONAL BOUNDARY VALUE PROBLEMS

In this case our unknown function depends only on two variables, either x and t or x and y. We may use the Laplace transformation with respect to only one of the variables (t or x), to reduce the partial differential equation to an ordinary differential equation. Then we solve the resulting differential equation to find the Laplace transform of the original equation. By inverting the solution of the ordinary differential equation, using the Laplace inverse formulas and methods, we will obtain the solution.

Example

To solve the heat equation

$$u_t = u_{xx} \text{ for } x > 0, t > 0 \tag{8}$$

with the boundary-value

$$u(x,0) = 1,\ u_x(0,1) = -u(0,1),\ |u(x,t)| < M \tag{9}$$

we take the Laplace transform of (8) with respect to t

$$\mathcal{L}\{u_t\} = \mathcal{L}\{u_{xx}\}, \tag{10}$$

to obtain the initial value problem

$$sU - 1 = \frac{d^2 U}{dx^2}, \tag{11}$$

$$U_x(0,s) = -U(0,s), \text{ and } U(x,s) \text{ is bounded.} \tag{12}$$

The equation (11) is a second order differential equation with the general solution

$$U(x,s) = c_1 e^{\sqrt{s}x} + c_2 e^{-\sqrt{s}x} + \frac{1}{s}. \tag{13}$$

From the boundedness condition of (12), we have $c_1 = 0$. Therefore the solution (13) is simplified to

$$U(x,s) = c_2 e^{-\sqrt{s}x} + \frac{1}{s}. \tag{14}$$

From the first condition of (12) we find $c_2 = \dfrac{1}{s(\sqrt{s}-1)}$,

$$U(x,s) = \frac{1}{s(\sqrt{s}-1)} e^{-\sqrt{s}x} + \frac{1}{s} \tag{15}$$

Therefore by the inverse Laplace transformation we have

$$u(x,t) = \mathcal{L}^{-1}\left\{\frac{1}{s(\sqrt{s}-1)} e^{-\sqrt{s}x} + \frac{1}{s}\right\}$$

$$= 1\mathcal{L}^{-1}\left\{\frac{1}{s(\sqrt{s}-1)} e^{-\sqrt{s}x}\right\} \tag{16}$$

8.6.2 TWO-DIMENSIONAL BOUNDARY VALUE PROBLEM

In this case our unknown function depends on three variables x, y, and t. We may use Laplace transformation to solve our boundary value problem by taking Laplace transformation of our equation twice. For example, apply Laplace transformation with respect to x variable then to t variable. Then our equation will reduce to an ordinary differential equation. Solving the resulting ordinary differential equation we obtain a solution, such that, by applying Laplace inverse transformation to it twice, we will obtain the solution of our boundary value problem. This method is called **Iterated Laplace transformation**.

CHAPTER 9

TABLES

9.1 TABLE OF GENERAL PROPERTIES

Original Function $f(t)$ $f(t) = \mathcal{L}^{-1}\{F(s)\}$	Transformed Function $F(s)$ $F(s) = \mathcal{L}\{f(t)\}$ $= \int_0^\infty e^{-st} f(t)\,dt$
Inversion formula $\dfrac{1}{2\pi i}\int_{c-i}^{c+i} e^{is} F(s)\,ds$	$F(s)$
Linearity property $a\,f(t) + b\,g(t)$	$a\,F(s) + b\,G(s)$
Differentiation $f'(t)$	$s\,F(s) - f(0+)$
General Differentiation $f^{(n)}(t)$; $n = 1, 2, \ldots$	$s^n F(s) - s^{n-1} f(0+)$ $- s^{n-2} f'(0+) - \ldots - f^{(n-1)}(0+)$

E-104

$f(t)$	$F(s)$
Integration $$\int_0^t f(u)\, du$$	Division by s $$\frac{1}{s} F(s)$$
Integration $$\int_0^t \int_0^u f(v)\, dv\, du$$	Division by s^2 $$\frac{1}{s^2} F(s)$$
Convolution $$\int_0^t f(t-u) g(u)\, du = f*g(t)$$	Product $$F(s) \cdot G(s)$$
Multiplying by t^n $$(-1)^n t^n f(t);\ n = 1, 2, \ldots$$	Differentiation $$F^{(n)}(s)$$
Dividing by t $$\frac{1}{t} f(t)$$	Integration $$\int_s^\infty F(x)\, dx$$
$e^{at} f(t)$	Shifting $F(s-a)$
$\frac{1}{c} f\left(\frac{t}{c}\right);\ c > 0$	$F(cs)$
$\frac{1}{c} e^{(b/c)t} f\left(\frac{t}{c}\right);\ c > 0$	$F(cs - b)$

$f(t)$	$F(s)$
Shifting $f(t-b)\,h_1(t-b);\quad b>0$	$e^{-bs}F(s)$
Periodic Functions $f(t+a)=f(t)$	$\dfrac{\int_0^a e^{-st}f(t)\,dt}{1-e^{-as}}$
$f(t+a)=-f(t)$	$\dfrac{\int_0^a e^{-st}f(t)\,dt}{1+e^{-as}}$
Half-Wave Rectification of $f(t)$ $f(t)\sum_{n=0}^{\infty}(-1)^n h_1(t-na)$	$\dfrac{F(s)}{1-e^{-as}}$
Full-Wave Rectification of $f(t)$ $\lvert f(t)\rvert$	$F(s)\coth\dfrac{as}{2}$
Heaviside Expansion $\sum_{n=1}^{m}\dfrac{P(a_n)}{Q'(a_n)}e^{a_n t}$	$\dfrac{P(s)}{Q(s)}$ where $P(s)$ is of degree $<m$ $Q(s)=(s-a_1)(s-a_2)\ldots(s-a_m)$
$e^{at}\sum_{n=1}^{r}\dfrac{P^{(r-n)}(a)}{(r-n)!}\dfrac{t^{n-1}}{(n-1)!}$	$\dfrac{P(s)}{(s-a)^r}$ P is a polynomial of degree $<r$

9.2 TABLE OF MORE COMMON LAPLACE TRANSFORMS

$f(t) = \mathcal{L}^{-1}\{F(s)\}$	$F(s) = \mathcal{L}\{f(t)\}$
1	$\dfrac{1}{s}$
t	$\dfrac{1}{s^2}$
$\dfrac{t^{n-1}}{(n-1)!}\ ;\ n = 1, 2, \ldots$	$\dfrac{1}{s^n}$
$\dfrac{1}{\sqrt{\pi t}}$	$\dfrac{1}{\sqrt{s}}$
$2\sqrt{\dfrac{t}{\pi}}$	$s^{-3/2}$
$\dfrac{2^n t^{n-1/2}}{1\ 3\ 4\ 5\ldots(2n-1)\sqrt{\pi}}$	$s^{-(n+1/2)}\ ;\ n = 1, 2, 3, \ldots$
$t^r\ ;\ r > -1$	$\dfrac{\Gamma(r+1)}{s^{r+1}}$
e^{at}	$\dfrac{1}{s-a}$

$f(t) = \mathcal{L}^{-1}\{F(s)\}$	$F(s) = \mathcal{L}\{f(t)\}$
$t\,e^{at}$	$\dfrac{1}{(s-a)^2}$
$\dfrac{t^{n-1} e^{-at}}{(n-1)!}$	$\dfrac{1}{(s+a)^n}$; $n = 1, 2, \ldots$
$t^r e^{-at}$; $r > -1$	$\dfrac{\Gamma(r+1)}{(s+a)^{r+1}}$
$\dfrac{e^{-at} - e^{-bt}}{b-a}$; $a \neq b$	$\dfrac{1}{(s+a)(s+b)}$
$\dfrac{a\,e^{-at} - b\,e^{-bt}}{a-b}$; $a \neq b$	$\dfrac{s}{(s+a)(s+b)}$
$\sin st$	$\dfrac{a}{s^2 + a^2}$
$\cos at$	$\dfrac{s}{s^2 + a^2}$
$\sinh at$	$\dfrac{a}{s^2 - a^2}$
$\cosh at$	$\dfrac{s}{s^2 - a^2}$

$f(t) = \mathcal{L}^{-1}\{F(s)\}$	$F(s) = \mathcal{L}\{f(t)\}$
$\dfrac{1}{a^2}(1 - \cos at)$	$\dfrac{1}{s(s^2 + a^2)}$
$\dfrac{1}{a^3}(at - \sin at)$	$\dfrac{1}{s(s^2 + a^2)}$
$\dfrac{1}{2a^3}(\sin at - at \cos at)$	$\dfrac{1}{(s^2 + a^2)^2}$
$\dfrac{t}{2a}\sin at$	$\dfrac{s}{(s^2 + a^2)^2}$
$\dfrac{1}{sa}(\sin at + at \cos at)$	$\dfrac{s^2}{(s^2 + a^2)^2}$
$\dfrac{1}{b}e^{-at}\sin bt$	$\dfrac{1}{(s+a)^2 + b^2}$
$e^{-at}\cos bt$	$\dfrac{s+a}{(s+a)^2 + b^2}$
$e^{-at} - e^{at/2}\left(\cos\dfrac{at\sqrt{3}}{2} - \sqrt{3}\sin\dfrac{at\sqrt{3}}{2}\right)$	$\dfrac{3a^2}{s^3 + a^3}$

$f(t) = \mathcal{L}^{-1}\{F(s)\}$	$F(s) = \mathcal{L}\{f(t)\}$
$\sin at \cosh at - \cos at \sinh at$	$\dfrac{4a^3}{s^4 + 4a^4}$
$\dfrac{1}{2a^2}\sin at \sinh at$	$\dfrac{s}{s^4 + 4a^4}$
$(1 + a^2 t^2)\sin at - at \cos at$	$\dfrac{8a^3 s^2}{(s^2 + a^2)^3}$
$\dfrac{1}{2\sqrt{\pi t^3}}(e^{-bt} - e^{-at})$	$\sqrt{s+a} - \sqrt{s+b}$
$\dfrac{1}{\sqrt{\pi t}} - ae^{a^2 t}\,\text{erfc}\,a\sqrt{t}$	$\dfrac{1}{\sqrt{s}+a}$
$\dfrac{1}{\sqrt{\pi t}} + ae^{a^2 t}\,\text{erfc}\,a\sqrt{t}$	$\dfrac{\sqrt{s}}{s - a^2}$
$\dfrac{1}{a}e^{a^2 t}\,\text{erf}\,a\sqrt{t}$	$\dfrac{1}{\sqrt{s}(s - a^2)}$
$e^{a^2 t}\,\text{erfc}\,a\sqrt{t}$	$\dfrac{1}{\sqrt{s}(\sqrt{s}+a)}$
$\dfrac{1}{\sqrt{b-a}}e^{-at}\,\text{erf}\left[\sqrt{(b-a)t}\right]$	$\dfrac{1}{(s+a)\sqrt{s+b}}$

$f(t) = \mathcal{L}^{-1}\{F(s)\}$	$F(s) = \mathcal{L}\{f(t)\}$
$a e^{-at}[I_1(at) + I_0(at)]$	$\dfrac{\sqrt{s+2a}}{\sqrt{s}} - 1$
$e^{-(a+b)t/2} I_0\left(\dfrac{a-b}{2} t\right)$	$\dfrac{1}{\sqrt{(s+a)(s+b)}}$
$\sqrt{\pi}\left(\dfrac{t}{a-b}\right)^{k-\frac{1}{2}} e^{-\frac{1}{2}(a+b)t} I_{k-\frac{1}{2}}\left(\dfrac{a-b}{2} t\right)$	$\dfrac{\Gamma(k)}{(s+a)^k (s+b)^k}\;;\; k > 0$
$t e^{-\frac{1}{2}(a+b)t}\left[I_0\left(\dfrac{a-b}{2}t\right) + I_1\left(\dfrac{a-b}{2}t\right)\right]$	$\dfrac{1}{(s+a)^{\frac{1}{2}} (s+b)^{\frac{1}{2}}}$
$\dfrac{1}{t} e^{-at} I_1(at)$	$\dfrac{\sqrt{s+2a} - \sqrt{s}}{\sqrt{s+2a} + \sqrt{s}}$
$\dfrac{k}{t} e^{-\frac{1}{2}(a+b)t} I_k\left(\dfrac{a-b}{2} t\right)$	$\dfrac{(a-b)^k}{\left(\sqrt{s+a} + \sqrt{s+b}\right)^{2k}}\; k > 0$
$\dfrac{1}{a^k} e^{-\frac{1}{2}at} I_k\left(\dfrac{1}{2} at\right)$	$\dfrac{\left(\sqrt{s+a} + \sqrt{s}\right)^{-2k}}{\sqrt{s}\sqrt{s+a}}\; v > -1$

$f(t) = \mathcal{L}^{-1}\{F(s)\}$	$F(s) = \mathcal{L}\{f(t)\}$
$J_0(at)$	$\dfrac{1}{\sqrt{s^2 + a^2}}$
$a^n J_n(at); \quad n > -1$	$\dfrac{(\sqrt{s^2+a^2} - s)^n}{\sqrt{s^2+a^2}}$
$\dfrac{\sqrt{\pi}}{\Gamma(k)} \left(\dfrac{t}{2a}\right)^{k-1/2} J_{k-1/2}(at)$	$\dfrac{1}{(s^2 + a^2)^k}; \quad k > 0$
$\dfrac{ka^k}{t} J_k(at); \; k > 0$	$\left(\sqrt{s^2 + a^2} - s\right)^k$
$a^n I_n(at); \quad n > -1$	$\dfrac{(s - \sqrt{s^2 - a^2})^n}{\sqrt{s^2 - a^2}}$
$\dfrac{\sqrt{\pi}}{\Gamma(k)} \left(\dfrac{t}{2a}\right)^{k-1/2} I_{k-1/2}(at)$	$\dfrac{1}{(s^2 - a^2)^k}; \quad k > 0$
$h_1(t - a)$	$\dfrac{1}{s} e^{-as}$

$f(t) = \mathcal{L}^{-1}\{F(s)\}$	$F(s) = \mathcal{L}\{f(t)\}$
$(t-a)\, h_1(t-a)$	$\dfrac{1}{s^2} e^{-as}$
$\dfrac{(t-a)^{r-1}}{\Gamma(r)} h_1(t-a);\ r > 0$	$\dfrac{1}{s^r} e^{-as}$
$h_1(t) - h_1(t-a)$ (rectangular pulse of height 1 from 0 to a)	$\dfrac{1 - e^{-as}}{s}$
Step Function $\sum_{n=0}^{\infty} h_1(t-na)$ (staircase with steps at $a, 2a, 3a, 4a$)	$\dfrac{1}{s(1 - e^{-as})}$
$\sum_{n=1}^{\infty} c^{n-1} h_1(t-na)$ (steps of heights $1, c, c^2$ at $a, 2a, 3a$)	$\dfrac{1}{s(e^{as} - c)}$

$f(t) = \mathcal{L}^{-1}\{F(s)\}$	$F(s) = \mathcal{L}\{f(t)\}$
Square Wave Function $$h_1(t) + 2\sum_{n=1}^{\infty}(-1)^n h_1(t-2na)$$	$\dfrac{1}{2}\tanh as$
Triangular Wave Function $$h_1(t) + 2\sum_{n=1}^{\infty}(-1)^n (t-2na)h_1(t-2na)$$	$\dfrac{1}{s^2}\tanh as$
$$2\sum_{n=0}^{\infty} h_1[t-(2n+1)a]$$	$\dfrac{1}{s\sinh as}$

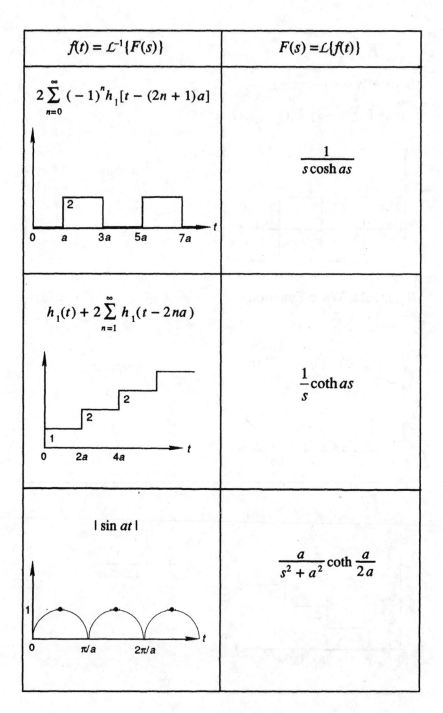

$f(t) = \mathcal{L}^{-1}\{F(s)\}$	$F(s) = \mathcal{L}\{f(t)\}$
$2\sum_{n=0}^{\infty}(-1)^n h_1[t-(2n+1)a]$	$\dfrac{1}{s\cosh as}$
$h_1(t) + 2\sum_{n=1}^{\infty} h_1(t-2na)$	$\dfrac{1}{s}\coth as$
$\lvert \sin at \rvert$	$\dfrac{a}{s^2+a^2}\coth\dfrac{a}{2a}$

$f(t) = \mathcal{L}^{-1}\{F(s)\}$	$F(s) = \mathcal{L}\{f(t)\}$
$\sum_{n=0}^{\infty} (-1)^n h_1(t - n\pi) \sin t$	$\dfrac{1}{(s^2 + 1)(1 - e^{-\pi s})}$
$J_0(2\sqrt{rt})$	$\dfrac{1}{s} e^{-\frac{r}{s}}$
$\dfrac{1}{\sqrt{\pi t}} \cos 2\sqrt{at}$	$\dfrac{1}{\sqrt{s}} e^{-\frac{a}{s}}$
$\dfrac{1}{\sqrt{\pi t}} \cosh 2\sqrt{at}$	$\dfrac{1}{\sqrt{s}} e^{\frac{a}{s}}$
$\dfrac{1}{\sqrt{\pi a}} \sin 2\sqrt{at}$	$\dfrac{1}{s^{3/2}} e^{-\frac{a}{s}}$
$\dfrac{1}{\sqrt{\pi a}} \sinh 2\sqrt{at}$	$\dfrac{1}{s^{3/2}} e^{\frac{a}{s}}$
$\left(\dfrac{t}{k}\right)^{\frac{r-1}{2}} J_{r-1}(2\sqrt{kt})$	$\dfrac{1}{s^r} e^{-\frac{k}{s}}; \ r > 0$
$\left(\dfrac{t}{k}\right)^{\frac{r-1}{2}} I_{r-1}(2\sqrt{kt})$	$\dfrac{1}{s^r} e^{\frac{k}{s}}; \ r > 0$

$f(t) = \mathcal{L}^{-1}\{F(s)\}$	$F(s) = \mathcal{L}\{f(t)\}$
$\dfrac{a}{2\sqrt{\pi t^3}}\exp\left(-\dfrac{k^2}{4t}\right)$	$\dfrac{1}{s}e^{-k\sqrt{s}}$; $k > 0$
$\operatorname{erfc}\dfrac{k}{2\sqrt{t}}$	$\dfrac{1}{s}e^{-k\sqrt{s}}$; $k \geqq 0$
$\dfrac{1}{\sqrt{\pi t}}\exp\left(-\dfrac{k^2}{4t}\right)$	$\dfrac{1}{\sqrt{s}}e^{-k\sqrt{s}}$; $k \geqq 0$
$2\sqrt{t}\, i\operatorname{erfc}\dfrac{k}{2\sqrt{t}}$	$\dfrac{1}{s^{3/2}}e^{-k\sqrt{s}}$; $k \geqq 0$
$-\gamma - \ln t$ γ = Euler's constant = .577215...	$\dfrac{1}{s}\ln s$
$\ln t$	$-\dfrac{(\gamma + \ln s)}{s}$ γ = Euler's constant = .577215...
$\cos t\, si(t) - \sin t\, ci(t)$	$\dfrac{\ln s}{s^2 + 1}$
$-\sin t\, si(t) - \cos t\, ci(t)$	$\dfrac{s\ln s}{s^2 + 1}$
$\dfrac{1}{t}(e^{-bt} - e^{-at})$	$\ln\dfrac{s+a}{s+b}$

$f(t) = \mathcal{L}^{-1}\{F(s)\}$	$F(s) = \mathcal{L}\{f(t)\}$
$-2\,ci\!\left(\dfrac{t}{k}\right)$	$\dfrac{1}{s}\ln(1 + k^2 s^2)\,;\quad k > 0$
$\dfrac{2}{t}(1 - \cos at)$	$\ln\dfrac{s^2 + a^2}{s^2}$
$\dfrac{2}{t}(1 - \cosh at)$	$\ln\dfrac{s^2 - a^2}{s^2}$
$\dfrac{1}{t}\sin kt$	$\arctan\dfrac{k}{s}$
$\dfrac{1}{k\sqrt{\pi}}\exp\!\left(-\dfrac{t^2}{4k^2}\right)$	$e^{k^2 s^2}\operatorname{erfc} ks\,;\quad k > 0$

9.3 TABLE OF SPECIAL FUNCTIONS

Gamma Function
$$\Gamma(r) = \int_0^\infty x^{r-1} e^{-x} dx \,; \quad r > 0$$
Bessel Function
$$J_n(t) = \sum_{k=0}^\infty \frac{(-1)^k \left(\frac{1}{2} t\right)^{2k+n}}{k!\, \Gamma(k+n+1)}$$
Modified Bessel Function
$$I_n(t) = (-i)^{-n} J_n(it)$$
The Error Function
$$\operatorname{erf}(t) = \frac{2}{\sqrt{\pi}} \int_0^t e^{-u^2} du$$
The Complementary Error Function
$$\operatorname{erfc}(t) = 1 - \operatorname{erf}(t)$$
Sine Integral Function
$$si(t) = \int_0^t \frac{\sin u}{u} du$$
Cosine Integral Function
$$ci(t) = \int_0^\infty \frac{\cos u}{u} du$$

Exponential Integral Function

$$ei(t) = \int_t^\infty \frac{e^{-u}}{u} \, du$$

The Unit Step Function or Heaviside Function

$$h_1(t - t_0) = \begin{cases} 0 & \text{if } t < t_0 \\ 1 & \text{if } t \geq t_0 \end{cases}$$

The Unit Impulse Function

$$\delta(t) = \lim_{\varepsilon \to 0} f(t) \qquad f(t) = \begin{cases} \dfrac{1}{\varepsilon} & \text{if } 0 \leq t \leq \varepsilon \\ 0 & \text{if } t > \varepsilon \end{cases}$$

Beta Function

$$B(m, n) = \int_0^1 u^{m-1}(1-u)^{n-1} \, du = \frac{\Gamma(m) \, \Gamma(n)}{\Gamma(m+n)}$$

Handbook of Electrical Engineering

SECTION F

Automatic Control Systems/Robotics

CHAPTER 1

SYSTEM MODELING: MATHEMATICAL APPROACH

1.1 ELECTRIC CIRCUITS AND COMPONENTS

The equations for an electric circuit obey Kirchoff's laws which can be stated as follows:

A) Σ potential differences around a closed circuit = 0

B) Σ currents at a junction or node = 0

Voltage drops across three basic electrical elements:

A) $v_R = Ri$

B) $v_L = L \frac{di}{dt} = LDi$

C) $v_C = \frac{q}{c} = \frac{1}{c} \int_0^t i\,d\tau + \frac{Q_o}{C} = \frac{1}{c} \cdot \frac{i}{D}$

where operator D is defined as the following:

$$DX = \frac{dx}{dt}$$

and $\quad \dfrac{X}{D} \equiv \displaystyle\int_{-\infty}^{t} x\, d\tau$

SERIES R-L CIRCUIT

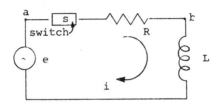

Fig. Simple R-L circuit

$$Ri + \dfrac{L di}{dt} = e = \dfrac{R}{L} \cdot \dfrac{v_L}{D} + v_L$$

(because $i_{inductor} = \dfrac{1}{LD} v_L$)

$$\dfrac{v_b - v_a}{R} + \dfrac{1}{L} \cdot \dfrac{v_b}{D} = 0 \ \ldots \ \text{Kirchoff's second law}$$

$$\left(\dfrac{1}{R} + \dfrac{1}{LD}\right) v_b - \dfrac{v_a}{R} = 0$$

SERIES R-L-C CIRCUIT

$$v_R + \dfrac{L}{R} D v_R + \dfrac{1}{RC} \cdot \dfrac{v_R}{D} = e$$

MULTILOOP ELECTRIC CIRCUITS

Loop method:

 A loop current is drawn in each closed loop; then Kirchoff's voltage equation is written for each loop. These equations are solved simultaneously to obtain output (voltage) in terms of input (voltage) and the circuit parameters.

Node Method:

 The rules for writing the node equations:

A) The number of equations required is equal to the number of unknown node voltages.

B) An equation is written for each node.

C) The equation includes the following terms: the node voltage multiplied by the sum of all the admittances that are connected to this node, and the node voltage of the other end of each brand multiplied by the admittance connected between the two nodes.

1.2 MECHANICAL TRANSLATION SYSTEMS

The mechanical translation system is characterized by mass, elastance and damping.

Representation of the basic elements:

A) Mass

    a) It is the inertial element.

    b) Reaction force f_M = M x acceleration = Ma

$$= M \frac{dv}{dt} = M \frac{d^2s}{dt^2}.$$

    c) Network representation:

    a has the motion of the mass

    b has the motion of the reference

    f_M is a function of time

B) Elastance (or stiffness, k):

    a) Representation:

b) Reaction force $f_k = k(x_c - x_d)$; x_c is the position of c and x_d is the position of d.

C) Damping (viscous friction, B):

a) Representation:

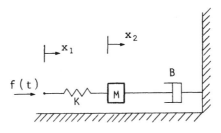

b) The reaction damping force $f_B = B(v_e - v_f)$
$$= B(Dx_c - Dx_f)$$

SIMPLE TRANSLATION SYSTEM

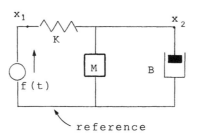

$f = f_k = k(x_a - x_b)$

$f_k = f_M + f_B = MD^2 x_2 + BDx_2$

These can be solved for displacements x_1 and x_2 and Dx_1 and Dx_2.

Fig: Mechanical Network

The system is initially at rest. To draw the mechanical method, x_1 and x_2 and the reference are located.

MULTI-ELEMENT SYSTEM

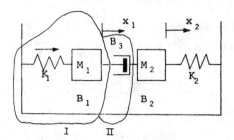

The system equations must be written in terms of two displacements x_1 and x_2. These are the nodes in the equivalent mechanical network:

$$(M_1 D^2 + B_1 D + B_3 D + K_1)x_1 - (B_3 D)x_2 = f \ldots$$

because the forces at node x_1 must add to zero. At node x_2, the equations are written by observing the above pattern:

$$(M_2 D^2 + B_2 D + K_2)x_2 - (B_3 D)x_1 = 0$$

Using these, an equivalent mechanical network can be drawn using the following table of electrical and mechanical analogies.

1.3 MECHANICAL AND ELECTRICAL ANALOGS

Mechanical Element	Electrical Element
M - mass	C - capacitance
f - force	i - current
$v = \frac{dv}{dt}$ - velocity	e or v - voltage
B - damping coefficient	$G = \frac{1}{R}$ - conductance
K - stiffness coefficient	$\frac{1}{L}$ = reciprocal inductance

An equivalent electrical network is drawn using the table of electrical and mechanical analogs.

These two networks have the same mathematical forms.

1.4 MECHANICAL ROTATIONAL SYSTEMS

Network elements of mechanical rotational systems:

   J: Moment of inertia

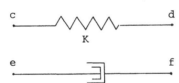

A) The reaction torque = T_J = $J\alpha$ = $JD\omega$ = $JD^2\theta$

   where θ is the angular displacement.

B) Reaction spring torque = T_k = $K(\theta_c - \theta_d)$

   where θ_c and θ_d are the positions of the two ends of a spring $(\theta_c - \theta_d)$ = the angle of twist.

C) Damping torque T_B = $B(\omega_e - \omega_f)\ldots$

   where $\omega_e - \omega_f$ = relative angular velocity of the ends of the dashpot.

SIMPLE ROTATIONAL SYSTEM

The governing equation is

$$JD^2\theta + BD\theta + K\theta = T(t)\ldots$$

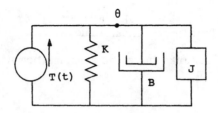

Only one equation is necessary because the system has only one node. The actual system consists of a shaft with fins of a moment of inertia, J, which is immersed in oil. The fluid has a damping factor of B.

1.5 THERMAL SYSTEMS

Only a few thermal systems are represented by linear differential equations. The basic requirement is that the temperature of the body should be assumed to be uniform.

NETWORK ELEMENTS

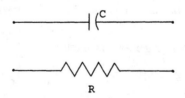

C: Thermal capacitance
R: Thermal resistance

A) $h = $ Heat stored $= \frac{q}{D} C(T_2 - T_1)$ due to change in temperature $(T_2 - T_1)$

B) $q = CD(T_2 - T_1)$ in terms of rate of heat flow.

C) $q = $ Rate of heat flow $= \frac{T_3 - T_4}{R}$ where T_3 and T_4 are two boundary temperatures.

Temperature is analogous to potential.

SIMPLE THERMAL SYSTEM: MERCURY THERMOMETER

A) Network representation

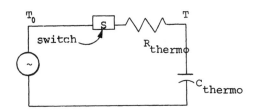

Fig: Simple Network

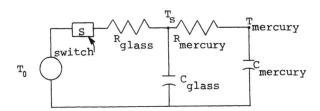

Fig: Exact Network

B) q = flow of heat = $(T_0 - T)/R$ where T_0 is the temperature of the bath and T is the temperature before immersing into the bath.

C) h = Heat entering the thermometer = $C(T - T_1)$

D) $RCDT + T = T_0$ is the governing equation.

E) More exact analysis:

T_s = The temperature at the inner surface between the glass and the mercury

For node s: $\left(\dfrac{1}{R_g} + \dfrac{1}{R_m} + C_g D \right) T_s - \dfrac{T_m}{R_m} = \dfrac{T_o}{R_g}$

For node m: $\dfrac{-T_s}{R_m} + \left(\dfrac{1}{R_m} + DC_m \right) T_m = 0$

F-8

$$\left[C_g C_m D^2 + \left(\frac{C_g}{R_m} + \frac{C_m}{R_g} + \frac{C_m}{R_m} \right) D + \frac{1}{R_g R_m} \right] T_m$$

$$= \frac{T_o}{R_g R_m} \quad \text{the governing equation.}$$

The latter equation is from the form

$$(A_2 D^2 + A_1 D + A_o) T_m = k T_o.$$

1.6 POSITIVE-DISPLACEMENT ROTATIONAL HYDRAULIC TRANSMISSION

The hydraulic transmission is used when a large torque is required. It contains a variable displacement pump driven at a constant speed. It is assumed that the transmission is linear over a limited range.

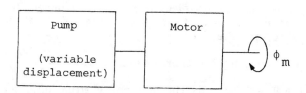

A) $q_p = q_m + q_l + q_c$ Since the fluid flow rate from the pump must equal the sum of the flow rates.

$q_p = x d_p \cdot \omega_p$

$q_m = d_m \cdot \omega_m$, $q_l = L P_L$ and $q_c = Dv = \dfrac{v D P_L}{K_B}$

B) Torque at the motor shaft = $T = n_T \cdot d_m \cdot P_L$

$\qquad \qquad = C P_L$

where P_L is the load-induced pressure-drop across motor.

DIFFERENT CASES

A) Inertia load

$$T = J \cdot D^2 \phi_m = CP_L$$

B) Inertia load coupled through a spring

$$T = K(\phi_m - \phi_L) = JD^2\phi_L = CP_L$$

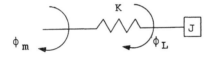

C) In case of inertia load, value of P_L obtained from equation $q_p = q_m + q_l + q_c$ is substituted into $T = CP_L$. The resulting equation can be solved for ϕ_m in terms of x. The same procedure is adopted in case of a spring-coupled inertia load.

1.7 D-C AND A-C SERVOMOTOR

D-C SERVOMOTOR

$T(t)$ = The torque = $Ka \phi \, i_m \ldots i_m$ = armature current,

ϕ = the flux, K_a = constant of proportionality

Modes of Operation

A) An adjustable voltage is applied to the armature while the field current is held constant.

B) An adjustable voltage is applied to the field while the armature current is kept constant.

Armature Control

A constant field current is obtained by exciting the field from a fixed Dc source.

$T(t) = k_T i_m$ where ϕ is a constant
(since field current is a constant)

Back emf = $e_m = k_i \phi \omega_m = k_b \omega_m = k_b \cdot D\theta_m \ldots \omega_m$ = speed

Armature-controlled d-c motor:

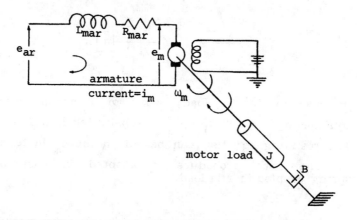

$e_{ar} = L_m(L_{mar} + R_{mar}) + e_m$

$JD\omega_m + B\omega_m = T(t) \ldots$ Torque equation with motor load.

$$\frac{L_{mar} J}{k_T} D^3 \theta_m + \frac{L_{mar} B + R_{mar} J}{k_T} D^2 \theta_m + \frac{R_{mar} B + K_b K_T}{k_T} D\theta_m = e_{ar}$$

The system equation.

Field Control

In this case the armature current i_{mar} is constant so that T(t) is proportional only to the flux ϕ:

$$T(t) = K_3 \cdot \phi \cdot i_{mar} = K_3 K_2 \cdot i_{mar} \cdot i_{field} = k_f \cdot i_{field}$$

where i_{mar} is constant.

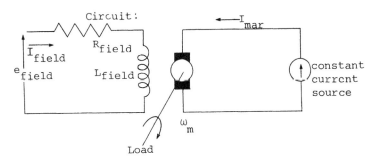

$$e_{field} = L_{field} D_{field} + R_{field} \cdot i_{field}$$

$$(K_f \cdot D + R_{field})(JD + B)\omega_m = k_f \cdot e_{field}$$

The system equation.

A-C SERVOMOTOR

This may be considered to be a two-phase induction motor having two field coils positioned 90 electrical degrees apart.

In a two-phase induction motor, the speed which is a little below the synchronous speed is constant; however when the unit is used as a servomotor, the speed is proportional to the input voltage.

The two-phase induction motor shown below can be used as a servomotor by applying an ac voltage e to one of the windings. Thus e is fixed and when e_{c_0} is varied, the torque and speed are a function of this voltage.

Because these curves are non-linear, we must approximate them by straight lines in order to obtain linear differential equations.

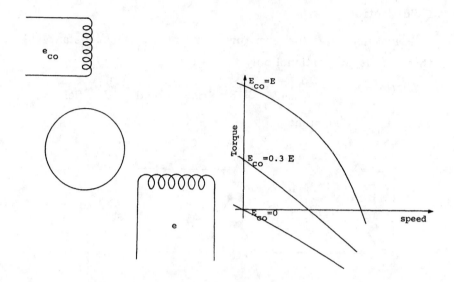

From these curves, it is noted that the generated torque is a function of the speed ω and voltage e_{c_0}.

The torque equation is given by

$$\boxed{e_{c_0}\frac{\partial T}{\partial e_{c_0}} + \frac{\partial T}{\partial \omega}\omega = T(e_{c_0},\omega)} \qquad (a)$$

Because of the straight line approximation we are using, it is justifiable to let

$$\frac{\partial T}{\partial e_{c_0}} = K_{c_0} \quad \text{and} \quad \frac{\partial T}{\partial \omega} = K_\omega$$

Suppose we assume a load consisting of inertia and damping, then

$$T_L = JD\omega + B\omega \qquad (b)$$

However, the generated torque must be equal to the load torque. Therefore, equating equations (a) and (b) we get

$$K_{c_0}e_{c_0} + K\omega = JD\omega + B\omega$$

$$\boxed{JD\omega + (B - K)\omega = K_{c_0}e_{c_0}}$$

1.8 LAGRANGE'S EQUATION

$$\frac{d}{dt}\left(\frac{\partial T}{\partial \dot{q}_n}\right) - \frac{\partial T}{\partial q_n} + \frac{\partial D}{\partial \dot{q}_n} + \frac{\partial V}{\partial q_n} = Q_n$$

where $n = 1, 2, 3, \ldots$ are the independent coordinates or degrees of freedom which exist in the system and

- T = total kinetic energy of system
- D = dissipation function of system
- V = total potential energy of system
- Q_n = generalized applied force at the coordinate n
- q_n = generalized coordinate
- $\dot{q}_n = dq_n/dt$ (generalized velocity)

Energy functions for translational mechanical elements

Table 1.1

Element	Kinetic energy T	Potential energy V	Dissipation factor D	Forcing function Q
Force, f	—	—	—	f (force)
Mass, M	$\frac{1}{2}Mv^2 = \frac{1}{2}M\dot{x}^2$	—	—	—
K (spring, x_1, x_2)	—	$\frac{K}{2}\left[\int (v_1 - v_2)dt\right]^2$ $= \frac{1}{2}K(x_1 - x_2)^2$ where x is the displacement	—	—
Damping, B (v_1, v_2)	—	—	$\frac{1}{2}B(v_1 - v_2)^2$ $= \frac{1}{2}K(\dot{x}_1 - \dot{x}_2)^2$	—

Energy functions for electric circuits based on the loop or mesh analysis

Table 1.2

Element	Kinetic energy T	Potential energy V	Dissipation function D	Forcing function Q
Voltage source, e	—	—	—	e
Inductance, L	$\tfrac{1}{2}Li^2 = \tfrac{1}{2}L\dot{q}^2$ where q is the charge	—	—	—
Capacitance, C	—	$\dfrac{(\int i\,dt)^2}{2C} = \dfrac{q^2}{2C}$	—	—

CHAPTER 2

SOLUTIONS OF DIFFERENTIAL EQUATIONS: SYSTEM'S RESPONSE

2.1 STANDARDIZED INPUTS

Sinusoidal function: $r(t) = \cos\omega t$

Power-series function: $r = a_0 + a_1 t + a_2 t^2 + \ldots$

Unit step function: $r = u(t)$

Unit ramp function: $r = tu(t)$

Unit parabolic function: $r = t^2 u(t)$

Unit impulse function: $r = \delta(t)$

2.2 STEADY STATE RESPONSE

SINUSOIDAL INPUT

Input: $r(t) = A\cos(\omega t + \alpha)$

This is generally the form of input.

$$= \text{real part of } (Ae^{j(\omega t+\alpha)}) = \text{Re}(Ae^{j(\omega t+\alpha)})$$

$$= \text{Re}(Ae^{j\alpha} e^{j\omega t}) = \text{Re}(\mathbf{A}e^{j\omega t})$$

using Euler's identity.

Form of the equation to be solved:

$$A_m D^m C + A_{m-1} D^{m-1} C \ldots + A_{-n} D^{-n} C = r$$

$\bar{R} = Re^{j\alpha}$ phasor representation of the input.

$\bar{C}$ = Phasor representation of the output

$$= \frac{\bar{R}}{A_m (j\omega)^m + A_{m-1}(j\omega)^{m-1} \ldots + \underline{A}_n (j\omega)^{-n}}$$

Time response = $c(t) = |\bar{C}| \cos(\omega t + \phi)$

Steady-State Sinusoidal Response of Series RLC Circuit:

$$LDi + Ri + \frac{1}{CD} i = e$$

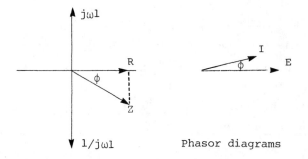

Phasor diagrams

The equation obtained after application of Kirchoff's voltage law.

$$e = Re[\sqrt{2}\, \bar{E}(j\omega) e^{j\omega t}]$$

The input to the RLC circuit.

$$\bar{I}(j\omega) = \text{Phasor current} = \frac{\bar{E}(j\omega)}{j\omega L + R + \dfrac{1}{j\omega c}}$$

$$\bar{Z}(j\omega) = j\omega L + R + \frac{1}{j\omega c}$$

Frequency transfer function:

$$\overline{G}(j\omega) = \frac{\overline{V}_R(j\omega)}{\overline{E}(j\omega)} = \frac{R}{j\omega L + R + \frac{1}{j\omega C}}$$

$\overline{V}_R(j\omega)$ is the voltage (output) across R in phasor form.

POWER SERIES INPUT

Consider the general differential equation:

$$A_m D^m C + \ldots + A_0 C + A_{-1} D^{-1} C + \ldots + A_n D^{-n} C = r$$

where $r(t) = a_0 + a_1 t + a_2 t^2 + \ldots + a_k t^k$.

To find the particular (steady state) solution of the response $c(t)$, we assume a solution of the form

$$(H) = b_0 + b_1 t + b_2 t^2 + \ldots + b_v t^v$$

Substituting in the differential equation and equating coefficients on both sides will enable us to find the constants $b_0, b_1, \ldots$.

Note: 1) k is the highest power of t on the right side. Therefore t^k must also appear on the left side of the equation.

2) The highest power of t on the left side will result from the lowest order derivative. Let x be the order of the lowest derivative.

$$v = k + x \qquad v \geq 0$$

Special Cases of Power Series Input

(i) Step-Function Input:

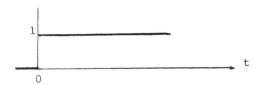

A) Step-function input:

Voltage equation: $A_2 D^2 \omega_m + A_1 D \omega_m + A_0 \omega_m = u(t)$
where $u(t)$ is the input.

The response is of the form $\omega_m = b_o$ because, for step f_n, the highest exponent of t is $k = 0$ for $u(t)$.

$$D\omega_m = 0 = D^2 \omega_m$$

$$\boxed{b_o = \frac{1}{A_o}}$$

B) Ramp-function:

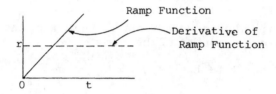

Assume we have a system whose equation is

$$C \cdot D\theta + A = m$$

By integrating, the response is

$$C \cdot \theta + \left(ns + \frac{1}{R}\right) D^{-1}\theta = q$$

One form of the response is

$$\boxed{\theta = b_o}$$

because the highest exponent of t in the input $n = t$ is $k = 1$.

After integrating,

$$D^{-1}\theta = b_o t + C_o$$

$$b_o = \frac{1}{A}, \quad \theta = \frac{1}{A} \text{ at a steady state.}$$

C) Parabolic-function input:

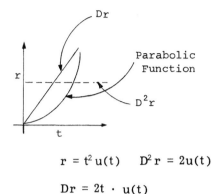

$$r = t^2 u(t) \quad D^2 r = 2u(t)$$

$$Dr = 2t \cdot u(t)$$

Example: Given a system whose equation is

$$AD^2 x_2 + BDx_2 + Kx_2 = Kx_1$$

k = 2 because the highest exponent of t in the input is 2.

Order of the lowest derivation in the system equation

$$x = 0$$

The form of the steady-state response is

$$\boxed{x_2(t) = b_0 + b_1 t + b_2 t^2}$$

$$Dx_2 = b_1 + 2b_2 t \quad \text{and} \quad D^2 x_2 = 2b_2$$

Steady-state solution:

$$x_2(t) = \frac{-2A}{K} + \frac{2B^2}{K^2} - \frac{2Bt}{K} + t^2.$$

2.3 TRANSIENT RESPONSE

Transient response of a differential equation:

REAL ROOTS

Steps:

1) Write a homogeneous equation by equating the given

differential equation to zero.

Differential equation:

$$b_m D^m c + b_{m-1} D^{m-1} c + \ldots + b_o D^o c + \ldots + b_{-n} D^{-n} c = r$$

Homogeneous equation:

$$b_m D^m c_t + b_{m-1} D^{m-1} c_t + \ldots + b_{-n} D^{-n} c_t = 0$$

2) Solution of the homogeneous equation gives the general expression for the transient response.
Assume a solution $C_t = e^{kt}$.

3) Substituting $C_t = e^{kt}$ in the equation results in the characteristic equation:

$$b_m k^m + b_{m-1} k^{m-1} + \ldots + b_o + \ldots + b_{-n} k^{-n} = 0$$

Its roots are $k_1, k_2, \ldots$.

4) So if there are no multiple roots the transient response is $C_t = A_1 e^{k_1 t} + A_2 e^{k_2 t} \ldots$.

Short-cut method:

We can obtain the characteristic equation by substituting into the homogeneous equation: k for DC_t, k^2 for $D^2 C_t$ etc. For the general equation we are using, our characteristic equation will consist of m + n constants. We must consequently have m + n initial conditions in order to set up m + n equations which will enable us to determine the transient response.

COMPLEX ROOTS

When the roots of the characteristic equation are complex, the above method cannot be used; instead, the response takes the form

$$A e^{\sigma t} \sin(\omega_d t + \phi) \quad \text{(For 2 complex roots)}$$

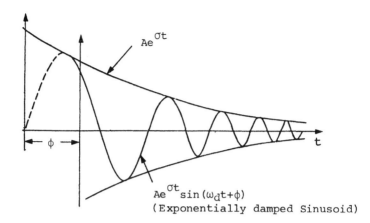
(Exponentially damped Sinusoid)

Damping ratio and undamped national frequency:

$$\text{Damping ratio } \zeta = \frac{\text{Actual damping}}{\text{Critical damping}}$$

If the characteristic equation has a pair of complex conjugate roots, the form of the quadratic factor is $b_2 m^2 + b_1 m + b_0$.

b_1 represents the effective damping constant of the system.

$$\zeta = \frac{\text{Actual damping}}{\text{Critical damping}} = \frac{b_1}{b_1^*} = \frac{b_1}{2\sqrt{b_2 b_0}}$$

If $\zeta < 1$, the response is said to be underdamped.
If $\zeta > 1$, the response is said to be overdamped.
If $\zeta = 0$,

A) $\omega_m = \sqrt{\dfrac{b_0}{b_2}}$ = undamped natural frequency

B) the zero damping constant means that the transient response is a sine wave of constant amplitude.

We can rewrite the above equation $b_2 m^2 + b_1 m + b_0$ as follows:

$$\frac{b_2}{b_0} m^2 + \frac{b_1}{b_0} m + 1 = \frac{1}{\omega_n^2} m^2 + \frac{2\zeta m}{\omega_n} + 1$$

where

$$\omega_m = \sqrt{\frac{b_0}{b_2}}$$

$$\rightarrow \quad m^2 + 2\zeta\omega_n m + \omega_n^2$$

This is the general form of the characteristic equation whose roots are:

$$m_{1,2} = -\zeta\omega_n \pm j\omega_n\sqrt{1-\zeta^2}$$

Transient response for underdamped case:

$$Ae^{-\zeta\omega_n t}\sin(\omega_n\sqrt{1-\zeta^2}\,t + \phi)$$

ω_d - the damped frequency of oscillation

$$\omega_d = \omega_n\sqrt{1-\zeta^2}$$

Time constant:

The transient terms have the form Ae^{kt}. The value of time that makes the exponent of e equal to -1 is called the time constant T.

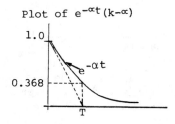

Plot of $e^{-\alpha t}$ $(k-\alpha)$

In case of the damped sinusoid

$$T = \frac{1}{|\sigma|}, \quad \text{for } k = \sigma + \omega_d$$

$$T = \frac{1}{\zeta\omega_n}$$

where the larger the $\zeta\omega_n$, the greater the rate of decay of the transient.

F-23

2.4 FIRST AND SECOND-ORDER SYSTEM

FIRST-ORDER SYSTEM

An example:

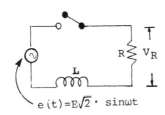

Applying KVL to series RL circuit,

$$e = \frac{L}{R} Dv_R + v_R$$

The characteristic equation is therefore

$$\frac{L}{R} x + 1 = 0 \quad \text{or,} \quad x = -\frac{R}{L}$$

So the transient solution is

$$v_{R,t} = Ae^{-(R/L)t}$$

The steady-state solution in phasor form is,

$$\overline{V}_{R,SS}(j\omega) = \frac{E(j\omega)}{1 + \left(\frac{L}{R}\right) j\omega} = \frac{E(j\omega)}{\left[1 + \left(\frac{\omega L}{R}\right)^2\right]} \underline{/-\tan^{-1} \frac{\omega L}{R}}$$

Therefore, the steady-state voltage in time domain is

$$V_{R,SS} = \frac{E\sqrt{2}}{\left[1 + \left(\frac{\omega L}{R}\right)^2\right]^{\frac{1}{2}}} \sin\left(\omega t - \tan^{-1} \frac{\omega L}{R}\right)$$

The complete solution is:

$$V_R = \frac{E\sqrt{2}}{1 + \left(\frac{\omega}{L}\right)^2} \sin\left(\omega t - \tan^{-1} \frac{\omega L}{R}\right) + Ae^{-\frac{R}{L}t}$$

|← steady-state solution →| |← transient →|
                                              solution

Evaluation of A using initial conditions:

initial conditions: $V_R = 0$ at $t = 0$

$$A = \frac{\omega \cdot R \cdot L \cdot E\sqrt{2}}{R^2 + (\omega L)^2}$$

obtained after putting $V_R = 0$ and $t = 0$ in the complete solution.

SECOND-ORDER SYSTEM

An example

$$E = i_2 \left[R_1 + \frac{R_1 R_2}{R_3} + \frac{R_1}{R_3} LD + \frac{R_1}{CDR_3} + R_2 + LD + \frac{1}{CD} \right]$$

$$10 = \left[2D + 40 + \frac{200}{D} \right] i_2 \quad \text{after application of KVL.}$$

The steady-state output is: $i_{2,ss} = 0$...since the branch contains a capacitor.

The characteristic equation: $m + 20 + \frac{100}{m} = 0$.

Its roots are $m_{1,2} = -10$...so the circuit is critically damped.

The output current:

$$i_2(t) = A_1 e^{-10t} + A_2 t e^{-10t}$$

$$Di_2(t) = -10A_1 e^{-10t} + A_2(1 - 10t)e^{-10t}$$

Initial conditions: $i_1(0^-) = i_2(0^-) = 0$

$$i_2(0^+) = i_2(0^-) = 0$$

because i_2 can't change instantly in the inductor.

A) $Di_2(t) = 10i_1(t) - 25i_2(t) - vi(t)$

B) $i_1(0^+) = 0.5$ because $i_2(0^+) = 0$

C) $v_c(0^-) = v_c(0^+)$ = steady-state value of = 10 capacitor voltage for $t < 0$

Hence, $Di_2(t) = -5$

$A_1 = 0$, $A_2 = -5$, so $i_2(t) = -5t\, e^{-10t}$

Second-order transients:

Simple second-order equation:

$$\boxed{\frac{D^2 c}{\omega_n^2} + \frac{2\zeta}{\omega_n} Dc + c = r}$$

because there are no derivatives of r on the right-hand side.

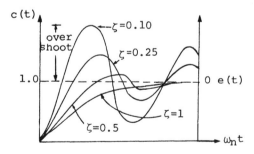

Representation of the Transient Response
of a Second Order System

$$c(t) = 1 - \frac{e^{-\zeta \omega_n t}}{\sqrt{1 - \zeta^2}} \sin(\omega_n \sqrt{1 - \zeta^2}\, t + \cos^{-1}\zeta)$$

This is the response to a unit step with initial conditions set to zero.

$e(t)$ - error in the system

$e = r - c$, where $r = u(t)$

Representation of the transient response of a second-order system.

Conclusion: The amount of overshoot (i.e. beyond 1.0) depends on the damping ratio ζ.

For the overdamped ($\zeta > 1$) and critically damped case ($\zeta = 1$), there is no overshoot.

For underdamped ($\zeta < 1$), the system oscillates around the steady state value before it settles down at the steady state.

$$\boxed{t_p = \pi/\omega_n \sqrt{1-\zeta^2} \quad = \text{The time at which the peak overshoot occurs.}}$$

$$\boxed{c_p = 1 + e^{-\zeta\pi/\sqrt{1-\zeta^2}} \quad \text{The value of the peak overshoot.}}$$

$$\boxed{M_o = \frac{c_p - c_{ss}}{c_{ss}} \quad \text{per unit overshoot.}}$$

Peak overshoot v/s ζ and Frequency of Oscillation v/s Damping ratio

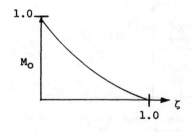
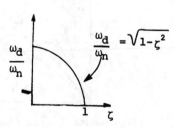

Error equation:

$$e = r - c = \frac{e^{-\zeta\omega_n t}}{\sqrt{1-\zeta^2}} \sin(\omega_n \sqrt{1-\zeta^2} \cdot t + \cos^{-1}\zeta)$$

RESPONSE CHARACTERISTICS

One of the characteristics of the system's response is the setting time. Setting time is the time required for the envelope of the transient to die out or to reduce to an insignificantly small value. This time is a function of a number of time constants involved in the transfer function of the system, and can be presented mathematically as follows:

$$T_s = \frac{\text{no. of time constants in the system's transfer function}}{(\text{undamped natural frequency})(\text{the damping constant})}$$

2.5 TIME-RESPONSE SPECIFICATIONS

The following terms are used as specifications to evaluate the performance of a system. These are:

A) Maximum overshoot (Cp): the magnitude of the first overshoot as shown.

B) t_p (Time to reach the maximum overshoot): the time required to reach the first overshoot.

C) Duplicating time (t_o): the time when the response of the system has the final value for the first time.

D) Settling time (t_s): the time taken by the control system to reach the final specified value and thereafter remain within a specified tolerance.

E) Frequency of oscillation (ω_d): the frequency of oscillation of the tansient response.

CHAPTER 3

APPLICATIONS OF LAPLACE TRANSFORM

3.1 DEFINITION OF LAPLACE TRANSFORM

$$L[f(t)] = \int_0^\infty f(t)e^{-st}dt = F(s)$$

provided f(t) is piecewise continuous over every finite interval and is of exponential order.

Laplace Transform of Some Simple Functions:

$L[\text{step function } u(t)] = \frac{1}{s}$ ... if $\sigma > 0$

$L[e^{-\alpha t}] = \frac{1}{s+\alpha}$ .... if $\sigma > -\alpha$

$L[\cos\omega t] = \frac{s}{s^2 + \omega^2}$ ...... if $\sigma > 0$

$L[\text{Ramp function } f(t) = t] = \frac{1}{s^2}$ ... if $\sigma > 0$

where s is a complex variable and has the form $\sigma + j\omega$.

Table 3.1   Table of Laplace-transform pairs

F(s)	f(t) $0 \leq t$
1. 1	$u_1(t)$ unit impulse at $t = 0$
2. $\dfrac{1}{s}$	1 or $u(t)$ unit step at $t = 0$
3. $\dfrac{1}{s^2}$	$tu(t)$ ramp function
4. $\dfrac{1}{s^n}$	$\dfrac{1}{(n-1)!} t^{n-1}$ n is a positive integer
5. $\dfrac{1}{s} e^{-as}$	$u(t-a)$ unit step starting at $t = a$
6. $\dfrac{1}{s}(1 - e^{-as})$	$u(t) - u(t-a)$ rectangular pulse
7. $\dfrac{1}{s+a}$	e^{-at} exponential decay
8. $\dfrac{1}{(s+a)^n}$	$\dfrac{1}{(n-1)!} t^{n-1} e^{-at}$ n is a positive integer
9. $\dfrac{1}{s(s+a)}$	$\dfrac{1}{a}(1 - e^{-at})$
10. $\dfrac{1}{s(s+a)(s+b)}$	$\dfrac{1}{ab}\left[1 - \dfrac{b}{b-a} e^{-at} + \dfrac{a}{b-a} e^{-bt}\right]$
11. $\dfrac{s+\alpha}{s(s+a)(s+b)}$	$\dfrac{1}{ab}\left[\alpha - \dfrac{b(\alpha-a)}{b-a} e^{-at} + \dfrac{a(\alpha-b)}{b-a} e^{-bt}\right]$
12. $\dfrac{1}{(s+a)(s+b)}$	$\dfrac{1}{b-a}(e^{-at} - e^{-bt})$
13. $\dfrac{s}{(s+a)(s+b)}$	$\dfrac{1}{a-b}(ae^{-at} - be^{-bt})$
14. $\dfrac{s+\alpha}{(s+a)(s+b)}$	$\dfrac{1}{b-a}[(\alpha-a)e^{-at} - (\alpha-b)e^{-bt}]$
15. $\dfrac{1}{(s+a)(s+b)(s+c)}$	$\dfrac{e^{-at}}{(b-a)(c-a)} + \dfrac{e^{-bt}}{(c-b)(a-b)} + \dfrac{e^{-ct}}{(a-c)(b-c)}$
16. $\dfrac{s+\alpha}{(s+a)(s+b)(s+c)}$	$\dfrac{(\alpha-a)e^{-at}}{(b-a)(c-a)} + \dfrac{(\alpha-b)e^{-bt}}{(c-b)(a-b)} + \dfrac{(\alpha-c)e^{-ct}}{(a-c)(b-c)}$
17. $\dfrac{\omega}{s^2+\omega^2}$	$\sin \omega t$
18. $\dfrac{s}{s^2+\omega^2}$	$\cos \omega t$
19. $\dfrac{s+\alpha}{s^2+\omega^2}$	$\dfrac{\sqrt{\alpha^2+\omega^2}}{\omega} \sin(\omega t + \phi)$ $\phi = \tan^{-1}\dfrac{\omega}{\alpha}$
20. $\dfrac{1}{s(s^2+\omega^2)}$	$\dfrac{1}{\omega^2}(1 - \cos \omega t)$
21. $\dfrac{s+\alpha}{s(s^2+\omega^2)}$	$\dfrac{\alpha}{\omega^2} - \dfrac{\sqrt{\alpha^2+\omega^2}}{\omega^2} \cos(\omega t + \phi)$ $\phi = \tan^{-1}\dfrac{\omega}{\alpha}$
22. $\dfrac{1}{(s+a)(s^2+\omega^2)}$	$\dfrac{e^{-at}}{a^2+\omega^2} + \dfrac{1}{\omega\sqrt{a^2+\omega^2}} \sin(\omega t - \phi)$ $\phi = \tan^{-1}\dfrac{\omega}{\alpha}$
23. $\dfrac{1}{(s+a)^2+b^2}$	$\dfrac{1}{b} e^{-at} \sin bt$
24. $\dfrac{1}{s^2+2\zeta\omega_n s+\omega_n^2}$	$\dfrac{1}{\omega_n\sqrt{1-\zeta^2}} e^{-\zeta\omega_n t} \sin \omega_n\sqrt{1-\zeta^2}\, t$
25. $\dfrac{s+a}{(s+a)^2+b^2}$	$e^{-at}\cos bt$
26. $\dfrac{1}{s^2(s+a)}$	$\dfrac{1}{a^2}(at - 1 + e^{-at})$
27. $\dfrac{1}{s(s+a)^2}$	$\dfrac{1}{a^2}(1 - e^{-at} - ate^{-at})$
28. $\dfrac{s+\alpha}{s(s+a)^2}$	$\dfrac{1}{a^2}[\alpha - \alpha e^{-at} + a(a-\alpha)t e^{-at}]$

LAPLACE TRANSFORM THEOREMS

A) Linearity

$$L[a \cdot f(t)] = a \cdot L[f(t)] = a \cdot F(s) \quad \ldots \text{ if a is a constant}$$

B) Superposition

$$L[f_1(t) \pm f_2(t)] = F_1(s) \pm F_2(s)$$

C) Translation in time

$$L[f(t - a)] = e^{-as} F(s) \quad \ldots \text{ if a is a positive real, and } f(t - a) = 0 \text{ for } 0 < t < a.$$

D) Complex differentiation

$$L[t \cdot f(t)] = \frac{-d}{ds} F(s)$$

E) Translation in the s domain

$$L[e^{at} \cdot f(t)] = F(s - a)$$

F) Real differentiation

$$L[Df(t)] = sF(s) - f(o^+)$$

$f(o^+)$ is the value of the limit of $f(t)$ as the origin, $t = 0$ is approached from the right-hand side.

$$L[D^2 f(t)] = s^2 F(s) - sf(o) - Df(o)$$

$$L[D^n f(t)] = s^n F(s) - s^{n-1} f(o) \ldots - D^{n-1} f(o)$$

G) Real integration

$$L[D^{-n} f(t)] = \frac{F(s)}{s^n} + \frac{D^{-1} f(o)}{s^n} + \ldots + \frac{D^{-n} f(o)}{s}$$

H) Final value F

if $f(t)$ and $DF(t)$ are Laplace transformable, then

$$\lim_{s \to o} sf(s) = \lim_{t \to \infty} f(t)$$

According to the final value theorem we can find the

final value of the function f(t) by working in the s domain which saves us the work involved in taking the inverse Laplace transform.

Limitations of this theorem:

A) Whenever SF(s) has poles on the imaginary axis or in the right half s plane, there is no finite final value of f(t) since SF(s) becomes infinite and the theorem cannot be used.

B) Suppose the driving function is sinusoidal and is equal to $\sin\omega t$. The $L[\sin\omega t]$ has poles at $s = \pm j\omega$; furthermore, $\lim_{t \to \infty} \sin\omega t$ does not exist. Thus this theorem is invalid whenever the driving function is sinusoidal.

INITIAL VALUE THEOREM

$$\lim_{s \to \infty} S F(S) = \lim_{t \to 0} f(t) \ldots \text{ if } \lim S F(S) \text{ exists.}$$

COMPLEX INTEGRATION THEOREM

$$L\left[\frac{f(t)}{t}\right] = \int_{s}^{\infty} F(s)ds$$

providing the limit $f(t)/t$ exists
$t \to 0^+$

In words this means that division by the variable in the real time domain is equivalent to integration with respect to s in the s domain.

3.2 APPLICATION OF LAPLACE TRANSFORM TO DIFFERENTIAL EQUATIONS

An example

Step 1) $AD^2 x_2 + MDx_2 + kx_2 = kx_1$ ....

the differential equation where $x_1(t)$ is the input and $x_2(t)$ is called the response function. First of all take the Laplace transform of both sides:

$$L[AD^2 x_2 + MDx_2 + kx_2] = L[kx_1].$$

2) Then take the Laplace transform of each term; after substituting in the original equation, rearrange the equation:

$$kx_1(s) = (As^2 + Ms + k)x_2(s)$$
$$- [Asx_2(o) + ADx_2(o) + Mx_2(o)]$$

characteristic function $x_2(s)$ - the transform equation

$$x_2(s) = \frac{kx_1(s) + Asx_2(o) + Mx_2(o) + ADx_2(o)}{As^2 + Ms + k}$$

3) $x_2(t) = L^{-1}[x_2(s)]$ ... the response function

3.3 INVERSE TRANSFORM

$$F(s) = P(s)/Q(s) = [a_n s^n + a_{n-1} s^{n-1} + \ldots]/[s^m + b_{m-1} s^{m-1} + \ldots]$$

$$= P(s)/[(s - s_1)(s - s_2) \ldots (s - s_m)]$$

after breaking $Q(s)$ into linear and quadratic factors.

Application of partial expansion in finding inverse Laplace transform.

A) $F(s)$ has first order real poles:

$$F(s) = \frac{P(s)}{Q(s)} = \frac{P(s)}{s(s-s_1)(s-s_2)} = \frac{A_0}{s} + \frac{A_1}{s-s_1} + \frac{A_2}{s-s_2}$$

$$\boxed{f(t) = L^{-1}[F(s)] = A_0 + A_1 e^{s_1 t} + A_2 e^{s_2 t}}$$

Two inverse transforms.

Evaluation of coefficients A_k:

$$A_k = \left[(s - s_k) \frac{P(s)}{Q(s)}\right]_{s=s_k}$$

A_k is called the residue of $F(s)$.

B) Multiple-order real poles:

$$F(s) = \frac{P(s)}{Q(s)} = \frac{P(s)}{(s-s_1)^2(s-s_2)} = \frac{A_{12}}{(s-s_1)^2} + \frac{A_{11}}{s-s_1} + \frac{A_2}{s-s_2}$$

$$f(t) = A_{12} \cdot t \cdot e^{s_1 t} + A_{11} \cdot e^{s_1 t} + A_2 e^{s_2 t}$$

A general formula for finding coefficients associated with the repeated real pole of order n,

$$A_{rn} \left[(s - s_r)^n \frac{P(s)}{Q(s)}\right]_{s=s_r}$$

$$A_{r(n-k)} = \frac{1}{k!} \frac{d^k}{ds^k} \left[(s - s_r)^n \frac{P(s)}{Q(s)}\right]_{s=s_r}$$

C) Complex conjugate poles:

$$f(s) = \frac{P(s)}{Q(s)} = \frac{P(s)}{(s^2 + 2\zeta\omega_n s + \omega_n^2)(s-s_3)}$$

$$= \frac{A_1}{s-s_1} + \frac{A_2}{s-s_2} + \frac{A_3}{s-s_3}$$

$$= \frac{A_1}{s + \zeta\omega_n - j\omega_n\sqrt{1-\zeta^2}} + \frac{A_2}{s + \zeta\omega_n + j\omega_n\sqrt{1-\zeta^2}}$$

$$+ \frac{A_3}{s-s_3}$$

Hence, the inverse function of $F(s)$ is,

$$f(t) = A_1 e^{(-\zeta\omega_n + j\omega_n\sqrt{1-\zeta^2})t} + A_2 e^{(-\zeta\omega_n - j\omega_n\sqrt{1-\zeta^2})t} + A_3 e^{s_3 t}$$

$$= 2|A_1| e^{6t} \sin(\omega_d t + \phi) + A_3 e^{s_3 t}$$

where ϕ = angle of A_1 + 90°

Now, $A_1 = [(s - s_1)F(s)]_{s=s_1}$ and $A_3 = [(s - s_3)F(s)]_{s=s_3}$

HAZONY AND RILEY RULE

For a normalized ratio of polynomials:

A) If the denominator is one degree higher than the numerator, the sum of the residues is one.

B) If the denominator is two or more degrees higher than the numerator, the sum of the residues is zero.

GRAPHICAL METHOD

$$F(s) = \frac{P(s)}{Q(s)} = k \frac{\prod_{m=1}^{\omega}(s - z_m)}{\prod_{k=1}^{\nu}(s - p_k)} = \frac{A_1}{s - p_1} + \frac{A_2}{s - p_2}$$

Pole-zero plot of the function $F(s)$:

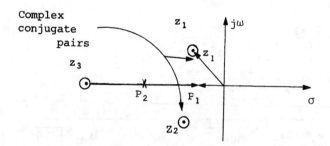

$$A_k = k \frac{\text{Product of directed distances from each zero to the pole } p_k}{\text{Product of directed distances from all other poles to the pole } p_k}$$

... except f ~ k = 0.

F-35

An example:

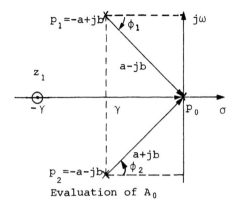

Evaluation of A_0

$$-F(s) = \frac{k(s+\gamma)}{s[(s+a)^2+b^2]} = \frac{k(s+\gamma)}{s(s+a-jb)(s+a+jb)} = \frac{k(s-z_1)}{s(s-p_1)(s-p_2)}$$

$$= \frac{A_0}{s} + \frac{A_1}{s+a-jb} + \frac{A_2}{s+a+jb}$$

$$A_0 = \frac{k\gamma}{(a-jb)(a+jb)} = \frac{k\gamma}{a^2+b^2}$$

$$A_1 = \frac{k[(\gamma-a)+jb]}{(-a+jb)(j2b)}$$

$$= \frac{k}{2b}\sqrt{\frac{(\gamma-a)^2+b^2}{a^2+b^2}}\; e^{j(\phi-\theta-\pi/2)}$$

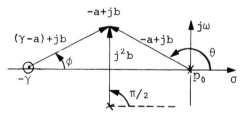

Fig. Evaluation of A_1.

where $\phi = \tan^{-1}\dfrac{b}{\gamma-a}$

$\theta = \tan^{-1}\dfrac{b}{-a}$

F-36

A_1 and A_2 are complex conjugate

$$\sigma - f(t) = A_0 + 2|A_1| e^{-\gamma t} \cdot \cos(bt + \phi - \theta - \pi/2)$$

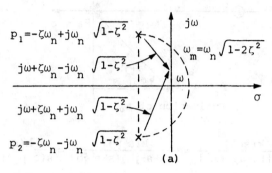

(a)

Pole-zero diagram.

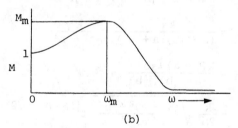

(b)

Magnitude of frequency response.

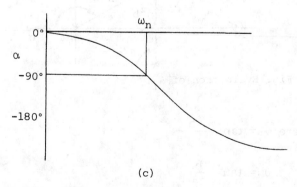

(c)

Angle of frequency response.

3.4 FREQUENCY RESPONSE FROM THE POLE-ZERO DIAGRAM

Frequency response: It is the steady-state response with a sine-wave forcing function for all values of frequency.

It is given by two curves:

A) M curve: the ratio of output amplitude to input amplitude as a function of frequency.

B) α curve: phase angle of the output α as a function of frequency.

Frequency-response characteristics:

A) at ω = 0, the magnitude is a finite value and the angle is 0°.

B) If the number of poles is more than the number of zeros, as ω → ∞, the magnitude approaches zero and the angle is -90° times the difference between the number of poles and zeros.

C) In order to have a peak value M_m, there must be present complex poles near the imaginary axis. More precisely, the damping ratio must be less than 0.707.

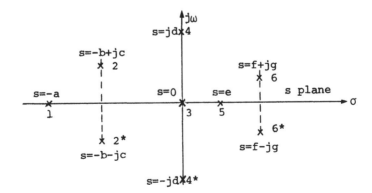

Location of poles in the s plane.

Table 3.1

Pole	Response	Characteristics
1	Ae^{-at}	Damped exponential
2-2*	$Ae^{-bt}\sin(ct + \phi)$	Exponentially damped sinusoid
3	A	Constant
4-4*	$A\sin(dt + \phi)$	Constant sinusoid
5	Ae^{et}	Increasing exponential (unstable)
6-6*	$Ae^{ft}\sin(gt + \phi)$	Exponentially increasing sinusoid (unstable)

3.5 ROUTH'S STABILITY CRITERION

Consider the characteristic equation

$$Q(s) = b_n s^n + b_{n-1} s^{n-1} + b_{n-2} s^{n-2} + \ldots + b_1 s + b_0 = 0$$

where all coefficients are real.

A) All powers of s from s^n to s^0 must be present in the characteristic equation. The system is unstable if any coefficients other than b_0 are zero or if any of the coefficients are negative. In this case, the roots are either imaginary or complex with positive real parts.

B) To determine the number of roots in the right half s plane:

The coefficients of the characteristic are arranged in the following Routhian array.

$$
\begin{array}{c|cccccc}
s^n   & b_n   & b_{n-2} & b_{n-4} & b_{n-6} & \cdots \\
s^{n-1} & b_{n-1} & b_{n-3} & b_{n-5} & b_{n-7} & \cdots \\
s^{n-2} & c_1   & c_2   & c_3   &   & \cdots \\
s^{n-3} & d_1   & d_2   &   &   & \cdots \\
\vdots &   &   &   &   &   \\
s^1   & j_1   &   &   &   &   \\
s^0   & k_1   &   &   &   &   \\
\end{array}
$$

c_1, c_2, c_3, etc., are evaluated as follows:

$$c_1 = \frac{b_{n-1}b_{n-2} - b_n b_{n-3}}{b_{n-1}}$$

$$c_2 = \frac{b_{n-1}b_{n-4} - b_n b_{n-5}}{b_{n-1}}$$

$$c_3 = \frac{b_{n-1}b_{n-6} - b_n b_{n-7}}{b_{n-1}}$$

This pattern is continued until the rest of the c's are all equal to zero. The d row is formed by using the s^{m-1} and s^{m-2} row. The constants are

$$d_1 = \frac{c_1 b_{n-3} - b_{n-1} c_2}{c_1}$$

$$d_2 = \frac{c_1 b_{n-5} - b_{n-1} c_3}{c_1}$$

$$d_3 = \frac{c_1 b_{n-7} - b_{n-1} c_4}{c_1}$$

When no more d terms are present, the other rows are formed in a similar way.

Routh's criterion: The number of roots of the characteristic equation with positive real parts is equal to the number of changes of sign of the coefficients in the first column of the Routhian array. Thus a system will be stable if all terms in the first column have identical signs.

An example

The Routh's array

s^4	1	2	9
s^3	10	3	
s^2	$\frac{17}{10}$	9	
s^1	-50		
s^0	9		

There are two changes of sign, so there are two roots in the right-half of the s plane.

C) Theorems: (For special cases)

Division of a row: The coefficients of any row may be multiplied or divided by the number without changing the signs of the first column.

A zero coefficient in the first column: The procedure below can be used whenever the first term in a row is zero, but the other terms are not equal to zero.

Method A: A small positive number δ is substituted for the zero. The rest of the terms in the array are evaluated as usual.

Method B: Substitute in the original equation $s = \frac{1}{x}$; then solve for the roots of x with positive real parts. The number of roots x with positive real parts will be the same as the number of s roots with positive real parts.

A ZERO ROW

A) An auxiliary equation is formed from the preceding row.

B) The array is completed by replacing the all-zero row by the coefficients obtained by differentiating the auxiliary equation.

C) The roots of the auxiliary equation are also roots of the original equation. These roots occur in pairs and are the negatives of each other.

3.6 IMPULSE FUNCTION: LAPLACE-TRANSFORM AND ITS RESPONSE

$f(t) = \dfrac{u(t) - u(t - a)}{a}$ Analytical expression

$f(s) = \dfrac{1 - e^{-as}}{as}$ its Laplace-transform

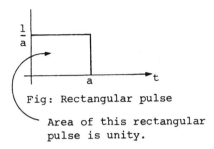

Fig: Rectangular pulse

Area of this rectangular pulse is unity.

A unit impulse: The limit of $f(t)$ as $a \to 0$ is termed a unit impulse and is designated $b\delta(t)$.

$$\delta(t) = \lim_{a \to 0} f(t) = \lim_{a \to 0} \frac{u(t) - u(t - a)}{a}$$

$\Delta(s)$ = The Laplace-transform of the unit impulse = 1.

SECOND-ORDER SYSTEM WITH IMPULSE EXCITATION

$$F(s) = \frac{\omega_n^2}{s^2 + 2\zeta\omega_n s + \omega_2^2}$$

Laplace-transform with an impulse input

The impulse function is the derivative of the step function, so the response to an impulse is the derivative of the response to a step function.

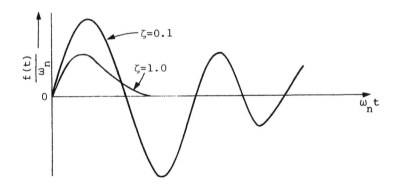

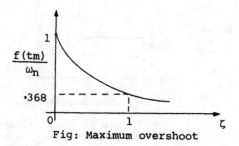

Fig: Maximum overshoot

$$t_m = \frac{\cos^{-1}\zeta}{\omega_n \sqrt{1-\zeta^2}}$$

where the maximum overshoot occurs.

$$f(t_m) = \omega_n e^{-(\zeta \cos^{-1}\zeta)/\sqrt{1-\zeta^2}}$$

CHAPTER 4

MATRIX ALGEBRA AND Z-TRANSFORM

4.1 FUNDAMENTALS OF MATRIX ALGEBRA

OPERATIONS WITH MATRICES

A) Matrix equality:

$\overline{A} = \overline{B}$ if and only if $a_{jk} = b_{jk}$ for $1 \leq j \leq p$ and $1 \leq k \leq q$ where the size of both matrices is $p \times q$.

B) Addition and subtraction:

If $\overline{A} = [a_{jk}]$ and $\overline{B} = [b_{jk}]$ are both $p \times q$ matrices, then $\overline{A} + \overline{B} = \overline{C}$ and $\overline{A} - \overline{B} = \overline{D}$ implies that $\overline{C} = [c_{jk}]$ and $\overline{D} = [d_{jk}]$ are also of the size $p \times q$ and $c_{jk} = a_{jk} + b_{jk}$ and $d_{jk} = a_{jk} - b_{jk}$ where $j = 1, 2, \ldots, p$ and $k = 1, 2, 3, \ldots, q$.

C) Matrix multiplication:

Scalar multiplication:

$$a\overline{B} = \overline{B}a = [ab_{jk}]$$

where a is a scalar.

Matrix multiplication:

$\overline{A} = [a_{jk}]$ of size $p \times q$ and $\overline{B} = [b_{jk}]$ of size $l \times m$. The product $\overline{A}\overline{B} = \overline{D}$ is defined only when $q = 1$, then $\overline{A}$ and $\overline{B}$ are conformable.

$$d_{jk} = \sum_{k'=1}^{a} a_{jk'} b_{k'k}, \text{ size of } \overline{D} \text{ is } p \times m.$$

D) Transpose, conjugate and the associate matrix:

If $\overline{B} = [b_{jk}]$ then the transpose of $\overline{B}$ is $\overline{B}^T = [b_{kj}]$. The matrix $\overline{B}$ is symmetric if $\overline{B} = \overline{B}^T$. If $\overline{B} = -\overline{B}^T$, then $\overline{B}$ is skew-symmetric. $(\overline{A}\,\overline{B})^T = B^T A^T$.

$A^* = $ conjugate of $A = [\overline{a}_{jk}]$

Associate matrix of $\overline{A}$ = conjugate transpose of $\overline{A}$.

Hermitian matrices: These are the matrices for which $\overline{B} = \overline{B^T}$, these are called skew-hermitian if $\overline{A} = -\overline{A^T}$.

E) Matrix inversion:

$\overline{B}\,\overline{A} = \overline{A}\,\overline{B} = \overline{I}$ then $\overline{A} = \overline{B}^{-1}$...A should be a square matrix.

$A^{-1} = \dfrac{\overline{C}^T}{|\overline{A}|}$ where $\overline{C}^T$ is the adjoint matrix denoted by Adj($\overline{A}$).

$A^{-1} = \text{Adj}(\overline{A})/|\overline{A}|$ Inverse of a non-singular matrix.

F) Special relationships:

$(\overline{A}\,\overline{B}\,\overline{C}\ldots\overline{F})^{-1} = \overline{F}^{-1}\ldots\overline{C}^{-1}\,\overline{B}^{-1}\,\overline{A}^{-1}$

If $\overline{A}^{-1} = A$, then $\overline{A}$ is called involutory.

If $\overline{B}^{-1} = \overline{B}^T$ then $\overline{B}$ is called orthogonal.

If $\overline{C}^{-1} = \overline{C}^{T*}$, then $\overline{C}$ is called unitary.

G) Cofactors and minors:

Minors: $\overline{B}$ is a $m \times m$ matrix, then minor M_{jk} is the determinant of $m-1 \times m-1$ matrix formed from $\overline{B}$ by eliminating the $j + n$ row and the kth column.

Cofactors:

The cofactors are given by $C_{jk} = (-1)^{j+k} M_{jk}$ where every element b_{jk} of $\bar{B}$ has a cofactor C_{jk}.

Determinants:

$\bar{B}$ is a m × n matrix, then $|\bar{B}| = \sum_{j=1}^{m} b_{ij} C_{ij}$ where i is any arbitrary row.

The Laplace expansion can be done with respect to any column j' to get $|\bar{B}| = \sum_{i=1}^{} b_{ij'} c_{ij'}$

H) Integration and differentiation of matrices:

If $\bar{C}(t) = [c_{jk}(t)]$, then $\frac{d\bar{c}}{dt} = \left[\frac{d}{dt} c_{jk}(t)\right]$ and

$$\int \bar{D} \, dt = \left[\int d_{ij}(t) dt\right]$$

Useful role for differentiating a determinant:

$$\frac{\partial |\bar{B}|}{\partial b_{ij}} = \bar{C}_{ij}$$

PARTITIONED MATRICES

$\bar{C}\,\bar{D} = \bar{E}$ can be divided as follows:

$$\begin{bmatrix} \bar{C}_1 \\ \bar{C}_2 \end{bmatrix} [D_1 \mid D_2] = \begin{bmatrix} \bar{C}_1 \bar{D}_1 & \bar{C}_1 \bar{D}_2 \\ \bar{C}_2 \bar{D}_1 & \bar{C}_2 \bar{D}_2 \end{bmatrix} = \begin{bmatrix} F_1 & F_2 \\ F_3 & F_4 \end{bmatrix}$$

$$\begin{bmatrix} \bar{C}_1 & \bar{C}_2 \\ \bar{C}_3 & \bar{C}_4 \end{bmatrix} \begin{bmatrix} D_1 \\ D_2 \end{bmatrix} = \begin{bmatrix} \bar{C}_1 \bar{D}_1 + \bar{C}_2 \bar{D}_2 \\ \bar{C}_3 \bar{D}_1 + \bar{C}_4 \bar{D}_2 \end{bmatrix} = \begin{bmatrix} E_1 \\ E_2 \end{bmatrix}$$

A position can be used to find the inverse of a matrix $\bar{C}$.

$$\bar{C}\,\bar{D} = \bar{I} \rightarrow \begin{bmatrix} \bar{C}_1 & \bar{C}_2 \\ \bar{C}_3 & \bar{C}_4 \end{bmatrix} \begin{bmatrix} \bar{D}_1 & \bar{D}_2 \\ \bar{D}_3 & \bar{D}_4 \end{bmatrix} = \begin{bmatrix} I & 0 \\ 0 & I \end{bmatrix}$$

This position results in two simultaneous equations $\bar{C}_1 \bar{D}_1 + \bar{C}_2 \bar{D}_3 = I$ and $\bar{C}_3 \bar{D}_1 + \bar{C}_4 \bar{D}_3 = 0$ from which $\bar{D}_1$ and $\bar{D}_3$ can be obtained. Two other equations give $\bar{D}_2$ and $\bar{D}_4$ and ultimately result into $\bar{C}^{-1}$.

Diagonal, block diagonal and triangular matrices:

Diagonal matrix: If all the non-zero elements of a square matrix are on the main diagonal, then it is called a diagonal matrix.

Triangular matrix: A square matrix, all of whose elements below (or above) the main diagonal are equal to zero, is called an upper (or lower) triangular matrix.

4.2 Z-TRANSFORMS

This is especially useful for linear systems with sampled or discrete input signals.

Block diagram:

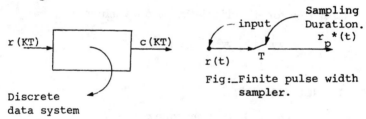

Fig:_Finite pulse width sampler.

The input for a discrete data system is given by

$$r^*(t) = \sum_{k=0}^{\infty} r(kT)\, \delta(t - kT)$$

where k varies from 0 to ∞ and these numbers are T seconds apart.

The output of the sampler is given by:

$$r_p^*(t) = r(t) \sum_{k=0}^{\infty} [u(t - kT) - u(t - kT - p)]$$

where u is the unit step function, T is the sampling period, and P is the sampling ration ($P \ll T$).

DEFINITION

The z-transform is defined as $z = e^{Ts}$, where s is the Laplace transform variable.

$$s = \frac{1}{T} \ln z$$

Table of z-transforms:

Table 4.1

Laplace Transform	Time Function	z-Transform
1	Unit impulse $\delta(t)$	1
$\dfrac{1}{s}$	Unit step $u(t)$	$\dfrac{z}{z-1}$
$\dfrac{1}{1-e^{-Ts}}$	$\delta_T(t) = \sum\limits_{n=0}^{\infty} \delta(t-nT)$	$\dfrac{z}{z-1}$
$\dfrac{1}{s^2}$	t	$\dfrac{Tz}{(z-1)^2}$
$\dfrac{1}{s^3}$	$\dfrac{t^2}{2}$	$\dfrac{T^2 z(z+1)}{2(z-1)^3}$
$\dfrac{1}{s^{n+1}}$	$\dfrac{t^n}{n!}$	$\lim\limits_{a \to 0} \dfrac{(-1)^n}{n!} \dfrac{\partial^n}{\partial a^n}\left(\dfrac{z}{z-e^{-aT}}\right)$
$\dfrac{1}{s+a}$	e^{-at}	$\dfrac{z}{z-e^{-aT}}$
$\dfrac{1}{(s+a)^2}$	te^{-at}	$\dfrac{Tze^{-aT}}{(z-e^{-aT})^2}$
$\dfrac{a}{s(s+a)}$	$1 - e^{-at}$	$\dfrac{(1-e^{-aT})z}{(z-1)(z-e^{-aT})}$
$\dfrac{\omega}{s^2+\omega^2}$	$\sin \omega t$	$\dfrac{z \sin \omega T}{z^2 - 2z\cos\omega T + 1}$
$\dfrac{\omega}{(s+a)^2+\omega^2}$	$e^{-at}\sin \omega t$	$\dfrac{ze^{-aT}\sin \omega T}{z^2 e^{2aT} - 2ze^{aT}\cos \omega T + 1}$
$\dfrac{s}{s^2+\omega^2}$	$\cos \omega t$	$\dfrac{z(z-\cos \omega T)}{z^2 - 2z\cos\omega T + 1}$
$\dfrac{s+a}{(s+a)^2+\omega^2}$	$e^{-at}\cos \omega t$	$\dfrac{z^2 - ze^{-aT}\cos \omega T}{z^2 - 2ze^{-aT}\cos \omega T + e^{-2aT}}$

INVERSE Z

Methods:

A) Partial-fraction expansion

F-48

B) Power-series

C) The inversion formula

POWER-SERIES METHOD

A) The z-transform is expanded into a power series in powers of z^{-1}.

B) The coefficient of z^{-k} is the value of $r(t)$ at $t = kT$.

INVERSION FORMULA

Inversion formula: $r(kT) = \dfrac{1}{2\pi j} \oint_L R(z) z^{k-1} dz$

where L is a circle of radius $|z| = t^{cT}$. The center of this is at the origin and c is such that all the poles of $R(z)$ are inside the circle.

THEOREMS

A) Addition and Subtraction

If $r_1(kT)$ and $r_2(kT)$ have z-transforms $R_1(z)$ and $R_2(z)$, respectively, then

$$\zeta[r_1(kT) \pm r_2(kT)] = R_1(z) \pm R_2(z)$$

B) Multiplication by a Constant

$$\zeta[ar(kT)] = a\zeta[r(kT)] = aR(z)$$

where a is a constant.

C) Real Translation

$$\zeta[r(kT - nT)] = z^{-n} R(z)$$
and
$$\zeta[r(kT + nT)] = z^n \left[R(z) - \sum_{k=0}^{n-1} r(kT) z^{-k} \right]$$

D) Complex Translation

$$\zeta[e^{\mp akT} r(kT)] = R(ze^{\pm aT})$$

E) Initial-Value Theorem

$$\lim_{k \to 0} r(kT) = \lim_{z \to \infty} R(z)$$

F) Final-value Theorem

$$\lim_{k \to \infty} r(kT) = \lim_{z \to 1} (1 - z^{-1})R(z)$$

if the function, $(1 - z^{-1})R(Z)$, has no poles on or outside the circle centered at the origin in the z-plane, $|z| = 1$.

CHAPTER 5

SYSTEM'S REPRESENTATION: BLOCK DIAGRAM, TRANSFER FUNCTIONS, AND SIGNAL FLOW GRAPHS

5.1 BLOCK DIAGRAM AND TRANSFER FUNCTION

Fundamentals:

A block diagram is a symbolic representation of:

A) the flow of information in a system, and

B) the functions performed by each component in the system.

The transfer function expresses the relationship between output and input of a system and is frequently given in terms of the Laplace variable s.

In finding the transfer function of a system, all initial conditions must be set to zero.

The transfer function may be expressed in terms of:

A) the operator D . . . . (D)

B) the Laplace-transform variable s . . . . C(j)

C) in phasor form . . . . C(jω)

This latter case is used in the sinusoidal steady-state analysis.

The denominator of the transfer function is the characteristic equation when it is set equal to zero.

Summation points in the block diagrams:

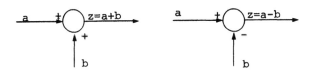

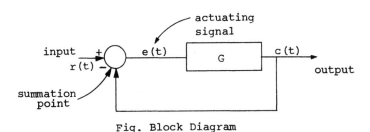

Fig. Block Diagram

Transfer function:

$$G(D) = \text{Transfer function} = \frac{V_{out}}{V_{in}}$$

G(S) = Laplace-transform of the transfer function

$$= \frac{E_2(s)}{E_1(s)} = RCs/1 + RCs$$

$\overline{G}(j\omega)$ = Frequency transfer function = $\dfrac{\overline{E}_2(j\omega)}{\overline{E}_1(j\omega)}$

Blocks in Cascade:

If the operation of an element or a component can be described independently, then a block can be used to represent that component.

Combination of cascade blocks:

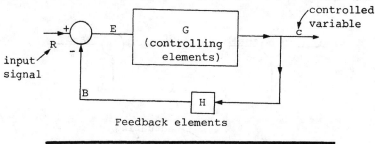

Fig. Equivalent Block diagram after cascading.

Determination of control ratio:

$$\text{Control ratio} = \frac{C(s)}{R(s)} = \frac{G(s)}{1 + G(s)H(s)}$$

$$= \text{Close-loop transfer function}$$

Characteristic equation: $1 + G(s)H(s) = 0$

Open-loop transfer function $= \frac{B(s)}{E(s)} = G(s)H(s)$

Forward transfer function $= \frac{C(s)}{E(s)} = G(s)$

5.2 TRANSFER FUNCTIONS OF THE COMPENSATING NETWORKS

Lag compensator:

$$\frac{E_1(s)}{E_2(s)} = \frac{1 + (R_1 + R_2)Cs}{1 + R_2Cs}$$

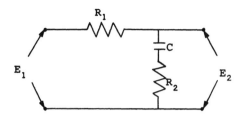

$$\frac{E_2(s)}{E_1(s)} = \frac{1 + R_2Cs}{1 + (R_1 + R_2)Cs} = G(s)$$

$$G(s) = \frac{1}{\alpha} \frac{s + 1/T}{s + 1/\alpha T}$$

$\ldots$ $\alpha = \dfrac{R_1 + R_2}{R_2}$ and $T = R_2C$

Lead compensator:

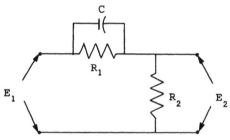

$G(s) = \dfrac{s + 1/T}{s + 1/\alpha T}$ $\qquad \alpha = \dfrac{R_2}{R_1 + R_2}$

and $T = R_1C$

Lag-lead compensator:

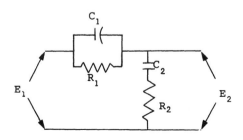

F-54

$$G(s) = \frac{E_2(s)}{E_1(s)} = \frac{1 + (T_1 + T_2)s + T_1 T_2 s^2}{1 + (T_1 + T_2 + T_{12})s + T_1 T_2 s^2}$$

$$\ldots T_1 = R_1 C_1, \quad T_2 = R_2 C_2 \quad T_{12} = R_1 C_2$$

$$= \frac{(1 + R_1 C_1 s)(1 + R_2 C_2 s)}{(1 + R_1 C_1 s)(1 + R_2 C_2 s) + R_1 C_2 s}$$

5.3 SIGNAL FLOW GRAPHS

This graph is the pictorial representation of a set of simultaneous equations. The nodes represent the system variable and a branch acts as a signal multiplier.

DEFINITION AND ALGEBRA

Node: A node performs two functions:

A) Addition of all the incoming signals.

B) Transmission or distribution of the total incoming signals of a node to all the outgoing branches.

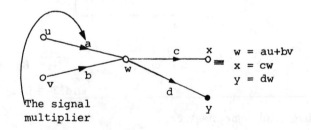

The signal multiplier

$w = au+bv$
$x = cw$
$y = dw$

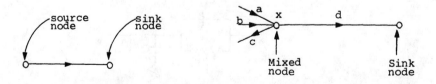

These are the equations represented by the adjacent flow graph.

FLOW-GRAPH ALGEBRA
Rules

A) Series paths (cascade nodes): Series paths are combined into a single path by multiplying the transmittances of the individual paths.

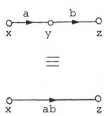

B) Parallel paths: Parallel paths are combined by adding the transmittance.

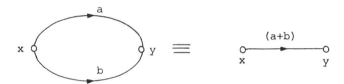

C) Elimination of a node: A node representing a variable can be eliminated as follows:

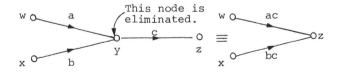

MASON'S RULE gives the overall transmittance of a system and is defined as follows:

$$T = \text{overall transmittance} = \frac{\Sigma T \Delta}{\Delta} \text{ where}$$

A) Δ represents the determinant of the graph given by

$$\Delta = 1 - \Sigma L_1 + \Sigma L_2 - \Sigma L_3 + \ldots \text{ and}$$

a) L_1 is defined as the transmittance of each loop, thus ΣL_1 would be the sum of the individual transmittances of all loops.

b) ΣL_2 = sum of L_2 where L_2 is the product of the transmittances of 2 non-touching loops. Two loops are non-touching if they do not share a common node.

c) ΣL_3 = sum of all possible combinations of the product of transmittances of non-touching loops taken three at a time.

B) T denotes the transmittance of each forward path between a source and a sink node.

C) If we remove the path which has transmittance T, the determinant of the subgraph produced is denoted by Δ.

CHAPTER 6

SERVO CHARACTERISTICS: TIME-DOMAIN ANALYSIS

6.1 TIME-DOMAIN ANALYSIS USING TYPICAL TEST SIGNALS

Following signals are used for analysis:

A) Step function: This function is defined as follows.

(i) Step Function

$$r(t) = \begin{cases} R_0 & \text{for } t > 0 \\ 0 & \text{for } t < 0 \end{cases}$$

$$= R_0 u(t)$$

where $u(t)$ is the unit step function.

B) Ramp function:

$$r(t) = R_0 t \quad \text{for } t \geq 0 \text{ and zero elsewhere}$$

$$= R_0 t u(t)$$

ii) Ramp Function

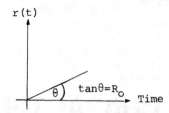

C) Parabolic function:

(iii) Parabolic Function

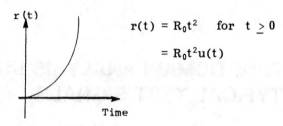

$$r(t) = R_0 t^2 \quad \text{for} \quad t \geq 0$$
$$= R_0 t^2 u(t)$$

Any random input signal can be considered as being composed of these signals.

CONCLUSION

If any linear system is analyzed mathematically or experimentally for each of these input signals then it can be said that the response of this system to these basic test signals will be the representation of the actual system response to any random signal.

6.2 TYPES OF FEEDBACK SYSTEMS

The standard form of the transfer function is

$$G(s) = \frac{k_n(1 + a_1 s + a_2 s^2 + \ldots + a_m s^m)}{s^n(1 + b_1 s + b_2 s^2 + \ldots + b_p s^p)}$$

where $a_1, a_2, \ldots, b_1, b_2, \ldots$ are constant coefficients.

k_n = overall gain of transfer function $G(s)$

$n = 0, 1, 2$ denotes the type of the transfer function.

Characteristics of different types of systems:

A) Type 0: A constant error signal $E(s)$ results in a constant value of the output signal $C(s)$.

B) Type 1: A constant $E(s)$ signal results in a constant rate of change of the output signal.

C) Type 2: A constant $E(s)$ signal will produce a constant D^2C of the output variable.

Note: A type of the given system can be readily known from $G(j\omega)$ and log $G(j\omega)$ versus ω plots.

6.3 GENERAL APPROACH TO EVALUTION OF ERROR

IMPORTANT FACTS

A) Final-value theorem:

$$\lim_{t \to \infty} f(t) = \lim_{s \to 0} sF(s)$$

B) Differentiation theorem:

$$L[D^n C(t)] = s^n C(s)$$

when all the initial conditions are zero.

C) If the input signal to a unit feedback, stable system is a power series, then the steady-state output will have the same form as the input.

Steady-state error function: General Form

$$G(s) = \frac{C(s)}{E(s)} = \frac{k_n(1 + T_1 s)(1 + T_2 s)\ldots}{s^n(1 + T_a s)(1 + T_b s)\ldots}$$

$$E(s) = \frac{(1 + T_a s)(1 + T_b s)\ldots s^n C(s)}{s^n (1 + T_a s)(1 + T_b s)(1 + T_c s)}$$

This is obvious after rearranging the top equation.

$$e(t)_{ss} = \lim_{s \to 0} [SE(s)] = \lim_{s \to 0} \left[\frac{s(1+T_a s)(1+T_b s)\ldots s^n C(s)}{k_n (1+T_1 s)(1+T_2 s)} \right]$$

$$= \lim_{s \to 0} \frac{s[s^n C(s)]}{K_n}$$

This is the steady-state error.

This can be written as follows:

$$\boxed{e(t)_{ss} = \frac{D^n C(t)_{ss}}{k_n}}$$

This equation is useful when $D^n C(t)_{ss}$ = constant. Then $e(t)_{ss}$ will be constant and is equal to E_o. So $k_n E_o = D^n C(t)_{ss}$ = constant = C_n.

$$E(s) = \frac{s^n (1 + T_a s)(1 + T_b s)\ldots R(s)}{s^n (1 + T_a s)(1 + T_b s) + \ldots + k_n (1 + T_1 s)(1 + T_2 s)}$$

When $H(s) = 1$, i.e. unity feedback.

$$\boxed{e(t)_{ss} = \lim_{s \to 0} s \left[\frac{s^n (1 + T_a s)(1 + T_b s)\ldots R(s)}{s^n (1+T_a s)(1+T_b s) + \ldots + k_n (1+T_1 s)(1+T_2 s)} \right]}$$

The general expression.

ERROR SERIES: CONCEPTS

This error series is useful when the input to a feedback system is an arbitrary function of time.

Mathematical representation:

$$E(s) = R(s)/1 + G(s)H(s)$$

$$e(t) = \int_{-\infty}^{t} A(\tau)r(t-\tau)d\tau$$

where $A(q)$ is the inverse Laplace transform of $1/(1 + G(s)H(s))$.

Error series:

A) $$e(t) = r(t) \int_0^t A(\tau)d\tau - \dot{r}(t) \int_0^t \tau A(\tau)d\tau$$

$$+ \ddot{r}(t) \int_0^t \frac{\tau^2}{2!} A(\tau)d\tau$$

where $r(t-\tau) = r(t) - \tau \dot{r}(t) + \frac{\tau^2}{2!} \ddot{r}(t)$ Taylor expansion

B) $$e_s(t) = C_0 r_s(t) + C_1 \dot{r}_s(t) + \frac{C_2}{2!} \ddot{r}_s(t)$$

Where C_0, C_1 are called the error coefficients and $e_s(t)$ is known as the error series.

$$C_n = (-1)^n \int_0^0 \tau^n A(\tau)d\tau$$

Evaluation of the error coefficients from $A(s)$:

A) $$C_0 = \lim_{s \to 0} A(s) = \lim_{s \to 0} \int_0^\infty A(s)e^{-\tau s}d\tau$$

B) $$C_1 = \lim_{s \to 0} \frac{d}{ds} A(s)$$

C) $$C_n = \lim_{s \to 0} \frac{d^n}{ds^n} A(s)$$

6.4 ANALYSIS OF SYSTEMS: UNITY FEEDBACK

TYPE 0 SYSTEM (n = 0)

Step input $(r(t) = R_0 u(t))$:

$$\boxed{\begin{aligned} e(t)_{ss} &= \frac{R_0}{1 + k_0} = \text{constant} \\ &= E_0 \end{aligned}}$$

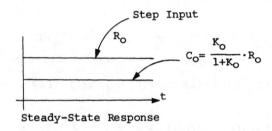

Steady-State Response

This is obtained from the general expression for $e(t)_{ss}$. Since

$$R(s) = \frac{R_0}{s}$$

Ramp input:

$$\boxed{e(t)_{ss} = \infty}$$

since $r(t) = R_1 t\, u(t)$

$$R(s) = R_1/s^2$$

Parabolic input:

$$\boxed{\begin{aligned} e(t)_{ss} &= r(t)_{ss} - C(t)_{ss} \quad \text{which approaches a value of infinity} \\ &= \infty \quad\quad\quad\quad\quad\quad \text{since } r(t) = R_2 t^2 u(t) \end{aligned}}$$

Conclusions:

A) A constant input (i.e. step input) produces a constant value of the output with a constant error signal.

B) When a ramp-function input produces a ramp output with a smaller slope, there is an error which approaches a value of ∞ with increasing time.

C) Type 0 system cannot follow a parabolic input.

TYPE 1 SYSTEM (n = 1)

Step input:

$$\boxed{e(t)_{ss} = 0}$$

This is obtained from the general equation by putting n = 1 and

$$R(s) = \frac{R_0}{s}$$

Ramp input:

$$\boxed{e(t)_{ss} = \infty}$$
since $r(t) = R_2 t^2 u(t)$
$R(s) = 2R_2/s^3$

Parabolic input:

$$\boxed{e(t)_{ss} = \frac{R_1}{k_1} = \text{constant} = E_0 \neq 0}$$
since $R(s) = R_1/s^2$

Conclusions:

A) Type 1 system with a constant input produces a

steady-state constant output of value equal to the input, i.e. zero steady-state error.

B) Type 1 system with a ramp input produces a ramp output with a constant error signal.

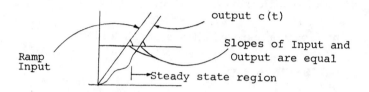

Ramp Input — output c(t) — Slopes of Input and Output are equal — Steady state region

C) With a parabolic input, a parabolic output is produced with an error which increases with time $e(\infty) = \infty$.

TYPE 2 SYSTEM (n = 2)

Step input:

$$\boxed{e(t)_{ss} = 0}$$

This is obtained by putting n = 2 and $R(s) = R_0/s$.

Ramp input:

$$e(t)_{ss} = 0 \qquad \text{since } R(s) = R_1/s^2$$

Parabolic input:

$$e(t)_{ss} = \frac{2 \cdot R_2}{k_2} = \text{constant} \quad \text{since } R(s) = \frac{2R_2}{s^3}$$
$$= E_0 \neq 0$$

F-65

Summary:

Table 6.1

Steady-state characteristics: Unity feedback systems

System type n	$r(t)_{ss}$ Steady-state (input)	(Steady-state error function)	(Steady-state output)	Value of $e(t)$ at $t = \infty$
0	Step	$\dfrac{R_0}{1+K_0}$	$\dfrac{K_0}{1+K_0}R_0$	$\dfrac{R_0}{1+K_0}$
0	Ramp	$\dfrac{R_1}{1+K_0}t - C_0$	$\dfrac{K_0 R_1}{1+K_0}t + C_0$	∞
0	Parabolic	$\dfrac{R_2}{1+K_0}t^2 - C_1 t - C_0$	$\dfrac{K_0 R_2}{1+K_0}t^2 + C_1 t + C_0$	∞
1	Step	0	R_0	0
1	Ramp	$\dfrac{R_1}{K_1}$	$R_1 t - \dfrac{R_1}{K_1}$	$\dfrac{R_1}{K_1}$
1	Parabolic	$-C_1 t - C_0$	$R_2 t_2 + C_1 t + C_0$	∞
2	Step	0	R_0	0
2	Ramp	0	$R_1 t$	0
2	Parabolic	$2\dfrac{R_2}{K_2}$	$R_2 t^2 - \dfrac{2R_2}{K_2}$	$\dfrac{2R_2}{K_2}$

CHAPTER 7

ROOT LOCUS

7.1 ROOTS OF CHARACTERISTIC EQUATIONS

Basic approach:

$$G(s) = \frac{C(s)}{E(s)} = \frac{k}{s(s+a)}$$

forward transfer function

STATIC LOOP SENSITIVITY

This is the value of k, when the transfer function is expressed in such form that the coefficients of s in both numerator and denominator are equal to unity.

$$\frac{C(s)}{R(s)} = \frac{k}{s^2 + 2s + k} = \frac{k}{s^2 + 2\zeta\omega_n s + \omega_n^2} \quad \text{for } a = 2$$

Roots of the characteristic equation:

$$s = -\zeta\omega_n \pm \omega_n \sqrt{\zeta^2 - 1}$$

where $\omega_n = \sqrt{k}$ and $\zeta = \dfrac{1}{\sqrt{k}}$

Plot of the roots of the characteristic equation:

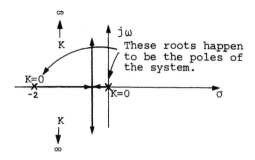

Roots are obtained for different values of k and these are plotted as shown.

A smooth curve is drawn (as shown by thick lines) through those points.

System performace v/s sensitivity k: salient points from the root locus.

An increase in the gain of the system results in:

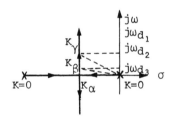

Fig. A sample plot.

A) a decrease in the damping ratio ζ, so the overshoot of the time response increases.

B) increase in ω_n (i.e., undamped natural frequency).

C) increase in ω_d (i.e., damped natural frequency)...ω_d is the imaginary component of the complex root.

D) The rate of decay, σ, is unchanged.

E) Root locus is a vertical line for $k \geq k_\alpha$ and
$$\zeta\omega_n = \sigma = \text{const.}$$

F-68

7.2 IMPORTANT PROPERTIES OF THE ROOT LOCI

EFFECT OF ADDITION OF POLES

When a pole is added to the function $G(s)H(s)$ in the left half of the s-plane, the net effect is of pushing the original locus towards the right half plane.

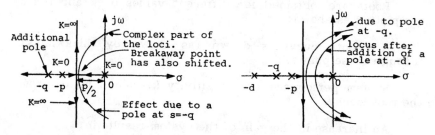

Fig. Original plot

Fig. Addition of pole at -d

ADDITION OF ZERO

Adding zeros has the effect of moving the root locus towards the left half of the s-plane.

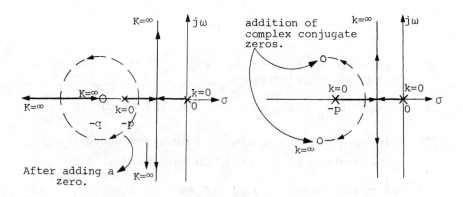

STEPS IN PLOTTING THE ROOT-LOCUS

A) The open loop transfer function $G(s)H(s)$ of the system is first determined.

B) The numerator and the denominator of the transfer function are then factorized into factors of the form (s + p).

C) Locate zeros and poles of the open-loop transfer function in the s-plane.

D) Determine the roots of the characteristic of the closed loop system, i.e. [1 = G(s)H(s) = 0]. The locus of the roots of the characteristic equation is then found from the plotted zeros and poles of the open loop function.

E) The locus is calibrated in terms of k. If the gain of the open-loop system is predetermined, the location of the exact roots of HG(s)H(s) is known. If the location of the roots is specified, k can be known.

F) Now that the roots are known, the system's response is the inverse Laplace transform.

G) If the specifications are not satisfied, then the shape that meets the desired specifications is determined and compensating networks are introduced in the system to meet these requirements.

Poles of the control ratio $\frac{C(s)}{R(s)}$:

$$\boxed{\frac{C(s)}{R(s)} = M(s) = \frac{N_1 D_2}{D_1 D_2 + N_1 N_2} = \frac{P(s)}{Q(s)}}$$

where $G(s) = \frac{N_1(s)}{D_1(s)}$, $H(s) = \frac{N_2(s)}{D_2(s)}$

and $B(s) = 1 + G(s)H(s) = D_1 D_2 + N_1 N_2 / D_1 D_2$

i.e. the zeros of B(s) are the poles of $\frac{C(s)}{R(s)}$ and these zeros determine the transient response.

Factors of Q(s) fall into the following categories:

Pole of C(s)	Corresponding inverse of the form
S	u(t) - unit step
$S + \frac{1}{T}$	$e^{-t/T}$
$s^2 + 2\zeta\omega_n s + \omega_n^2$	$e^{-\zeta\omega_n t} \sin(\omega_n \sqrt{1-\zeta^2}\, t + \phi)$, $\zeta < 1$

F-70

$P(s)$ only modifies the constant multiplier of these transients.

Conditions to plot the root-locus for $k > 0$:

$$|G(s)H(s)| = 1 \quad \text{Magnitude condition}$$
$$\angle G(s)H(s) = (1 + 2m)180° \quad \text{for } m = 0, \pm 1, \pm 2 \ldots \text{for } k > 0.$$
$$\text{Angle condition}$$

Conditions for negative values of k:

$$|G(s)H(s)| = 1 \quad \text{Magnitude condition}$$
$$\angle G(s)H(s) = (m)360° \ldots \text{for } m = 0, \pm 1, \pm 2 \ldots \text{for } k < 0$$
$$\text{Angle condition}$$

The magnitude and angle conditions:

APPLICATION

$$G(s)H(s) = \frac{k(s - z_1)}{s^n(s - p_1)(s - p_2)} \quad \text{General form of the open-loop transfer function.}$$

Note: $s - p_1$, $s - p_2$, ..., $s - z_1$ etc., are the complex numbers representing the directed line segments.

$$1 + G(s)H(s) = 0 \quad \text{The characteristic equation.}$$

Application of the magnitude and angle conditions results in

$$\frac{|k||s - z_1| \ldots |s - z_n|}{|s^n||s - p_1||s - p_2|\ldots} = 1$$

$$-\zeta = \angle s - z_1 \ldots + \angle s - z_w - n\angle s - \angle s - p_1$$

$$= \begin{cases} (1 + 2m)180° & \ldots \text{if } k > 0 \\ (m)360° & \ldots \ldots \text{if } k < 0 \end{cases}$$

So modification of these equations results in:

A) $|k|$ = static loop sensitivity = $\dfrac{|s^n||s - p_1|\ldots|s - p_q|}{|s - z_1|\ldots|s - z_m|}$

F-71

> B) $+\zeta$ = -(angles of numerator terms) + (angles of denominator terms)
>
> $$= \begin{cases} -(1+2m)180° & \text{...if } k > 0 \\ -(m)360° & \text{........if } k < 0 \end{cases}$$
>
> Note: These two conditions are used in the graphical construction of the root locus.

Construction rules for plotting the root locus:

A) Total number of branches in the complete root loci:

   The number of branches in a root loci is equal to the number of poles of the open-loop transfer function.

   If the equation is expressed in the following form

   $$s^q + b_1 \ldots s^{q-1} \ldots b_n + k(s^p + a_1 s^{p-1} \ldots) = 0$$

   then the number of branches of the root loci is greater or equal to q and p.

B) Locus on the real axis:

   If the total number of poles and zeros to the right of the search point on the real axis is odd, then this point is on the root locus.

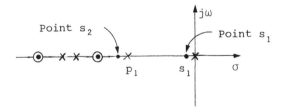

Note: Point s_1 lies on the real axis root locus but not s_2.

The following is the required angle condition so that the selected search point will be on the root locus.

$$(R_p - R_z)180° = (1 + 2m)180°$$

Where R_p = number of poles to the right of the search point on the real axis and R_z = number of zeros.

C) End points of the locus:

The starting points of the root-locus (i.e. for $k = 0$) are the open-loop poles and the end points (for $k = \infty$) are the open-loop zeros (∞ is an equivalent zero).

Asymptotes of the root loci (behavior as $s \to \infty$):

$$\theta = \frac{(1 + 2m)180°}{[\text{number of poles of } G(s)H(s)] - [\text{number of zeros of } G(s)H(s)]}$$

= This is the angle of straight lines or asymptotes.

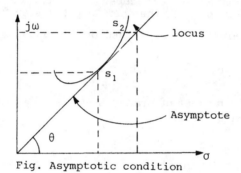

Fig. Asymptotic condition

The behavior of the root-locus near ∞ are important because when number of poles of $G(s)H(s) \neq$ number of zeros of $G(s)H(s)$, then $2\,|\text{number of poles } G(s)H(s) - \text{number of zeros of } G(s)H(s)|$ will tend to infinity in the s-plane.

Intersection of the asymptotes (centroid):

$$\sigma_0 = \text{Real axis interception of the asymptotes} = \frac{\sum_{c=1}^{v} \text{Re}(P_c) - \sum_{m=1}^{w} \text{Re}(Z_m)}{v - w}$$

since the proportions of the locus away from the asymptotes and near the axes are important. σ_0 is also the centroid of the pole zero plot where

$$A(s) = \frac{\prod_{c=1}^{v} (s - P_c)}{\prod_{m=1}^{w} (s - z_m)} = -k$$

Note that there are 2|number of poles of $G(s)H(s) = v -$ number of zeros ϕ of $G(s)H(s) = w$|.

Asymptotes whose intersection lies on the real axis.

Also note that the point of intersection is always on the real axis.

Real axis break-away points (saddle points):

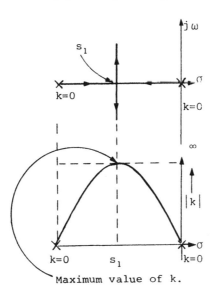

Maximum value of k.

The break-away point: Since k starts with a value of zero at the poles and increases in value, as the locus moves away from the poles, there is a point somewhere in between where the k's for the two branches simultaneously reach a maximum value. This is the break-away point.

Break-away points on the root-loci of an equation correspond to multiple-order roots of the equation.

A root locus can have more than one break-away point and these need not always be on the real axis; however the break-away points may be real or complex conjugate pairs.

The break-in point: This is the value of σ for which $|k|$ is a minimum between 2 zeros. This is shown in the diagram.

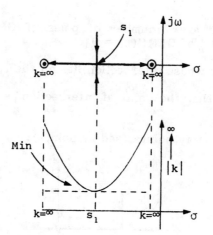

Break-in points

Break-away points: These can be determined by taking the derivative equating to zero and then determining its roots. The root that occurs between the poles (or the zeros) is the break-away (or break-in) point.

Example: $G(s)H(s) = k/s\,(s+1)(s+2)$

so $W(s) = s(s+1)(s+2) = -k = s^3 + 3s^2 + 2s$

$$\frac{dW(s)}{ds} = 3s^2 + 6s + 2 = 0$$

Hence, the roots are:

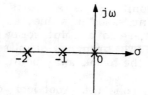

Fig. Pole-zero plot.

$s_{1,2} = -1 \pm 0.6$ i.e. -0.4 and -1.57

The break-away point for $k > 0$ lie between $s = 0$ and $s = -1$ in order to satisfy the angle condition, $s_1 = -0.4$. The other point, $s_2 = -1.57$ is the break-in point for $k < 0$.

so,
$$-k = (-0.4)^3 + 3(-0.4)^2 + 2(-0.4)$$
$$k = 0.38$$

(This is the value of k at the break-away point for $k > 0$.)

The necessary angular condition near the break-away point: A break-away point is determined as follows:

Step 1: A check point is selected between two poles on the real axis at a distance d; then the angular contribution due to all poles and zeros which are on the real axis is $\frac{d}{L_i}$, where L_i is the distance between the check point and the pole or zero.

Step 2: The angular contribution towards this check point due to complex poles or zeros is
$$\Delta\phi = 2b\delta/b^2 + a^2 \qquad \text{as shown}$$

Step 3: All the angular contributions with proper signs are summed to zero. The point which satisfies this zero condition is the break-away point.

Complex pole (or zero): Angle of departure

For $k > 0$, the angle of departure from a complex pole is equal to 180° (1 + 2m) minus the sum of the angles from the other poles plus the sum of the angles from the zeros. Any of these angles may be tve or -ve. For $k < 0$, the departure angle is 180° from that obtained for $k > 0$.

Imaginary-axis intercepting point:

The points where the complete root locus intersects the imaginary axis of the s-plane, and the corresponding values of k can be determined by means of the Routh-Hurwitz criterion:

Example: $s^3 + as^2 + bs + kd = 0$

The closed-loop characteristic equation.

A Routhian array is formed

$$\begin{array}{c|cc} s^3 & 1 & b \\ s^2 & a & \\ s^1 & (ab-kd)a & kd \\ s^0 & kd & \end{array}$$

The Routhian array formed from the closed-loop characteristic equation.

Undamped oscillation occurs if the s^1 row is zero. Thus, the auxiliary equation obtained from the s^2 row is:

$$as^2 + kd = 0 \text{ and its roots } s_1 = +j\sqrt{\frac{kd}{a}} \text{ and } s_2 = -j\sqrt{\frac{kd}{a}}$$

i.e. $k = \dfrac{ab}{d}$

Root locus branches (non-intersection and intersection):

Properties

A) Any point on the root locus satisfies the angle condition. There are no root locus intersections at a point on the root locus if $\dfrac{dW(s)}{ds} \neq 0$ at this point. In this case there is only one branch of the root locus through the point.

B) A point on the root locus will have branches through it (i.e. it is an intersection point) if the derivatives of $W(s)$ vanish at this point. Thus if the first $y-1$ derivatives of $W(s)$ vanish at a given point on the root locus, there will be y branches approaching and y branches leaving this point. The angle between 2 approaching branches is

$$\lambda_y = \pm \frac{360°}{y}$$

while the angle between 2 branches (one leaving and the other approaching the same point) is

$$\phi_y = \pm \frac{180°}{y}$$

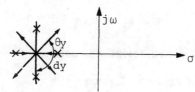

Fig. Root locus.

Conservation of the sum of the system of roots:

Grant's rule states that the sum of the closed-loop roots is equal to the sum of the open-loop poles. This is applicable when the open-loop transfer function is such that $v - z \geq W$.

Determination of roots on the root locus:

Application of Grant's rule

Grant's rule which is stated as follows, is used to find one real or two complex roots of the system provided the dominant roots of the characteristic equations is known.

$$\sum_{k=1}^{v} p_k = \sum_{l=1}^{v} r_l$$

7.3 FREQUENCY RESPONSE

DEFINITION

The frequency response gives the ratio of the phasor output to the phasor input for any inputs over a range of frequencies. The combined plots of the magnitude (M) and the angle (α) of $\frac{C(j\omega)}{R(j\omega)}$ versus the angular frequency is the frequency response of a control system. How to obtain the frequency response:

a) To determine the frequency response, the closed-loop control ratio should be known.

b) Then the control ratio is expressed as a function of frequency by substituting $s = j\omega$ as follows.

$$\frac{C(j\omega)}{R(j\omega)} = \frac{k(j\omega + a)}{(j\omega + b - cj)(j\omega + b + c)(j\omega + d - ej)(j\omega + d + ej)}$$

Please note that this is the most generalized form.

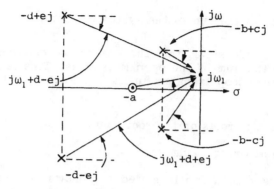

Fig. Frequency response from the plot.

c) For any frequency ω_1, a point ω_1 is chosen on the $j\omega$ axis, and the directed lines are drawn from all poles and zeros to this point. Then the lengths of the directed lines and angles these lines make with horizontal lines are determined as shown in the figure.

d) When the magnitudes and angles for each term (i.e. for each of the directed lines) of the $\frac{C(j\omega)}{R(j\omega)}$ equation are obtained, the value of $C(j\omega)/R(j\omega)$ is obtained for that particular frequency ω_1. Note that the clockwise angles are -ve and the anticlockwise angles are +ve.

e) This procedure is repeated for sufficient numbers of angular frequencies and a smooth curve is drawn as shown below.

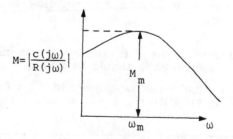

CHAPTER 8

SPECIAL POLE-ZERO TOPICS: DOMINANT POLES AND THE PARTITION METHOD

8.1 TRANSIENT RESPONSE: DOMINANT COMPLEX POLES

The conditions which are necessary for the time response to be nominated by only one pair of complex poles, require a pole-zero pattern with the following characteristics:

A) All other poles in the pole-zero diagram must be far to the left of the dominant poles, so the transient response due to these poles are small in amplitude.

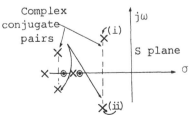

Pattern whose time response is dominated by complex poles.

B) In the pole-zero diagram, any pole which is not far away from the dominant complex poles must be near a zero, so that the magnitude of the transient response is small; since its effect will be modified by the zero. Note: The dominant poles are drawn darker.

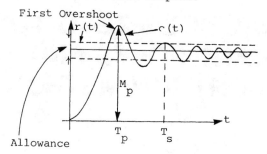

The Time Response

FIGURES OF MERIT

A) M_p (peak overshoot): The amplitude of the first overshoot.

B) t_p (peak time): The time to reach the peak overshoot from the initial starting time.

C) t_s (setting time): The time for the response to first reach and thereafter remain within the allowable limit (usually 2% of the final value).

D) n: The number of oscillations in the response up to the setting time. Not that there are two oscillations in a complete cycle.

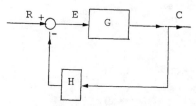

Non-Unity Feedback System

Figures of merit for the non-unity feedback system:

$$G(s)H(s) = \frac{k_G k_H \prod_{m=1}^{\omega} (s - z_m)}{\prod_{i=1}^{v} (s - p_i)}$$

where product $k_G k_H = k$ is the static loop sensitivity.

$$\frac{C(s)}{R(s)} = \frac{k_G \prod_{m=1}^{n} (s - z_n)}{\prod_{i=1}^{v} (s - p_i)}$$

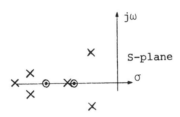

The desired Pole-Zero pattern

Here k_G is not k unless the system has unity feedback.

$$= \frac{A_0}{s} + \frac{A_1}{s - p_1} + \ldots$$

$$c(t) = \begin{matrix}\text{Time solution if the} \\ \text{system has a dominant} \\ \text{pole } p_1 = \sigma + j\omega_d\end{matrix} = \frac{P(o)}{Q(o)} + 2 \left| \frac{k_G \prod_{m=1}^{n} (p_1 - z_m)}{p_1 \prod_{i=2}^{v} (p_1 - p_i)} \right|$$

$$e^{\sigma t} \cdot \cos[\omega_d t + \underline{/P(p_1)} - \underline{/p_1} - \underline{/Q'(p_1)}]$$

$$+ \sum_{k=3}^{v} \frac{P(p_k) e^{p_k t}}{p_k Q'(p_k)}$$

where

$$Q'(p_k) = \left. \frac{Q(s)}{s-p_k} \right]_{s=p_k}$$

$$T_p = \frac{1}{\omega_d} \left\{ \frac{\pi}{2} - \begin{bmatrix} \text{sum of angles} \\ \text{from zeros of } \frac{C(s)}{R(s)} \\ \text{to } P_1, \text{ the dominant} \\ \text{pole} \end{bmatrix} + \begin{bmatrix} \text{sum of angles from} \\ \text{all other poles of } \frac{C(s)}{R(s)} \\ \text{to dominant pole } p_1 \\ \text{(including conjugate} \\ \text{pole)} \end{bmatrix} \right\}$$

$$M_p = \text{The peak overshoot} = \underbrace{\frac{P(o)}{Q(o)}}_{\text{The final value}} + \underbrace{\frac{2\omega_d}{\omega_n^2} \left| \frac{k_G \prod_{m=1}^{\omega^1} (p_1 - z_m)}{\prod_{i=2}^{v} (p_1 - p_i)} \right| e^{\sigma T_p}}_{\text{The overshoot } M_0}$$

$$\boxed{\begin{array}{l} t_s = \dfrac{m}{|\sigma|} = \dfrac{m}{\zeta \omega_n} = \text{Four time constants for 2\% error, i.e., } m = 4 \\[1em] n = \dfrac{\text{Settling time}}{\text{Period}} = \dfrac{t_s}{2\pi/\omega_d} = \dfrac{2\omega_d}{\pi|\sigma|} = \dfrac{2}{\pi} \dfrac{\sqrt{1-\zeta^2}}{\zeta} \end{array}}$$

Additional Significant Poles:

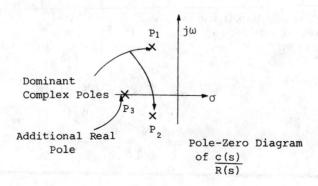

Pole-Zero Diagram of $\frac{c(s)}{R(s)}$

$$c(t) = 1 + 2|A_1|e^{-\zeta\omega_n t}\sin(\omega_n\sqrt{1-\zeta^2}\,t + \phi) + A_0 e^{P_0 t}$$

The time response to a unit step input for a system,

$$\frac{C(s)}{R(s)} = \frac{k}{(s^2 + 2\zeta\omega_n s + \omega_n^2)/(s + P_0)}$$

The effects of an additional real pole can be given as follows:

Time Responses As A Function of Real-Pole Location

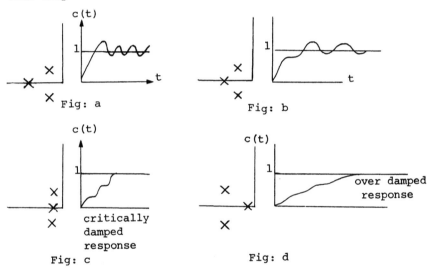

Fig: a

Fig: b

Fig: c — critically damped response

Fig: d — over damped response

A) Peak overshoot M_p is reduced.

B) The setting time t_s is increased because $A_0 e^{P_0 t}$ is the transient term due to P_0 and A_0 is negative.

C) $|A_0|$ depends on the relative location of P_0 with respect to dominant pole of the distance between P_0 and complex poles is large, then A_0 is small.

Effects of additional real pole and zero:

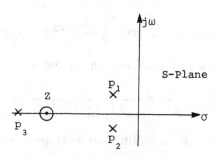

Fig. Pole-zero plot.

A) The complete time response to a unit-step function has the same form as in the case of an additional real pole.

B) Sign of A_o depends on the relative locations of the real pole and the real zero. A_o is negative if the zero is to the left of p_o and is positive if the zero is to the right of p_o. A_o is proportional to the distance from p_o to z.

C) If the zero is close to the pole, A_o is small, then the contribution of this transient term is small.

8.2 POLE-ZERO DIAGRAM AND FREQUENCY AND TIME RESPONSE

CHARACTERISTICS

A) Fig. a:
    a) Frequency response curve has a single peak M_m and $1.0 < M < M_m$ in the frequency range $0 < \omega < 1.0$.
    b) Time response: The first maxima of c(t) due to the oscillatory term is greater than $c(t)_{ss}$ and the c(t) response after this maxima oscillates around the value $c(t)_{ss}$.

Relation: An illustration.

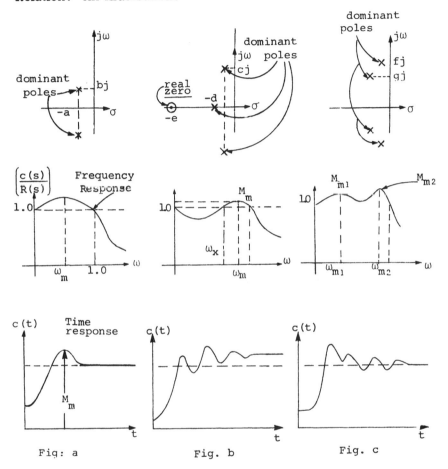

Fig:-Comparison of frequency and time responses

B) Fig. b: Time response: The first maximum of c(t) due to oscillatory term is less than $c(t)_{ss}$.

C) Fig. c: Time response: The first maxima of c(t) in the oscillation is greater than $c(t)_{ss}$, and the oscillatory portion of c(t) does not oscillate about a value of $c(t)_{ss}$.

The system's time-response can be predicted to a

greater extend from the shape of the frequency-response plot.

Table 8.1

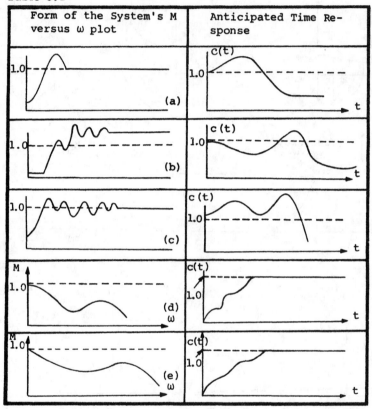

Correlation between Frequency and Time Response

8.3 FACTORING OF POLYNOMIALS USING ROOT-LOCUS

Partition method: The application of the root-locus method for factoring polynomials is called the partition method.

An example:

$$s^3 + es^2 + fs + g = 0$$

where e, f and g are constants.

A) This can be partitioned at s^3, s^2 and s.

B) Partition at s^3:

$$s^3 = -(es^2 + fs + g)$$

$$-1 = \frac{e(s^2 + \frac{f}{e}s + \frac{g}{e})}{s^3} = I_1(s)$$

$$= \frac{e(s + \beta)(s + \gamma)}{s^3}$$

C) Partition at s^2:

$$s^3 + es^2 = -(fs + g)$$

$$-1 = \frac{f(s + \frac{g}{f})}{s^2(s + e)} = I_2(s)$$

D) $s^3 + es^2 + fs = -g$

$$-1 = \frac{g}{s(s^2 + es + f)} = \frac{g}{s(s + \alpha)(s + \sigma)} = I_3(s)$$

E) Each resulting equation after partitioning has a form $I(s) = -1$. It should satisfy the angle and magnitude conditions. Since the resulting equation looks like $G(s)H(s) = -1$, the root-locus method can be utilized to determine the roots of a polynomial.

F) For a third-degree polynomial, partition at s^2 is preferred.

CHAPTER 9

SYSTEM ANALYSIS IN THE FREQUENCY-DOMAIN: BODE AND POLAR PLOTS

9.1 THE FREQUENCY RESPONSE AND THE TIME RESPONSE: RELATIONSHIP

If the frequency response of a system is known, its time-response can be known using the fourier integral.

FOURIER INTEGRAL

Definition

If $f(t)$ is any time-variant input signal, its Fourier integral is given by

$$\boxed{\overline{F}(j\omega) = \int_{-\infty}^{\infty} f(t)e^{-j\omega t}dt.}$$

For a control-system shown below, the frequency response is given by taking Fourier integral of C(t) where

$$\overline{C}(j\omega) = \frac{\overline{G}(j\omega)}{1 + \overline{G}(j\omega)\overline{H}(j\omega)} \overline{R}(j\omega)$$

where $R(j\omega)$ is the Fourier integral of the input signal (t).

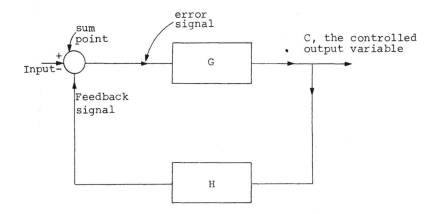

So now taking the inverse Fourier integral of $\overline{C}(j\omega)$ results in

$$c(t) = \text{Time-response of the system} = \int_{-\infty}^{\infty} \frac{\overline{C}(j\omega)}{2\pi} e^{j\omega t} d\omega.$$

9.2 FREQUENCY RESPONSE PLOTS

For a non-unity feedback control system, the control ratio is given by

$$\frac{\overline{C}(j\omega)}{\overline{R}(j\omega)} = \frac{\overline{G}(j\omega)}{1 + \overline{G}(j\omega)\overline{H}(j\omega)}.$$

For each value of $\omega = \omega_1$ these are two important quantities which are frequently used in the frequency response plot; these are

A) $\left|\dfrac{\overline{C}(j\omega_1)}{\overline{R}(j\omega_1)}\right|$ ... at $\omega = \omega_1$.

B) α = The angle between $\overline{C}(j\omega)$ and $\overline{R}(j\omega)$, where $\overline{C}(j\omega)$ and $\overline{R}(j\omega)$ are phasors.

SYSTEM CHARACTERISTICS

A) Ideal characteristics: $\alpha = 0$
   $R = C$ i.e. input = output.

   Note: α is the angle between phasors R and C.

B) Actual characteristics:

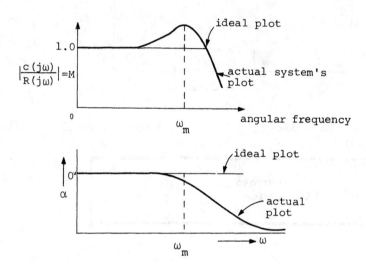

Note: The actual plot is different from the ideal plot due to presence of energy-storage devices and energy dissipation.

C) The bandwidth: It is the frequency range from 0 to ω_{band}, where

$$\left| \frac{C(j\omega)}{R(j\omega)} \right| = 0.707 \text{ times the value at } \omega = 0.$$

For $\omega = \omega_{band}$,

$$\left| \frac{C(j\omega)}{R(j\omega)} \right| = 0.707 \left| \frac{C(o)}{R(o)} \right|.$$

D) $M \to 1$ and $\alpha \to 0°$ as ω approaches 0.

E) As $\omega \to \infty$ (or very large angular frequencies), $M \to 0$ and $\alpha \to -k\pi/2$ radians where -k is equal to the degree of M.

9.3 POLAR PLOT

A) This method is used to represent the open-loop steady-state sinusoidal response. Using this polar plot, the stability and frequency response of the closed-loop system can be estimated.

B) The forward transfer function $= \overline{G}(j\omega) = \dfrac{\overline{C}(j\omega)}{\overline{E}(j\omega)}$.

   Note that the bar represents the phasor.

   For $\omega = \omega_1$, the phasor $G(j\omega)$ has the magnitude

   $$\left| \dfrac{C(j\omega_1)}{E(j\omega_1)} \right|$$

   and the angle $\theta(\omega_1)_j$. In this way many phasors are plotted for different values of the angular frequency and then a smooth curve is drawn passing through the tips of these phasors.

C) A sample polar plot:

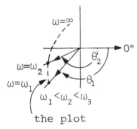

the plot

Note: The length of the arrow is $\left| \dfrac{\overline{C}(j\omega_1)}{\overline{E}(j\omega_1)} \right|$ for $\omega = \omega_1$.

9.4 LOGARITHMIC PLOTS

DEFINITIONS OF VARIOUS TERMS

Decibel: The log (always to the base 10) of the magnitude of a transfer function $\overline{G}(j\omega)$ is expressed in decibels as follows:

$$20 \log |\overline{G}(j\omega)| \quad (db)$$

$$\ln |\overline{G}(j\omega)| \, e^{j\phi(\omega)} = \ln |\overline{G}(j\omega)| + j\phi(\omega)$$

$$\log |\overline{G}(j\omega)| \, e^{j\phi(\omega)} = \log |\overline{G}(j\omega)| + j 0.434 \phi(\omega)$$

(since $\log_{10} e = 0.434.$).

OCTAVE AND DECADE

These are the units used to express the frequency bands or the frequency ratios.

An octave is not a fixed frequency band but it can be defined as a band of frequencies with lower limit defined as f' and upper limit defined as f" such that f" = 2f'. Mathematically it is

$$\boxed{\frac{\log(f''/f')}{\log 2} = 3.32 \log \frac{f''}{f'} \text{ octaves}}$$

(i.e. any arbitrarily frequency can be substituted in the above formula to determine the length of the frequency band in octaves).

Log $\left(\frac{f''}{f'}\right)$ in decades, gives the number of decades from f' to f".

So $\boxed{1 \text{ octave} = 3.32 \text{ decade.}}$

Decibels of some common numbers:

Number	Decibels
0.1	-20
0.5	-6
1.0	0
2.0	6
10.0	20
100.0	40

The following are deductions drawn from the characteristics of logarithms:

A) When a number is doubled, its decimal value increases by 6 db.

B) When a number increases by a factor of 10, its decibal value increases by 20.

9.5 LOG-MAGNITUDE AND PHASE DIAGRAM: BASIC APPROACH

The transfer function is usually written in the following generalized form:

$$\overline{G}(j\omega) = \frac{K(1+j\omega T')(1+j\omega T'')^n}{(\omega j)^n \left[1+j\omega T_1)(1+j\frac{2\zeta}{\omega_n}\omega + \frac{-\omega^2}{\omega_n^2}\right]} = KG(j\omega)$$

where k represents the gain of the system.

The basic approach in drawing curves of log magnitude and angle versus log frequency is the following: curves are drawn for each factor and a complete curve is obtained with the help of asymptotic approximations.

Curves for the basic factors found in the numerator and denominator of the transfer function.

A) Constants:

The curve of $Lmk = 20\log k(db)$ is a horizontal straight line.

The angle is 0 if k is positive; it is 180° if the value of k is negative.

This constant has the effect of raising or lowering the log-magnitude curve of the transfer function along the vertical axis.

B) $j\omega$:

The curve of $Lm \frac{1}{j\omega} = 20\log\left|\frac{1}{j\omega}\right| = -20\log\omega$ is a straight line with a slope of $-20db/decade$ or $-6db/octave$.

The value of the angle is (always) $-90°$.

When $j\omega$ appears in the numerator, the curve is a straight line with a slope of +6db/octave or +20db/decade.

Note: a) Both curves pass through the 0 db point at $\omega = 1$.

b) The curves for $j\omega$ and $1/j\omega$ differs very little; there is a change in the sign of the slope of the log-magnitude curve and a change in the sign of the angle of the phase curve.

For $(j\omega)^m$ term, the curve is a straight line of slope ±6m (depending on whether it is in the numerator or the denominator) and angle is $\pm n\,90°$.

C) $1 + j\omega T$:

$$Lm(1 + j\omega T) = 20\log |1 + j\omega T|$$
$$= 20\log \sqrt{1 + \omega^2 T^2}.$$

Since magnitude of $1 + j\omega T$ is $1 + \omega^2 T^2$.

Plot of $1 + j\omega T$:

A) When the frequency of the operation is very large, i.e. $\omega \gg 1/T$,

$$Lm(1 + j\omega T) = 20\log \omega T$$

where f is the frequency.

B) At very small values of frequencies, i.e. $\omega \ll 1/T$,

$$Lm(1 + j\omega T) = 0$$

because $\log 1 = 0$

Plot of $1/(1 + j\omega T)$;

The plot for low frequencies is a 0 db horizontal line and for very high frequencies the plot is a straight line with a slope of -6db per octave. This plot is shown below.

ASYMPTOTES For the log-magnitude plot shown below, i.e. $Lm[1/1+j\omega T]$, there are two asymptotes.

A) One is of slope = 0 below $\omega = 2\pi f = \frac{1}{T}$.

B) The other of slope -6db/octave above $\omega = \frac{1}{T}$.

Corner frequency: It is the point of intersection of the asymptotes of the log magnitude plot.

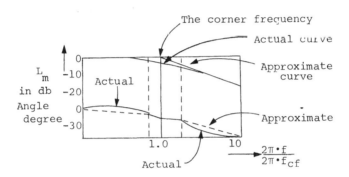

For $\frac{1}{1 + j\omega T}$ function the angular corner frequency is

$$\omega_g = \frac{1}{T}$$

as shown in the figure.

C) $\omega = \omega_n$ is called the corner frequency; it is the point where the asymptotes intersect. At this frequency resonance occurs. ω_n is also called the undamped natural frequency.

D) The phase curve is also a function of ζ. The following are noted:
   a) the angle is 0° at zero frequency;
   b) the angle is -90° at the corner frequency;
   c) the angle is -180° at infinite frequency.

9.6 RELATION BETWEEN SYSTEM TYPE, GAIN AND LOG-MAGNITUDE CURVES

TYPE 0 SYSTEM

A type 0 system is characterized by the transfer function

Log-magnitude plot of the transfer function

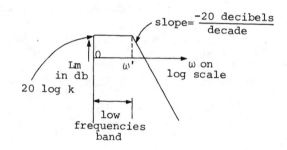

$$G = \frac{K}{1 + ST}$$

i.e. $\quad G(j\omega) = \dfrac{K}{1 + j\omega T} \quad$ since $s = j\omega$.

Facts:

A) For $2\pi f = \omega < \dfrac{1}{T}$, $\text{Lm}G(j\omega) = 20\log k = $ constant.

B) Corner frequency $= \omega^1 = \dfrac{1}{T}$.

C) K = static error coefficient for the step input

TYPE 1 SYSTEM

This system is characterized by $G = \dfrac{k}{s(1 + sT)}$, i.e.

$$G(j\omega) = \frac{k}{j\omega(1 + j\omega T)}.$$

Log-magnitude plot:

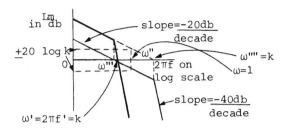

Fig. —curve for k>corner frequency
    —curve for k<corner frequency.

Facts:

A) For the low frequencies slope = -20 db/decade.

B) The gain k = static error coefficient for the ramp input.

Type 2 system:

This type of system is characterized by

$$G(s) = \frac{k}{s^2(1+sT)}.$$

The log-magnitude plot of $G(j\omega)$:

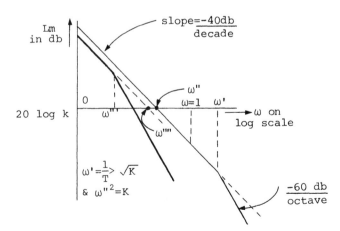

Fig: — for $\omega''' = \frac{1}{T} < \sqrt{k}$ and $\omega''''^2 = k$
     — for $\omega' = \frac{1}{T} > \sqrt{k}$ and $\omega''^2 = k$

A) The gain k = static error coefficient for the parabolic input.

B) For low frequencies, the slope = -40 db/decade.

9.7 DIRECT POLAR PLOTS

DIRECT POLAR PLOTS: ILLUSTRATIVE EXAMPLES

Lag compensator:

A lag compensator is a simple series R-L circuit.

Its transfer function is:

$$G = \frac{R}{R + SL} \quad \text{i.e.} \quad G(j\omega) = \frac{R}{R + j\omega L}$$

$$G(j\omega) = \frac{1}{1 + j\omega \frac{L}{R}} = \frac{1}{1 + j\omega T}$$

where $T = \frac{L}{R}$ = time constant of the circuit.

Polar plot:

$$G(j\omega) = \frac{1}{1 + j\omega T}$$

For $\omega = 0$ radian,

$$G(j\omega) = \left|\frac{1}{1 + 0}\right| \underline{/\text{-tan}^{-1}T\omega}$$

$$= 1 \underline{/0°} \quad \text{because } \tan^{-1} 0 = 0°$$

For $\omega = \infty$,

$$G(j\omega) = 0 \underline{/\text{-}90°}$$

The plot is shown below:

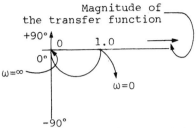

Note: Polar plot is a semicircle of diameter 1.

Lag-lead network: This is an R-C network.

$$G(s) = \frac{V_o(s)}{V_i(s)} = \frac{1 + (T_1 + T_2)s + T_1 T_2 s^2}{1 + (T_1 + T_2 + T_{12})s + T_1 T_2 s^2}$$

where $T_1 = R_2 C_2$, $T_2 = R_1 C_1$ and $T_{12} = R_2 C_1$.

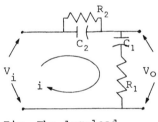

Fig. The lag-lead network.

The circuit acts as a lag network in the lower frequency range $0 - \omega_1$ and as a lead network when ω is beyond ω_1.

The polar plot:

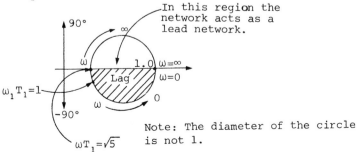

Note: The diameter of the circle is not 1.

9.8 INVERSE POLAR PLOTS

It is a plot between $G^{-1}(j\omega) = \dfrac{\overline{E}(j\omega)}{\overline{C}(j\omega)}$ and the angular frequency. The direct polar plots have some drawbacks when used for feedback systems.

Polar plots for different systems:

For a lag network:

The transfer function for the lag network at $\omega = 0$, $\overline{G}(j\omega) = 1$.

$$G(j\omega) = \dfrac{1}{1 + j\omega T}$$

hence,

$$\overline{G}^{-1}(j\omega) = 1 + j\omega T.$$

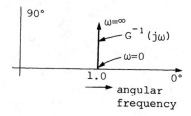

For a lead network:

$$G^{-1}(j\omega) = 1 + \dfrac{1}{j\omega T} \qquad \text{at } \omega = 0,\ G^{-1}(j\omega) = 1.$$

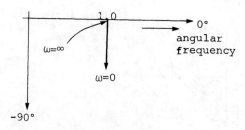

Type 0 and type 1 feedback systems:

$$G^{-1}(j\omega) = \frac{(1 + j\omega T_1)(1 + j\omega T_2)}{k} \text{ for type 0 system.}$$

$$= \frac{j\omega(1 + j\omega T')(1 + j\omega T'')(1 + j\omega T''')}{k}$$

for type 1 system.

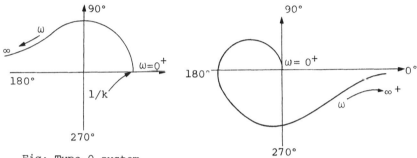

Fig: Type 0 system.

Fig. Type 1 system.

Characteristics:

A) When $\omega = \omega_1$, $\omega_1^2 T_1 T_2 = \omega_1^2 R_1 C_1 R_2 C_2 = 1$ so by proper choice of R_1, C_1, R_2 and C_2 this network can be made to act like a lead or a lag network.

B) Minimum value of the transfer function =

$$\frac{T_1 + T_2}{T_1 + T_2 + T_{12}} \quad \text{at } \omega = \omega_1 T_1.$$

From the polar plot, the following can be seen:

a) for frequencies below $\omega_1 T_1 = \sqrt{5.0}$, the circuit acts like an RL circuit, i.e. it is a lag compensator.

b) for frequencies greater than $\omega_1 T_1 = \sqrt{5.0}$, the circuit acts like an RC circuit. In this case it acts as a lead compensator.

Please note that this value is needed for designing the control system and for eliminating certain noisy frequencies.

C) Polar plots for different types of systems:
Table 9.1

Type of System	Forward transfer function of the feedback system.	Polar Plot	Comments/characteristics
0	$G(j\omega) = \dfrac{C(j\omega)}{E(j\omega)}$ $= \dfrac{k}{(1+j\omega T_1)(1+j\omega T_2)}$	(polar plot)	i) The exact shape of the plot is determined by the time constants T_1 & T_2. ii) If the time constants T_1 & T_2 are nearly equal, the curve gets pulled more into the 3rd quadrant.
	$G(j\omega) = \dfrac{k(1+j\omega T')^2}{(1+j\omega T'')\times(1+j\omega T''')\times(1+j\omega T'''')^2}$	(polar plot)	i) T'' & $T''' > T'$ and $T' > T''''$
1	$G(j\omega) = \dfrac{k}{j\omega(1+j\omega T_1)\times(1+j\omega T_2)\times(1+j\omega T_3)}$	(polar plot)	i) $R = \lim_{\omega \to 0} \text{Re}[G(j\omega)]$ $= -k(T_1+T_2+T_3)\ldots$ The true asymptote. ii) $\omega_y = \dfrac{1}{[T_1 T_2+T_2 T_3+T_3 T_1]^{\frac{1}{2}}} \ldots$ $\ldots$ can be obtained by the equation $I_w(G(j\omega)) = 0$ iii) R is always a direct function of the ramp error coefficient.
2	$G(j\omega) = \dfrac{k}{(j\omega)^2 \times(1+j\omega T_1)\times(1+j\omega T_2)}$	(polar plot)	i) This polar plot represents an unstable feedback constant system.

D) $\left[1 + \dfrac{2\zeta}{\omega_n} j2\pi f + \dfrac{-(2\pi f)^2}{(2\pi f_n)^2}\right]^{-1}$

There are many curves for this factor depending on the value of ζ.

a) For $\zeta > 1$:

the function can be factored into two real factors. The log-magnitude curve is plotted for the individual factors and these curves are then combined..

b) For $\zeta < 1$:

In this case no factorizing is necessary.

$$Lm[function]^{-1} = -20\log \sqrt{\left(\frac{2\zeta\omega}{\omega_n}\right)^2 + \left(1 - \frac{\omega^2}{\omega_n^2}\right)^2}$$

$$angle = -\tan^{-1}\left[\frac{2\zeta\omega/\omega_n}{1 - \omega^2/\omega_n^2}\right]$$

Plot

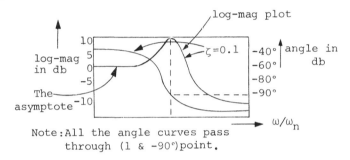

Note: All the angle curves pass through (1 & -90°) point.

Note: All the angle curves pass through (1 and -90°) point.

FACTS

A) For very small values of ω, the asymptote is log-magnitude = 0, line as shown in the figure.

B) For higher values of ω, the high frequency asymptote has a slope of -40db per decade. The log-magnitude is approximately

$$-20\log \frac{\omega^2}{\omega_n^2} = -40\log \frac{\omega}{\omega_n}.$$

HOW TO PLOT AN INVERSE POLAR PLOT

A) First of all the system type, i.e. whether the system is type 0 or type 1, etc., is determined from the forward transfer function G(s).

B) Then $\lim_{\omega \to 0} G^{-1}(j\omega)$ is determined after substituting $s = j\omega$ in the transfer function and evaluating the value of

$G^{-1}(j\omega)$ by putting $\omega = 0$. This zero frequency point is always approached in the clockwise direction.

C) $\lim_{\omega \to \infty} \overline{G}'(j\omega) = \infty \underline{/(a+b-c)90°}$ is determined. This gives the high-frequency end of the polar plot. In this expression, b is the number of zeros, ω' is the number of poles and a is given as follows:

$$G^{-1}(s) = \frac{s^n(1 + T's)(1 + T''s)\ldots(1 + T_s^b)}{K(1 + T_1 s)\ldots(1 + T_c s)}$$

This is the most general form of $\overline{G}'(s)$.

D) Points of intersection:

$I_m[\overline{G}'(j\omega)] = 0$ this gives point on the negative real axis

Exact shape of the plot near negative real axis is very important and this particular point plays an important role in the system behavior.

$Re[G^{-1}(j\omega)] = 0$ on the imaginary axis.

E) If there are no time constants in the denominator, the curve is a smooth curve and the angle increases continuously as the angular frequency increases from 0 to ∞.

The presence of time constants results in dents or up and down (i.e. the angle will not vary continuously, i.e. it will not increase or decrease continuously) in the plot.

9.9 DEAD TIME

Some elements produce no output or may produce an output after a considerable time lag in response to the input signal to the control system. These are known as transport-lag elements; an anologous example in the electrical system is the D-type flip-flop.

$$V_{out}(t) = V_{in}(t - \gamma)u(t - \gamma)$$

where γ is called the dead time.

γ is represented by non-linear characteristics called dead-time characteristics given by the following curves.

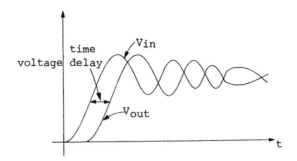

$$\bar{G}(j\omega) = e^{-j\omega\gamma} = 1\underline{/-\omega\gamma}$$

is the transfer function of a dead time element.

Polar plot:

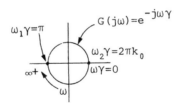

$k_0 = 1, 2, 3\ldots$ as ω increases.

CHAPTER 10

NYQUIST STABILITY CRITERION

10.1 DETERMINING AND ENHANCING THE SYSTEM'S STABILITY

Necessary condition for system's stability in terms of poles and zeros:

A) No zeros of T(s) should be present on the imaginary axis or in the right half of the s-plane as shown below. Note that the poles of T(s) are the poles of the open-loop transfer function.

B) There must be no zeros on the jω axis; zeros present on the jω axis result in a quasistable system and should there be any slight variation in the gain k, the system should become unstable.

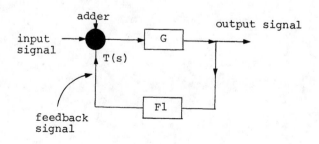

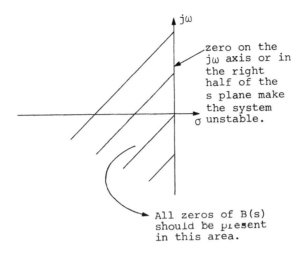

Note the following about Nyquist plots:

A) Unlike the polar plots dealt with in the preceding chapter, Nyquist's plots are drawn for a range of frequencies from $-\infty$ to $+\infty$.

B) Nyquist's plots are plots of the open-loop transfer function $G(s)H(s)$.

C) The curve for negative frequencies is symmetrical to the curve for positive frequencies.

Nyquist's Criterion: A generalized statement

A) $$N = \frac{\text{Phase change of } 1 + G(s)H(s)}{2\pi}$$

$= P_{R.\bar{H}.} \; Z_{R.H.}$ Note that $T(s) = 1 + G(s)H(s)$

where,

$Z_{R.H.}$ = Number of zeros in the right half of the s-plane
$P_{R.H.}$ = Number of poles in the right half of the s-plane
N = The number of revolutions of $1 + G(s)H(s)$ about the origin.

Note: N may be positive, negative, or zero. Counterclockwise revolutions around T(s) are positive and clockwise revolutions are negative.

These are the conventions used in applying this criterion.

B) A pictorial and a mathematical approach: concept of revolutions

Suppose that $T(s) = 1 + G(s)H(s)$ is written in the following generalized form:

$$1 + G(j\omega)H(j\omega) = \frac{(s - z')(s - z'')}{(s - p')(s - p'')}$$

zeros and z' and p' are the poles of the characteristic equation.

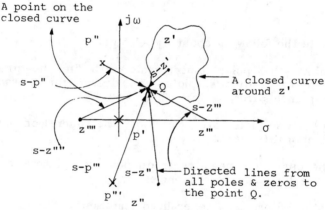

Note: The length of the directed lines are given by s-z'...& s-p'...

Fig. Pole-zero diagram for B(s)

Note that if the point Q rotates clockwise,

A) S-Z phasor turns through complete 360°.

B) Rest of the directed lines rotate through a net angle of 0°.

CONCLUSION

All the poles and zeroes which are outside the closed path, which itself lies in the right half of the s-plane, give an individual contribution to the net angular rotation from 0° to T(s) as an arbitrary point moves around the closed path.

$$\begin{bmatrix} \text{The total number of} \\ \text{rotations of} \\ 1 + G(s)H(s) \end{bmatrix} = \begin{bmatrix} \text{Number of} \\ \text{poles en-} \\ \text{closed, P} \end{bmatrix} - \begin{bmatrix} \text{Number of} \\ \text{zeros en-} \\ \text{closed, Z} \end{bmatrix}$$

Note the rotations result due to clockwise movement of an arbitrary point on the closed curve which is drawn so as to enclose p poles and z zeros.

Stable system and the revolutions:

Table 10.1

Type of System	The Transfer Function $G(s)H(s)$ Of The System	Direct Polar Plot	Comments/Characteristics Important Points
Type 0	$\dfrac{K}{S(1+T's)(1+T''s)}$ i) No zeros ii) Two poles which are located on the left hand side		i) Nyquist criterion: $N=0$... since number of poles in the right half of S-plane ii) $P_{R.H.}=0$, hence $N=0=R_{R.H.}-Z_{R.H.}$ so the system is stable for all values of K.
Type 1	$\dfrac{K}{S(1+T's)(1+T''s)}$		i) If the gain K is increased beyond a certain value, then the system becomes unstable. ii) $N=-2=$ the rotations of $1=G(S)H(S)$ in the direction of increasing value of 0.
Type 2	$\dfrac{K(1+T_1S)}{S^2(1+T_2s)(1+T_3s)(1+T_4s)}$ $T_1 > T_2$, T_3 and T_4 $P_{R.H.}=0$		i) $P_{R.H.}=0$, $N=0$ so $Z_{R.H.}=0$ hence the system is stable ii) If gain is increased, then the system can become unstable (because $-1+j0$ point is on the R.H. side of the intersection part of the curve and the -ve real axis).

A) For a stable automatic control system, the total number of revolutions of $1 + G(s)H(s)$ around the origin of the σ, $j\omega$ axis must be clockwise and

Total number of revolutions = Number of poles that are located in the right half of s-plane.

F-110

B) When there are no poles on the right-hand side of $j\omega$ axis, the necessary condition is

N_1 = The total number of revolutions about $(-1 + j0)$ point

= 0.

10.2 INVERSE POLAR PLOTS: APPLICATION OF NYQUIST'S CRITERION

A MATHEMATICAL APPROACH

A) Clockwise revolutions of

$$T'(s) = \frac{1}{G(s)H(s)} + 1 = \text{Number of zeros on the right half of s-plane} = Z$$

B) Total number of anticlockwise revolutions of

$$\frac{1}{G(s)H(s)} + 1 \text{ due to poles} = \text{Number of poles located in the right half of s-plane} = P$$

C) N' = Total number of revolutions of

$$\frac{1}{G(s)H(s)} + 1 \text{ about the origin of the s-plane} = P - Z$$

CRITERION FOR THE SYSTEM'S STABILITY

A) The net number of revolutions of $\frac{1}{G(s)H(s)} + 1$ about the origin point 0, must be counterclockwise and,

B) the number of revolutions should be equal to the number of poles that are located in the right half of the s-plane.

Note: The clockwise revolutions are treated as negative and the counterclockwise revolutions are treated as positive.

C) When there are no poles located in the right half of the s-plane, the stability criterion for the system is

$$\left[\begin{array}{c}\text{Total number of revolutions}\\ \text{about } -1 + \text{jo}\end{array}\right] = 0.$$

D) If the system's G(s)H(s) has some zeros in the right side, then the number of poles which may be present on the right half plane are found by using Routh's criterion after which point the criterion is applied.

CONCEPTS OF REVOLUTIONS OF T'(s)

Suppose that there are some poles and zeros of $T'(s) = 1 + \left[\dfrac{1}{G(s)H(s)}\right]$ which lie in the right half of the s-plane.

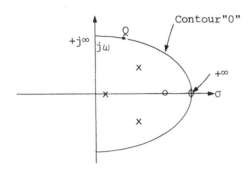

A closed contour O is drawn which can enclose all these poles and zeros. Such a contour should be semicircle of ∞ radius as shown.

Consider a point Q on this contour. Then, one revolution of T'(s) is defined as follows:

A) The point Q should go from +0 to +∞ on the imaginary jω axis,

B) then from +∞ to -∞ on the semicircle of ∞ radius,

C) then from -∞ to -0 on -jω axis.

10.3 PHASE MARGIN AND GAIN MARGIN: DEFINITIONS

The gain crossover:

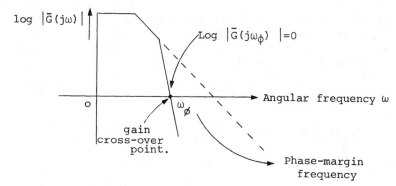

It is the point on $\log|\bar{G}(j\omega)|$ versus ω curve where $\log|\bar{G}(j\omega)| = 0$ or where $|\bar{G}(j\omega)| = 1$.

Phase margin frequency: ($\omega\phi$)
It is the angular frequency where $\log|\bar{G}(j\omega)| = 0$ or $|\bar{G}(j\omega)| = 1$.

PHASE MARGIN

Phase margin = 180° + (negative of the angle of $\bar{G}(j\omega)$ at the gain crossover frequency).

Phase crossover

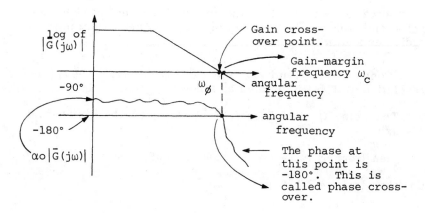

GAIN MARGIN

It is the extra gain which makes the system just unstable. It is given as:

$$\text{log magnitude of gain margin} = -\text{Lm}\overline{G}(j\omega_c).$$

10.4 SYSTEM STABILITY

A) For a stable system, the phase margin must be positive. A negative phase margin means an unstable system.

B) Gain margin expressed db should be a positive number for a stable system.

Negative gain margin in db implies an unstable system.

C) Phase margin gives the better estimate of damping ratio ζ than the gain margin.

Log magnitude-angle diagram, stability of the system log-magnitude-angle diagram

A) This is a plot between angle α of the torque for function $\overline{G}(j\omega)$ and its magnitude $|\overline{G}(j\omega)|$.

B) This is plotted from the following two plots:
   a) $\log|\overline{G}(j\omega)|$ versus ω and,
   b) angle α of $\overline{G}(j\omega)$ versus ω.

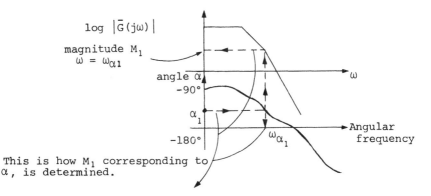

This is how M_1 corresponding to α_1 is determined.

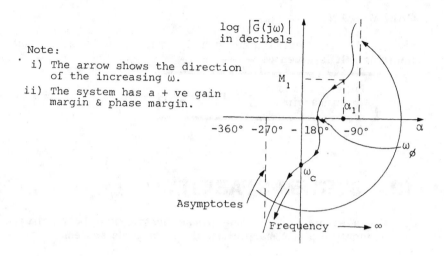

Note:
i) The arrow shows the direction of the increasing ω.
ii) The system has a + ve gain margin & phase margin.

10.5 STABILITY

Effects of changing the gain and the phase margin

Table 10.2

Value of gain k	Value of the phase margin	Value of the gain margin	Effect on system
—	positive	positive	The system is stable
changing	—	—	The log-magnitude curve is elevated or lowered without changing angle characteristics.
increased	—	—	The curve is raised and both gains and phase margins are decreased so the system becomes unstable. High values of gain k are necessary to reduce steady-state errors.
decreased	—	—	The system stability is increased because both the gain and phase margins are increased.

F-115

A simple rule to determine the stability.

A curve is drawn for $\bar{G}(j\omega)$ (i.e. the log magnitude-angle curve) for $\omega > 0$ and $\omega < \infty$.

The curve is traced in the direction of increasing value of the angular frequency.

The system is stable if the (0db, -180°) point is lying on the right-hand side of the curve.

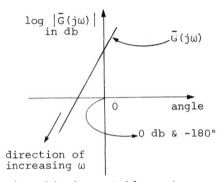

Fig. This is a stable system.

Conditionally stable system: This system is characterized by the curve shown below.

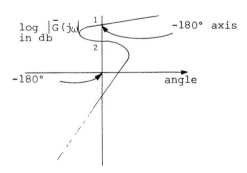

The characteristic is that the curve crosses $\log|\bar{G}(j\omega)|$ db axis at more than one point as shown.

10.6 EFFECT OF ADDING A POLE OR A ZERO: EFFECT ON THE POLAR PLOTS

A) If an additional zero is inforced in the pole-zero pattern of the system, it results into a counterclockwise rotation of the system's direct polar plot. So the system's stability is increased.

B) An extra pole rotates the polar plot clockwise, making the sytem less stable.

CHAPTER 11

PERFORMANCE EVALUATION OF A FEEDBACK CONTROL SYSTEM IN THE FREQUENCY-DOMAIN

11.1 PERFORMANCE EVALUATION USING DIRECT POLAR PLOT

This mathematical analysis of the system's direct polar plot gives us additional information regarding the steady-state error. This information is helpful in the design of compensating networks.

Feedback system: A block diagram.

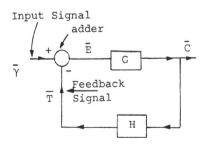

Analysis: Consider for the sake of analysis, H = 1 and let the following be the $\bar{G}(j\omega)\bar{H}(j\omega)$ curve.

A) The vector $\overline{AB}$ as shown in the figure gives

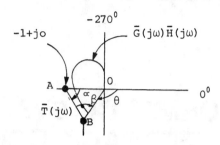

$$\overline{T}(j\omega) = 1 + \overline{G}\,\overline{H}$$

$$\overline{T}(j\omega) = |\underbrace{\overline{T}(j\omega)}_{\text{length AB}}| \, e^{j\alpha(\omega)}$$

B) $\dfrac{\overline{C}(j\omega)}{\overline{R}(j\omega)} = \dfrac{\overline{G}(j\omega)}{1 + \overline{G}(j\omega)}$ The closed-loop transfer function.

The vector $\overline{OB}$ in the figure is $\overline{G}(j\omega)$ because $\overline{H}(j\omega) = 1$. So,

$$\dfrac{\overline{C}(j\omega)}{\overline{R}(j\omega)} = \dfrac{\text{length of } \overline{OB}}{\text{length of } \overline{AB}} = \dfrac{OB}{AB}\, e^{j\beta} \quad \text{because } \theta - \alpha = \beta.$$

Note that both α and θ are negative and angle β is positive according to the convention followed.

C) $\dfrac{\overline{E}}{\overline{R}} = \dfrac{\text{Error signal phasor}}{\text{Input signal phasor}} = \dfrac{1}{AB \underline{/\alpha}}$.

Conclusion: The error between the input signal and the feedback signal is very very small if the length of the vector $\overline{AB}$ is very large. This is further explained as follows:

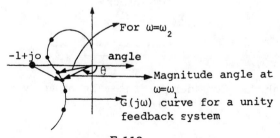

F-119

The frequency of operation ω_0 is chosen in such a way that the length of the vector will result in an error which is in the specified limit. This can be achieved by keeping $\omega =$ constant and by changing ζ, the damping ratio (this point is stressed in the next article).

11.2 RESONANT FREQUENCY AND THE MAXIMUM MAGNITUDE OF C/R OF A SECOND ORDER SYSTEM

SECOND-ORDER SYSTEM

Characteristics

A second order system is characterized by the following transfer functions:

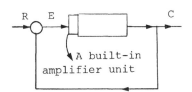

$$\frac{\overline{C}(s)}{\overline{R}(s)} = \frac{\omega_n^2}{S^2 + 2\omega_n \zeta S + \omega_n^2}$$

where

$$\zeta = \frac{D}{2}(KA)^{-\frac{1}{2}}$$

$$\omega_n = \left(\frac{K}{A}\right)^{\frac{1}{2}}$$

and

$$\frac{\overline{C}(s)}{\overline{E}(s)} = \frac{K}{S(As + D)}$$

$$\frac{\overline{C}(j\omega_1)}{\overline{R}(j\omega_1)} = \text{Mag}(\omega_1) e^{j\beta(\omega_1)}.$$

This is applicable for $\omega = \omega_1$.

A) The maximum value of Mag is found from the following equation

$$\frac{d}{d\omega}(\text{Mag})^2 = 0.$$

B) The angular frequency at which $\text{Mag} = \text{Mag}_{max}$ is

$$\omega_m = \begin{pmatrix} \text{undamped natural} \\ \text{frequency} \end{pmatrix} [1-2(\text{damping ratio})^2]^{\frac{1}{2}}$$

$$\text{Mag}_{max} = \frac{1}{2\zeta\sqrt{1-\zeta^2}} \quad \text{and}$$

$$M_p = \frac{\text{Peak overshoot}}{\text{value of the step signal}} = 1 + e^{\frac{-\zeta\pi}{\sqrt{1-\zeta^2}}}$$

Plot: The following is a set of plots which highlights some of the facts regarding the frequency and line response. It is quite clear that for the smaller values of ζ, the values of Mag_{max} and M_p are larger.

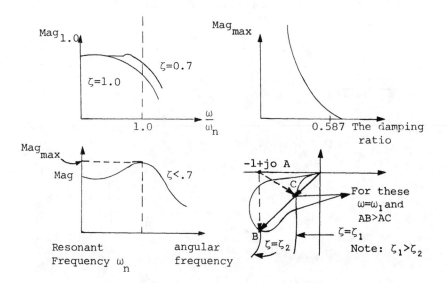

The steady-state or the error at $\omega = \omega_1$ can be altered by changing the damping constant. This is quite clear from the figure.

The time response can be made faster by increasing the resonance frequency which in turn depends on the damping ratio ζ. So, the lower the value of ζ, the higher the value of resonant frequency.

In order to make Mag_{max} higher, the $\overline{G}(j\omega)$ plot for this unity feedback system should be close to $-1 + jo$ point.

A) Type 0 system: If the $\overline{G}(jo) = k$ point on the $\overline{G}(j\omega)\overline{H}(j\omega)$ curve is away from the origin, then for a step input there will be an exact steady-state time response. This exactness depends on how far this point is.

B) Type 1 system: Ramp input steady-state time response. In this case the $\omega \to o$ asymptote should be away from the imaginary axis of the plot.

11.3 PLOTTING MAXIMUM MAGNITUDE AND RESONANT FREQUENCY ON THE COMPLEX PLANE

These curves which represent $\text{Mag}(\omega)$ and $\beta(\omega)$ = constant values of the closed-loop transfer function are plotted on the direct polar plot of the system.

CONSTANT MAGNITUDE CIRCLES

Plotting procedures

A) First of all the $G(j\omega)H(j\omega)$ curve is plotted as shown in the following figure, and two vectors $\overline{OA}$ and $\overline{BA}$ are drawn.

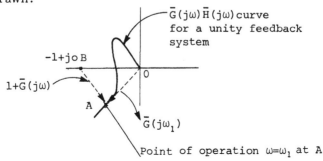

F-122

B) From this plot:

$$\frac{\bar{C}(j\omega_1)}{\bar{R}(j\omega_1)} = \frac{\text{output}}{\text{input}} = \frac{OA}{BA} = \text{Mag}(\omega_1) = \frac{|\bar{G}(j\omega_1)|}{|1 + \bar{G}(j\omega_1)|}.$$

C) Now if the curve $\bar{G}(j\omega)$ is plotted on an x-yj plane, then

$$\bar{G}(j\omega) = x + jy$$

and

$$\text{Mag}(\omega_1) = \frac{|x + jy|}{|1 + x + jy|}$$

so

$$\text{Mag}^2 = x^2 + y^2/(1 - x)^2 + y^2.$$

This equation for Mg represents a circle which is shown as follows:

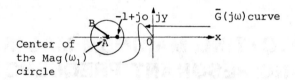

Note: $\quad OA = \dfrac{-\text{Mag}^2}{\text{Mag}^2 - 1}\quad$ and $\quad AB = \left|\dfrac{\text{Mag}}{\text{Mag}^2 - 1}\right|.$

Characteristics of Mag(ω_1) circle:

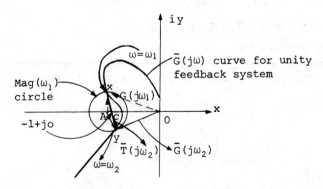

A)

For all points on Mag circle:

F-123

$$\frac{Oy}{Cy} = \frac{Ox}{Cx} = \text{Magnitude of } \overline{G}(j\omega) \text{ for } \omega = \omega_1.$$

B) If at x and y_1 the angular frequencies are ω_1 and ω_2 then

$$\left|\frac{\overline{G}(j\omega_1)}{\overline{B}(j\omega_1)}\right| = \left|\frac{\overline{G}(j\omega_2)}{\overline{B}(j\omega_2)}\right|$$

C)

[Figure: Nyquist plot showing $\overline{G}(j\omega)$ with concentric circles labeled Mag_1, Mag_2, Mag_3, points $\overline{G}(j\omega_1)$, $\overline{G}(j\omega_2)$, $\overline{G}(j\omega_3)$, and $-1+j0$. Note: $Mag_1 > Mag_2 > Mag_3$. Circle Mag_2 is tangent to $\overline{G}(j\omega)$.]

Circle Mag_2 is tangent to $\overline{G}(j\omega)$.

In this case for only the tangent point ω_2

$$\left|\frac{\overline{G}(j\omega_2)}{\overline{B}(j\omega_2)}\right| = Mag_2.$$

Locus of constant Mag curves:

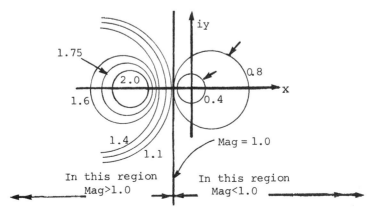

In this region $Mag > 1.0$

In this region $Mag < 1.0$

Constant angle contours: Plotting Procedure

A) $\dfrac{\bar{C}}{\bar{R}} = \text{Mag}(\omega) e^{i\beta(\omega)}$

$= \dfrac{\bar{G}(j\omega)}{\bar{G}(j\omega) + 1}$

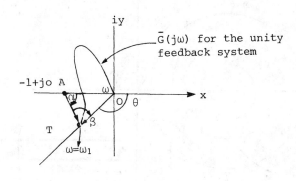

$\beta = \alpha - \theta$

$\overline{OB} = \bar{G}(j\omega_1)$ and $\overline{AB} = 1 + \bar{G}(j\omega_1) = \bar{B}(j\omega_1)$

This is the closed-loop transfer function.

B) Now a constant angle contour is a curve for which the value of the angle β is always constant.

C) $\text{Mag}(\omega) e^{i\beta(\omega)} = \dfrac{x + iy}{(1 + x) + iy}$

$= \dfrac{(x + iy)[(1 + x) - iy]}{(1 + x^2)^2 - y^2}$

Hence

$\tan\beta = \dfrac{y}{(x + 1)x + y^2} = N$

It represents a circle of radius $= \dfrac{1}{2}\sqrt{\dfrac{N^2 + 1}{N^2}}$

and its center is $\left(-\dfrac{1}{2}, \dfrac{1}{2N}\right)$.

Characteristics:

A)

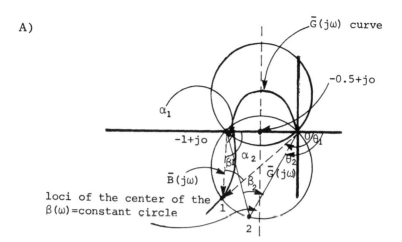

Note: The circle always passes through $(-1+j0)$ and 0.

For points 1 and 2:

$$\boxed{\beta_1 = \beta_2 = [\text{value of the } \beta(\omega) = \text{constant}].}$$

B)

Note: (i) length a = length b if $\beta_1 = -\beta_2$

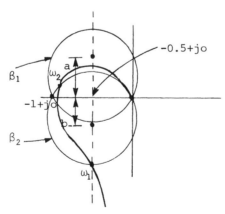

$\beta_0 = \underline{/\bar{A}(j\omega_1)} - \underline{/\bar{B}(j\omega_1)} = 180° + \beta_2$.

This $\beta(\omega)$ at $\omega = \omega_1$

β' = value of $\beta(\omega)$ at $\omega = \omega_2$ $= \beta_2$

F-126

C) Locus of constant β circles

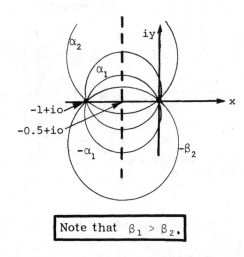

Note that $\beta_1 > \beta_2$.

D) Intersection with constant magnitude curve:

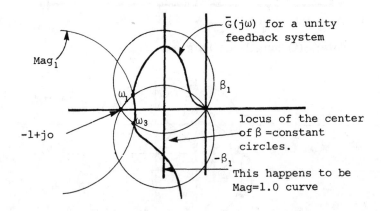

For $\omega = \omega_1$, $\quad \dfrac{\bar{C}(j\omega_1)}{\bar{R}(j\omega_1)} = \text{Mag}_1 \, e^{i(180° + \beta_1)}$.

For $\omega = \omega_3$, $\quad \dfrac{\bar{C}(j\omega_3)}{\bar{R}(j\omega_3)} = \text{Mag}_1 \, e^{i(180° - \beta_1)}$.

Gain adjustments using Mag circles:

F-127

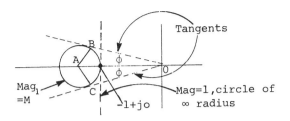

$$\sin \phi = \frac{AB}{AO} = \frac{M/(M^2 - 1)}{M^2/(M^2 - 1)} = \frac{1}{M}.$$

This adjustment procedure is given in the next article.

11.4 MAGNITUDE AND ANGLE CURVES IN THE INVERSE POLAR PLANE

MATHEMATICAL ANALYSIS

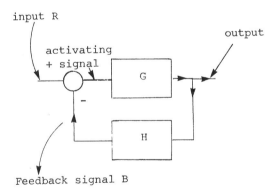

$$\frac{\overline{C}(j\omega)}{\overline{R}(j\omega)} = \overline{G}(j\omega) = |\overline{G}(j\omega)| \; \underline{/\theta(\omega)}$$
$$= \text{Mag}\underline{/\theta(\omega)}$$

$$\frac{\overline{R}(j\omega)}{\overline{C}(j\omega)} = \frac{1}{\text{Mag}} \; \underline{/-\theta(\omega)}$$

$$= \frac{1}{\overline{G}(j\omega)} + \overline{H}(j\omega)$$

PHASOR REPRESENTATION

Unity feedback system

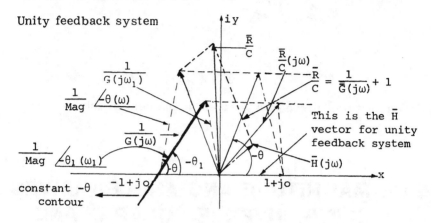

Note: The dot-dash-dot representation is for a nonunity feedback system.

CONCLUSIONS FROM PHASOR DIAGRAM

A) The curves of constant values of 1/Mag is a circle of radius 1/Mag with its center located at $-1 + jo$.

B) The contour for constant $-\theta$ is a radial line passing through $-1 + jo$ point, i.e. the center of the set of 1/Mag circles.

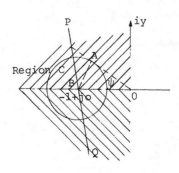

a) $\sin \psi = \dfrac{AB}{OB} = \dfrac{1/\text{Mag}}{1} = \dfrac{1}{\text{Mag}}$ compare this with that for direct

In region C and along the line PQ, the angle is $-\theta_1$ and in the D region, the angle is $-\theta_1 + 180°$.

CHARACTERISTICS

A)

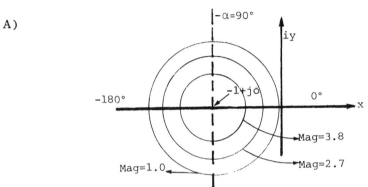

Fig: Inverse polar plot

Note: The Mag = 2.7 → $\dfrac{1}{\text{Mag}} = \dfrac{1}{2.7}$.

B) $\dfrac{1}{G(j\omega)}$ curve and Mag and θ curve:

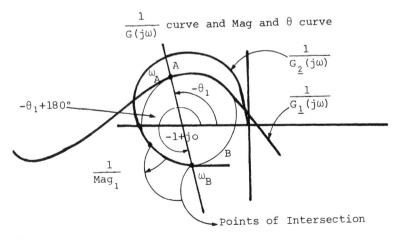

POINTS TO BE NOTED

F-130

A) At points A and B, the magnitude is the same but angles differ by 180°.

B) For large values of Mag_1, the circles are small and do not cross the $1/G_1(j\omega)$ curve.

C) Non-unity feedback case:

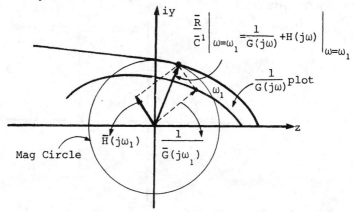

Mag Circle

Note: The center in this case is located at the origin.

11.5 GAIN ADJUSTMENT FOR A DESIRED MAXIMUM MAGNITUDE USING A DIRECT POLAR PLOT

Steps:

1: Express and plot the transfer function $G(j\omega)$ of the given system in the following form:

$$G(\omega) = \frac{K(1 + j\omega T')(1 + j\omega T'') \cdots}{(j\omega)^n (1 + j\omega T_1)(1 + j\omega T_2) \cdots}.$$

2: Then the curve for the following modified form of $G(x)$ is plotted.

$$G(\omega)_{mod} = \frac{(1 + j\omega T')(1 + j\omega T'') \cdots}{(j\omega)^n (1 + j\omega T_1) \cdots} = \frac{G(\omega)}{k}$$

3: Let the circle which is tangent to the $G(j\omega)$ curve be Mag_{max}. This is found out by trial and error.

4: A line is drawn whose slope is

$$\tan\psi = \tan\left[\sin^{-1}\frac{1}{\text{Mag}_1}\right].$$

5: Now a circle is found out by a trial and error method which is tangent to both $G(x)_{mod}$ and the line drawn in step number 4 is as follows.

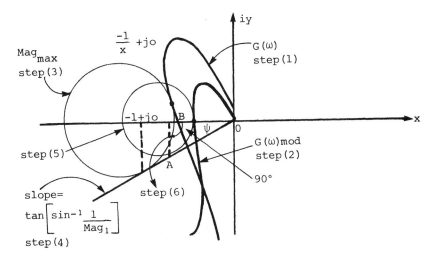

Please note the steps written beside each curve to understand the method.

6: Draw a line AB which is perpendicular to the x-axis.

7: The value of $K' = \frac{1}{OB}$, so the initial gain K has changed by a factor equal to $\frac{K'}{K}$ = length of line segment OB.

8: So the additional gain required to produce the given value of Mag_1 is given by $\frac{K'}{K}$ = length OB.

11.6 NICHOL'S CHART

These are the set of curves representing constant Mag and β on a log-magnitude versus angle plot.

MATHEMATICAL ANALYSIS

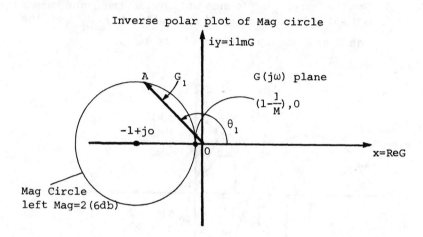

The equation of the circle:

$$(OA)^2 Mag^2 + 2(OA)(Mag)^2 \cos(\angle Aox) - 1 + Mag^2 = 0.$$

$$|\overline{OA}| = -\cos\angle Aox \pm \sqrt{\cos^2(Aox) - \frac{Mag^2 - 1}{Mag^2}}$$

$$|\angle Aox| = \cos^{-1}\left[\frac{1 - Mag^2 - OA^2 Mag^2}{2(OA)(Mag)^2}\right].$$

MODIFICATIONS

A) A log magnification-angle diagram shows the relation between the magnitude of $G(j\omega)$ and angle.

B) An inverse plot depicts $G^{-1}(j\omega)$ of the system.

C) So following modifications are required:

a) $1/|\overline{OA}| = r$ and $-|\angle Aox| = \theta$
b) The magnitude is represented in decibels.

Action Plot

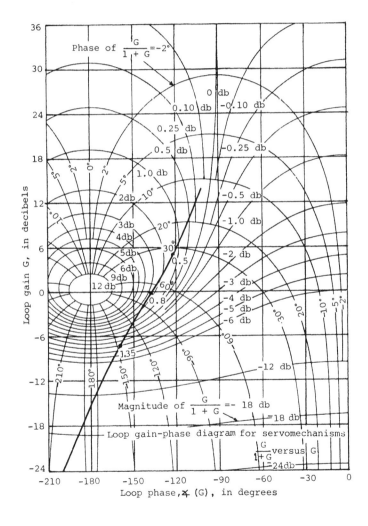

F-134

IMPORTANT POINTS OF NICHOL'S CHART

A) Constant Mag and angle curves repeat for every 360°.

B) There is symmetry at every 180° interval.

C) Mag = 1(0 db) curve is asymptotic to $\theta = -90°$ and $\theta = -270°$ line.

D) When Mag < -6db, the curves are negative because their magnitudes are negative as shown in the figure.

E) Mag > 1.0 curves are the closed curves and are bounded within $-90°$ and $-270°$

Note: Some of these points are shown by the corresponding number.

CHAPTER 12

SYSTEM STABILIZATION: USE OF COMPENSATING NETWORKS AND THE ROOT LOCUS

12.1 FUNCTION OF A COMPENSATING NETWORK

STABILIZED SYSTEM

Definition

The stabilized system must meet the following minimum requirements:

A) It ought to have a satisfactory transient response.

B) A large gain (k) to keep the steady-state error in the specified limit.

Compensating networks: location and function

Selection of location:

A) The selection depends on the type of the system and the modifications that are required.

B) Series cascading: The network is placed at a low-energy point so the dissipation is the lowest. An additional buffer amplifier is required to avoid the loading of the compensating network.

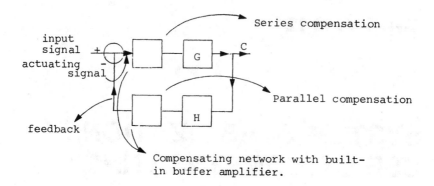

Compensating network with built-in buffer amplifier.

Function:

A) The compensation is done by introducing poles and zeros. Each additional pole increases the number of roots of the system by 1.

B) The compensating network effectively reshapes the root locus of the system.

12.2 TYPES OF COMPENSATIONS

12.2.1 PI (INTEGRAL AND PROPORTIONAL) CONTROL

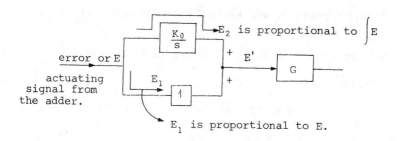

E_2 is proportional to $\int E$

E_1 is proportional to E.

Increase in type of system: The type of system is increased by a suitable network when 1_{ss} is very large, but the transient response is okay. This is done by adjusting the roots which are close to jω axis (i.e. the dominant roots).

PI controller effectively does this. This controller works as follows:

Transfer function:

$$E' = E_1 + E_2 = (1 + \frac{k_0}{s})E$$

$$\boxed{G = \frac{E'}{E} = (1 + \frac{k_0}{s}).}$$

Pole-zero pattern:

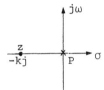

$$G = \frac{s + k_0}{s}.$$

Note: k_0 is selected very small.

Effects:

A) Effect due to the presence of pole P only: It shifts the root locus to the right, so the time response slows down.

B) Zero alone: The distance k_0 should be minimum so as to reduce the increase in response time.

C) The static error coefficient is ∞ because the type of system has increased.

D) $e_1(t)$ increases as long as there is an error signal $e(t)$.

12.2.2 LAG COMPENSATOR

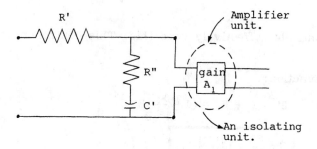

This is nothing more than an integral compensation.

Transfer function: (with amplifier)

$$G = A_1 \cdot \frac{(s + \frac{1}{T'})}{\left(s + \frac{1}{\beta T'}\right)\beta} \quad , \quad \beta > 1.$$

Steady-state accuracy:

A) $\quad G_{or} = \dfrac{k' \prod\limits_{i=1}^{n}(s - z_i)}{\prod\limits_{j=1}^{n}(s - p_j)}$.

The general form of system's original open-loop transfer function.

B) $\quad G' = G_{com} \times G_{or} = \dfrac{A_1 k'(s + \frac{1}{T'})}{\beta[(s + \frac{1}{\beta T'})]} \times \dfrac{\prod\limits_{i=1}^{n}(s - z_i)}{\prod\limits_{j=1}^{n}(s - p_j)}$.

$$\boxed{k'' = \text{Static loop-sensitivity for new root } S_1 = \frac{A_1 k'}{\beta}}$$

$$k_a = \prod_{i=1}^{n}(-z_i) \Big/ \prod_{j=1}^{n}(-p_j) k' \text{ for type } 0$$

$$k_b = \frac{\prod_{i=1}^{m}(-z_i)}{\prod_{j=1}^{n}(-p_j)} \beta k_0.$$

Design Procedure

A) Pole $S_p = \frac{-1}{\beta T'}$ and zero $S_2 = \frac{-1}{T'}$ are placed very close by varying R', R" and C' (i.e. the root locus is slightly affected).

B) Then the angle contribution of the poles and zeros of this compensator is determined at the dominant root of the original closed-loop system.

C) If the angle contribution is < s, the new root locus will be changed slightly and the new pole s, will be slightly displaced from the uncompensated value S_0, the transient response will not change much, i.e.

$$k'' \approx k' \text{ and } k_b \approx \beta k_a.$$

D) The gain for new root s_1 increased by

$$\beta = \frac{\text{compensator zero}}{\text{compensator pole}}.$$

12.2.3 PROPORTIONAL PLUS DERIVATIVE (PD) COMPENSATOR

Block diagram

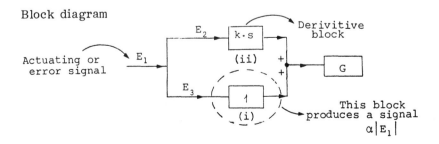

Function:

A) A zero is produced by block (i) and (ii).

B) Transfer function G = (1 + ks).

12.2.4 LEAD COMPENSATION

This is essentially a derivative compensator.

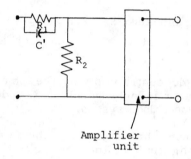

Amplifier unit

Transfer function

$$G = A'\beta \frac{1 + T's}{1 + \beta T's} = A' \frac{s - z_1}{s - p_1}$$

where $z_1 = -1/T'$ and $p_1 = -1/\beta T'$.

Pole-zero pattern:

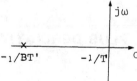

Note:

A) If β is very small then the location of the pole will be further away from the origin and it will have less effect on the root-locus.

B) The location of the zero is adjusted for the desired performance.

Characteristics:

A) The static loop sensitivity is proportional to the ratio

$$\frac{|s + 1/\beta T'|}{|s + 1/T'|}.$$

B) In order to increase the sensitivity, β must be small.

12.2.5 LEAD-LAG COMPENSATION

Advantages

A) A large increase in the gain, and

B) a large increase in the ω_n.

Transfer function:

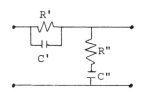

$$G = K_0 \frac{(s + \frac{1}{T'})(s + \frac{1}{T''})}{(s + \frac{1}{\beta T'})(s + \frac{\beta}{T''})}$$

$\beta > 1$ and $T' > T''$.

Relationship:

$$T' = R'C', \quad T'' = R''C''$$

$$\beta T' + \frac{T''}{\beta} = R'C' + R''C'' + R'C''.$$

Pole-zero pattern

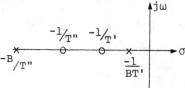

Application:

$$\text{Uncompensated forward transfer function} = G_{un} = \frac{k_0}{(s+5)(s+1)s}.$$

Block-diagram of the system:

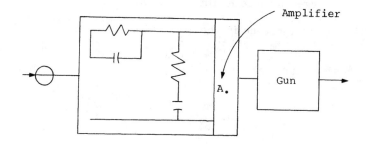

$$\text{New transfer } G_{NEW} = \frac{Ak_0(s+\frac{1}{T'})(s+\frac{1}{T''})}{(s+5)s(s+1)(s+\frac{1}{\beta T'})(s+\frac{\beta}{T''})}.$$

Selection of poles and zeros: Selecting β

The poles and zeros of this compensator are selected as follows:

A) Integral or lag element:
  a) $S_z = \frac{-1}{T'}$ and the pole $S_p = \frac{1}{\beta T'}$ are selected close to each other, with $\beta = 10$.

  b) To improve the gain, p and z are located closed to the origin and to the left.

B) Lead element:

$S_z = \frac{-1}{T''}$ is placed on the pole of the original system, so

$$\beta = 10 \text{ and,}$$

$$\frac{1}{T'} = \frac{1}{10}, \quad T'' = 1$$

$$\beta T' = 100 \text{ and } \frac{T''}{\beta} = 0.1 \text{ and}$$

$$G = \frac{k_0(s + \frac{1}{10})}{s(s + \frac{1}{100})(s + 5)(s + 10)} .$$

Root-locus:

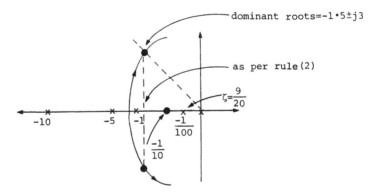

dominant roots = $-1 \cdot 5 \pm j3$

as per rule (2)

$\zeta = \frac{9}{20}$

K_1 = Static loop sensitivity at the dominant roots

$$= \frac{|S| \times |s + \frac{1}{100}| |s + 5| |s + \frac{1}{.1}|}{|s + \frac{1}{10}|}$$

$$= 143$$

ω_n = Undamped frequency $\cong$ 3 rad/sec

Error coefficient for the ramp input $= \dfrac{k_1 (\frac{1}{10})}{(\frac{1}{100})(s)(\frac{1}{0.1})} \cong 28'/\text{sec}.$

F-144

12.2.6 COMPARISON OF COMPENSATORS
Table 12.1

Type of compensator	Effect on		Comments
	Gain k	Undamped natural frequency ω_n	
Lag	The gain increases by a factor = β, so the e_{ss} reduces considerably.	Decrease in ω_n so the settling time of the original system increases slightly	
Lead	k increases slightly so the steady-state error reduces slightly	Large increase in ω_n	An amplifier unit with gain A is required such that A > gain of original system.
Lag-lead	k_n increases considerably.	Large increase in ω_n	

Frequency response:

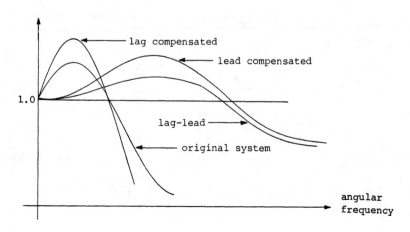

CHAPTER 13

FREQUENCY-RESPONSE PLOTS OF CASCADE COMPENSATED SYSTEMS

13.1 SELECTING A PROPER COMPENSATOR

A basic approach:

Let G_1 = The forward transfer function of the original feedback system.

Let G_2 = The desired transfer function, which means that the function will result in specified stability and e_{ss}.

Then $G = \dfrac{G_2}{G_1}$ is the equation from which the network of the compensator can be determined.

Characteristics of the fundamental compensators: The features of the basic types of compensators are given as follows:

Polar plots:

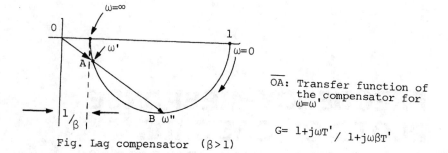

Fig. Lag compensator ($\beta > 1$)

$\overline{OA}$: Transfer function of the compensator for $\omega = \omega'$

$G = 1+j\omega T' / 1+j\omega\beta T'$

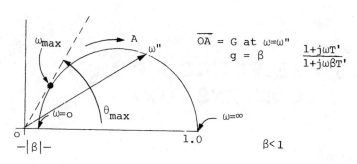

Fig. Lead network

$\overline{OA} = G$ at $\omega = \omega''$
$g = \beta$
$\dfrac{1+j\omega T'}{1+j\omega\beta T'}$

$\beta < 1$

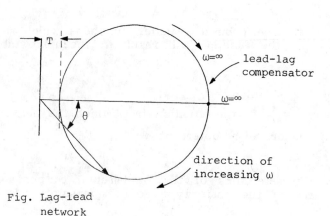

Fig. Lag-lead network

lead-lag compensator

direction of increasing ω

F-147

Lag-lead network:

$$T = \frac{T' + T''}{T' + T'' + T'''}$$

$$G = \frac{1 + j\omega(T' + T'') - \omega^2 T'T''}{1 + j\omega(T' + T'' + T''') - \omega^2 T'T''}$$

Table 13.1

Cascade Compensation: Effects on the System.							
Type of Compensator	Polarplot: Characteristics				Phasor $G(j\omega_1)$	Effects on the Original System Due to Addition of a Compensator	
	$\|G\| \omega = \omega_1$		$G(j\omega)$			Value of New ω_m	Over-all gain
	$\omega_1=0$	$\omega_1=\infty$	increasing ω	and/or increasing ω			
lag	1	$\frac{1}{\beta}$	decreases	angle varies from $0 - {}^-90°$	The phasor revolves clockwise	Lower	There is a very large increase. The gain is multiplied by β.
lead	β	1	increases	angle varies from $0 - {}^+90°$	The phasor revolves counterclockwise	Higher	A small increase.

Mathematical approach:

$$G = |G| \underline{/\theta}$$

The transfer function of a lag and lead compensator.

$$\theta = \tan^{-1} \omega T' - \tan^{-1} \omega \beta T'$$

$$= \tan^{-1}\left[\frac{\omega T' - \omega\beta T'}{1 + \omega^2 \beta T'^2}\right].$$

$$\tan\theta = \omega T' - \omega\beta T'/1 + \omega^2 \beta T'^2.$$

The maximum value of the angle θ at an angular frequency

$$\omega = 1/T'\sqrt{\beta} \quad \text{and} \quad \theta_{max} = \sin^{-1}\left[\frac{\frac{1}{\beta} - 1}{\frac{1}{\beta} + 1}\right].$$

For given values of β, θ and ω, the value of T' can be determined which is required for the compensator.

13.2 ANALYSIS OF A LAG NETWORK

When to use a lag network:

If the transient response of a feedback system is reasonably okay but the e_{ss} is very large, then the gain of the system is increased by introducing a lag network.

How to increase the gain of the system:

$$G_{com} = A_0 \frac{1 + T's}{1 + \beta T's} \qquad \text{The transfer function of the compensator.}$$

Type 0 system: The e_{ss} can be reduced by increasing step error coefficient k.

$$k = \lim_{j\omega \to 0} G(j\omega) \qquad \text{initial value}$$

$$k' = \lim_{j\omega \to 0} G_{com} G = A_0 k$$

So there must be a built-in amplifier of gain $A_0 = \frac{k'}{k}$.

Note: The change in the gain must not affect the transient response of the system.

Characteristics:

Note:

A) The phase margin at $\omega = \omega_{\theta_1}$ has been reduced.

B) The compensator is designed such that the log-magnitude β occurs at $\omega = \omega_{\theta_1}$.

C) The gain can now be increased so that the Lm plot have value of odb at $\omega = \omega_{\theta_2}$.

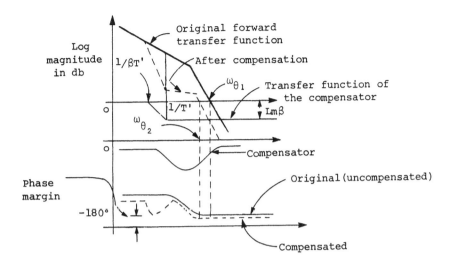

Effects:

A) Gain adjustment for specified Mag_{max}.

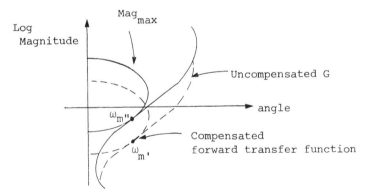

a) Increase in gain to obtain same Mag_{max} = The amount by which dotted curve must be raised to make it tangent to the Mag_{max} curve.

b) Resonant frequency $\omega_{m'} < \omega_{m''}$.

The reason is the negative angle introduced by the network.

B) The lag network acts as a low-pass filter.

13.3 ANALYSIS OF A LEAD NETWORK

Characteristics:

A) Transfer function:

$$G(j\omega) = \beta \; \frac{1 + j\omega T'}{1 + j\omega \beta T'} \; .$$

Note that $\beta < 1$ and the amplifier is not considered.

B) Lm and angle:

$\mathrm{Lm}\, G(j\omega) = \mathrm{Lm}\,\beta + \mathrm{Lm}(1 + j\omega T') - \mathrm{Lm}(1 + j\omega \beta T')$

$\underline{/G(j\omega)} = \underline{/1 + j\omega T'} - \underline{/1 + j\omega \beta T'} \; .$

C) Plots:

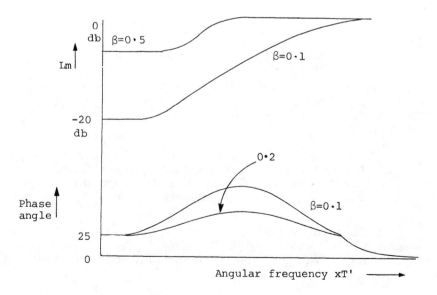

Application to a Feedback System: Curves and Important Facts

A)

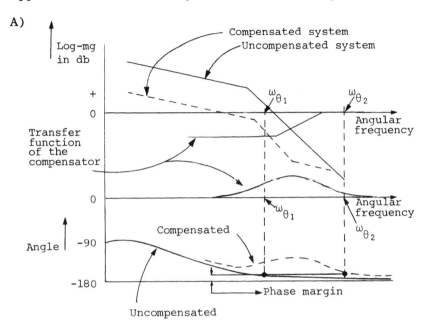

B) The lead network is a high-pass filter. For the frequencies below $\omega = 1/T'$, this lead network introduces an attenuation of value log-magnitude β.

C) The main function of the lead compensator is to increase ω_θ (the phase-margin frequency). By properly choosing T', the phase-margin frequency can be increased from ω_{θ_1} to ω_{θ_2}.

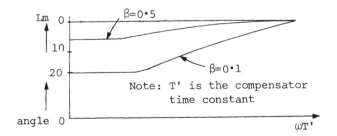

This figure shows that for a constant β, if T' is made smaller, then the new gain of the system is very large.

D) Selection of T':

    a) Type 0 system:

        $T' \leq$ Second largest value of the time constant of the uncompensated system

    b) Type 1 or higher order:

        $T' \leq$ The largest time constant of the original forward transfer function

13.4 ANALYSIS OF A LAG-LEAD COMPENSATOR

Characteristics:

A) Transfer function:

$$G(j\omega) = \frac{A_0(1 + j\omega T')(1 + j\omega T'')}{(1 + j\omega \beta T')(1 + \frac{j\omega T''}{\beta})} \quad \beta > 1 \text{ and } T' > T''$$

$$= A_0 \underbrace{\left[\frac{1 + j\omega T'}{1 + j\omega \beta T'}\right]}_{\text{lag effect}} \underbrace{\left[\frac{(1 + j\omega T'')}{(1 + \frac{j\omega T''}{\beta})}\right]}_{\text{lead effect}}$$

B) Plot:

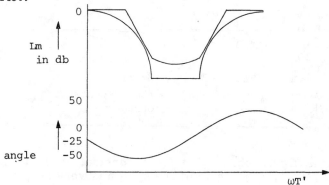

C) Uncompensated and compensated plots:

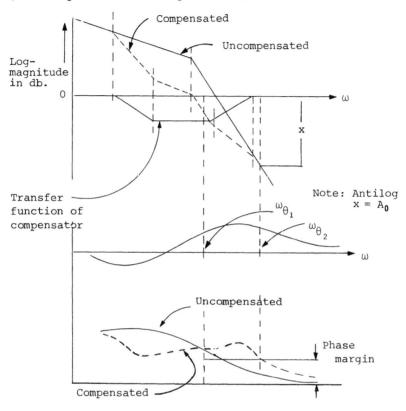

CHAPTER 14

FEEDBACK COMPENSATION: PARALLEL COMPENSATION

14.1 PARALLEL COMPENSATION: PROS AND CONS OF SELECTING A FEEDBACK COMPENSATOR

A faster time of response can be obtained with an introduction of a feedback compensator. The components of the compensator can be adjusted to obtain a small closed-loop time constant. Sometimes this time is an important factor.

Sometimes situations solely dictate the choice of a parallel compensation for greater stability.

The design procedure for a feedback compensator is more complicated than the series compensator.

Parameter variations: Effect on the overall stability

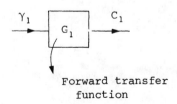

Fig.1 Open-loop system

F-155

CASE STUDY 1

Open-loop system:
$$C_1 = \gamma_1 G_1$$

$$dC_1 = \frac{dG_1}{G_1} C_1.$$

So a change in G can cause a corresponding change in the output, C_1.

CASE STUDY 2

$$C_2 = \frac{\gamma_2 G_2}{1 + G_2 H_2}.$$

$$dC_2 = \frac{\gamma_2 dG_2}{(1 + G_2 H_2)^2} = \left(\frac{1}{(1 + G_2 H_2)} \frac{dG_2}{G}\right) C_2.$$

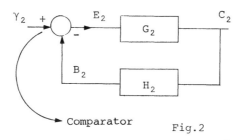

Fig.2

Thus when going from open-loop to closed-loop control, the effect of parameter changes upon the output is reduced by a factor of $1/(1 + GH)$. Note: In case of a series compensator, the value of $1 + G_2 H_2$ is high compared to $1 + G$ which results due to a parallel compensator. This is the plus point of a parallel compensator.

$$-dC_2 = \frac{\gamma_2 G_2^2 C_1 H_2}{(1 + G_2 H_2)^2} \quad \text{Assuming } G_2 \text{ is fixed.}$$

$$\approx -C_2 \frac{dH_2}{H_2} \quad \text{This is valid if } |G_2 H_2| > 1.$$

This also points out the advantages of a parallel compensator.

Sensitivity function of a system:

$$S = \frac{\Delta M}{\Delta \sigma}\bigg|$$ for given variations in the parameter value.

Note:

A) M is the response and σ represents a set of parameters in the open-loop of the system.

B) $|S| \leq 1$.

C) $|S| \to 1$ as the system becomes more sensitive to the variations of a parameter.

D) Open-loop poles have very small effects on C (i.e. the output response) if and only if $S \leq 1$.

E) Parallel compensator makes a system less sensitive as compared to a series compensator.

14.2 EFFECTS OF THE DIFFERENT TYPES OF FEEDBACK ON THE SYSTEM'S TIME RESPONSE

A comparison

A) Specifications:

$$\text{input } R(t) = \frac{u(t)}{0.1}.$$

B) $C_{ss} t = 1$; System with no feedback

$$C(s) = \frac{A}{S + 10} R(s)$$

$$= \frac{10A}{S(S + 10)}.$$

```
        input   ┌─────────┐  output
        ──────→ │    A    │ ──────→
                │  s + 10 │
                └─────────┘
```

The value of gain A can be obtained using the final value theorem.

$$C_{ss}(t) = \lim_{S \to 0} S(LS) = \frac{10A}{10} = 1.$$

So $A = 1$ and

$$C(s) = \frac{1}{(1 + \frac{1}{10}S)S}$$

so the system time constant is 1/10 second.

Original system with unity feedback:

$$C(s) = \frac{A_0}{\left[\frac{1}{10}s + 1 + \frac{A_0}{10}\right]s} \qquad \text{because } R(s) = \frac{10}{S}.$$

Applying the final value theorem, $A_0 = 1.11$ the desired output is

$$C(s) = \frac{1}{s(1 + 0.09)s}.$$

System with series compensation:

```
r(t)=         ┌──────┐   ┌──────┐
1/·10 u(t) →○→│  A₀  │──→│ s+10 │────────┬──→
           ↑  │(10+s)│   │s+100 │        │
           │  └──────┘   └──────┘        │
           │    G₁          G₂   Lead    │
           │                    Compensator
           └──────────────────────────────┘
```

$$\frac{C(s)}{R(s)} = \frac{G_1 G_2}{1 + G_1 G_2} = \frac{\frac{A_0}{s + 10} \cdot \frac{s + 10}{s + 100}}{1 + \frac{A_0}{s + 100}}$$

$\Rightarrow \qquad C(s) = \dfrac{10 A_0}{[(s + 100) + A_0]s} \qquad$ since $R(s) = \dfrac{10}{S}$.

F-158

$$C(t)_{ss} = 1 = \lim_{s \to 0} SC(s) = \frac{10A_0}{A_0 + 100} \quad \text{from use of final value theorem.}$$

$\Rightarrow \quad A_0 + 100 = 10A_0$

$\qquad A_0 = 11.$

Hence,
$$C(s) = \frac{110}{S(S+11)} = \frac{1}{S\left(\frac{111}{110} + \frac{S}{110}\right)} = \frac{1}{S(1 + 0.009S)}.$$

Therefore the system's time constant is 0.009 second; note the improvement.

Original system with parallel compensation:

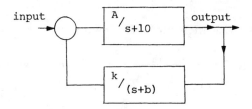

Let the closed-loop response be given by

$$C(s) = \frac{10}{s}\left[\frac{A(s+b)}{s^2 + (10+b)s + (10b+Ak)}\right]$$

$$C_{ss} = \lim_{S \to 0} SC(s) = \lim_{S \to 0}\left[\frac{10A(s+b)}{s^2 + (10+b)s + (10b+Ak)}\right]$$

$\qquad = 1.$

So simplifying this we get

$$1 = \frac{10Ab}{[10b + Ak]} \quad \text{i.e.} \quad 10b + Ak = 10Ab.$$

The system time constant in the previous case was .009. So the compensator can be designed for a time constant < .009. It is possible to get a faster time of response using a feedback compensator.

Location of a feedback compensator in the system:

A) The location of a feedback compensator totally depends on the type of system to be compensated.

B) There is always a unity feedback from the output to the input besides the feedback compensating loop. These two loops make the overall system complicated as compared to the system with a series compensator.

14.3 APPLICATION OF LOG-MAGNITUDE CURVE: FOR FEEDBACK COMPENSATION

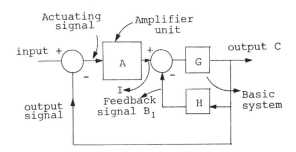

$$G_1 = \frac{C}{I} = \frac{G}{1 + GH}.$$

Note all of them are functions of $j\omega$ and G_1 is the transfer function of the system.

Let $|GH| \ll 1$, then $G_1 \approx G$.

Let $|GH| \gg 1$, then the following approximation is good

$$G_1 \approx \frac{1}{H}.$$

where G_1 and H are phasors and $j\omega$ dependent quantities.

Note: $|GH| \approx 1$ case is neglected.

Direct Application:

$$G = \frac{k'}{S(1 + ST')}$$

Forward transfer function of a system whose $|H(j\omega)| = 1$ and angle $\angle H(j\omega) = 0°$.

A) Log-magnitude plot of GH:

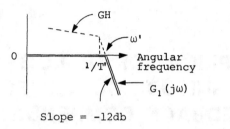

Slope = -12db

B) Observations from the plot:

For $\omega < \omega'$, $|G_1(j\omega)|$ can be given by $\frac{1}{H(j\omega)}$.

For $\omega > \omega'$, $|G_1(j\omega)H(j\omega)| < 1$.

C) So $G_1(j\omega)$ is represented by a thick line as shown above.

D) From the above observations, it can be said that the $G_1(j\omega)$ should look like

$$G_1(j\omega) = \frac{1}{\left[\frac{2\omega j \zeta}{\omega'} + \frac{-\omega^2}{\omega'^2} + 1\right]}$$

To obtain exact curve of $G_1(j\omega)$: Steps

A) The forward transfer function (overall) is expressed in the following form:

$$G_1(j\omega) = \frac{G''(j\omega)}{H(j\omega) + G''(j\omega)H(j\omega)}$$

where $G'' = G \cdot H$ and is a function of ω.

F-161

B) Next, the angle and log-magnitude of $G'' = H(j\omega)G_1(j\omega)$ are plotted on the Nichol's chart.

C) The points of intersection of $G''(j\omega)$ plot with constant magnitude and constant angle curves are determined.

D) These points are used to plot the curve $\dfrac{G''(j\omega)}{1 + G''(j\omega)}$.

E) $1/H(j\omega)$ curve is next plotted and it is combined to produce the entire $G_1(j\omega)$ curve.

CHAPTER 15

SYSTEM SIMULATIONS: USE OF ANALOG COMPUTERS

15.1 ANALOG COMPUTER: BASIC COMPONENTS

OPERATIONAL AMPLIFIER

Schematic diagram:

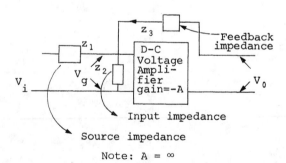

Note: $A = \infty$

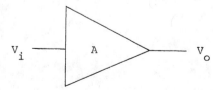

Fig. Symbolic Representation

$$\frac{V_o}{V_i} = \frac{-Z_3}{Z_1} \left[\frac{1}{1 + \frac{1}{A}[1 + Z_3(Z_1+Z_2)/Z_1 Z_2]} \right].$$

For an ideal opamp, $Z_2 \gg Z_1$ and Z_3 and $A \to \infty$, so

$$\frac{V_o}{V_i} = \frac{-Z_3}{Z_1}$$

Opamp as a summer:

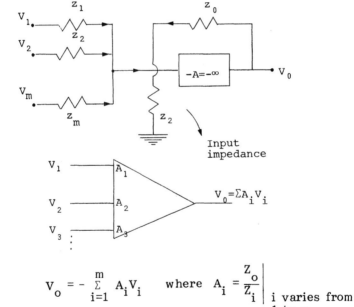

$$V_o = - \sum_{i=1}^{m} A_i V_i \quad \text{where} \quad A_i = \left. \frac{Z_o}{Z_i} \right| \text{ i varies from 1 to m.}$$

When the input and feedback impedances are pure resistors, this scheme is known as a summer.

Integrator:

$$\frac{V_o(s)}{V_i(s)} = \frac{-1}{RCs}$$

F-164

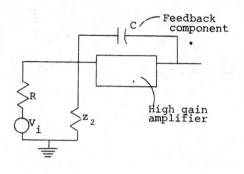

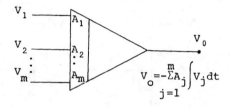

and in the time-domain this relation is

$$V_o(t) = \frac{-1}{RC} \int V_i(t)dt$$

$\frac{1}{RC}$ represents a change in gain.

Potentiometer:

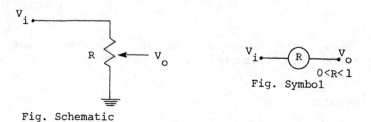

Fig. Schematic   Fig. Symbol

15.2 SIMULATIONS USING ANALOG COMPUTER

Second-order differential equation:

$$D^2x + \frac{a_1}{a_2}Dx + \frac{ax}{a_2} = f(t)/a_2 \quad \text{A second-order equation.}$$

This equation is solved for the highest derivative.

$$a_2 D^2 x = f(t) - ax - a_1 Dx$$

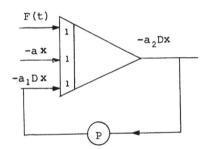

because

$$\int a_2 D^2 x = \int F(t) + \int -ax + \int -a_1 Dx$$

$$a_2 Dx = \int F(t) + \int -ax + \int -a_1 Dx,$$

(The output of integrator is $-a_2 Dx$ because of negative unity gain.)

$$-\int a_2 Dx = -a_2 x \quad \text{so}$$

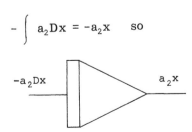

So the scheme for simulating the equation is:

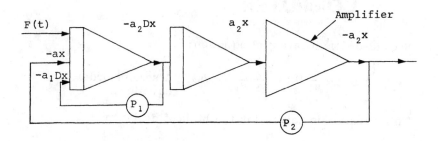

Simulation of a lag network:

Transfer function and the governing differential equation:

$$\frac{V_{out}(s)}{V_{in}(s)} = \frac{A}{1 + ST'} \qquad \text{The transfer function.}$$

$$(T'D + 1)V_{out}(t) = AV_{in}(t).$$

Computer simulation:

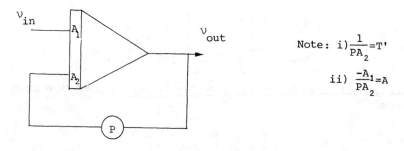

Note: i) $\frac{1}{PA_2} = T'$

ii) $\frac{-A_1}{PA_2} = A$

Simulation of a lead network:

Transfer function and the governing differential equation:

$$\frac{V_{out}(s)}{V_{in}(s)} = \frac{1 + T's}{1 + \beta T's} \qquad \text{Transfer function.}$$

$$V_{in} - V_{out} + T' Dv_{in} = \beta T' DV_{out} \qquad \text{Differential equation.}$$

Mathematical manipulation:

$$\frac{V_{out}(s)}{V_{in}(s)} = \frac{1}{\beta}\left[\frac{\beta - 1}{1 + \beta T's} + 1\right].$$

Case #1: $\beta < 1$

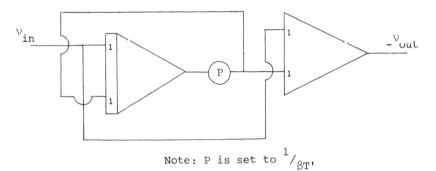

Note: P is set to $1/\beta T'$.

Analog computer simulations of special curves:

Use of ideal diodes and relays

Saturation curve:

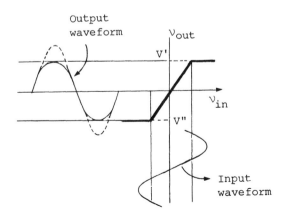

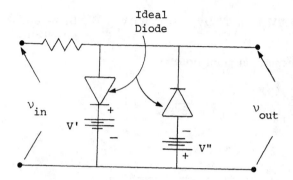

Fig. Shunt-type simulating circuit.

Hysteresis curve:

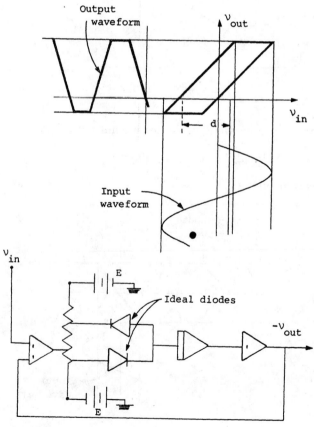

Fig. Back-lash circuit

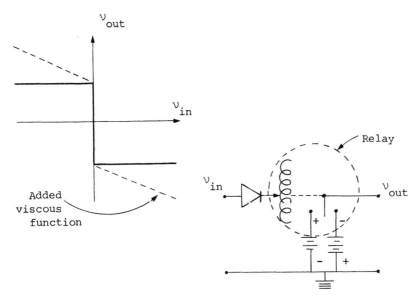

Fig.: Coulomb-friction circuit

Simulation of a continuous function:

The continuous function is represented as follows on a x-y plane:

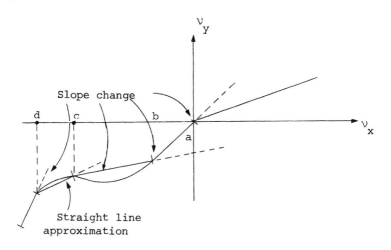

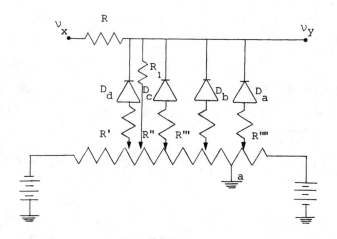

Procedure:

A) The number of ideal diodes required in the circuit is equal to number of changes in the slope.

B) When $V_x < d$, then the diode D_d conducts and V_y depends on R and R'. For $V_x > d_1$ the diode enters in the cutoff region.

C) When the input voltage is in the range d-c, the slope from d to c depends solely on R and R_1 since all the diodes are non-conducting.

D) For $V_x > c$, the diode D_c conducts and the slope from c to b depends on R/R".

E) Tap point a on the potentiometer is reached by trial and error.

15.3 APPLICATION OF ANALOG COMPUTERS FOR SYSTEM TUNING

Unity feedback system: Tuning

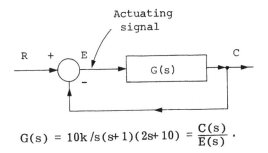

$$G(s) = 10k/s(s+1)(2s+10) = \frac{C(s)}{E(s)}.$$

So the corresponding differential equation is

$$D^3C + 6D^2C + 5DC = 5ke$$

$$D^3C = 5ke - 6D^2 + 5DC.$$

This is given by integrator #1.

The analog computer can be set up as follows:

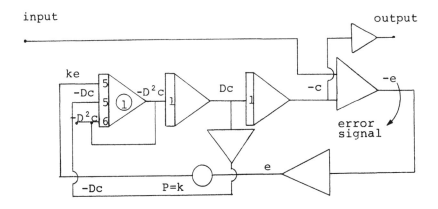

System response and adjusting potentiometer P:

The system gain k is adjusted by means of a potentiometer until the desired response is achieved.

Compensated unity feedback system:

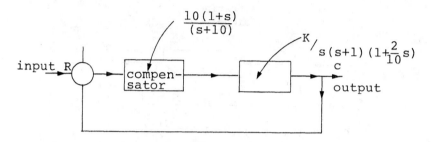

The overall transfer function is

$$\frac{C(s)}{E(s)} = \frac{10k}{s(s+10)(1+\frac{2}{10}s)}$$

So the corresponding differential equation is

$$D^3C + 15D^2C + 50DC = 50ke$$

i.e. $D^3C = 50ke - 15D^2C - 50DC$

The computer set-up is:

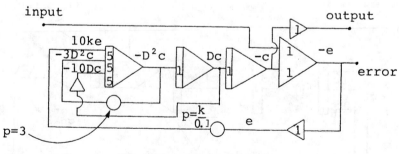

Time-response

Note: It can be seen that the settling time has been reduced considerably.

15.4 SETTING OF A CONTROL SYSTEM: CONTROLLER SETTING

ZIEGLER-NICHOLS METHOD

Salient features:

A) It is widely used for single-loop process control systems.

B) Procedure:

The loop is opened just after the controller and,

a small step input (Δs) is applied to the final control element.

The feedback signal is determined at the point where it is compared with the input signal.

The open-loop response to a small step signal is plotted and following quantities are determined from the curve.

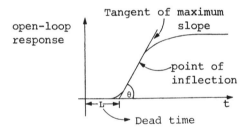

C) Calculations of the controller constants:

Proportional control

$$k_p = \Delta s / L \tan \theta$$

$$T_i = 3.33L \text{ minutes.}$$

PID control (proportional plus integral plus derivative control):

$$k_p = 1.2 \Delta s / L \tan \theta$$

$$T_i = 2L \text{ minutes}$$

$$T_d = 0.5L \text{ minutes.}$$

F-174

ULTIMATE-CYCLE METHOD

Following two parameters are required:

A) Ultimate gain k_u: It is the gain at which the closed-loop system just begins to oscillate.

B) P_u (ultimate period): It is the time period associated with frequency of oscillation ω_0.

Controller settings:

Proportional control
$$k_p = 0.5 k_u.$$

PI control
$$k_p = k_u/2.22$$
$$T_i = 0.83 P_u.$$

PID controller
$$k_p = 0.6 k_u$$
$$T_i = 0.5 P_u$$
$$T_d = 0.125 P_u.$$

Handbook of Electrical Engineering

SECTION G

Mathematics for Engineers

CHAPTER 1

VECTORS, MATRICES, AND EQUATION SYSTEMS

Note: Vectors are indicated by **bold** type.

1.1 VECTORS IN THREE DIMENSIONS

FAMILIAR IDEAS

    Vector **X** = (a, b, c) (See Figure 1.1)

        Vector Components a, b, c

        Equality of Vectors

        Scalar

        Zero Vector **0** = $(0, 0, 0)$

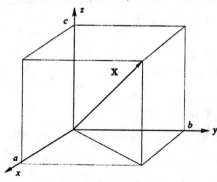

FIGURE 1.1 — Vector Components

DEFINITION 1.1 Norm (Length, Magnitude) of $X = (a, b, c)$

$|X| = \{a^2 + b^2 + c^2\}^{1/2}$.

DEFINITION 1.2 Unit Vector

A vector whose norm is 1, $|X| = 1$.

DEFINITION 1.3   Basis Vectors in x, y, z Space

$i = (1, 0, 0), j = (0, 1, 0), k = (0, 0, 1)$.

REMARK 1.1 Representing X in Terms of Basis Vectors:

$X = (a, b, c) = ai, bj + ck$

DEFINITION 1.4 Direction Cosines of $X = (a, b, c)$

The cosines of the angles α, β, Γ between X and the x, y, z axes (Figure 1.2).

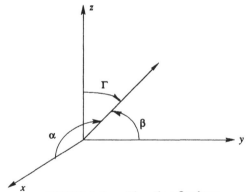

FIGURE 1.2 — Direction Cosines

THEOREM 1.1 $\Sigma = (\cos \alpha, \cos \beta, \cos \Gamma)$ is a Unit Vector

$1 = \cos^2\alpha + \cos^2\beta + \cos^2\Gamma$.

DEFINITION 1.5 Collinearity of X, Y

X, Y are parallel or have opposite directions.

DEFINITION 1.6 Sum of $X = (a, b, c)$, $Y = (A, B, C)$

$X + Y = (a + A, b + B, c + C)$.

DEFINITION 1.7 Product δX for δ = Scalar, X = Vector
$\delta X = (\delta a, \delta b, \delta c)$.

THEOREM 1.2 Properties of δX

1. $|\delta X| = |\delta| |X|$;
2. δX is parallel to X if $\delta > 0$,

   opposite to X if $\delta < 0$.

DEFINITION 1.8 Linear Combination of X, Y

$Z = AX + BY$, A, B = scalars.

REMARK 1.2 (Parallelogram Law)

$X + Y$ is diagonal of parallelogram of X, Y (Figure 1.3).

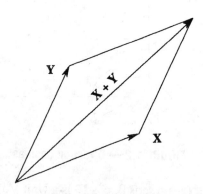

FIGURE 1.3 — Sum of Vectors

DEFINITION 1.9 Inner (Scalar) Product of $X = (a, b, c)$, $Y = (A, B, C)$

$(X, Y) = X \cdot Y = aA + bB + cC$.

THEOREM 1.3 Properties of $X \times Y$

For any X, Y and scalar δ,

$X \times Y = Y \times X$ (Commutative)

$(\delta X) \times Y = \delta(X \times Y)$ (Associative)

$|X \times Y| \leq |X| |Y|$ (Cauchy-Schwarz Inequality)

(= if and only if (iff) X and Y are collinear).

THEOREM 1.4 (Triangle Inequality) For any X, Y,

$|X + Y| \leq |X| + |Y|$

$||X| - |Y|| \leq |X - Y|$.

THEOREM 1.5 The Angle θ between X, Y (Figure 1.4)

$\cos(\theta) = (X \times Y) / (|X| |Y|)$.

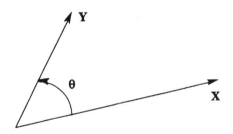

FIGURE 1.4 — Angle Between Vectors

REMARK 1.3 A Unit Vector Parallel to X

For $X \neq 0$, $Y = X / |X|$ is parallel to X and $|Y| = 1$.

REMARK 1.4 Direction Cosines of X

$\cos \alpha = i \times (X / |X|)$, $\cos \beta = j \times (X / |X|)$,

$\cos \Gamma = k \times (X / |X|)$.

DEFINITION 1.10 X, Y Orthogonal

The angle between them is 90 or $\pi/2$.

> **REMARK 1.5** Condition for Orthogonality of **X, Y**
>
> $\mathbf{X} \times \mathbf{Y} = \theta$.

DEFINITION 1.11 Orthonormal Vectors

Orthogonal unit vectors.

> **DEFINITION 1.12** Determinants of Second and Third Order
>
> $$\begin{vmatrix} a & b \\ c & d \end{vmatrix} = ad - bc \qquad (1.1)$$
>
> $$\begin{vmatrix} a & b & c \\ d & e & f \\ g & h & k \end{vmatrix} = a\begin{vmatrix} e & f \\ h & k \end{vmatrix} - b\begin{vmatrix} d & f \\ g & k \end{vmatrix} + c\begin{vmatrix} d & e \\ g & h \end{vmatrix} \qquad (1.2)$$

DEFINITION 1.13 Vector (Cross) Product $\mathbf{X} \times \mathbf{Y}$

Vector with *Right Hand Rule Direction* (Figure 1.5), norm equal to area of parallelogram defined by **X, Y**,

$$\mathbf{X} \times \mathbf{Y} = (|\mathbf{X}| \, |\mathbf{Y}| \sin(\theta)) \, \mathbf{N}.$$

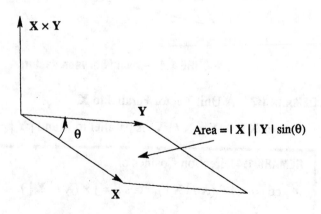

FIGURE 1.5 — Vector Product

THEOREM 1.6 Vector Products of the Basis Vectors

$$i \times j = k, \quad j \times i = -k, \quad j \times k = i,$$
$$k \times j = -i, \quad k \times i = j, \quad i \times k = -j.$$

THEOREM 1.7 Vector Product $X \times Y$ of $X = (a, b, c)$, $Y = (d, e, f)$,

$$X \times Y = \begin{vmatrix} i & j & k \\ a & b & c \\ d & e & f \end{vmatrix} = i \begin{vmatrix} b & c \\ e & f \end{vmatrix} - j \begin{vmatrix} a & c \\ d & f \end{vmatrix} + k \begin{vmatrix} a & b \\ d & e \end{vmatrix}$$

THEOREM 1.8 Vector Product Properties

For any X, Y, Z, A,

A. $(AX) \times Y = A(X \times Y) = X \times (AY)$;

B. $X \times (Y + Z) = (X \times Y) + (X \times Z)$ (Distributive);

C. $(X + Y) \times Z = (X \times Z) + (Y \times Z)$ (Distributive).

REMARK 1.6 Vector Product Non-Properties

Vector product is usually *not* commutative and *not* associative,
$$X \times Y \neq Y \times X, \quad X \times (Y \times Z) \neq (X \times Y) \times Z$$

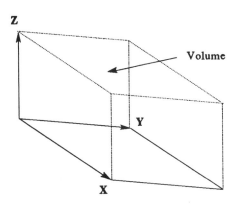

FIGURE 1.6 — {x y z} = Volume

DEFINITION 1.14 Scalar Triple Product $\{X\ Y\ Z\}$

$X \times (Y \times Z) = X, Y, Z$ parallelepiped volume (Figure 1.6).

1.2 VECTORS IN N-DIMENSIONS

DEFINITION 1.15   Norm (Magnitude, Length)

For $X = (x_1, x_2, ..., x_N)^{1/2}$

$|X| = \{x_1^2 + x_2^2 + ... + x_N^2\}^{1/2}$.

DEFINITION 1.16 Zero Vector

$0 = (0, 0, ..., 0)$.

DEFINITION 1.17 Unit Vector

Vector having norm 1.

DEFINITION 1.18 Unit Basis

$i_1 = (1, 0, ..., 0)$, $i_2 = (0, 1, 0, ..., 0)$

$i_{N-1} = (0, 0, ..., 1, 0)$, $i_N = (0, 0, ..., 1)$.

REMARK 1.7 Notation

$X = (x_1, x_2, ..., x_N)$ is often written as $X = \{x_j\}$.

DEFINITION 1.19 Sum of $X = \{x_j\}$, $Y = \{y_j\}$

$Z = X + Y = \{x_j + y_j\}$.

DEFINITION 1.20 Product δX of $X = \{x_j\}$, δ = Scalar

$\delta X = \{\delta x_j\}$.

REMARK 1.8 Representing $X = \{x_j\}$

$X = x_1 i_1 + x_2 i_2 + ... + x_N i_N$.

DEFINITION 1.21 X, Y Collinear

$Y = aX$, for some a.

DEFINITION 1.22   Scalar (Inner) Product

$X \times Y = x_1 y_1 + x_2 y_2 + \ldots + x_N y_N.$

THEOREM 1.9 Cauchy-Schwarz Inequality

$|X \times Y| \le |X| \; |Y|$

($=$ iff X, Y are collinear).

THEOREM 1.10 Triangle Inequality

$|X + Y| \le |X| + |Y|,$

$||X| - |Y|| \le |X - Y|$

($=$ iff X, Y are collinear).

DEFINITION 1.23   Angle θ Between X, Y

$\cos(\theta) = (X \times Y) / (|X| \; |Y|).$

DEFINITION 1.24 X, Y are

A. *Orthogonal* (Normal) if $X \times Y = 0$;

B. *Orthonormal* if they are orthogonal unit vectors;

C. *Parallel* if $\cos(\theta) = 1$.

REMARK 1.9   A Unit Vector Parallel to X

$Z = X / |X|$ is parallel to X, $|Z| = 1$.

DEFINITION 1.25 Direction Cosines of X

Components of $X / |X|$.

1.3 MATRICES AND SYSTEMS OF EQUATIONS

FAMILIAR IDEAS

Columns of a Matrix

Components (Elements) of a Matrix

$M \times N$ Matrix, M, N = Number of Rows, Columns

Rows of a Matrix

REMARK 1.10 Notation

$M \times N$ matrix $A = \{a_{ij}\}$ for i = row, j = column, and

$$A = \{a_{ij}\} = \begin{pmatrix} a_{11} & a_{12} & \cdots & a_{1N} \\ a_{21} & a_{22} & \cdots & a_{2N} \\ \cdots & & & \\ a_{M1} & a_{M2} & \cdots & a_{MN} \end{pmatrix}$$

DEFINITION 1.26 The $M \times N$ Matrix A is

I. a *Row Matrix* if it is a single row, $M = 1$;

II. a *Column Matrix* if it is a single column, $N = 1$;

III. a *Square Matrix of Order M* if $M = N$;

IV. a *Zero Matrix* 0 if all its elements are zero.

DEFINITION 1.27 Equality of $A = \{a_{ij}\}$ and $B = \{b_{ij}\}$

$a_{ij} = b_{ij}$, $i = 1, ..., M$; $j = 1, ..., N$.

DEFINITION 1.28 Diagonal of Square Matrix $A = \{a_{ij}\}$

The elements $a_{11}, a_{22}, ..., a_{NN}$; moreover, the a_{ij}, $i \neq j$ are the *Off-Diagonal Elements*.

DEFINITION 1.29 Diagonal Matrix

Off-diagonal elements are zero.

DEFINITION 1.30 Identity Matrix I of Order N

Diagonal matrix all of whose diagonal elements are 1.

DEFINITION 1.31 Sum of $M \times N$ Matrices A, B

$$C = A + B = \{a_{ij} + b_{ij}\}.$$

DEFINITION 1.32 Product αA, α = Scalar

$$\alpha A = \{\alpha \, a_{ij}\}.$$

DEFINITION 1.33 $-A = (-1)A.$

THEOREM 1.11 Properties of Arithmetic Operations

For any A, B, C and scalars α, β,

A. $A + B = B + A$ (Commutativity)

B. $(A + B) + C = A + (B + C)$ (Associativity)

C. $\alpha(A + B) = \alpha A + \alpha B$ (Distributivity)

D. $(\alpha + \beta)A = \alpha A + \beta A$ (Distributivity)

E. $A + 0 = A$

F. $A + (-A) = 0$

DEFINITION 1.34 Transpose A^T

Matrix obtained by interchanging rows and columns of A,

$$A^T = \{c_{ij}\},\ c_{ij} = a_{ji}.$$

EXAMPLE 1.1 Transpose A^T

For $A = \begin{pmatrix} 3 & 2 & 1 \\ 4 & 5 & 6 \end{pmatrix}$, $A^T = \begin{pmatrix} 3 & 4 \\ 2 & 5 \\ 1 & 6 \end{pmatrix}$.

DEFINITION 1.35 Symmetric Matrix

$A = A^T$.

DEFINITION 1.36 Skew-Symmetric Matrix

$A^T = -A$.

DEFINITION 1.37 Products of Matrices

Let $A = \{a_{ij}\}$, $B = \{b_{jk}\}$ be $L \times M$ and $M \times N$ matrices. Their product $C = \{c_{ik}\} = AB$ is

$$c_{ik} = \sum_{j=1}^{M} a_{ij} b_{jk}$$

A is *Post-Multiplied* by B, B is *Pre-Multiplied* by A.

THEOREM 1.12 Properties of Matrix Multiplication

For any A, B, C and scalar α,

A. $(A + B)C = AC + BC$ \hfill (Distributivity)

B. $A(B + C) = AB + AC$ \hfill (Distributivity)

C. $A(BC) = (AB)C$ \hfill (Associativity)

D. $\alpha(AB) = (\alpha A)B = A(\alpha B)$

E. $IA = AI$ for I the identity matrix

EXAMPLE 1.2 Non-Properties of the Matrix Product

For

$$A = \begin{pmatrix} 1 & -1 \\ 1 & -1 \end{pmatrix}, \quad B = \begin{pmatrix} 1 & 1 \\ 1 & 1 \end{pmatrix}$$

$$AB = \begin{pmatrix} 1 & -1 \\ 1 & -1 \end{pmatrix} \begin{pmatrix} 1 & 1 \\ 1 & 1 \end{pmatrix} = \begin{pmatrix} 0 & 0 \\ 0 & 0 \end{pmatrix}$$

$$BA = \begin{pmatrix} 1 & 1 \\ 1 & 1 \end{pmatrix} \begin{pmatrix} 1 & -1 \\ 1 & -1 \end{pmatrix} = \begin{pmatrix} 2 & -2 \\ 2 & -2 \end{pmatrix}$$

Therefore,

A. Matrix multiplication *may not be Commutative.*

B. The product of non-zero matrices *may be zero.*

THEOREM 1.13 $(AB)^T = B^T A^T$.

REMARK 1.11 Vectors Are Matrices

An N-dimensional vector X is an $N \times 1$ column matrix. For an $M \times N$ matrix A, AX is an M-dimensional vector.

EXAMPLE 1.3

$$\begin{pmatrix} 1 & 2 & 3 \\ 4 & 5 & 6 \end{pmatrix} \begin{pmatrix} 3 \\ 2 \\ 1 \end{pmatrix} = \begin{pmatrix} 10 \\ 28 \end{pmatrix}$$

DEFINITION 1.38 Mappings of Linear Transformations

The $M \times N$ matrix A maps N-dimensional vectors $\mathbf{X}$ into M-dimensional vectors $\mathbf{Y}$ by premultiplication of $\mathbf{X}$,

$$\mathbf{Y} = A\mathbf{X}.$$

Y is the *Image of* $\mathbf{X}$ under the mapping.

EXAMPLE 1.4 A Rotation

$$A = \begin{pmatrix} \cos(\theta) & -\sin(\theta) \\ \sin(\theta) & \cos(\theta) \end{pmatrix} \text{ maps } \mathbf{X} = \begin{pmatrix} x \\ y \end{pmatrix} \text{ onto}$$

$$\mathbf{Y} = \begin{pmatrix} u \\ v \end{pmatrix} = \begin{pmatrix} \cos(\theta) & -\sin(\theta) \\ \sin(\theta) & \cos(\theta) \end{pmatrix} \begin{pmatrix} x \\ y \end{pmatrix} = \begin{pmatrix} x\cos(\theta) & -y\sin(\theta) \\ x\sin(\theta) & +y\cos(\theta) \end{pmatrix}$$

$\mathbf{Y}$ is obtained by rotating $\mathbf{X}$ through the angle θ in the counterclockwise direction (Figure 1.7).

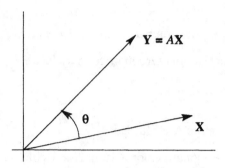

FIGURE 1.7 — Rotation by an Angle θ.

REMARK 1.12 A System of Two Equations in Two Unknowns

$$ax + by = A$$
$$cx + dy = B.$$

x, y are the *Unknowns*; a, b, c, d are the known *Coefficients*; A, B are the known *Right Hand Sides*. The system is *Homogeneous* if $A = B = 0$; if not, it is *Inhomogeneous*. The homogeneous system has the *Trivial Solution* $x = y = 0$. If the determinant

$$D = \begin{vmatrix} a & b \\ c & d \end{vmatrix} = ad - bc \neq 0,$$

then a unique *Solution* to the system is given by

$$x = (1/D) \begin{vmatrix} A & b \\ B & d \end{vmatrix} = (1/D)(Ad - Bb)$$

$$y = (1/D) \begin{vmatrix} a & A \\ c & B \end{vmatrix} = (1/D)(aB - cA)$$

If $D = 0$ then there are two possibilities:

A. $Ac = aB$: there are infinitely many solutions;

B. $Ac = aB$: there are no solutions.

> **DEFINITION 1.39** Inverse of $A = \{a_{ij}\}$
>
> The $N \times N$ matrix A^{-1} obeying
>
> $$A^{-1}A = AA^{-1} = I$$
>
> for I the $N \times N$ identity matrix. A is *Nonsingular* if it has an inverse, and *Singular* if it does not.

THEOREM 1.14 Properties of the Inverse

If A, B are nonsingular then

A. $(A^{-1})^{-1} = A$

B. $(AB)^{-1} = B^{-1}A^{-1}$

DEFINITION 1.40 Determinant of Any Order

The *Determinant of First Order* $D = a_{11}$ is the value of a_{11}. Determinants of 2nd and 3rd order have been defined in (1.12). Assuming that determinants of order $N-1$ are defined, the *Determinant of N-th Order*

$$D = \begin{vmatrix} a_{11} & a_{12} & \cdots & a_{1N} \\ a_{21} & a_{22} & \cdots & a_{2N} \\ \cdots & & & \\ a_{N1} & a_{N2} & \cdots & a_{NN} \end{vmatrix}$$

is defined in terms of the *Cofactors* C_{ij} as

[ROW] $\quad D = a_{i1}C_{i1} + a_{i2}C_{i2} + \ldots + a_{iN}C_{iN}$

[COLUMN] $\quad D = a_{1j}C_{1j} + a_{2j}C_{2j} + \ldots + a_{Nj}C_{Nj}$

The *Cofactor* C_{ij} of a_{ij} is the value of the determinant obtained by removing the i-th row and j-th column of D, multiplied by $(-1)^{i+j}$.

DEFINITION 1.41 Determinant det(A) of a Matrix A

The determinant of the array of rows and columns of A.

THEOREM 1.15 Properties of Determinants

A. $\det(A^T) = \det(A)$;

B. $\det(A) = 0$ if any row or column of A has only zeros;

C. $\det(A)$ is preserved under addition of rows or columns;

D. $\det(AB) = \det(A)\det(B)$.

THEOREM 1.16 Inverse of a Nonsingular Matrix $A = \{a_{ij}\}$

$$A = (1/\det(A)) = \begin{pmatrix} C_{11} & C_{21} & \cdots & C_{N1} \\ C_{12} & C_{22} & \cdots & C_{N2} \\ \cdots & & & \\ C_{1N} & C_{2N} & \cdots & C_{NN} \end{pmatrix}$$

for C_{ij} the cofactor of a_{ij}.

THEOREM 1.17 Cramers Rule

If $D = \det(A) \ne 0$ for A the coefficient matrix of

$$a_{11}x_1 + a_{12}x_2 + \ldots + a_{1N}x_N = b_1$$
$$a_{21}x_1 + a_{22}x_2 + \ldots + a_{2N}x_N = b_2$$
$$\ldots$$
$$a_{N1}x_1 + a_{N2}x_2 + \ldots + a_{NN}x_N = b_N$$

then the system has a unique solution

$$x_1 = D_1/D, \quad x_2 = D_2/D, \ldots, x_N = D_N/D.$$

Here D_j is obtained by replacing the j-th column of D by the right-hand side values $b_1, b_2, \ldots, b_n$.

REMARK 1.13 How to Solve $AX = B$ for Nonsingular A

Premultiply the equation by A^{-1} giving $X = A^{-1}B$.

1.4 COMPLEX VECTORS AND MATRICES

FAMILIAR IDEAS

Complex Number, $z = x + iy$

Imaginary Unit, i, such that $i^2 = -1$

Complex Conjugate, $\bar{z} = x - iy$

Complex Vector, $Z = \{z_j\} = (z_1, z_2, ..., z_N)$

DEFINITION 1.42 Complex Inner Product

For $Z = \{z_j\}$, $W = \{w_j\}$,

$Z \times W = z_1 w_1 + z_2 w_2 + ... + z_N w_N$.

DEFINITION 1.43 Hermitian Matrix

$\overline{A^T} = A$.

REMARK 1.14 Real Valued Hermitian Matrix is Symmetric.

DEFINITION 1.44 Skew-Hermitian Matrix

$\overline{A^T} = -A$.

DEFINITION 1.45 Eigenvalues and Eigenvectors

μ, X satisfying $AX = \mu X$, $X \neq 0$.

REMARK 1.15 Uniqueness of Eigenvector

Up to a multiplicative constant.

DEFINITION 1.46 Characteristic Equation

$\det(A - \mu I) = 0$. $\qquad(1.4)$

DEFINITION 1.47 Spectrum of A

The set of its eigenvalues.

G-16

DEFINITION 1.48 Unitary Matrix

$A^T = A^{-1}$.

DEFINITION 1.49 Orthogonal Matrix

A real unitary matrix.

THEOREM 1.18 Some properties of Eigenvalues:

A. The eigenvalues of a Hermitian matrix are real;

B. The eigenvalues of a skew-Hermitian matrix are either zero or pure imaginary;

C. The eigenvalues of a unitary matrix have absolute value equal to one.

CHAPTER 2

ESSENTIALS OF CALCULUS

2.1 CALCULUS OF ONE VARIABLE

FAMILIAR IDEAS

Closed Interval (a,b): Set of all t with $a \leq t \leq b$

Number Sequence (a_n): Limit $\lim_{n \to \infty} a_n$, convergence

Variable x, y: Independent, Dependent

Function $f(x)$: Bounded, Continuous, Uniformly Continuous, Domain of Definition, Range, Maximum, Minimum, Limit $\lim_{x \to a} f(x)$

Function Sequence $\{f_j(x)\}$: Pointwise, Uniform Convergence

Infinite Series $\sum_{n=0}^{\infty} a_n$: Limit, Absolute and Conditional Convergence, Convergence Tests

Infinite Series of Functions $\sum_{n=0}^{\infty} f_n(x)$: Pointwise, Uniform Convergence

Integral $\int_a^b f(x)dx$: Definition as $\lim_{n \to \infty} \sum_{n=0}^{N} f(x_n)\Delta x$

Derivative $f'(x)$: Definition as $\lim_{h \to \infty}[f(x+h) - f(x)]/h$ of Sum, Product, Quotient of Functions Geometric Meaning as Slope of Curve $y = f(x)$

THEOREM 2.1 Mean Value Theorem

For f, f' continuous on $[a, b]$, for some c on $[a, b]$ (Figure 2.1)

$f(b) - f(a) = f'(c)(b - a)$.

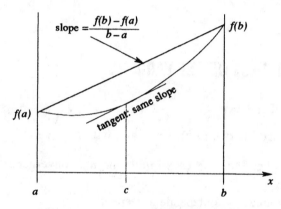

FIGURE 2.1 — Mean Value Theorem

DEFINITION 2.1 Differential $df[x, dx]$

$df[x, dx] = f'(x)dx$.

THEOREM 2.2 Justification for the Differential

If $f'(x)$ is continuous, then

$f(x + dx) \approx f(x) + df[x, dx]$.

THEOREM 2.3 L'Hospitals Rule

If $f(x), g(x) \to 0$ as $x \to a$, then

$\lim_{x \to a} f(x)/g(x) = \lim_{x \to a} f'(x)/g'(x)$.

THEOREM 2.4 The Taylor Series of $f(x)$ about $x = a$

$$f(x) = f(a) + f'(a)(x-a) + (1/2!)f''(a)(x-a)$$
$$+ \ldots + (1/n!)f^{(n)}(a)(x-a)^n + \ldots$$

DEFINITION 2.2 Antiderivative of f

$F(x)$ such that $F'(x) = f(x)$.

THEOREM 2.5 Fundamental Theorem of the Calculus

For continuous $f(x)$, $F'(t) = f(t)$ for

$$F(t) = \int_a^t f(x)dx.$$

THEOREM 2.6 Definite Integral

For any antiderivative $F(x)$ of $f(x)$,

$$\int_a^b f(x)dx = F(x)\Big|_{x=a}^{b} = F(b) - F(a).$$

REMARK 2.1 Elementary Functions (Figure 2.2)

A. Linear: $y = ax + b$

B. Trigonometric: $y = \cos(x)$, $y = \sin(x)$,

$y = \tan(x)$, $\sec(x) = 1/\cos(x)$,

$\mathrm{cosec}(x) = 1/\sin(x)$

C. Exponential/Log: $y = e^x$,

$y = \ln(x) = \int_1^x (1/t)\,dt$

D. Hyperbolic: $\cosh(x) = \frac{1}{2}(e^x + e^{-x})$,

$\sinh(x) = \frac{1}{2}(e^x - e^{-x})$

$\tanh(x) = \sinh(x)/\cosh(x)$

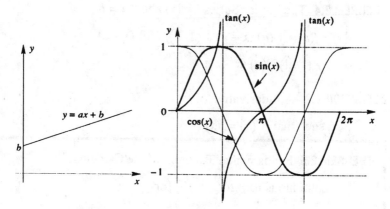

FIGURE 2.2 — Line and Trigonometric Functions

THEOREM 2.7 Basic Properties

A. For $x \approx 0$, $\sin(x) \approx x$, $\cos(x) \approx 1 - \frac{1}{2} x^2$

B. sin, cos, tan are periodic with period 2π

C. $\sin(x)/x \to 1$ for $x \to 0$

D. $\sin^2(x) + \cos^2(x) = 1$

E. $\sinh^2(x) + 1 = \cosh^2(x)$

F. $\sin(x + y) = \sin(x)\cos(y) + \cos(x)\sin(y)$

G. $1 + \tan^2(x) = \sec^2(x)$

DEFINITION 2.3 Compound Function

$z = H(x) = g(f(x))$ for $y = f(x)$, $z = g(y)$.

THEOREM 2.8 Chain Rule for Compound Function

If $H(x) = g(f(x))$, then $H'(x) = g'(f(x)) f'(x)$

DEFINITION 2.4 Inverse $x = g(y) = f^{-1}(y)$ of $y = f(x)$
Function such that $f(g(y)) = y$ and $g(f(x)) = x$.

THEOREM 2.9 Existence of the Inverse

If $f'(x) \neq 0$ then its inverse exists.

THEOREM 2.10 Derivative of the Inverse

$df^{-1}(y)/dy = 1/f'(x)$, for $y = f(x)$.

THEOREM 2.11 Properties of Definite Integral

A. $\int_a^b [Af(x) + Bg(x)]dx = A\int_a^b f(x)dx + B\int_a^b g(x)dx$

B. $(b-a) \min_{a \leq x \leq b} |f(x)| \leq \left|\int_a^b f(x)dx\right| \leq (b-a) \max_{a \leq x \leq b} |f(x)|$

C. Mean Value Theorem: For f continuous on $[a, b]$ there is some c for which

$$f(c)(b-a) = \int_a^b f(x)\,dx$$

D. Generalized Mean Value Theorem: For f, g continuous on $[a, b]$ and $g(x) \geq 0$ there is some c for which

$$\int_a^b f(x)g(x)\,dx = f(c)\int_a^b g(x)\,dx$$

REMARK 2.2 Improper Integrals — Typical Case

$$\int_0^1 \sqrt{x}\,dx = \lim_{\varepsilon \to 0} \int_\varepsilon^1 \sqrt{x}\,dx$$
$$= \lim_{\varepsilon \to 0} \{2 - 2\sqrt{\varepsilon}\} = 2$$

THEOREM 2.12 Techniques of Finding Integrals

A. Integration by Parts:

$$\int_a^b f(x)g'(x)dx = f(x)g(x)\Big|_{x=a}^b - \int_a^b g(x)f'(x)\,dx$$

B. Substitution: If $x = g(t)$, $g'(t) = 0$, then

$$\int_a^b f(x)dx = \int_{g^{-1}(a)}^{g^{-1}(b)} f(g(t))g'(t)dt.$$

2.2 VECTOR FUNCTIONS OF ONE VARIABLE

FAMILIAR IDEAS

Vector Function, $\mathbf{V}(t) = (x(t), y(t), z(t))$

Calculus of Vector Functions of One Variable Continuity, Derivative

REMARK 2.3 Parametric Representation of a Curve

For $x(t)$, $y(t)$, $z(t)$ continuous on $[a, b]$, the vector function $\mathbf{V}(t) = (x(t), y(t), z(t))$ is a *Parametric Representation* of a curve C, written as $C : \mathbf{X} = \mathbf{V}(t)$ (Figure 2.3).

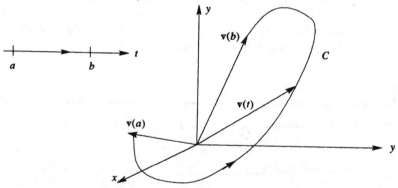

FIGURE 2.3 — Parametric Representation of $C: \mathbf{V} = \mathbf{V}(t)$

DEFINITION 2.5 Arc Length of C

If $x'(t)$, $y'(t)$, $z(t)$ are continuous, then the length of the arc of $C: \mathbf{X} = \mathbf{V}(t)$ from $\mathbf{V}(a)$ to $\mathbf{V}(t)$ for $a \leq t \leq b$ is a function $s(t)$ given by

$$s(t) = \int_a^t |\mathbf{V}'(\tau)| d\tau \qquad (2.1)$$

(See Figure 2.4.) If $\mathbf{V}'(t) \neq \mathbf{0}$, then s can replace t as the independent variable in the parametric representation.

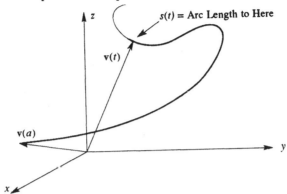

FIGURE 2.4 — Length of Arc $S(t)$ as Measured from $V(a)$

DEFINITION 2.6 Rectifiable Curve

A curve of finite length.

DEFINITION 2.7 Orientation

The direction of increasing arc length.

DEFINITION 2.8 Unit Tangent of C: $\mathbf{X} = \mathbf{V}(s)$ for Arc Length s:

$\mathbf{T}(s) = \mathbf{V}'(s)$

THEOREM 2.13 Frenet-Serret Formulas

$\mathbf{T}'(s) = \kappa(s)\mathbf{N}(s)$,

$\mathbf{N}'(s) = -\kappa(s)\mathbf{T}(s) + \tau(s)\mathbf{B}(s)$,

$\mathbf{B}'(s) = -\tau(s)\mathbf{N}(s)$,

for $\mathbf{T}, \mathbf{N}, \mathbf{B}$ the *tangent, normal, and binormal* to C and $\kappa(s), \tau(s)$ its *curvature* and *torsion*.

REMARK 2.4 Curves in Higher Dimensions

The parametric representation of a curve C in the space $(x, y, z, \ldots)$ is $\mathbf{V}(t) = (x(t), y(t), z(t), \ldots)$.

If $x'(t)$, $y'(t)$, $z'(t)$, ... are continuous, then the arc length from $\mathbf{V}(a)$ to $\mathbf{V}(t)$ is a function $s(t)$ again given by (2.1). If $\mathbf{V}'(t) \neq \mathbf{0}$, then the arc length s can serve as the independent variable defining C in place of t. C is *rectifiable* if it has finite length. s can replace t in the representation of C. C is a *smooth curve* if $\mathbf{V}'(x)$ is continuous.

2.3 FUNCTIONS OF TWO OR MORE INDEPENDENT VARIABLES

FAMILIAR IDEAS

Variables: $x, y, z, ...$

Function $z = f(x, y, z, ...)$: Domain, Range, Independent, Dependent Variables, Limits, Continuity

Partial Derivatives: $f_x(x, y, z, ...), ...$

Higher Order Derivatives: $f_{xx}(x, y, z, ...), ...$

Surface S: $z = f(x, y)$, Defined by $f(x, y)$ (Figure 2.5).

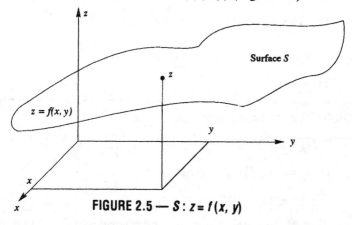

FIGURE 2.5 — $S: z = f(x, y)$

REMARK 2.5 Equation of a Plane

Plane normal to $\mathbf{N} = (a, b, c)$, passing through $(0, 0, d/c)$ is

$$ax + by + cz = d.$$

DEFINITION 2.9 Differential $df[x, y, z, ..., dx, dy, dz, ...]$

$df[x, y, z, ..., dx, dy, dz, ...]$
$= f_x(x, y, z, ...) dx + f_y(x, y, z, ...)dy + ...$

REMARK 2.6 The Differential in Approximation

$f(x + dx, y + dy, z + dz, ...) \approx f(x, y, z, ...)$
$+ df[x, y, z, ..., dx, dy, dz, ...]$

THEOREM 2.14 Taylor Series for $f(x, y)$

$f(a + h, b + k) = f(a, b) + [hf_x(a, b) + kf_y(a, b)]$
$+ (1/2!)[h^2 f_{xx}(a, b) + 2hk f_{xy}(a, b) + k^2 f_{yy}(a, b)]$
$+ (1/3!)[h^3 f_{xxx}(a, b) + 3h^2 k f_{xxy}(a, b)$
$+ 3hk^2 f_{xyy}(a, b) + k^3 f_{yyy}(a, b)]$
$+ ...$

DEFINITION 2.10 Gradient $\nabla f(x, y, z, ...) = \text{grad}. f(x, y, z, ...)$

$\nabla f(x, y, z, ...) = (f_x(x, y, z, ...), f_y(x, y, z, ...), ...)$

DEFINITION 2.11 Directional Derivative

Derivative in the direction of the unit vector **U** is

$D_U f = \mathbf{U} \cdot \nabla f$.

DEFINITION 2.12 Level Surfaces of $f(x, y, z, ...)$

Surfaces along which f is constant.

THEOREM 2.15 Direction of the Gradient

∇f is normal to the level surfaces of f and points in the direction of maximum rate of increase f.

DEFINITION 2.13 Differential $df[x, y, dx, dy]$

$df[x, y, dx, dy] = f_x(x, y)dx + f_y(x, y)dy$

THEOREM 2.16 Differential in Approximation

$$f(x + dx, y + dy) \approx f(x, y) + df[x, y, dx, dy]$$

DEFINITION 2.14 Compound Function

For $w = f(x, y, z, \ldots)$, $x = x(t)$, $y = y(t)$, $z = z(t)$, $\ldots$,

$$w = g(t) = f(x(t), y(t), z(t), \ldots). \tag{2.2}$$

THEOREM 2.17 Chain Rule for (2.2)

$$g'(t) = \nabla f \cdot \mathbf{V}',$$
$$\mathbf{V}'(t) = (x'(t), y'(t), z'(t), \ldots).$$

DEFINITION 2.15 Compound Function

For $w = f(x, y, z, \ldots)$, $x = x(u, v)$, $y = y(u, v)$, $z = z(u, v)$, $\ldots$,

$$g(u, v) = f(x(u, v), y(u, v), z(u, v), \ldots). \tag{2.3}$$

THEOREM 2.18 Chain Rule for (2.3)

$$g_u(u, v) = f_x x_u + f_y y_u + f_z z_u + \ldots$$
$$g_v(u, v) = f_x x_v + f_y y_v + f_z z_v + \ldots$$

2.4 VECTOR FIELDS AND DIVERGENCE

DEFINITION 2.16 Vector Field

A vector function defined in the space $x, y, z, \ldots$,

$$\mathbf{V}(x, y, z, \ldots).$$

DEFINITION 2.17 Divergence $\nabla \cdot \mathbf{V} = \text{div. } V$

For

$$\mathbf{V}(x, y, z, \ldots) = (A(x, y, z, \ldots), B(x, y, z, \ldots), \ldots),$$
$$\nabla \cdot \mathbf{V} = A_x + B_y + C_z + \ldots.$$

DEFINITION 2.18 Curl of a 3-Dimensional Vector Field

For
$$\mathbf{V}(x, y, z) = (A(x, y, z), B(x, y, z), C(x, y, z)),$$

$$\text{curl. } \mathbf{V}(x,y,z) = \nabla \times \mathbf{V} = \begin{vmatrix} \mathbf{i} & \mathbf{j} & \mathbf{k} \\ \partial/\partial x & \partial/\partial y & \partial/\partial z \\ A & B & C \end{vmatrix}$$

$$= (C_y - B_z)\mathbf{i} + (A_z - C_x)\mathbf{j} + (B_x - A_y)\mathbf{k}$$

THEOREM 2.19 div(curl. f) = 0.

DEFINITION 2.19 LaPlacian $\Delta f(x, y, z, ...)$

$$\Delta f(x, y, z, ...) = \nabla^2 f(x, y, z, ...)$$
$$= f_{xx} + f_{yy} + f_{zz} + ... \ .$$

2.5 THE DOUBLE INTEGRAL

FAMILIAR IDEAS (IN ANY NUMBER OF DIMENSIONS)

Point Set: A set of points (See Figure 2.6).

Connected Point Set: Any two points can be joined by a rectifiable curve.

Bounded Point Set: Can be contained in a sufficiently large circle (See Figure 2.7).

Closed Set: A set containing its boundary curve (See Figure 2.8).

Region: A closed, bounded, connected point set.

Non-Simply Connected Region: A region containing the interior of any circle contained in it (See Figure 2.9).

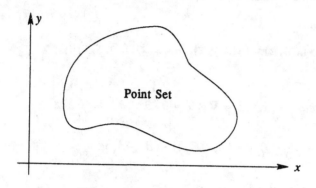

FIGURE 2.6 — A Point Set

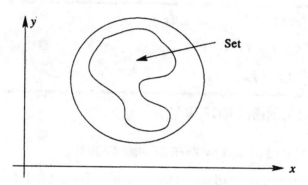

FIGURE 2.7 — Bounded Point Set

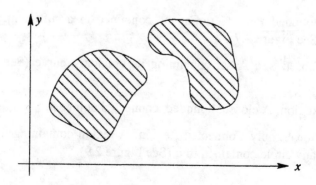

FIGURE 2.8 — Closed Set (Has 2 Parts)

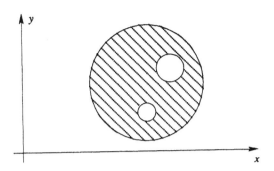

FIGURE 2.9 — A Non-Simply Connected Set

DEFINITION 2.20 The Integral of $f(x, y)$

For $f(x, y)$ continuous on a region D,

$$\iint_D f(x,y)dxdy = \lim_{N \to \infty} \sum_{i=1}^{N} f(x_i, y_j)\delta A$$

(Figure 2.10).

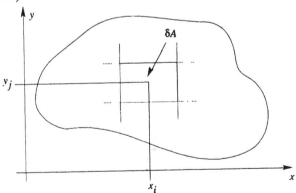

FIGURE 2.10 — Definition of Integral

DEFINITION 2.21 Iterated Integral

If D is the region between $y = y_1(x)$, $y = y_2(x)$, $y_1(x) \leq y_2(x)$, $a \leq x \leq b$ (see Figure 2.11), then

$$\iint_D f(x,y)dxdy = \int_a^b \left\{ \int_{y_1(x)}^{y_2(x)} f(x,y)dy \right\} dx$$

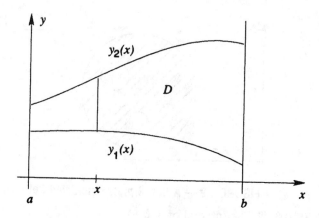

FIGURE 2.11 — Iterated Integral

REMARK 2.7   Mean Value Theorem

For continuous $f(x, y)$,

$$\iint_D f(x,y)dxdy = f(a,b)A(D),$$

where $A(D)$ is the area of D and (a, b) is some point of D.

DEFINITION 2.22 Jacobian

$$J = \partial(u,v)/\partial(x,y) = \begin{vmatrix} u_x & u_y \\ v_x & v_y \end{vmatrix} \qquad (2.4)$$

THEOREM 2.20 Change of Variables in an Integral

If $u = u(x, y)$, $v = v(x, y)$ maps the region D of the x, y plane onto the region E of the u, v plane (see Figure 2.12), then

$$\iint_D f(x,y)dxdy = \iint_E f(x(u,v),y(u,v))Jdudv$$

for J the Jacobian (2.4).

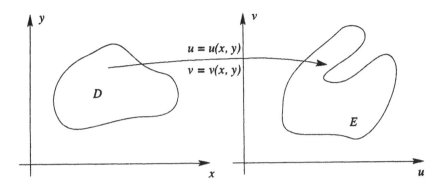

FIGURE 2.12 — Mapping of *D* onto *E*

2.6 LINE INTEGRALS

DEFINITION 2.23 Line Integral (Figure 2.13)

$$\int_C f(x,y,z,\ldots)ds = \lim_{N\to\infty} \sum_{j=1}^{N} f(P_j)\delta s$$

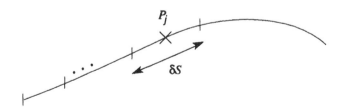

FIGURE 2.13 — Defining the Line Integral

REMARK 2.8 Properties of Line Integrals

Properties of ordinary integrals hold for line integrals (Theorem 2.11).

REMARK 2.9 Representation of Line Integrals

Commonly, the integrands of line integrals appear as

$h(x, y, z, ...)dx/ds$, $h(x, y, z, ...)dy/ds$,

$h(x, y, z, ...)dz/ds$, ...

for s the arc length. In this case we write

$$\int_C h(x,y,z,...)\{dx/ds\}ds = \int_C h(x,y,z,...)dx$$

DEFINITION 2.24 First-Order Differential Form

For given f, g, h, the expression

$fdx + gdy + hdz$

DEFINITION 2.25 Differential of $f(x, y, z)$

The function of the variables x, y, z, dx, dy, dz, defined as

$dF = F_x(x, y, z,)dx + F_y(x, y, z)dy + F_z(x, y, z)dz$

DEFINITION 2.26 Exact Differential Form

The first-order differential form if it is the differential of a function F:

$f = F_x, g = F_y, h = F_z.$

THEOREM 2.21 Independence of Path

The line integral

$$\int_C (fdx + gdy + hdz) \qquad (2.5)$$

along a curve joining P, Q (see Figure 2.14) is independent of path iff the differential form $fdx + gdy + hdz$ is exact.

THEOREM 2.22 Independence of Path

The integral of (2.5) is independent of path iff

$h_y = g_z, f_z = h_x, g_x = f_y$

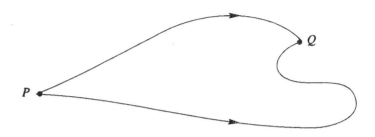

FIGURE 2.14 — Paths from P to Q

2.7 GREEN'S THEOREM

THEOREM 2.23 Green's Theorem:

$$\iint_D (f_x + g_y)dxdy = \int_C (fdy - gdx)$$

D is a region with boundary C, f, g, f_x, g_y are continuous in D, and C is smooth.

THEOREM 2.24 Area of a Region

The area $A(D)$ of the region bounded by the rectifiable, non-self-intersecting curve C (see Figure 2.15) is

$$A = \tfrac{1}{2}\int_C (xdy - ydx).$$

THEOREM 2.25 Integral of the Laplacian

$$\iint_D (f_{xx} + f_{yy})dxdy = \int_C (\partial f / \partial n)ds$$

for $\partial/\partial n$ the outer normal directional derivative to C.

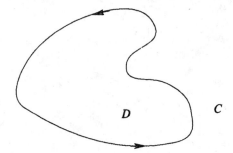

FIGURE 2.15 — The Area of *D*

2.8 SURFACES IN 3 DIMENSIONS

REMARK 2.10 Non-Parametric Representation of a Surface

$$S: F(x, y, z) = 0. \qquad (2.6)$$

THEOREM 2.26 Unit Normal Vector to S of (2.6)

$$n = \nabla F / |\nabla F|.$$

REMARK 2.11 Parametric Representation of a Surface

$$\mathbf{V}(u, v) = (x(u, v), y(u, v), z(u, v)) \qquad (2.7)$$

for (u, v) varying over a domain R of the u, v plane (see Figure 2.16); u, v are the *parameters* of the representation (2.7).

DEFINITION 2.27 Smooth Surface

A surface having a unique, continuous normal vector.

REMARK 2.12 A Curve on a Surface

Let S be given by (2.7) and let $C: u = u(t), v = v(t), a \leq t \leq b$ be a curve in R (see Figure 2.17). Then

$$\mathbf{V}[t] = (x(u(t), v(t)), y(u(t), v(t)), z(u(t), v(t)))$$

defines a curve C' on S.

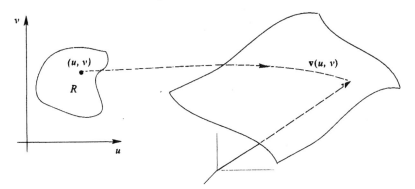

FIGURE 2.16 — Parametric Representation of Surface

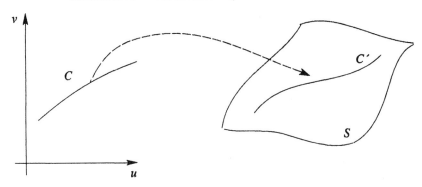

FIGURE 2.17 — A Curve on a Surface

THEOREM 2.27 The Length of a Curve on a Surface

The length of C' of Remark 2.12 is

$$L = \int_a^b \{Eu'^2 + 2Fu'v' + Gv'^2\}^{1/2} dt,$$

where E, F, G are the elements of the *First Fundamental Form of S*,

$$E = \mathbf{V}_u \times \mathbf{V}_u, \; F = \mathbf{V}_u \times \mathbf{V}_v, \; G = \mathbf{V}_v \times \mathbf{V}_v \qquad (2.9)$$

THEOREM 2.28 Surface Area Integral

The area of the surface S of Remark 2.12 is

$$A(S) = \iint_D \{EG - F^2\}^{1/2} du\, dv.$$

DEFINITION 2.28 Surface Integral of $f(x, y, z)$ over S

$$\iint_S f(x,y,z)dA = \lim_{N \to \infty} \sum_{n=0}^{N} f(x_i, y_j, z_k)\delta A.$$

THEOREM 2.29 Parametric Representation of the Surface Integral

For $\mathbf{V}(u, v) = (x(u, v), y(u, v), z(u, v))$ the parametric representation of S on the u, v domain D

$$\iint_S f(x,y,z)dA = \iint_D f(x(u,v), y(u,v), z(u,v))\{EG - F^2\}^{1/2} du dv.$$

THEOREM 2.30 Non-Parametric Representation

For $z = g(x, y)$ the non-parametric representation of S on the (x, y) domain D

$$\iint_S f(x,y,z)dA$$
$$= \iint_D f(x,y,g(x,y))\{1 + (g_x(x))^2 + (g'(y))^2\}^{1/2} dxdy.$$

DEFINITION 2.29 Orientable Surface

There is a unique, continuous normal direction (Figure 2.18).

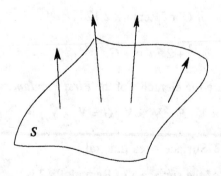

FIGURE 2.18 — Orientable Surface

REMARK 2.13 Integrals Over a Surface

For α, β, Γ the direction cosines of the unit normal to S,

$$\iint_S f(x,y,z)\alpha dA = \iint_S f(x,y,z)dydz$$
$$\iint_S f(x,y,z)\beta dA = \iint_S f(x,y,z)dzdx$$
$$\iint_S f(x,y,z)\Gamma dA = \iint_S f(x,y,z)dxdy$$

THEOREM 2.31 Stoke's Theorem

Let S be an oriented surface bounded by a smooth, closed, non-self-intersecting curve C (Figure 2.19). Let $\mathbf{V}(x, y, z)$ have continuous partial derivatives of first order in a region of x, y, z space containing S. Then

$$\iint_S (curl.\,\mathbf{V}) \cdot \mathbf{N} dA = \int_C \mathbf{V} \cdot \mathbf{T} ds,$$

where $\mathbf{T}$ is the unit tangent vector of C.

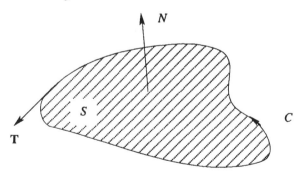

FIGURE 2.19 — Stoke's Theorem

2.9 VOLUME INTEGRALS

DEFINITION 2.30 Volume Integral of $f(x, y, z)$ (See Figure 2.20)

$$\iiint_D f(x,y,z)dV = \lim_{N \to \infty} \sum_{n=0}^{N} f(x_i, y_j, z_k)\delta V$$

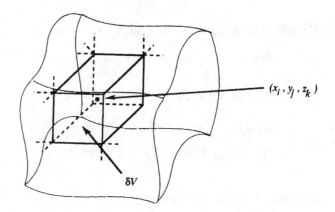

FIGURE 2.20 — Volume Integral

THEOREM 2.32 Divergence Theorem

If $\mathbf{V}(x, y, z)$ and its first partial derivatives are continuous in a region D bounded by an orientable surface S, then

$$\iiint_D \text{div. } \mathbf{V} dV = \iint_S \mathbf{V} \cdot \mathbf{N} \, dA \tag{2.10}$$

for $\mathbf{N}$ the unit vector in the outer normal direction.

REMARK 2.14 Divergence Theorem

(2.10) can be restated as

$$\iiint_D (u_x + v_y + w_z) dxdydz = \iint_S (udydz + vdzdx + wdxdy).$$

REMARK 2.15 Integrating the Laplacian

For any f, g

$$\iiint_D [f\Delta g + (\nabla f) \cdot (\nabla g)] dV = \iint_S f(\partial g / \partial n) dA.$$

CHAPTER 3

COMPLEX FUNCTIONS

3.1 BASIC CONCEPTS

FAMILIAR IDEAS

Imaginary Unit i with $i^2 = -1$

Complex Numbers $z = x + iy$, x, y = Real Numbers

Real, Imaginary Parts of z: $x = Re(z)$, $y = Im(z)$

Complex Conjugate $\bar{z} = x - iy$

Absolute Value or Modulus $|z| = (x^2 + y^2)^{1/2}$

Arithmetic of Complex Numbers

ADDITION

$$(x_1 + iy_1) + (x_2 + iy_2) = (x_1 + x_2) + i(y_1 + y_2)$$

MULTIPLICATION

$$(x_1 + iy_1)(x_2 + iy_2) = (x_1 x_2 - y_1 y_2) + i(x_1 y_2 + x_2 y_1)$$

MULTIPLICATION BY COMPLEX CONJUGATE

$$(x + iy)(x - iy) = x^2 + y^2$$

DIVISION OF z_1 BY z_2

$z_1 / z_2 = (z_1 \bar{z}_2) / (z_2 \bar{z}_2)$

Complex Plane or Argand Diagram (See Figure 3.1)

Parallelogram Law (See Figure 3.2)

Triangle Inequality $|z_1 + z_2| \leq |z_1| + |z_2|$

Polar Form $z = r \cos \theta + ir \sin \theta$ (See Figure 3.3)

Euler's Formula $e^{i\theta} = \cos \theta + i \sin \theta$

Exponential Form $z = re^{i\theta}$, $r = \bmod z$, $\theta = \arg z$

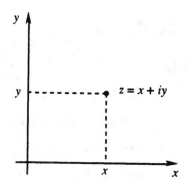

FIGURE 3.1 — Complex Plane

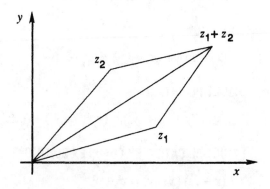

FIGURE 3.2 — Parallelogram Law

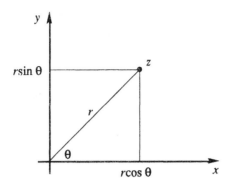

FIGURE 3.3 — Polar Form

THEOREM 3.1 DeMoivre's Formula

For $n = 0, \pm 1, \pm 2, \ldots,$

$(\cos \theta + i \sin \theta)^n = \cos(n\theta) + i \sin(n\theta)$

REMARK 3.1 The Circle of Radius R about $z = 0$

$z = z_1 + Re^{i\theta}, \quad -\infty < \theta < \infty$

3.2 SETS IN THE COMPLEX PLANE

FAMILIAR IDEAS

Set, Point Set in the Plane

Subset

Interior, Boundary Points of a Set

Boundary of a Set

Bounded Set

DEFINITION 3.1 Open Set

A set not containing any of its boundary points.

DEFINITION 3.2 Closed Set

A set containing all of its boundary points.

DEFINITION 3.3 Closure of a Set

The set together with its boundary points.

DEFINITION 3.4 Connected Set

Any two of its points can be linked by a finite number of straight line segments in the set.

DEFINITION 3.5 Domain

An open connected set.

3.3 FUNCTIONS OF A COMPLEX VARIABLE

FAMILIAR IDEAS

Complex Variable

Complex Function $w = f(z)$

Dependent, Independent Variables

Domain of Definition of $f(z)$

Range of $f(z)$

Single-Valued Function

Multi-Valued Function

DEFINITION 3.6 Polynomial of Order N

$$P(z) = \sum_{j=1}^{N} a_j z^j$$

DEFINITION 3.7 Rational Function

Ratio of polynomials $R(z) = P(z) / Q(z)$.

DEFINITION 3.8 Mapping (Transformation)

$w = f(z)$ (Figure 3.4).

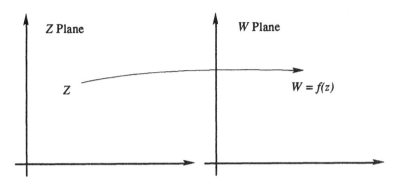

FIGURE 3.4 — Mapping w = f(z)

DEFINITION 3.9 Inverse

$f^{-1}(w)$ (Figure 3.5).

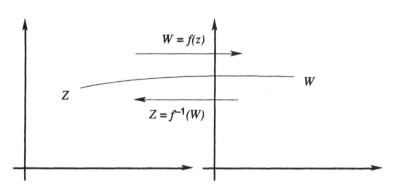

FIGURE 3.5 — Inverse Mapping

DEFINITION 3.10 Translation

$w = z + A$.

DEFINITION 3.11 Rotation

$w = ze^{i\theta}$, θ = Rotation Angle.

DEFINITION 3.12 Compound Function

$$F(z) = g(f(z))$$

3.4 LIMITS, CONTINUITY, AND DERIVATIVES

FAMILIAR IDEAS

Limit: $w_1 = \lim\limits_{z \to z_1} f(z)$

Continuity of $f(z)$

Uniform Continuity of $f(z)$

Properties of Continuity:

    Sum, Product, and Quotient are Continuous

    Composite Function is Continuous

Derivative of $f(z)$ at $z = z_1$:

$$f'(z_1) = \lim_{z \to z_1} [f(z) - f(z_1)]/[z - z_1]$$

Differentiability: Derivative Exists

Properties of Derivatives:

    Derivatives of Sum, Product, and Quotient

THEOREM 3.2 Differentiability = Cauchy-Riemann

If $f(z) = u(x, y) + iv(x, y)$ is differentiable at $z = x + iy$, then u, v satisfy the *Cauchy-Riemann Equations*

$$u_x(x_1, y_1) = v_y(x_1, y_1),$$
$$u_y(x_1, y_1) = -v_x(x_1, y_1).$$

If u_x, u_y, v_x, v_y satisfy (3.1) then, $f = u + iv$ is differentiable.

THEOREM 3.3 Formula for the Derivative

$$f'(z) = u_x(x, y) + iv_x(x, y) = v_y(x, y) - iu_y(x, y).$$

EXAMPLE 3.1 Derivative of e^z

For $f(z) = e^z$, $f'(z) = e^z$.

DEFINITION 3.13 Analytic (or Holomorphic) at a Point

Differentiable in some circle about the point.

DEFINITION 3.14 Entire Function

Analytic at all points.

DEFINITION 3.15 Singular Point z_0

Not analytic at z_0 but at a point in every circle about it.

THEOREM 3.4 Analyticity of Sum and Product

The sum and product of analytic functions is analytic.

THEOREM 3.5 Analyticity of Quotient

The quotient $f(z)/g(z)$ of analytic functions is analytic wherever g does not vanish.

THEOREM 3.6 Analyticity of Composite Functions

If $w = f(z)$, $u = g(w)$ are analytic, then so is $G(z) = g(f(z))$ wherever it is defined.

3.5 HARMONIC FUNCTIONS

DEFINITION 3.16 Harmonic Function $A(x, y)$

Solves Laplace equation: $A = A_{xx}(x, y) + A_{yy}(x, y) = 0$

REMARK 3.2 Laplace Equation in Polar Coordinates

$r^2 A_{rr}(r, \theta) + r A_r(r, \theta) + A_{\theta\theta}(r, \theta) = 0$.

THEOREM 3.7 Cauchy-Riemann = Laplace

If $u(x, y)$, $v(x, y)$ satisfy the Cauchy-Riemann equations, then they are both harmonic.

THEOREM 3.8 Analytic Implies Harmonic

For $f(z) = u(x, y) + iv(x, y)$ analytic, u, v are harmonic.

DEFINITION 3.17 Harmonic Conjugates

Harmonic functions u, v satisfying Cauchy-Riemann equations.

3.6 THE ELEMENTARY COMPLEX FUNCTIONS

FAMILIAR IDEAS

Branches of the Function $y = \tan^{-1}(x)$ (See Figure 3.6)

Zeroes of $f(z)$: Points where f Vanishes

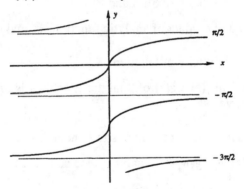

FIGURE 3.6 — Branches of tan⁻¹

DEFINITION 3.18 Branch of Multivalued Function $f(z)$

Any single-valued function $w = F(z)$, analytic in a domain in which it takes on one of the values at each point.

DEFINITION 3.19 Branch Cut

Boundary of the branch domain of a multivalued function.

EXAMPLE 3.2 $f(z) = z^{1/2}$

For $z = Re^{i\theta}$, $F(z) = \sqrt{R}e^{i\theta/2}$, $G(z) = -\sqrt{R}e^{i\theta/2}$ are branches if we choose the branch cut as the ray $\theta = 0$.

DEFINITION 3.20 $e^z = e^{x+iy} = e^x(\cos y + i \sin y)$

REMARK 3.3 Properties

A. entire with $(d/dz)e^z = e^z$

B. periodic with period $2\pi i$

C. never vanishes

D. $|e^z| = e^x$

DEFINITION 3.21 $\log(z)$

Inverse of e^z defined by $e^{\log(w)} = w$.

REMARK 3.4 $\log(z)$ is Multivalued

Since e^z is periodic,

$$w = \log(r) + i(\theta \pm 2n\pi), n = 0, 1, 2, \ldots, z = re^{i\theta}.$$

REMARK 3.5 Branches of $\log(z)$

A branch maps the z-plane onto any strip of height 2π in the w-plane; $\theta = \pi$ is a branch cut for the *Principal Branch* $\text{Log}(z)$ mapping z-plane onto strip $-\pi < v < \pi$.

THEOREM 3.9 Properties of Logarithm

A. $\log(z_1 z_2) = \log(z_1) + \log(z_2)$

B. $\log(z_1/z_2) = \log(z_1) - \log(z_2)$

C. $\log(z^\alpha) = \alpha \log(z)$, $\alpha = 0$

D. $\log(z)$ is analytic and

E. $(d/dz) \log(z) = 1/z$

F. $e^{\log(z)} = z$

DEFINITION 3.22 Complex Powers

$$z^\alpha = e^{\alpha \log(z)}$$

DEFINITION 3.23 Trigonometric Functions

$$\sin(z) = [e^{iz} - e^{-iz}]/2i$$
$$\cos(z) = [e^{iz} - e^{-iz}]/2$$

THEOREM 3.10 Derivatives of Trigonometric Functions

$$(d/dz)\sin(z) = \cos(z), \quad (d/dz)\cos(z) = -\sin(z)$$

THEOREM 3.11 Properties

A. $\sin^2(z) + \cos^2(z) = 1$

B. $\sin(z + 2\pi) = \sin(z), \cos(z + 2\pi) = \cos(z)$

C. $\sin(-z) = -\sin(z), \cos(-z) = \cos(z)$

D. $\sin(z_1 + z_2) = \sin(z_1)\cos(z_2) + \cos(z_1)\sin(z_2)$

E. $\cos(z_1 + z_2) = \cos(z_1)\cos(z_2) - \sin(z_1)\sin(z_2)$

F. $\sin(2z) = 2\sin(z)\cos(z)$

G. $\cos(2z) = \cos^2(z) - \sin^2(z)$

H. $\sin(z + \pi/2) = \cos(z)$

I. $\cos(z) = 0$ for $z = \pi/2 \pm n\pi, \ n = 0, 1, 2, \ldots$

J. $\sin(z) = 0$ for $z = \pm n\pi, \ n = 0, 1, 2, \ldots$.

DEFINITION 3.24 Tangent

$$\tan(z) = \sin(z) / \cos(z)$$

DEFINITION 3.25 Hyperbolic Functions

$$\sinh(z) = (1/2)[e^z - e^{-z}]$$
$$\cosh(z) = (1/2)[e^z + e^{-z}]$$

REMARK 3.6 Properties of Hyperbolic Functions

A. $(d/dz)\sinh(z) = \cosh(z)$

B. $(d/dz)\cosh(z) = \sinh(z)$

C. $\sin(z) = -\sinh(iz)$, $\cos(z) = \cosh(iz)$

D. $\sinh(z) - i\sin(iz)$, $\cosh(z) = \cos(iz)$

E. $\cosh^2(z) - \sinh^2(z) = 1$

F. $\sinh(-z) = -\sinh(z)$, $\cosh(-z) = \cosh(z)$

G. $\sinh(z_1 + z_2) = \sinh(z_1)\cosh(z_2) + \cosh(z_1)\sinh(z_2)$

H. $\cosh(z_1 + z_2) = \cosh(z_1)\cosh(z_2) + \sinh(z_1)\sinh(z_2)$

I. $\sinh(z), \cosh(z)$ are periodic with period $2\pi i$

J. $\cosh(z) = 0$ for $z = (1/2 \pm n)\pi i$, $n = 0, 1, 2, \ldots$

K. $\sinh(z) = 0$ for $z = n\pi i$, $n = 0, 1, 2, \ldots$

L. $\sinh(z) = \sinh(x)\cos(y) + i\cosh(x)\sin(y)$

M. $\cosh(z) = \cosh(x)\cos(y) + i\sinh(x)\sin(y)$

N. $|\sinh(z)|^2 = \sinh^2(x) + \sin^2(y)$

O. $|\cosh(z)|^2 = \sinh^2(x) + \cos^2(y)$.

REMARK 3.7 More on the Trigonometric Functions

A. $\sin(z) = \sin(x)\cosh(y) + i\cos(x)\sinh(y)$

B. $\cos(z) = \cos(x)\cosh(y) - i\sin(x)\sinh(y)$

C. $\sin(z), \cos(z)$ are not bounded:

$|\sin(z)|^2 = \sin^2(x) + \sinh^2(y)$

$|\cos(z)|^2 = \cos^2(z) + \sinh^2(y)$

DEFINITION 3.26 Hyperbolic Tangent

$\tanh(z) = \sinh(z)/\cosh(z)$.

DEFINITION 3.27 Inverse Trigonometric Functions

$\sin^{-1}(z) = -i\log[iz + (1-z^2)^{1/2}]$,

$\cos^{-1}(z) = -i\log[z + i(1 - z^2)^{1/2}]$,

$\tan^{-1}(z) = (1/_{2i}) \log[(1 + iz) / (1 - iz)]$.

THEOREM 3.12 Derivatives of Inverse Functions

$(d / dz) \sin^{-1}(z) = 1 / (1 - z^2)^{1/2}$,

$(d / dz) \cos^{-1}(z) = -1 / (1 - z^2)^{1/2}$,

$(d / dz) \tan^{-1}(z) = 1 / (1 + z^2)$.

DEFINITION 3.28 Inverse Hyperbolic Functions

$\sinh^{-1}(z) = \log[z + (z^2 + 1)^{1/2}]$

$\cosh^{-1}(z) = \log[z + (z^2 - 1)^{1/2}]$

$\tanh^{-1}(z) = \frac{1}{2}\log[(1 + z) / (1 - z)]$

3.7 INTEGRALS OF COMPLEX FUNCTIONS

FAMILIAR IDEAS

Continuous Curve

Rectifiable Curve

Oriented (Directed) Curve

Line Integral

Simple (Jordan) Curve: Does Not Intersect Itself

Simple (Jordan) Closed Curve: Intersects Only at Endpoints

Positive (Negative) Orientation: Counterclockwise (Clockwise) Orientation (See Figure 3.7)

Parametric Representation $z = z(t)$ of curve C

Tangent to a Curve

Smooth Curve: Has a Continuously Varying Tangent

Contour: A Finite Number of Smooth Curves

Simply Connected Domain (See Figure 3.8)

Antiderivative F of f: $F' = f$

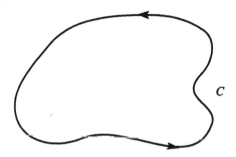

FIGURE 3.7 — Positive Orientation

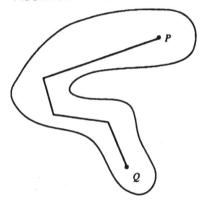

FIGURE 3.8 — Simply Connected Domain

THEOREM 3.13 Jordan Curve Theorem

A simple closed curve C divides the plane into three sets: points of C, interior to C, and exterior to C.

DEFINITION 3.29 Line Integral of $f(z)$ Along C

If $f = u + iv$ and C: $z = z(t)$, $a \leq t \leq b$, then

$$I = \int_C f(z)dz$$

$$= \int_a^b f[z(t)]z'(t)dt$$

$$= \int_a^b [(ux' - vy')dt + i\int_a^b [uy' + vx']dt.$$

THEOREM 3.14 Cauchy Integral Theorem

If f is analytic within and on the simple closed rectifiable curve C, then

$$\int_C f(z)dz = 0.$$

THEOREM 3.15 Independence of Path

The integral of an analytic function is independent of path (see Figure 3.9):

$$\int_{C_1} f(z)dz = \int_{C_2} f(z)dz.$$

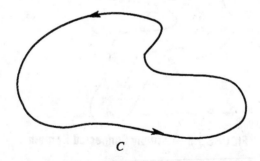

FIGURE 3.9 — Simple Closed Curve C

THEOREM 3.16 Fundamental Theorem of the Calculus

If $f(z) = F'(z)$, then for any curve C between P, Q in a domain of analyticity of f

$$\int_C f(z) = F(Q) - F(P).$$

THEOREM 3.17 Independence of Path Implies Antiderivative

If the line integrals of $f(z)$ are independent of path in a domain D then $f(z)$ has an antiderivative in D.

THEOREM 3.18 Analyticity Implies Antiderivative

If $f(z)$ is analytic, then it has an antiderivative $F(z)$.

THEOREM 3.19 Analyticity Implies Integral Formula

If $f(z)$ is analytic within a simple closed curve C, then for any z_0 within C

$$f(z_0) = (1/2\pi i)\int_C [f(s)/(s-z_0)]ds$$

THEOREM 3.20 Derivative in Terms of Boundary Data

$$f'(z_0) = (1/2\pi i)\int_C [f(s)/(s-z_0)^2]ds$$

THEOREM 3.21 Derivatives of Analytic Function are Analytic

$$f^{(n)}(z_0) = (n!/2\pi i)\int_C [f(s)/(s-z_0)^{n+1}]ds$$

THEOREM 3.22 Morera — Converse to Cauchy Integral Theorem

If $f(z)$ is continuous in D and its line integral around every closed curve C in D is zero, then $f(z)$ is analytic.

THEOREM 3.23 Maximum Modulus Theorem

If f is analytic in the closed curve C and M is its maximum modulus on C, then either $|f(z)| < M$ within C or $|f(z)| \equiv M$.

THEOREM 3.24 Cauchy Inequality

If C is the circle of radius R about z_0, $f(z)$ is analytic within and on C and $M(R)$ is its maximum modulus on C, then

$$|f^{(n)}(z_0)| \leq n!M(R)/R^n.$$

THEOREM 3.25 Louville's Theorem

If $f(z)$ is entire and bounded everywhere in the complex plane, then it is constant.

3.8 NUMBER SEQUENCES AND SERIES

FAMILIAR IDEAS

Number Sequence $\{z_j\}$

Bounded Sequence

Convergence of a Sequence to a Limit: $\lim_{j \to \infty} z_j$

Null Sequence

Cauchy Sequence: $|z_j - z_k| \to 0, j, k \to \infty$

Infinite Series: $\sigma^* = \sum_{n=1}^{\infty} S_n$

Tests for Convergence of Series

    Root Test, Integral Test, Ratio Test

EXAMPLE 3.3 Geometric Series

For $-1 < \alpha < 1$, $\sum_{n=0}^{\infty} \alpha^n = 1/[1-\alpha]$

DEFINITION 3.30 Absolute Convergence of a Series

$$\sum_{n=1}^{\infty} |s_n| \text{ converges.}$$

DEFINITION 3.31 Conditional Convergence

The series converges but not absolutely.

THEOREM 3.26 Absolute Convergence Implies Convergence

3.9 FUNCTION SEQUENCES AND SERIES

FAMILIAR IDEAS

    Function Sequence $\{f_n(z)\}$

    Pointwise Convergence

    Uniform Convergence

    Infinite Function Series $\sum_{n=0}^{\infty} f_n(z)$

    Power Series $\sum_{n=0}^{\infty} a_n(z-z_0)^n$

THEOREM 3.27 Limit of Uniformly Convergent Series is Continuous

THEOREM 3.28 Uniformly Convergent Series Can be Integrated Term by Term

$$\int \left[\sum_{n=0}^{\infty} f_n(z)\right] dz = \sum_{n=0}^{\infty} \int f_n(z) dz.$$

THEOREM 3.29 Convergent Power Series

    A convergent power series can be integrated and differentiated term by term.

DEFINITION 3.32 Taylor Series

$$f(z) = \sum_{n=0}^{\infty} [f^{(n)}(z_o)/n!](z-z_o)^n.$$

DEFINITION 3.33 MacLaurin Series

    The Taylor Series for $z_0 = 0$.

DEFINITION 3.34 Laurent Expansion

For $f(z)$ analytic between circles C_0, C_1 (See Figure 3.10)

$$f(z) = \sum_{n=0}^{\infty} a_n(z-z_0)^n + \sum_{n=1}^{\infty} b_n/(z-z_0)^n;$$

b_1 is the *Residue* of f at z_0. The *Principal Part* is the latter sum of negative powers.

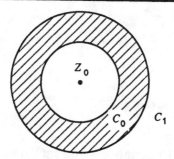

FIGURE 3.10 — Laurent Expansion

3.10 POLES AND RESIDUES

FAMILIAR IDEAS

Singular Point of $f(z)$

Isolated Singular Point z_0 of $f(z)$: The Only Singular Point in Some Circle About z_0

THEOREM 3.30 Value of Integral Around Closed Curve

If z_0 is an isolated singularity within C,

$$\int_C f(z)dz = 2\pi i b_1.$$

THEOREM 3.31 Integral Given by the Sum of the Residues

If f has finitely many isolated singularities in C, then its line integral is $2\pi i$ times the sum of its residues.

DEFINITION 3.35 Pole of Order N

The principal part has non-zero terms up to the power $-N$. For $N = 1$ we have a *Simple Pole*.

DEFINITION 3.36 Essential Singularity

The principal part has infinitely many terms.

THEOREM 3.32 Picard's Theorem

In any circle about an essential singularity $f(z)$ assumes every value except possibly for one, infinitely often.

3.11 ELEMENTARY MAPPINGS AND THE MOBIUS TRANSFORMATION

FAMILIAR IDEAS

Mapping by a Function $w = f(z)$

Domain D of f

Image S' of a Set S (See Figure 3.11)

Level Lines of $u(x, y)$

Expansion $(C > 1)$, Contraction $(C < 1)$, $f(z) = Cz$

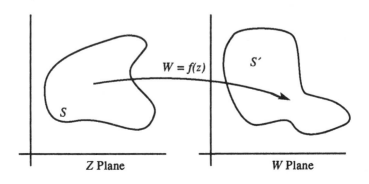

FIGURE 3.11 — Image S' of S

Rotations $w = f(z) = e^{i\theta}z$

Translations $w = fz = z + z_0$

Compound Mappings $w = f(z) = F(G(z))$

THEOREM 3.33 A Line is a Circle of Infinite Radius

THEOREM 3.34 Rotations, Contractions, Expansions, Translations

Preserve rectangles, lines, and circles.

DEFINITION 3.37 Linear Transformation

$w = Az + B$

DEFINITION 3.38 Inversion

$w = 1/z$

THEOREM 3.35 Inversion Preserves Orientation

DEFINITION 3.39 Mobius = Bilinear = Linear Fractional Transformation

$w = f(z) = [az + b] / [cz + d]$.

THEOREM 3.36 Decomposition of Linear Fractional Transformation

The composition of inversion and linear transformation.

THEOREM 3.37 Mobius Transformation

Maps lines and circles into lines and circles.

THEOREM 3.38 Determining a Unique Mobius Transformation

There is a unique Mobius transformation mapping 3 distinct points into 3 distinct points.

THEOREM 3.39 Fixed Point z_0

$z_0 = f(z_0)$

THEOREM 3.40 Fixed Points of Mobius Transformation

There are at most two.

THEOREM 3.41 Composition of Mobius Transformations

Composition of two Mobius transforms is a Mobius transform.

3.12 CONFORMAL MAPPINGS AND HARMONIC FUNCTIONS

FAMILIAR IDEAS

Angle Between Two Smooth Arcs (See Figure 3.12)

Preservation of Angle (See Figure 3.13)

Preservation of Orientation or Sense (See Figure 3.14)

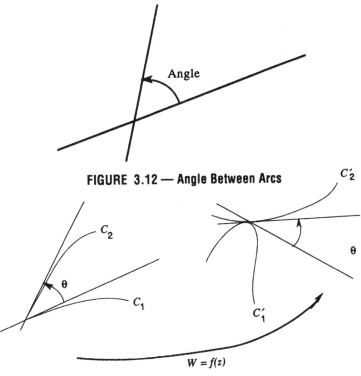

FIGURE 3.12 — Angle Between Arcs

FIGURE 3.13 — f Preserves Angle θ

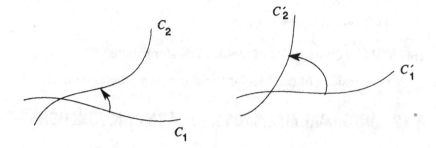

Figure 3.14 — *f* Preserves Orientation

THEOREM 3.42 Analytic Function Preserves Angle and Orientation

DEFINITION 3.40 Conformal Mapping

f is analytic and $f'(z) = 0$.

DEFINITION 3.41 Critical Point

f is analytic and f' vanishes.

THEOREM 3.43 Image of Angle Under Analytic Mapping

If for $m = 2, 3, \ldots,$

$$f'(z_0) = f''(z_0) = \ldots = f^{(m-1)}(z_0) = 0, f^{(m)}(z_0) \neq 0,$$

then the angle between any smooth curves meeting at z_0 is multiplied by m under the mapping $f(z)$.

> **DEFINITION 3.42** Harmonic Function $u(x, y)$
>
> Obeys the *Laplace Equation* $u_{xx} + u_{yy} = 0$.

THEOREM 3.44 Analytic Function Implies Cauchy-Riemann

If $f(z) = u + iv$ is analytic, then u, v obey the Cauchy-Riemann equations and are harmonic.

DEFINITION 3.43 u, v are Harmonic Conjugates

u, v obey the Cauchy-Riemann equations.

THEOREM 3.45 Level Lines of Harmonic Conjugates are Perpendicular

---
DEFINITION 3.44 Poisson Integral Formula

For u harmonic in the circle $C: z = Re^{i\phi}, 0 \leq \phi \leq 2\pi$

$$u(r,\theta) = (1/2\pi)\int_0^{2\pi} u(R,\phi)[(R^2-r^2)/(r^2 + R^2 - 2rR\cos(\theta - \phi))]d\phi$$

THEOREM 3.46 Mean Value Property

$$u(z=0) = (1/2\pi)\int_0^{2\pi} u(R,\phi)d\phi.$$

DEFINITION 3.45 Dirichlet Problem

Find a harmonic function in domain D attaining given values on the boundary.

3.13 ZEROS AND SINGULAR POINTS OF ANALYTIC FUNCTIONS

FAMILIAR IDEAS

Zero of a Function $f(z)$: $f(z) = 0$

Zero of Order N: $f(z) = f'(z) = \ldots f^{N-1}(z) = 0, f^N(z) \neq 0$

THEOREM 3.47 Zeros and Poles

A zero of order n of $f(z)$ is a pole of order n of $1/f(z)$.

DEFINITION 3.46 Removable Singular Point

The Laurent expansion contains no negative powers.

THEOREM 3.48 The Numbers of Zeros and Poles

For f analytic in C with N zeros and P poles,

$$(1/2\pi i)\int [f'(z)/f(z)]dz = N - P.$$

THEOREM 3.49 Argument Principle

$$(1/2\pi) \Delta \text{carg}(f) = N - P$$

THEOREM 3.50 Rouche's Theorem

For f, g analytic in C and $|f(z)| > |g(z)|$ on C, $f(z)$ and $f(z) + g(z)$ have the same number of zeros in C.

3.14 RIEMANN MAPPING THEOREM

DEFINITION 3.47 Conformal Equivalence of Domains D, D'

There exists a conformal mapping of D onto D'.

THEOREM 3.51 Riemann Mapping Theorem

For D founded by a simple closed Jordan curve C, there is an analytic function mapping D conformally onto $|w| < 1$.

CHAPTER 4

ORDINARY DIFFERENTIAL EQUATIONS

FAMILIAR IDEAS

Ordinary Differential Equation (ODE): An Equation Relating a Function $y(x)$ and Its Derivatives Up to Some Order N

Order of the ODE: The Highest Order of a Derivative in the Equation

Solution of the ODE: A Function Satisfying the Equation

Linear ODE: All Derivatives and $y(x)$ Itself Appear Linearly

4.1 ORDINARY DIFFERENTIAL EQUATIONS OF FIRST ORDER

FAMILIAR IDEAS

First Order ODE: $y'(x) = f(x, y(x))$

Initial Condition (IC): $y(x_0) = y_0$

Initial Value Problem (IVP): Find a Solution to the ODE Satisfying the IC

Separable ODE: Equation Can Be Written as $F(y)dy = G(x)dx$

DEFINITION 4.1 Exact ODE $A(x, y)dx + B(x, y)dy = 0$

There is a $U(x, y)$ such that $A = U_x, B = U_y$.

THEOREM 4.1 The Solution of an Exact Equation

The solution $y(x)$ obeys $U(x, y(x)) = $ constant.

THEOREM 4.2 When is $Adx + Bdy = 0$ Exact?

iff $A_y = B_x$.

DEFINITION 4.2 Integrating Factor $\Phi(x, y)$ for $Adx + Bdy = 0$

$\Phi(x, y) A(x, y)dx + \Phi(x, y)B(x, y)dy = 0$ is exact.

REMARK 4.1 Most General Linear First Order ODE

$$y'(x) + a(x)y(x) = b(x). \tag{4.1}$$

DEFINITION 4.3 Homogeneity of (4.1)

$b(x) \equiv 0$.

DEFINITION 4.4 IVP for (4.1)

Find $y(x)$ satisfying equation (4.1) and the IC.

IC: $y(x_0) = y_0$

REMARK 4.2 Solution of the IVP

Multiply equation (4.1) by the integrating factor

$$\Phi(x) = \exp(\int_{x_0}^{x} a(s)ds)$$

giving us

$$\{\Phi(x)y(x)\}' = b(x)\Phi(x)$$

and integrate both sides from x_0 to x.

THEOREM 4.3 Solution to Homogeneous Equation $y' = c_0 y$

$y(x) = A \exp(c_0 x)$, $A = $ any constant

REMARK 4.3 IVP for the Most General ODE of First Order

Find $y(x)$ such that

ODE: $y'(x) = f(x, y(x))$

IC: $y(x_0) = y_0$.

REMARK 4.4 Equivalent Integral Equation Formulation

$$y(x) = y_0 + \int_{x_0}^{x} f(s, y(s))\,ds \qquad (4.2)$$

DEFINITION 4.5 Picard Iteration for Equation (4.2)

A method for finding approximate

$$Y_0(x) \equiv y_0$$

$$Y_{n+1}(x) = y_0 + \int_{x_0}^{x} f(s, Y_n(s))\,ds$$

THEOREM 4.4 Fundamental Existence and Uniqueness Theorem

If $f(x, y)$ and $f_y(x, y)$ are continuous, then the IVP has a unique solution $y = y(x)$ in some interval $\alpha \leq x \leq \beta$.

4.2 LINEAR ORDINARY DIFFERENTIAL EQUATION

FAMILIAR IDEAS

Second Order Linear ODE: $y''(x) + a(x)y'(x) + b(x)y(x) = c(x)$

Homogeneous Equation: $c(x) \equiv 0$

Coefficients: $a(x), b(x)$

Linear Independence (Dependence) of $y_1(x), y_2(x)$: Neither is a Constant Multiple of the Other

THEOREM 4.5 Superposition Principle

If u, v solve the homogeneous equation

$$y''(x) + a(x)y'(x) + b(x)y(x) = 0, \tag{4.3}$$

then so does any linear combination $\alpha u(x) + \beta v(x)$.

REMARK 4.5 Homogeneous Equation with Constant Coefficients

To find the most general solution of

$$y''(x) + a_0 y'(x) + b_0 y(x) = 0$$

set $y = \exp(\Gamma x)$, giving the *Characteristic Equation (CE)*

$$\Gamma^2 + a_0 \Gamma + b_0 = 0.$$

If the CE has two distinct roots Γ_1, Γ_2, then

$$y(x) = Ae^{\Gamma_1 x} + Be^{\Gamma_2 x}$$

If the CE has one double root Γ^*, then

$$y(x) = \{A + Bx\}\, e^{\Gamma^* x}.$$

REMARK 4.6 Complex Roots of the CE

If $\Gamma_1 = \alpha + i\beta$, $\Gamma_2 = \alpha - i\beta$, then

$$y(x) = e^{\alpha x}\{A\cos(\beta x) + B\sin(\beta x)\}.$$

DEFINITION 4.6 Solution with Non-Constant Coefficients

$$y(x) = Ay_1(x) + By_2(x)$$

for A, B constant, y_1, y_2 any linearly independent pair of solutions of equation (4.3).

DEFINITION 4.7 Wronskian of $y_1(x), y_2(x)$

$$W = \begin{vmatrix} y_1(x) & y_2(x) \\ y_1'(x) & y_2'(x) \end{vmatrix} = y_1(x)y_2'(x) - y_2(x)y_1'(x)$$

THEOREM 4.6 ODE for the Wronskian

For solutions y_1, y_2 of equation (4.3), $W'(x) = -a(x)W(x)$, or

$$W(x) = A \exp(-\int_{x_0}^{x} a(s)ds)$$

for some x_0 and constant A. $W(x) \equiv 0$ or is never zero.

THEOREM 4.7 Linear Dependence (Independence)

Two solutions y_1, y_2 of equation (4.3) are linearly independent if their Wronskian is never zero. The solutions y_1, y_2 of equation (4.3) are dependent if their Wronskian vanishes identically.

DEFINITION 4.8 IVP for the Second Order Equation (4.3)

Find y(x) satisfying equation (4.3) and the initial conditions

$$y(x_0) = y_0, \quad y'(x_0) = y_0'$$

THEOREM 4.8 Existence Theorem for the IVP

For $a(x)$, $b(x)$ continuous in an interval about x_0 the IVP for equation (4.3) has a unique solution.

REMARK 4.7 *N*-th Order Linear Equation

The results for second order linear ODE's hold for equations of *N*-th order,

$$y^N(x) + a_1(x) y^{N-1}(x) + \ldots + a_N y(x) = 0 \qquad (4.4)$$

Linear Dependence (Independence) of functions $y_1, y_2, \ldots, y_N$ means that one of these functions can (cannot) be expressed as a linear combination of the others. Their *Wronskian* is the determinant

$$W = \begin{vmatrix} y_1 & y_2 & \ldots & y_N \\ y_1' & y_2' & \ldots & y_N' \\ \ldots & & & \\ y_1^N & y_2^N & \ldots & y_N^N \end{vmatrix}$$

For *N* solutions of equation (4.4) *W either vanishes identically or never, according as they are linearly dependent or independent.* Any solution of equation (4.4) can be written as a linear combination of N linearly independent solutions:

$$y(x) = A_1 y_1(x) + A_2 y_2(x) + \ldots + A_N y_N(x).$$

THEOREM 4.9 Solution of the Inhomogeneous ODE

$$y''(x) + a(x) y'(x) + b(x) y(x) = c(x) \qquad (4.5)$$

The sum of any *particular solution* $Y_p(x)$ of equation (4.5) and the general solution of the homogeneous equation ($c(x) \equiv 0$) $Y_{gen}(x)$,

$$y(x) = Y_p(x) + Y_{gen}(x).$$

REMARK 4.8 Solving the IVP for Equation (4.5)

Find *any* particular solution y_p of (4.5) and the *general* solution of the homogeneous equation, add them,

$$y(x) = Y_p(x) + A y_1(x) + B y_2(x).$$

Then select A, B so that the initial conditions hold.

4.3 SYSTEMS OF FIRST ORDER ODE'S

DEFINITION 4.9 System of N First Order ODE's

$$y_1'(x) = F_1(x, y_1(x), y_2(x), \ldots, y_N(x))$$
$$y_2'(x) = F_2(x, y_1(x), y_2(x), \ldots, y_N(x)) \qquad (4.6)$$
$$\ldots$$
$$y_N'(x) = F_N(x, y_1(x), y_2(x), \ldots, y_N(x))$$

DEFINITION 4.10 Initial Value Problem (IVP)

Find $y_1(x)$, $y_2(x)$, ..., $y_N(x)$ satisfying system (4.6) and assuming N given values at the initial point x_0.

THEOREM 4.10 Existence Theorem

If $F_j(x, y_1, y_2, \ldots, y_N)$ ($j = 1, \ldots, N$) and their partial derivatives are continuous then a unique solution exists in some interval about x_0.

DEFINITION 4.11 Autonomous ODE System

The system (4.6) if the F_j are independent of x:

$$y_1'(x) = F_1(y_1(x), y_2(x), ..., y_N(x))$$
$$y_2'(x) = F_2(y_1(x), y_2(x), ..., y_N(x))$$
...
$$y_N'(x) = F_N(y_1(x), y_2(x), ..., y_N(x))$$

DEFINITION 4.12 Trajectory

The curves in space given by $(y_1(x), y_2(x), ..., y_N(x))$.

DEFINITION 4.13 Phase Space

The space of $y_1, y_2, ..., y_N$ containing the trajectories.

DEFINITION 4.14 Critical Point

A point in phase space where $F_j (j = 1, ..., N) = 0$.

DEFINITION 4.15 Isolated Critical Point (ICP)

One having a neighborhood containing no others.

DEFINITION 4.16 Stable ICP

If for some x_0 a trajectory is sufficiently close to the ICP, then it will remain close to it permanently.

DEFINITION 4.17 Unstable ICP

One which is not stable.

DEFINITION 4.18 Attractive ICP

If for some x_0 a trajectory is sufficiently close to the ICP, then it will converge to it as $x \to \infty$.

4.4 METHODS FOR SOLVING ODE'S

FAMILIAR IDEAS

Taylor Series: Can Be Differentiated and Integrated Term by Term

Analytic Real-Valued Function at $x = x_0$: A Function Expandable in a Taylor Series about $x = x_0$

THEOREM 4.11 Analytic Solution to ODE

For $a(x)$, $b(x)$, $c(x)$ analytic at $x = x_0$ every solution of

$$y''(x) + a(x)y'(x) + b(x)y(x) = c(x)$$

is analytic and can be expanded in powers of $x - x_0$.

DEFINITION 4.19 Laplace Transform of $f(x)$

$$\mathcal{L}\{f\} = F(s) = \int_0^\infty e^{-sx} f(x) dx$$

THEOREM 4.12 Laplace Transform of N-th Derivative of $f(x)$

$$\mathcal{L}\{f^N\} = s^N \mathcal{L}\{f\} - s^{N-1}f(0) - s^{N-2}f'(0) - \ldots - f^{N-1}(0) \qquad (4.7)$$

DEFINITION 4.20 Convolution $h = f*g$ of $f(x)$, $g(x)$

$$h(x) = (f*g)(x) = \int_0^x f(y)g(x-y)dy$$

THEOREM 4.13 Laplace Transform of Convolution

$$\mathcal{L}\{f*g\} = \mathcal{L}\{f\} \mathcal{L}\{g\}.$$

REMARK 4.9 Laplace Transform Solution of IVP's

To solve in IVP for a *Linear, Inhomogeneous ODE with Constant Coefficients,* we apply the transform to the equation and use equation (4.7) to replace the transform of every derivative by a term involving only the transform of the solution. The values of the derivatives of the solution at the initial point (chosen as $x = 0$ in (4.7)) are known from the initial conditions. The equation is solved for the Laplace Transform of the solution which is then found using tables of Laplace Transforms.

4.5 BOUNDARY VALUE PROBLEMS

DEFINITION 4.21 Boundary Value Problem (BVP) for the ODE

$$y''(x) = F(x, y(x), y'(x)) \tag{4.8}$$

For given $h_a, k_a, h_b, k_b, m_a, m_b$, find $y(x)$ satisfying (4.8) and the *Boundary Conditions*

$$h_a y(a) + k_a y'(a) = m_a \tag{4.9}$$
$$h_b y(b) + k_b y'(b) = m_b$$

REMARK 4.10 Some BVP's Have No Solution

If $y''(x) + y(x) = 0$, $0 \le x \le \pi$, then

$$y(x) = A \cos(x) + B \sin(x).$$

For the boundary conditions $y(0) = 0$, $y(\pi) = 1$, the BVP has no solution since $\sin(0) = 0$ implies $A = 0$ while $\sin(\pi) = 0$ implies that $y(\pi)$ vanishes.

DEFINITION 4.22 Sturm Liouville Equation

$$(p(x)y'(x))' + (q(x) + \mu r(x))y(x) = 0. \tag{4.10}$$

DEFINITION 4.23 Sturm Liouville Problem

Find constants μ and solutions to equation (4.10) not vanishing identically and satisfying the homogeneous BC obtained from system (4.9) by setting $m_a = m_b = 0$. Values of μ for which this can be done are *eigenvalues* while the solutions are *eigenfunctions*.

THEOREM 4.14 Orthogonality of Eigenfunctions $y_1(x), y_2(x)$

$$\int_a^b r(x) y_1(x) y_2(x) dx = 0.$$

THEOREM 4.15 Eigenvalues are Real Numbers

If $r(x) = 0$, then the eigenvalues are real numbers.

CHAPTER 5

FOURIER ANALYSIS AND INTEGRAL TRANSFORMS

5.1 BASIC IDEAS

FAMILIAR IDEAS

Periodic Function: $f(t) = f(t + T)$, T = Period

Harmonic Vibration: $f(t) = A \sin[\omega t - \delta]$, $A \cos[\omega t - \delta]$, A = Amplitude, ω = Angular Frequency

Frequency: $f = \omega/2\pi$

Left, Right Limits $f(t_0 - 0)$, $f(t_0 + 0)$ of $f(t)$ for $t \to t_0$ (Figure 5.1)

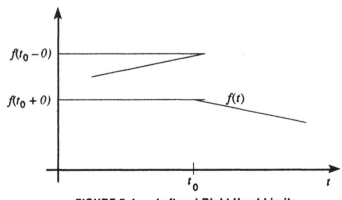

FIGURE 5.1 — Left and Right Hand Limits

Jump Discontinuity at a Point

Piecewise Continuous Function

Even, Odd Functions

DEFINITION 5.1 Mean Value at a Jump Discontinuity t_0

$$f_{mean}(t_0) = \tfrac{1}{2}\{f(t_0 + 0) + f(t_0 - 0)\}$$

DEFINITION 5.2 Phase Angle

The harmonic wave

$$f(t) = a\cos(\omega t) + b\sin(\omega t)$$

can be written as

$$f(t) = A\cos(\omega t - \delta),$$

for $\delta = \arctan(b/a)$ the *Phase Angle*, $A = [a^2 + b^2]^{1/2}$.

DEFINITION 5.3 Trigonometric Polynomial of Order N

A function

$$S_N(t) = \tfrac{1}{2}a_0 + \sum_{n=1}^{N}(a_n\cos(n\omega t) + b_n\sin(n\omega t))$$

$$= \tfrac{1}{2}a_0 + \sum_{n=1}^{N} A_n\cos(n\omega t - \delta_n)$$

DEFINITION 5.4 Superposition of Harmonic Waves

$$f(t) = \sum_{n=0}^{\infty} A_n\cos(\omega_n t - \delta_n),$$

with $0 < \omega_1 < \omega_2 < \ldots$.

DEFINITION 5.5 Fundamental and Higher Harmonics

$\cos(\omega_1 t - \delta_1)$, $\omega_n = n\omega_1$; ω_1, f_1 are the *Fundamental Angular Frequency* and *Fundamental Frequency*.

DEFINITION 5.6 Higher Harmonics

$\cos(\omega_2 t - \delta_2), \ldots, \cos(\omega_n - \delta_n)$ are the second, third, ..., n-th harmonics, respectively; $\omega_2, \omega_3, \ldots, \omega_n$ are the second, third, ..., n-th harmonic angular frequencies. The $f_1, f_2, \ldots$ are the fundamental, second, ..., n-th harnomic frequencies.

DEFINITION 5.7 Beats

Rhythmic amplitude changes due to harmonic interactions.

THEOREM 5.1 Euler's Formula and Implications

$e^{iz} = \cos(z) + i\sin(z),$

$e^{-iz} = \cos(z) - i\sin(z).$

$\cos(z) = \frac{1}{2}(e^{iz} + e^{-iz}),$

$\sin(z) = -\frac{1}{2}i(e^{iz} - e^{-iz}).$

REMARK 5.1 Harmonic Wave Superpositions

$$f(t) = \sum_{n=-\infty}^{\infty} B_n \exp(i\omega_n t)$$

5.2 FOURIER SERIES

FAMILIAR IDEAS

Pointwise Convergence

Uniform Convergence

DEFINITION 5.8 Fourier Series of $f(t)$ on $-\pi < x < \pi$

$$f(t) = \tfrac{1}{2} a_0 + \sum_{n=1}^{\infty} [a_n \cos(nt) + b_n \sin(nt)],$$

with the *Fourier Coefficients*

$$a_0 = (1/2\pi)\int_{-\pi}^{\pi} f(s)ds,$$
$$a_n = (1/\pi)\int_{-\pi}^{\pi} f(s)\cos(ns)ds,$$
$$b_n = (1/\pi)\int_{-\pi}^{\pi} f(s)\sin(ns)ds.$$

DEFINITION 5.9 Dirichlet Conditions on $f(t)$ for Convergence

A. f is periodic;

B. at all but a finite number of points f is differentiable;

C. $f(t), f'(t)$ are piecewise continuous;

D. at points of discontinuity $f(t)$ is its mean value.

DEFINITION 5.10 Piecewise Smooth Function $f(t)$

f obeys the Dirichlet conditions.

THEOREM 5.2 Convergence Theorem for Fourier Series

If the Dirichlet conditions hold, then the Fourier Series converges at every point where f is continuous; at a jump convergence is to the mean value; convergence is uniform on every closed interval of continuity.

DEFINITION 5.11 Gibbs Phenomenon

"Overshooting" of Fourier polynomials at a jump.

THEOREM 5.3 Riemann-Lebesque Theorem

If the Dirichlet conditions hold, then

$$\lim_{n\to\infty}\int_a^b f(t)\cos(nt)dt = \lim_{n\to\infty}\int_a^b f(t)\sin(nt)dt = 0$$

THEOREM 5.4 Bessel's Inequality

$$\tfrac{1}{2}a_0^2 + \sum_{n=1}^{N}[a_n^2 + b_n^2] \le (1/\pi)\int_{-\pi}^{\pi}(f(t))^2 dt$$

THEOREM 5.5 Parseval's Identity

$$\tfrac{1}{2}a_0^2 + \sum_{n=1}^{N} [a_n^2 + b_n^2] = (1/\pi)\int_{-\pi}^{\pi} (f(t))^2 dt$$

DEFINITION 5.12 Fourier Series in Complex Notation

$$f(t) = \sum_{n=-\infty}^{\infty} \alpha_n e^{int}$$

$$\alpha_n = (1/2\pi)\int_{-\pi}^{\pi} f(t)e^{-int} dt$$

THEOREM 5.6 Fourier Series for Even Functions

Can be written in terms of cosines.

THEOREM 5.7 Fourier Series for Odd Functions

Can be written in terms of sines.

DEFINITION 5.13 Multidimensional Fourier Series

$$f(s,t) = \sum_{m=0}^{\infty} \sum_{n=0}^{\infty} \gamma_{mn} \{a_{mn} \cos(ms)\cos(nt)$$

$$+ b_{mn} \sin(ms)\cos(nt) + c_{mn} \cos(ms)\sin(nt)$$

$$+ d_{mn} \sin(ms)\sin(nt)\}$$

$$\gamma_{mn} = \begin{cases} \tfrac{1}{4}, & \text{for } m = n = 0 \\ \tfrac{1}{2}, & \text{for } m > 0, n = 0 \text{ or } n > 0, m = 0 \\ 1, & \text{for } m, n > 0 \end{cases}$$

$a_{mn} = (1/\pi^2) \iint_D f(s, t) \cos(ms) \cos(nt)\, dsdt$

$b_{mn} = (1/\pi^2) \iint_D f(s, t) \sin(ms) \cos(nt)\, dsdt$

$$c_{mn} = (1/\pi^2) \iint_D f(s, t) \cos(ms) \sin(nt)\, dsdt$$

$$d_{mn} = (1/\pi^2) \iint_D f(s, t) \sin(ms) \sin(nt)\, dsdt$$

THEOREM 5.8 Termwise Integration of the Fourier Series

Fourier Series can be integrated term by term.

THEOREM 5.9 Termwise Differentiation of the Fourier Series

Fourier Series can be differentiated term by term.

5.3 FOURIER SERIES AND VECTOR SPACE CONCEPTS

FAMILIAR IDEAS

Vector in N Dimensions

Inner Product

Norm, Unit Vector

Cauchy-Schwarz Inequality

Angle Between Two Vectors

Triangle Inequality

Orthogonality of Vectors

Linear Combination of Vectors

DEFINITION 5.14 Vector Space

Vector collection containing the sum of any two and the product of any by a scalar.

THEOREM 5.10 Continuous Functions as Vectors

Piecewise continuous functions form a vector space.

DEFINITION 5.15 Inner Product of Continuous Functions on $[a,b]$

$$(f_1, f_2) = \int_a^b f_1(t) f_2(t) dt.$$

DEFINITION 5.16 Norm of $f(x)$

$$\|f\| = \{(f,f)\}^{1/2} = \left\{\int_a^b f(t)^2 dt\right\}^{1/2}$$

THEOREM 5.11 Cauchy-Schwarz Inequality

$$|(f, g)| \leq \|f\| \|g\|$$

DEFINITION 5.17 Angle Between f, g

$$\cos(\theta) = (f, g) / [\|f\| \|g\|]$$

DEFINITION 5.18 Orthogonality of f, g

$$(f, g) = 0.$$

THEOREM 5.12 Orthogonality of the Trigonometric Functions

For $a = -\pi$, $b = \pi$, $\cos(mx)$, $\sin(nx)$, $m,n = 0, 1, \ldots$ are orthogonal.

DEFINITION 5.19 Orthonormality

f, g are orthogonal and their norms are equal to one.

DEFINITION 5.20 Least-Square Distance Between f, g

$$d(f, g) = \|f - g\| \tag{5.1}$$

THEOREM 5.13 Fourier Polynomial is Best Approximation

The Fourier polynomial of order N is the closest linear combination of sin, cos functions in the sense of $\| \; \|$.

THEOREM 5.14 Generalized Fourier Series

Let $\{0_n\}$ be an orthonormal family of functions. Their nearest linear combination to a function f in the sense of (5.1) is found by choosing the coefficients of the 0_n as the *Fourier Coefficients*

$$a_n = (f, 0_n).$$

THEOREM 5.15 Bessel's Inequality

$$\|f\|^2 \geq \sum_{n=1}^{N} \alpha_n^2.$$

DEFINITION 5.21 Completeness of the Orthonormal Family $\{0_n\}$

Every function with a finite norm is the limit of a sequence of linear combinations of the family elements.

THEOREM 5.16 Parseval's Identity

If the $\{0_n\}$ are complete, then Bessel's inequality becomes an equality for $N \to \infty$.

DEFINITION 5.22 Weighted Inner Products and Norms

The above hold if the inner product is redefined as

$$(f,g) = \int_a^b W(t)f(t)g(t)dt$$

with $W(t) > 0$ a given continuous function.

DEFINITION 5.23 Inner Product for Complex-Valued Functions

$$(f,g) = \int_a^b f(t)\overline{g(t)}dt$$

5.4 FOURIER TRANSFORMS

DEFINITION 5.24 Absolutely Integrable Function $f(t)$

$\int_{-\infty}^{\infty} |f(t)| dt$ exists.

THEOREM 5.17 Fourier Integral Theorem

$$f(t) = (1/\pi) \int_0^{\infty} d\gamma \int_{-\infty}^{\infty} f(\gamma) \cos[s(t - \gamma)] ds.$$

THEOREM 5.18 Equivalent Forms of the Fourier Integral Theorem

$$f(t) = (1/2\pi) \int_0^{\infty} \int_{-\infty}^{\infty} f(\gamma) e^{is(t-\gamma)} d\gamma ds$$

$$f(t) = (1/2\pi) \int_{-\infty}^{\infty} e^{ist} ds \int_0^{\infty} f(\gamma) e^{-is\gamma} d\gamma.$$

DEFINITION 5.25 Fourier Transform $F(\gamma)$ of $f(t)$

$$F(s) = \int_{-\infty}^{\infty} f(\gamma) \exp(-is\gamma) d\gamma.$$

We write $F = \Phi\{f\}$ or $F(\gamma) = \Phi\{f\}[\gamma]$.

THEOREM 5.19 Inverse $f = \Phi^{-1}\{F\}$ of the Fourier Transform

$$f(t) = (1/2\pi) \int_{-\infty}^{\infty} F(s) e^{ist} ds.$$

THEOREM 5.20 Fourier Transform of an Odd Function $f(t)$

$$F_s(\gamma) = \int_0^{\infty} f(t) \sin(t\gamma) dt.$$

THEOREM 5.21 Fourier Transform of an Even Function $f(t)$

$$F_c(\gamma) = \int_0^{\infty} f(t) \cos(t\gamma) dt.$$

THEOREM 5.22 Special Properties of the Fourier Transform

$$\Phi\{f(t - t^*)\} = \Phi(f)\exp(-i\gamma t^*)$$
$$\Phi^{-1}\{F(\gamma - \gamma^*)\} = f(t)\exp(i\gamma t^*)$$
$$\Phi\{f(\alpha t)\} = (1/|\alpha|)F(\alpha)$$
$$\Phi\{f(-t)\} = F(-\gamma)$$
$$\Phi\{\Phi\{f(t)\}\} = 2\pi f(-t)$$

THEOREM 5.23 Convolution Theorem

The Fourier Transform of the convolution of two functions is the product of their Fourier Transforms.

THEOREM 5.24 Parseval's Identities

Let $f(t)$, $g(t)$ have Fourier Transforms $F(\gamma)$, $G(\gamma)$. Then

$$\int_{-\infty}^{\infty} f(t)g(t)\,dt = (1/2\pi)\int_{-\infty}^{\infty} F(\gamma)G(\gamma)\,d\gamma,$$

$$\int_{-\infty}^{\infty} f(s)G(s)\,ds = \int_{-\infty}^{\infty} F(s)g(s)\,ds,$$

$$\int_{-\infty}^{\infty} f(t)^2\,dt = (1/2\pi)\int_{-\infty}^{\infty} |F(\gamma)|^2\,d\gamma.$$

THEOREM 5.25 Continuity of the Fourier Transform

If f is absolutely continuous, then its Fourier Transform $F(\gamma)$ is continuous for all γ and tends to zero as $|\gamma| \to \infty$.

THEOREM 5.26 The Transform of the Derivative

$$\Phi(f')[\gamma] = \gamma i \Phi(f).$$

THEOREM 5.27 The Transform of the Indefinite Integral

If

$$g(t) = \int_0^t f(s)\,ds,$$

then $\Phi(g)[\gamma] = (1/i\gamma) \Phi(f) [\gamma]$.

DEFINITION 5.26 The Fourier Transform of $f(s, t)$

$$F(\sigma,\gamma) = \int_{-\infty}^{\infty} \int_{-\infty}^{\infty} f(s,t)e^{-i(\sigma s+\gamma t)}dsdt.$$

THEOREM 5.28 Inverse Fourier Transform

$$f(s,t) = 1/(2\pi)^2 \int_{-\infty}^{\infty} \int_{-\infty}^{\infty} F(\sigma,\gamma)e^{i(s\sigma+t\gamma)}d\sigma d\gamma.$$

DEFINITION 5.27 Magnitude Spectrum of $f(t)$

Magnitude Spectrum = $|F(\gamma)|$.

DEFINITION 5.28 Phase Spectrum of $f(t)$

Phase Spectrum = $\arg(F(\gamma))$.

DEFINITION 5.29 Energy Content of a Function $f(t)$

$$\int_{-\infty}^{\infty} |f(t)|^2 dt.$$

THEOREM 5.29 Energy Content in Terms of Fourier Transform

Energy Content = $(1/2\pi) \int_{-\infty}^{\infty} |F(\gamma)|^2 d\gamma.$

DEFINITION 5.30 Cross-Correlation of f, g

$$R_{fg}(\gamma) = \int_{-\infty}^{\infty} f(t) g(t - \gamma) dt.$$

DEFINITION 5.31 Uncorrelated Functions

Their cross correlation is zero for all arguments.

DEFINITION 5.32 Auto-Correlation of $f(t)$

$$R_f(\gamma) = \int_{-\infty}^{\infty} f(t) f(t - \gamma) dt.$$

THEOREM 5.30 Fourier Transforms of Correlations

$$\Phi\{R_{fg}\}\,[s] = \Phi\{f\}\,[s]\,\Phi\{g\}\,[-s]$$

$$\Phi\{R_f\}\,[s] = \Phi\{f\}\,[s]\,\Phi\{f\}\,[-s]$$

THEOREM 5.31 Wiener-Khintchine Theorem

$$|\Phi\{f\}\,[\gamma]|^2 = \Phi\{R_f\}\,[\gamma].$$

5.5 SPECIAL FUNCTIONS

FAMILIAR IDEAS

    Fourier Series

    Fourier Coefficients

    Complete Family of Functions

    Sturm Liouville Problem

        Eigenvalues and Eigenfunctions

DEFINITION 5.33 Gamma Functions

$$\Gamma(p) = \int_0^\infty e^{-t} t^{p-1}\,dt, \quad \text{for } p > 0.$$

THEOREM 5.32 Gamma Function

$$\Gamma(p+1) = p!\quad \text{for } p = 1, 2, \ldots.$$

DEFINITION 5.34 Bessel's Equation

$$t^2 y'' + t y' + (t^2 - m^2) y = 0, \text{ for } m = \text{constant}.$$

THEOREM 5.33 Bessel Functions

    For any m Bessel's Equation has a solution $J_m(t)$ having a finite limit as $t \to \infty$, and a solution $Y_m(t)$ which is unbounded as $t \to \infty$. J_m is a *Bessel Function of the First Kind of Order m*; Y_m is a *Bessel Function of the Second Kind of Order m*, or *Neumann Function*.

REMARK 5.2 Frobenius Method for Finding Bessel Functions

Look for a solution in the form

$$y = \sum_{n=0}^{\infty} a_n t^{n+\beta} ;$$

the unknown a_n are found by substitution.

THEOREM 5.34 Power Series for Bessel Functions of First Kind

$$J_m(t) = \sum_{n=1}^{\infty} \{(-1)^n (t/2)^{m+2n}\} / \{\Gamma(n+1)\Gamma(m+n+1)\}.$$

For $m = 1, 2, \ldots$

$$J_m(t) = \sum_{n=1}^{\infty} \{(-1)^n (t/2)^{2n+m}\} / \{n!(m+n)!\}.$$

For all m,

$$J_{-m}(t) = (-1)^m J_m(t).$$

THEOREM 5.35 Case of m Not an Integer

Then J_m and J_{-m} are linearly independent and the general solution of Bessel's equation is

$$y = a_1 J_m + a_2 J_{-m}.$$

THEOREM 5.36 Case of m an Integer

Then J_m and J_{-m} are linearly dependent and form an independent pair of solutions to Bessel's Equation, for which every solution can be represented as

$$y = a_1 J_m + a_2 Y_m.$$

THEOREM 5.37 Series Representation of the Neumann Function

For integer m,

$$Y_m(t) = (2/\pi) [\ln(t/2) + \gamma] J_m(t)$$

$$- (1/\pi) \sum_{k=0}^{m-1} (m - k - 1)! \, (t/2)^{2k-m} / k!$$

$$- (1/\pi) \sum_{k=0}^{\infty} (-1)^k \{W(k) + W(m + k)\} \, (t/2)^{2k+m} / [k!(m + k)!]$$

with $\gamma = 0.5772156 \ldots$ the *Euler's Constant* and

$W(k) = 1 + 1/2 + 1/3 + \ldots + 1/k.$

DEFINITION 5.35 Hankel Functions $H_m^{(1)}$, $H_m^{(2)}$

$$H_m^{(1)} = J_m + iY_m, \, H_m^{(2)} = J_m - iY_m.$$

THEOREM 5.38 Recurrence Relations for Bessel Functions

If C_m is J_m, Y_m, $H_m^{(1)}$ or $H_m^{(2)}$, then

$$C_{m+1}(t) = (2m/t) \, C_m(t) - C_{m-1}(t)$$
$$C_m' = \tfrac{1}{2}[C_{m-1} - C_{m+1}]$$
$$tC_m'(t) = mC_m(t) - tC_{m+1}(t)$$
$$tC_m'(t) = tC_{m-1}(t) - mC_m(t)$$
$$(d/dt) \, [t^m C_m(t)] = t^m C_{m-1}(t)$$
$$(d/dt) \, [t^{-m} C_m(t)] = -t^{-m} C_{m+1}(t)$$

THEOREM 5.39 Behavior for $t \approx 0$

$$J_m(t) \approx (\tfrac{1}{2}t)^m / m!$$
$$Y_0(t) \approx (2/\pi) \ln(t)$$
$$Y_m(t) \approx -(1/\pi) \, (m - 1)! \, (2/t)^m$$

THEOREM 5.40 Behavior for $t \to \infty$

$$J_m(t) \approx (2/\pi t)^{1/2} \cos\{t - \tfrac{1}{2} m\pi - \tfrac{1}{4}\pi\}$$

$$Y_m(t) \approx (2/\pi t)^{1/2} \sin\{t - \tfrac{1}{2} m\pi - \tfrac{1}{4}\pi\}$$
$$H_m^{(1)}(t) \approx (2/\pi t)^{1/2} \exp\{i(t - \tfrac{1}{2} m\pi - \tfrac{1}{4})\}$$
$$H_m^{(2)}(t) \approx (2/\pi t)^{1/2} \exp\{-i(t - \tfrac{1}{2} m\pi - \tfrac{1}{4})\}$$

DEFINITION 5.36 Bessel Inner Product $(f, g)_B$

$$(f, g)_B = \int_0^1 t f(t) g(t) dt.$$

DEFINITION 5.37 Bessel Norm

$$||f||_B^2 = \int_0^1 t f(t)^2 dt.$$

THEOREM 5.41 An Inner Product for Bessel Functions

For any α, β,

$$\int_0^1 t J_m(\alpha t) J_m(\beta t) dt$$
$$= \{\alpha J_m(\beta) J_m'(\alpha) - \beta J_m(\alpha) J_m'(\beta)\} / \{\beta^2 - \alpha^2\}$$

THEOREM 5.42 Orthogonality of the Bessel Functions

If either or both of α, β are zeros of J_m, then

$$f(t) = J_m(\alpha t), \ g(t) = J_m(\beta t)$$

are orthogonal in the Bessel inner product.

THEOREM 5.43 Zeros of $J_m(t)$

Bessel functions have infinitely many positive zeros that can be arranged in increasing order. The first four zeros of J_2 are 5.14, 8.42, 11.62, and 14.80.

THEOREM 5.44 Fourier-Bessel Series for $f(t)$

For $\{\alpha_n\}$ the zeros of J_m and $m \geq -\tfrac{1}{2}$, let

$$f_n(t) = J_m(\alpha_n t).$$

Then

$$f(t) = \sum_{j=1}^{\infty} c_j f_n(t)$$

$$c_j = (f, f_j)_B / ||f_j||_B^2.$$

DEFINITION 5.38 Legendre's Differential Equation

$(1 - t^2)y'' - 2ty' + m(m + 1)y = 0$ for any m.

DEFINITION 5.39 Legendre Polynomials P_m

The solutions to Legendre's ODE for $m = 1, 2, \ldots$.

THEOREM 5.45 Orthogonality of the Legendre Polynomials

$$\int_{-1}^{1} P_m(t)P_n(t)dt = \begin{cases} 0, & \text{for } m \neq n \\ 2/(2n+1), & \text{for } m = n \end{cases}$$

THEOREM 5.46 Rodrigue's Formula for P_m

$$P_m(t) = [1/(2^m m!)] d^m / dt^m \{(t^2 - 1)^m\}.$$

THEOREM 5.47 Recurrence Formulas for P_m

$$P_{m+1}(t) = [(2m+1)/(m+1)]tP_m(t) - [m/(m+1)]P_{m-1}(t)$$

$$P_{m+1}'(t) = 2(m+1)P_m(t) + P_{m-1}'(t)$$

THEOREM 5.48 Legendre-Fourier Series for $f(t)$

$$f(t) = \sum_{m=1}^{\infty} \alpha_m P_m(t)$$

$$\alpha_m = (f, P_m) / ||P_m||^2$$

$$||P_m||^2 = 2/(2m+1).$$

CHAPTER 6

PARTIAL DIFFERENTIAL EQUATIONS (PDE'S)

6.1 FUNDAMENTAL IDEAS

FAMILIAR IDEAS

Domain R in 1, 2, 3 Space Dimensions (See Figure 6.1)

Boundary Γ of a Domain in 1, 2, 3 Space Dimensions

REMARK 6.1 Time and Space

PDE's first arose in heat and mass transfer and wave propagation. Some processes are *Time-Dependent:* quantities of interest vary with time, such as changes of weather. Others are *Steady State*, not varying in time, such as a soap film spanning a wire frame (See Figure 6.2). Sometimes time is *Reversible*, with past and future interchangeable. Then the questions "what will a quantity be if it was given earlier" and "what was it if it is now some value" can both be answered. For other *Irreversible* processes this can't be done: there is a well defined sense of time that cannot be reversed, an example being the melting of an ice cube in hot coffee (See Figure 6.3).

REMARK 6.2 Number of Spatial Variables

We live in a three-dimensional world. However, the geometry

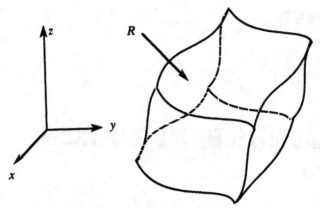

FIGURE 6.1 — Domain R

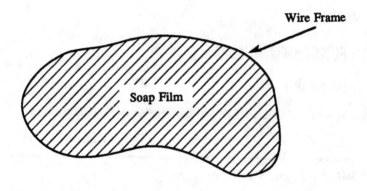

FIGURE 6.2 — Soap Film

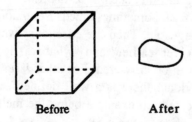

FIGURE 6.3 — Melting Ice Cube

of many processes permits us to represent quantities of interest in terms of functions of fewer than three spatial variables. A process may be axially symmetric, quantities of interest depending only on the depth along the cylinder axis and the distance from it (see Figure 6.4). Similarly, the radius of the cylinder may be so large that effectively the quantities of interest are independent of the distance from the axis, and depend only on the depth along it. Then the process is one dimensional.

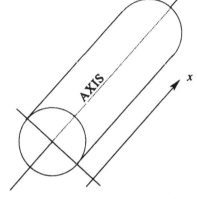

FIGURE 6.4 — Axially Symmetric Process

Note that a quantity may depend on space variables *and* other parameters of a process. We will then have more than three independent variables.

REMARK 6.3 Coordinate Systems

PDE's are represented in coordinate systems fitting the process being modeled. The most popular are the rectangular, cylindrical, and spherical, with the variables

Rectangular:	x, y, z	(See Figure 6.5)
Circular Cylindrical:	r, θ, z	(See Figure 6.6)
Spherical Polar:	$r, \theta, 0$	(See Figure 6.7)

Relations between them are given by:

Circular Cylindrical/Rectangular

$$x = r\cos(\theta), \quad y = r\sin(\theta), \quad z = z$$

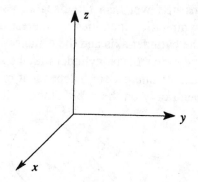

FIGURE 6.5 — Rectangular Coordinates

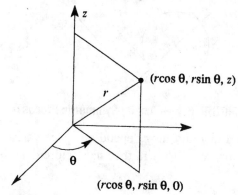

FIGURE 6.6 — Circular Cylindrical Coordinates

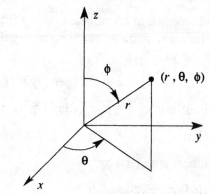

FIGURE 6.7 — Spherical Polar Coordinates

Spherical Polar/Rectangular

$$x = r\cos(\theta)\sin(\phi), \quad y = r\sin(\theta)\sin(\phi), \quad z = r\cos(\phi)$$

Where there is no dependence on z, the circular cylindrical reduce to the *Polar Coordinates*,

$$x = r\cos(\theta), \quad y = r\sin(\theta)$$

DEFINITION 6.1 PDE

A relation between a function $u(x, y, \ldots)$ and its partial derivatives with respect to $x, y, \ldots$ to various orders. We also say "PDE in u."

EXAMPLE 6.1 Three Classical PDE's

Three classical PDE's are

Heat Equation:	$u_t = \alpha u_{xx}$
Wave Equation:	$u_{tt} = c^2 u_{xx}$
Laplace Equation:	$u_{xx} + u_{yy} = 0$

DEFINITION 6.2 Qualities of a PDE in u

Solution of PDE: Any function obeying the PDE

Example: $u = e^x\cos(y)$ obeys $u_{xx} + u_{yy} = 0$

Linear PDE: Terms involving u are linear

Example: $u_t = \alpha u_{xx}$

Order of PDE: Order of highest order derivative

Example: $u_t = \alpha u_{xx}$, $u_{tt} = c^2 u_{xx}$ are 2nd order

Homogeneous PDE: No term that does not depend on u

Example: $u_{xx} + u_{yy} + 3u_x - 2u_y = 0$

Quasilinear PDE: Linear in highest order derivatives

Example: Minimal surface equation,

$$(1 + u_y^2)u_{xx} - 2u_x u_y u_{xy} + (1 + u_x^2)u_{yy} = 0$$

Nonlinear: Not linear

Example: $u_t + uu_x = 0$ and $u_t = \cos(u_x)$

Inhomogeneous: Not homogeneous

Example: Poisson's equation, $u_{xx} + u_{yy} = f$

DEFINITION 6.3 Superposition Principle

Any linear combination $au + bv$ of solutions u, v to a linear, homogeneous PDE is a solution.

DEFINITION 6.4 Types of $au_{xx} + bu_{xy} + cu_{yy} = 0$ (6.1)

Elliptic: $b^2 - 4ac < 0$

Hyperbolic: $b^2 - 4ac > 0$

Parabolic: $b^2 - 4ac = 0$

REMARK 6.4 Types of PDE's and Their Meaning

For hyperbolic and parabolic PDE's, one of the variables represents time while the other is space. These PDE's arise in time-dependent processes. In some cases, they are reversible; in others, they are not. Elliptic PDE's arise in steady-state processes. Each variable is spatial.

DEFINITION 6.5 Hadamard Well Posed Problem

A problem with a unique solution depending continuously on data.

REMARK 6.5 Well Posedness

A condition for the problem to be physically reasonable. In recent years, many *Non-Well-Posed* problems have become important in new areas of science and technology.

DEFINITION 6.6 Boundary Condition

A condition imposed on the PDE solution at the boundary.

DEFINITION 6.7 Initial Condition

A condition imposed on the PDE solution at an initial time.

DEFINITION 6.8 Separable Solution to a PDE

A solution that can be given as the product of functions of each of the variables.

DEFINITION 6.9 Similarity Solution to a PDE

A solution that can be given in terms of less variables.

6.2 THE LAPLACE EQUATION

DEFINITION 6.10 Laplace Equation in Rectangular Coordinates

$$u_{xx} + u_{yy} = 0.$$

REMARK 6.6 Type of Equation

Elliptic: $a = 1$, $b = 0$, $c = 1$, $b^2 - 4ac < 0$ in equation (6.1).

REMARK 6.7 Physical Meaning of Equation

Steady-state, equilibrium states.

REMARK 6.8 Equation in Higher Dimensions

$$u_{xx} + u_{yy} + u_{zz} = 0.$$

REMARK 6.9 Other Coordinate Systems

Circular Cylindrical:

$$(1/r)\,(ru_r)_r + (1/r^2)\,u_{\theta\theta} + u_{zz} = 0$$

Spherical Polar:

$(1/r^2)(r^2 u_r)_r + (1/[r^2 \sin(\phi)])(\sin(\phi) u_\phi)_\phi$
$+ (1/[r^2 \sin^2(\phi)]) u_{\theta\theta} = 0$

Polar:

$(1/r)(r u_r)_r + (1/r^2) u_{\theta\theta} = 0$

REMARK 6.10 Properties of Equation

Order 2, Homogeneous, Linear

REMARK 6.11 Properties of Solutions

Maximum Principle: In a domain R (see Figure 6.8), the maximum and minimum of u are attained on the boundary Γ.

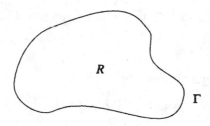

FIGURE 6.8 — Maximum Principle

REMARK 6.12 Form of Well Posed Problem

Dirichlet Problem: Find a solution u in R (see Figure 6.9) attaining given values on Γ.

EXAMPLE 6.2 A Well Posed Problem and Its Solution

For $f(\theta)$ continuous on $[0, 2\pi]$ ($f(0) = f(2\pi)$) find $u(r, \theta)$ obeying Laplace's equation in $0 \le r \le 1$ (see Figure 6.10) and B.C.

$u(1, \theta) = f(\theta)$.

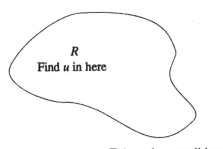

FIGURE 6.9 — Well Posed Problem

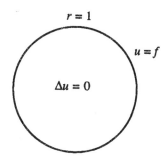

FIGURE 6.10 — Well Posed Problem in the Unit Circle

Separation of Variables for $u(r, \theta) = A(r)B(\theta)$ gives

$$(1/r)(rA'(r))'B(\theta) + (1/r^2)B''(\theta)A(r) = 0$$

or

$$[r^2 A''(r) + rA'(r)]/A(r) = -B''(\theta)/B(\theta). \qquad (6.2)$$

But the left and right sides of this equation are functions of independent variables. Thus, a change in θ must not induce a change in the left hand side and vice versa, or both sides must be constants. For $n = 1, 2, \ldots$, let this value be n^2. Equating the left hand side to n^2 yields

$$r^2 A''(r) + rA'(r) - n^2 A(r) = 0.$$

Seeking a solution $A(r)$ of the form $A(r) = r^\delta$ yields

$$\delta(\delta - 1) + \delta - n^2 = 0$$

or $\delta = \pm n$. A solution $A(r) = r^{-n}$ would not be defined at $r = 0$ and is unacceptable. Hence, $A(r) = r^n$. The right hand side of equation (6.2) gives the ODE

$$B''(\theta) + n^2 B(\theta) = 0$$

having the general solution

$$B(\theta) = a_n \cos(\theta) + b_n \sin(\theta).$$

Thus, for any $n = 1, 2, \ldots$, the separable solution u is

$$u(r, \theta) = r^n(a_n \cos(n\theta) + b_n \sin(n\theta)).$$

Similarly, any constant a_0 satisfies the Laplace equation. By superposition the *General Solution* is

$$u(r,\theta) = \sum_{n=0}^{\infty} r^n(a_n \cos(n\theta) + b_n \sin(n\theta)).$$

At the boundary $r = 1$,

$$u(1,\theta) = \sum_{n=0}^{\infty} (a_n \cos(n\theta) + b_n \sin(n\theta)).$$

Hence, the boundary condition becomes

$$f(\theta) = \sum_{n=0}^{\infty} (a_n \cos(n\theta) + b_n \sin(n\theta)),$$

which is met by choosing the coefficients a_n, b_n as the Fourier coefficients of $f(\theta)$, resulting in the solution.

REMARK 6.13 Related Equations

The steady-state with a heat source is Poisson's PDE

$$u_{xx} + u_{yy} = f(x, y).$$

REMARK 6.14 Relation to Complex Functions

For analytic $f = u + iv$, u, v obey the Cauchy-Riemann equations

$$u_x = v_y, \quad u_y = -v_x.$$

Differentiating the first equation with respect to x, the second with respect to y, and adding gives the Laplace PDE for u. Similarly, we obtain the Laplace equation for v.

There is an intimate relation between analyticity and solving the Laplace equation.

DEFINITION 6.11 Harmonic Functions

Solutions of Laplace's equation.

6.3 THE HEAT EQUATION

> **DEFINITION 6.12** Heat Equation in Rectangular Coordinates
>
> $$u_y = \alpha u_{xx}, \quad \alpha > 0. \tag{6.3}$$

REMARK 6.15 Type of Equation

Parabolic: $a = \alpha$, $b = 0$, $c = 0$, $b^2 - 4ac = 0$ in equation (6.1).

REMARK 6.16 Representation of Variables

The heat equation is parabolic; one of the variables is "time;" the equation models a "marching" process with time y. We usually write "t" in place of y to stress that it is not a spatial variable, giving us

$$u_t = \alpha u_{xx}.$$

REMARK 6.17 Physical Meaning

The heat equation gives the temperature rate of change in terms of the present value of T_{xx}. It is a tool for predicting future temperature from its history.

EXAMPLE 6.3 Heat Transfer in a Rod

Insulate a cylindrical metal rod along its length, leaving its faces (A, B in Figure 6.11) exposed. At any of its points the temperature is constant on any cross section and depends only on the distance from the faces. Placing the x-axis along the rod length, the rod temperature is $u(x, t)$. If there are no internal heat sources and the rod physical properties are constant, then u obeys (6.3).

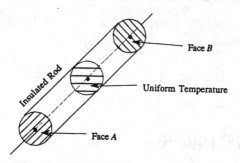

FIGURE 6.11 — Insulated Rod

REMARK 6.18 Equation in Higher Dimensions

$$u_t = \alpha[u_{xx} + u_{yy} + u_{zz}].$$

REMARK 6.19 Other Coordinate Systems

Circular Cylindrical

$$u_t = \alpha[(1/r)(ru_r)_r + (1/r^2)u_{\theta\theta} + u_{zz}]$$

Spherical Polar

$$u_t = \alpha[(1/r^2)(r^2 u_r)_r + (1/[r^2\sin(\phi)])(\sin(\phi)u_\phi)_\phi + (1/[r^2\sin^2(\phi)])u_{\theta\theta}]$$

Polar

$$u_t = \alpha[(1/r)(ru_r)_r + (1/r^2)u_{\theta\theta}]$$

REMARK 6.20 Properties of Equation (6.3)

Order 2, Homogeneous, Linear

REMARK 6.21 Properties of Solutions

Maximum Principle: The maximum and minimum values of any solution to the heat equation in the rectangle

$$R: 0 \le t \le t^*, 0 \le x \le a$$

of the x, t plane (see Figure 6.12) are achieved either at the time $t = 0$ or at the boundaries $x = 0, a$.

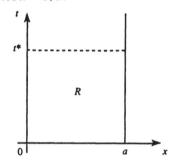

FIGURE 6.12 — Maximum Principle

REMARK 6.22 Form of Well Posed Problem

Solve equation (6.3) on $0 < x$, $t < \infty$ for the initial condition

$$u(x, 0) = f(x)$$

and the boundary condition

$$u(0, t) = g(t)$$

(Figure 6.13).

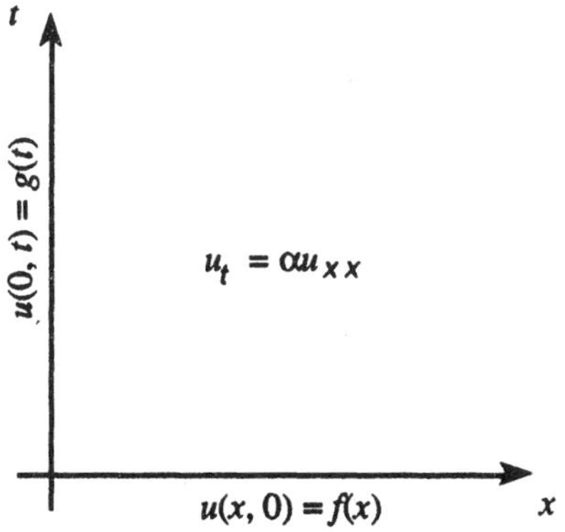

$u(0, t) = g(t)$

$u_t = \alpha u_{xx}$

$u(x, 0) = f(x)$

FIGURE 6.13 — Problem for Heat Equation

EXAMPLE 6.4 A Well Posed Problem and Its Solution

PDE $u_t(x, t) = \alpha u_{xx}(x, t)$

IC $u(x,0) = f_0$

BC $u(0, t) = g_0$

$u(x, t)$ is a similarity solution

$$u(x, t) = g_0 + (f_0 - g_0)\, \text{erf}(x/[2\sqrt{\alpha t}]).$$

Here is the error function erf is defined by

$$\text{erf}(s) = (2/\sqrt{\pi})\int_0^s \exp(-w^2)\,dw$$

satisfying the conditions: $\text{erf}(0) = 0$, $\text{erf}(\infty) = 1$.

REMARK 6.23 Related Equations

Laplace Equation: $u_{xx} + u_{yy} + u_{zz} = 0$

Steady-state form of the heat equation

Heat Source: $u_t = \alpha u_{xx} + W$

Nonlinear Case: For ρ, c the density and specific heat (taken as constants) and $k(u)$ the (temperature dependent) thermal conductivity, the heat transfer model is the nonlinear heat equation

$$\rho c u_t = (k(u) u_x)x$$

6.4 THE FIRST ORDER WAVE EQUATION

DEFINITION 6.13 In Rectangular Coordinates

$$u_t + c u_x = 0, \quad c = \text{given} \tag{6.4}$$

REMARK 6.24 Physical Meaning

Sound is a pressure wave moving through air causing the air density to vary with position x and time t; the wave moves at the sound speed c. If the direction of motion is that of increasing x then $c > 0$; if not, then $c < 0$. In either case, u will approximately obey the PDE (6.4).

REMARK 6.25 Equation in Higher Dimensions

In three space dimensions,

$$u_t + c_1 u_x + c_2 u_y + c_3 u_z = 0.$$

REMARK 6.26 Other Coordinate Systems

Circular Cylindrical

$$u_t + c_1 u_r + (c_2/r) u_\theta + c_3 u_z = 0$$

Spherical Polar

$$u_t + c_1 u_r + (c_2/[r\sin(\phi)]) u_\theta + (c_3/r) u_\phi = 0$$

Polar

$$u_t + c_1 u_r + (c_2/r) u_\theta = 0$$

REMARK 6.27 Properties of the PDE

Order 1, Homogeneous, Linear

REMARK 6.28 Properties of Solutions

$$u(x, t) = f(x - ct),$$

constant on all lines (Figure 6.14) $L: x = ct + x_0$.

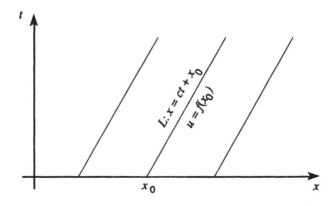

FIGURE 6.14 — $f(x - ct)$ Constant on L

DEFINITION 6.14 Characteristics of Equation (6.4)

The lines L of Remark 6.28.

REMARK 6.29 Form of Well Posed Problem

An initial condition and a single boundary condition at the upstream boundary.

EXAMPLE 6.5 A Well Posed Problem and Its Solution

PDE $u_t + u_x = 0$, ($c = 1$ in equation (6.4))

IC $u(x,0) = 0$, $x > 0$

BC $u(0, t) = 1$, $t > 0$

The solution is constant along all lines

$x = t + $ const.

$u = 1$ for $0 < x < t$ and $u = 0$ beyond (Figure 6.15).

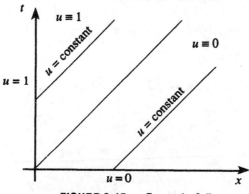

FIGURE 6.15 — Example 6.5

EXAMPLE 6.6 A Non-Well Posed Problem

For $u_t + u_x = 0$ signals move to the right; we cannot give initial conditions to the left of a boundary where we impose a condition (see Figure 6.16) since this would result in the impossibility of a line of constancy of the solution along which the solution has one value, meeting the boundary where a different value is imposed.

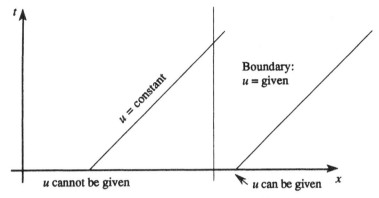

FIGURE 6.16 — Non-Well Posed Problem

6.5 THE WAVE EQUATION

DEFINITION 6.15 Wave Equation in Rectangular Coordinates

$u_{yy} = c^2 u_{xx}$, c = given

REMARK 6.30 Type of Equation

Hyperbolic: $a = c^2$, $b = 0$, $c = -1$, $b^2 - 4ac > 0$ in equation (6.1).

REMARK 6.31 Representation of Variables

The PDE is hyperbolic: one of the variables is time, and this models a "marching" process. The timelike variable is y. Accordingly, we write t in its place and have

$$u_{tt} = c^2 u_{xx}. \tag{6.5}$$

REMARK 6.32 Physical Meaning

The PDE describes wave propagation and vibrations. For example, if one subjects a string to vibrations $u(x, t)$ represents the time-dependent amplitude at the point x, time t. c is the wave speed.

REMARK 6.33 Equation in Higher Dimensions

$$u_{tt} = c^2[u_{xx} + u_{yy} + u_{zz}]$$

REMARK 6.34 Other Coordinate Systems

Circular Cylindrical

$$u_{tt} = c^2[(1/r)(ru_r)_r + (1/r^2)u_{\theta\theta} + u_{zz}]$$

Spherical Polar

$$u_{tt} = c^2[(1/r^2)(r^2 u_r)_r + (1/[r^2\sin(\phi)])(\sin(\phi)u_\phi)_\phi$$
$$+ (1/[r^2\sin^2(\phi)])u_{\theta\theta}]$$

Polar

$$u_{tt} = c^2[(1/r)(ru_r)_r + (1/r^2)u_{\theta\theta}]$$

REMARK 6.35 Properties of Wave Equation

Order 2, Homogeneous, Linear

REMARK 6.36 Properties of Solutions

General solution can be expressed as

$$u(x, t) = A(x - ct) + B(x + ct),$$

that is, the superposition of two waves moving at speed c, in opposite directions. The solution at any x_0 and time $t_0 > 0$ depends on the "past" within the *Domain of Dependence* bounded by the lines of slopes $\pm 1/c$ passing through the point (see Figure 6.17). Similarly, the solution

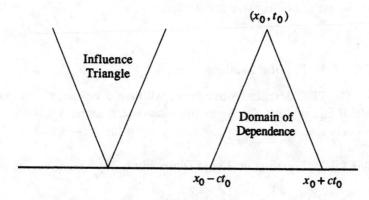

FIGURE 6.17 — Domains of Dependence and Influence

value at any initial time t_1 will effect the future only within the *Influence Triangle* bounded by the two lines of slope $\pm 1/c$ passing through this point.

DEFINITION 6.16 Characteristics

The families of lines of slope $\pm 1/c$ covering the plane.

REMARK 6.37 Form of Well Posed Problem

Consists of the PDE plus *two* initial conditions specifying u, u_t initially, and a single boundary condition at each boundary line. An example is the initial-value problem:

Find $u(x, t)$ satisfying equation (6.5) on $[a, b]$ and obeying the initial conditions

$$u(x, 0) = f(x), \quad u_t(x, 0) = g(x).$$

and the boundary conditions

$$u(a, t) = \alpha(t), \quad u(b, t) = \beta(t), t > 0,$$

for f, g, α, β given functions.

EXAMPLE 6.7 A Well Posed Problem and Its Solution

Solve the wave equation on the infinite line $-\infty < x < \infty$ for the initial conditions

$$u(x, 0) = f(x), \quad u_t(x, 0) = g(x).$$

The solution is given by D'Alembert's formula

$$u(x, t) = \tfrac{1}{2}[f(x - ct) + f(x + ct)] + (1/2c) \int_{x-ct}^{x+ct} g(s)ds.$$

REMARK 6.38 Related Equations

For many cases it is known that a vibrating system has a known angular frequency ω. For two space dimensions the solution will be of the form

$$u(x, y, t) = e^{-i\omega t} U(x, y).$$

Substitution into the wave equation

$$u_{tt} = c^2(u_{xx} + u_{yy})$$

yields the differential equation for $U(x, y)$,

$$U_{xx} + U_{yy} + (\omega/c)^2 U = 0.$$

This is the *Helmholt* or *Reduced Wave Equation*.

CHAPTER 7

CALCULUS OF VARIATIONS

Calculus of Variations is concerned with finding extreme values (maxima, minima). We examine this subject initially for functions and then for functions of functions or "functionals."

FAMILIAR IDEAS

Connected Set: A Set Each Pair of Whose Points Can Be Connected by Finitely Many Broken Lines (See Figure 7.1)

Boundary Point: Any Circle About the Point Contains Points Belonging to the Set and Points Not Belonging to It

Closed Domain: A Domain Containing All Its Boundary Points

Relative Minimum of $f(x)$: x^* such that $f(x) \geq f(x^*)$ for All $x \sim x^*$

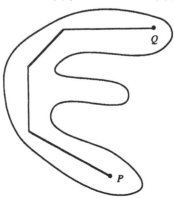

FIGURE 7.1 — Connected Set

Relative Maximum: x^* such that $f(x) \leq f(x^*)$ for All $x \sim x^*$

Relative Extremum: Relative Maximum or Minimum

Stationary Point of $f(x)$: x^* for which $f'(x^*) = 0$

Absolute Maximum (Minimum): x^* such that $f(x) < (>) f(x^*)$ for All $x = x^*$

Absolute Extremum: Absolute Maximum or Minimum

Relative Extremum is a Stationary Point

If x^* is a Stationary Point and $f'' < (>) 0$ then x^* is a Relative Maximum (Minimum)

7.1 BASIC THEORY OF MAXIMA AND MINIMA

PROBLEM 7.1 Fundamental Problem of Extrema of Functions

Find values $X = (x_1, ..., x_N)$ where a function $f(X)$ continuous in a closed domain of N dimensional space has relative and absolute extrema.

THEOREM 7.1 Weierstrass Theorem

A continuous function in a closed domain has a largest and smallest value either in the interior or on the boundary.

DEFINITION 7.1 Distance Between Points

For $X = (x_1, ..., x_n)$, $X^* = (x_1^*, ..., x_n^*)$,

$$|X - X^*| = (\sum_{i=1}^{n} (x_i - x_i^*)^2)^{1/2}$$

DEFINITION 7.2 Spherical Neighborhood of X^*

The points X for which $|X - X^*| < \varepsilon$ for some ε.

DEFINITION 7.3 Relative Minimum X^*

$f(X^*) \leq f(X)$ for all X in some spherical neighborhood.

DEFINITION 7.4 Relative Maximum X^*

$f(X^*) \geq f(X)$ for all X in some spherical neighborhood.

DEFINITION 7.5 Relative Extremum

Relative maximum or minimum.

DEFINITION 7.6 Absolute Minimum/Maximum X^* in R

$f(X^*)$ is least/greatest over all R.

EXAMPLE 7.1 Absolute Maximum and Minimum (Figure 7.2)

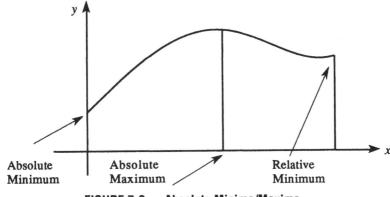

FIGURE 7.2 — Absolute Minima/Maxima

DEFINITION 7.7 Gradient of $f(X)$

$\nabla f(X) = (\partial f/\partial x_1, \ldots, \partial f/\partial x_n)$.

DEFINITION 7.8 Stationary Point X^* of $f(X)$

$\nabla f(X^*) = 0$.

REMARK 7.1 Stationary Point Need Not Be Extremum

$x = y = 0$ is stationary but not extremum for $f(x, y) = xy$.

REMARK 7.2 Constraints

The problem of finding minima or maxima is often constrained by the condition that the extremizing point X^* satisfy one or more auxiliary conditions.

DEFINITION 7.9 Constrained Extremizing Problem

Among all points X for which

$$g_j(X) = 0, \; j = 1, 2, \ldots, m$$

find a point X^* for which $f(X)$ has an extremum.

EXAMPLE 7.2 A Constrained Extremizing Problem

Among all bodies with given surface area the sphere has the largest volume.

REMARK 7.3 Lagrange Multipliers for Constrained Extrema

An extremum X^* of $f(x)$ is a stationary point for

$$F(X, \Gamma) = f(X) + \Gamma_1 g_1(X) + \ldots + \Gamma_n g_n(X):$$

$\partial F/\partial x_1 = 0$

$\partial F/\partial x_2 = 0$

...

$\partial F/\partial x_n = 0$

$\partial F/\partial \Gamma_1 = 0$

...

$\partial F/\partial \Gamma_n = 0$

There are $2n$ equations for the $2n$ values we seek.

EXAMPLE 7.3 The Right Cylinder of Given Area and Maximum Volume

Find the right cylinder of largest volume with given total surface area A_0.

Let r, h be the (unknown) radius and height of the cylinder. Its volume is $V = \pi r^2 h$ while the surface area constraint is $g(r, h) = 2\pi r^2 + 2\pi rh - A_0 = 0$. One way of solving this problem is to solve for h in terms of r from the constraint, substitute the result and then minimize V. We thus obtain

$$h = [A_0 - 2\pi r^2] / [2\pi r]$$
$$V = \pi r^2 [A_0 - 2\pi r^2] / [2\pi r]$$
$$= \frac{1}{2} r [A_0 - 2\pi r^2]$$

Differentiation of V with respect to r and setting the result equal to zero yields

$$r = [A_0 / 6\pi]^{1/2}, \; y = [2A_0 / 3\pi]^{1/2}.$$

The same result is obtained using a single Lagrange multiplier by seeking the stationary point of

$$F(r, h, \Gamma) = \pi r^2 h + \Gamma(2\pi r^2 + 2\pi rh - A_0).$$

Differentiating with respect to r, h, Γ and setting the results equal to zero yields the equations

$$F_r = 2\pi rh + \Gamma(4\pi r + 2\pi h) = 0,$$
$$F_\Gamma = 2\pi r^2 + 2\pi r\Gamma - A_0,$$
$$F_h = \pi r^2 + 2\pi r\Gamma = 0.$$

The third equation yields $\Gamma = -r/2$. The first then gives $r = h/2$ which, via the second, yields the earlier result.

REMARK 7.4 When Are Lagrange Multipliers Useful?

When elimination of variables becomes cumbersome or impossibly complex.

THEOREM 7.2 Direction of Gradient

That of the maximum rate of increase of the function. The negative of the gradient is in the direction of maximum decrease.

> **REMARK 7.5** Gradient Method ■ Method of Steepest Descent
>
> Starting at any point we move through small steps, each in the direction opposite that of the current gradient, until reaching a relative minimum. Starting at $\mathbf{X}_0$ we set
>
> $$\mathbf{X}_1 = \mathbf{X}_0 - t_1 \nabla f(X_0)$$
> $$\mathbf{X}_2 = \mathbf{X}_1 - t_2 \nabla f(X_1)$$
> ...
>
> for $t_1, t_2, \ldots$ sufficiently small.

REMARK 7.6 Shortcomings of the Gradient Method

The gradient method has two major limitations. The first is that the gradient may not exist, or be difficult to evaluate. A simple way to deal with this case is to replace the derivatives by finite differences to indicate a direction to be followed. The second limitation is that the method will lead to the nearest relative extremum instead of to the absolute extremum desired. Methods exist for "leaping" from a relative extremum in order to proceed to others. "Simulated Annealing" is one method, based on the use of tools of probability and giving the absolute extremum.

REMARK 7.7 Maximum Search

The gradient method finds relative maxima by moving in the direction of the gradient.

7.2 THE SIMPLEST PROBLEM OF VARIATIONAL CALCULUS

DEFINITION 7.10 Admissible Functions

For a given $f(x, y, z)$, $y = y(x)$ is *Admissible on* $[a, b]$ if

i) $y(x), y'(x), y''(x)$ are continuous on $[a, b]$;

ii) $f(x, y(x), y'(x))$ is defined and continuous on $a \le x \le b$;

iii) $y(a) = y_0, y(b) = y_1, y_0, y_1 =$ given.

DEFINITION 7.11 The Fundamental Problem

Let Ω be the collection of all admissible functions $y(x)$. Find $y = Y(x)$ in Ω giving the least value of

$$F[f] = \int_a^b f(x, y(x), y'(x))dx.$$

REMARK 7.8 Functionals

A *Functional* is a rule assigning a unique real number to each function $y(x)$ in Ω. $F[f]$ is a function over Ω.

REMARK 7.9 Notation

The third variable of f is the derivative y'. We write the partial derivative of f with respect to it as $f_{y'}$.

DEFINITION 7.12 First Variation $\delta F[f]$

$$\delta F[y] = \int_a^b \{\phi(x)f_y + \phi'(x)f_{y'}\}dx$$

for $\phi(x)$ vanishing at $x = a, b$.

DEFINITION 7.13 Stationary Function

A function $y(x)$ with vanishing first variation, $\delta F[y]$.

THEOREM 7.3 Euler Equation

If $Y(x)$ minimizes $F[y]$, then $Y(x)$ obeys the Euler ODE

$$F_y(x, Y(x), Y'(x)) = (d/dx)F_{y'}(x, Y(x), Y'(x)).$$

REMARK 7.10 On Euler's Equation

Euler's equation is a 2nd order ODE, for which the boundary conditions for admissibility at $x = a, b$ are suitable.

EXAMPLE 7.4 A Functional and Euler Equation

Let Ω be the set of all $y(x)$ with y', y'' continuous, and $y(0) = 0$, $y(1) = 1$. For the functional

$$F[y] = \int_0^1 [y(x)^2 + y'(x)^2]dx$$

$$f(x, y, z) = y^2 + z^2,$$

$$f_y = 2y, \quad f_{y'} = 2y'$$

and the Euler equation becomes

$$2y(x) = (d/dx)(2y'(x)) = 2y''(x)$$

or

$$y''(x) = y(x).$$

This equation has the general solution

$$y(x) = Ae^x + Be^{-x}.$$

Substitution into the boundary conditions yields

$$A + B = 0, \quad Ae + B/e = 1$$

or $A = 1/[e - 1/e]$, $B = -1/[e - 1/e]$. Hence, the solution to the minimum problem over Ω is

$$y(x) = (e^x - e^{-x}) / [e - 1/e]$$

$$= \sinh(x) / \sinh(1).$$

THEOREM 7.4 Fundamental Theorem of the Calculus of Variations

If $w(x)$ is any continuous function on $[a, b]$ such that

$$\int_a^b w(x)\phi(x)dx = 0$$

for all continuous $\phi(x)$ vanishing at $x = a, b$, then $w(x) \equiv 0$.

DEFINITION 7.14 A Functional of Many Functions

$$F[y_1, y_2, \ldots, y_N] = \int_a^b f(x, y_1(x), \ldots, y_N(x), y_1'(x), \ldots, y_N'(x))dx.$$

DEFINITION 7.15 Admissible Vector Function

A vector function $\mathbf{Y}(x) = (y_1, y_2, \ldots, y_N)$ defined on $[a, b]$, assuming given values at $x = a, b$, for which the integrand $F[y_1, y_2, ..., y_N]$ is defined.

DEFINITION 7.16 Minimum Problem for F

Among all admissible functions in Ω find one minimizing F.

THEOREM 7.5 Euler's Equation for F

If $Y^*(x)$ minimizes F, then it satisfies the system of N Euler equations

$\partial F/\partial y_i = (d/dx)\partial F/\partial y_i', i = 1, ..., N.$

DEFINITION 7.17 Functionals in Two Dimensions

Let R be a closed domain in the x, y plane and $f(x, y, u, v, w)$ be given. For any $u(x, y)$ define the functional $F[u]$ by

$$F[u] = \iint_R f(x, y, u(x, y), u_x(x, y), u_y(x, y))\, dxdy;$$

$u(x, y)$ is *admissible* if $F[u]$ is defined and u takes on given values at the boundary Γ of R. Ω is the set of all admissible functions.

DEFINITION 7.18 Minimum Problem for $F[u]$

Find $u^*(x, y)$ in Ω minimizing $F[u]$.

THEOREM 7.6 Euler's Equation for $F[u]$

$F_u(x, y, u(x, y), u_x(x, y), u_y(x, y))$
$= (\partial/\partial x) F_u x(x, y, u(x, y), u_x(x, y), u_y(x, y))$
$+ (\partial/\partial x) F_u y(x, y, u(x, y), u_x(x, y), u_y(x, y))$

EXAMPLE 7.5 Poisson's Equation as Euler's Equation

For given $g(x, y)$,

$$F(u) = \iint_R \{u_x^2 + u_y^2 + 2ug\}\, dxdy$$

Euler Equation: $g = u_{xx} + u_{yy}.$

7.3 SOME CLASSICAL PROBLEMS

Some classical Calculus of Variations Problems are

PROBLEM 7.2 Steiner's Problem (see Figure 7.3)

Find a road system of least total length joining N cities.

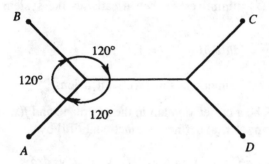

FIGURE 7.3 — Cities at A, B, C, D

PROBLEM 7.3 Brachistrochrone Problem

Find the path of least travel time for a particle, sliding without friction but under the influence of gravity, from a point A to a lower point B.

PROBLEM 7.4 Catenary Problem

What is the curve assumed by a string of given length with fixed end points, subject to the influence of gravity? The minimum problem is that of making the center of gravity of the string as low as possible.

PROBLEM 7.5 Plateau Problem

Find a surface of least area spanning a given closed wire frame in three dimensions.

PROBLEM 7.6 Hamilton's Principle

In progressing from a time t_0 to a time t_1, a mechanical system whose properties are determined by the kinetic energy T and the potential energy U will proceed in such a way as to make the integral

$$J = \int_{t_0}^{t_1} (T - U) dt$$

stationary over all possible motions.

7.4 CONTROL

DEFINITION 7.19 Control Process

A process governed by differential equations (ODE or PDE) including a forcing or *Control* term to be used to produce a desired performance for the system being modeled.

EXAMPLE 7.6 Temperature Control

Find a source $W(x, t)$ providing a desired temperature history for a rod subject to the conditions:

$T_t = \alpha T_{xx} + W(x, t)$ (PDE)

$T(a, t) = T(b, t) = 0$ (Boundary Conditions)

$T(x, 0) = f(x)$ (Initial Conditions)

EXAMPLE 7.7 A Least Time Control Problem

Find $u(t)$ such that $x''(t) = u(t)$, $x(0) = x'(0) = 0$,

$x(t^*) = 1$, $x'(t^*) = 0$, $-1 \leq u(t) \leq 1$ and

t^* is least.

This problem is one of finding the acceleration (positive or negative) minimizing the travel time from 0 to 1, and subject to the constraint of never exceeding 1 in absolute value. Moreover, the body is to begin and end its trip at rest.

This is a variational problem which cannot be handled via the Euler equation approach. The solution $u(t)$ is a *Bang-Bang* control, jumping between the constraining values ± 1. On some time interval $0 \le t \le t_{sw}$, $u(t) = 1$; it then switches to the value -1 on $t_{sw} \le t \le t^*$. The relation between t^* and t_{sw} is found by solving the problem

$$x''(t) = 1, \; x(0) = 0, \; x'(0) = 0$$

yielding the solution

$$x(t) = \tfrac{1}{2} t^2$$

with the velocity

$$v(t) = t.$$

This is the solution up to the (unknown) time t_{sw}. Beyond t_{sw} $x(t)$ is found via

$$x''(t) = -1, \; x(t^*) = 1, \; x'(t^*) = 0.$$

The solution to this problem is

$$x(t) = 1 - \tfrac{1}{2}(t - t^*)^2$$

with the velocity

$$v(t) = -(t - t^*).$$

At the time t_{sw} we require that $x(t)$ be continuous. Thus,

$$\tfrac{1}{2} t_{sw}^2 = 1 - \tfrac{1}{2}(t_{sw} - t^*)^2.$$

Solving for t^* in terms of t_{sw} yields

$$t^* = t_{sw} + [2 - t_{sw}^2]^{1/2}$$

From the calculus we find that t^* attains its least value at $t_{sw} = 1$. In turn $t^* = 2$ and $u(t)$ is found as the bang-bang control

$$u(t) = \begin{cases} 1, & 0 \le t \le 1 \\ -1, & 1 \le t \le 2. \end{cases}$$

7.5 DYNAMIC PROGRAMMING

DEFINITION 7.20 State of a System

A description of the system in terms of one or more quantities. Let the possible states of the system be denoted by S.

DEFINITION 7.21 Controller of a System

One or more functions representing our means for controlling the system. This can be represented as a function $u(t)$.

REMARK 7.11 On Dynamic Programming

Dynamic programming is a method for determining how to move a system from an initial state S_{init} at some starting time $t = a$ to a final state S_{end} at time $t = b$ at least total "cost."

DEFINITION 7.22 Cost of a Process

The cost of the system evolving from a state S_i to the state S_j over a time period $[a,b]$. It depends on the states, the times, and the controller and is written as $C[a, b, S_i, S_j, u]$.

DEFINITION 7.23 Least Cost of a Process

Given an initial state S_i and a final state S_j assume there is a controller U minimizing the cost

$$C[t_i, t_j, S_i, S_j, u]$$

over $[a, b]$. This least cost is denoted by $C^*[a, b, S_i, S_j]$.

THEOREM 7.7 Principle of Dynamic Programming

Let $a < c < b$ be any three times. Assume that we have found a least cost controller for moving from the initial state S_{init} to *any* possible state S at time c. Then the least cost process of going from S_{init} at time $t = a$ to S_{end} at time $t = b$ is found by minimizing the cost

$$\min. \{C^*[a, c, S_{init}, S] + C[c, b, S, S_{end}, u]\} \text{ all } u \text{ on } [c, b].$$

We can use this to incrementally find the least path in discrete time steps over the interval $[a, b]$ (Figure 7.4).

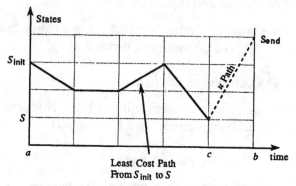

FIGURE 7.4 — Dynamic Programming Setting

7.6 LINEAR PROGRAMMING

The linear programming (LP) problem is best illustrated by the following simple problem:

EXAMPLE 7.8 An LP Problem

Maximize the function

$$y = -ax + b, \quad a, b > 0$$

subject to the conditions

$$y \leq -Ax + B, \quad A, B > 0 \tag{7.1}$$

$$x, y \geq 0.$$

There are four possibilities as seen when we compare the lines

$$L_1: y = -ax + b, L_2: y = -Ax + B$$

(see Figure 7.5). Let us refer to those points (x, y) lying below L_2 and on L_1 as *feasible*. In the first case, every point on L_1 lies below L_2 and the constraint (7.1) is not limiting. In case II, it is, and the solution is at the intersection of the two lines. For case III, the constraint is again not limiting. However, in case IV, there are no feasible points and no solution.

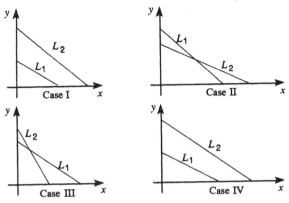

FIGURE 7.5 — The LP Problem

REMARK 7.12 Simplex Method

The solution will always be among the vertices of the region bounded by the lines and axes. This fact is the basis for the *Simplex Method* which is used for solving large-scale programming problems. The basic idea of the simplex method is to systematically move through vertices until reaching the solution vertex.

DEFINITION 7.24 General Linear Programming Problem

Maximize

$$f = c_1 x_1 + \ldots + c_n x_n$$

subject to the constraints

$$a_{11} x_1 + \ldots + a_{1n} x_n = b_1$$
$$a_{21} x_1 + \ldots + a_{2n} x_n = b_2$$
$$\ldots$$
$$a_{m1} x_1 + \ldots + a_{mn} x_n = b_m$$
$$x_j \geq 0, \quad j = 1, \ldots, n.$$

CHAPTER 8

NUMERICAL METHODS

8.1 SOLUTION OF EQUATIONS

DEFINITION 8.1 Roundoff Error

The error in replacing a number having infinitely many decimal places by one with a finite number of decimal places, as for example, $\pi \approx 3.14159$ and $\sqrt{2} \approx 1.414$.

DEFINITION 8.2 Approximation

A value x_{approx} that is to be used in place of the true value x_{true} of a quantity.

DEFINITION 8.3 Absolute Error of Approximation

$$\varepsilon_{abs} = |x_{true} - x_{approx}|.$$

DEFINITION 8.4 Relative Error of Approximation

$$\varepsilon_{rel} = e_{abs} / |x_{true}|.$$

DEFINITION 8.5 Fixed Point of a Function $\phi(x)$

An x^* for which

$$x^* = \phi(x^*)$$

(see Figure 8.1).

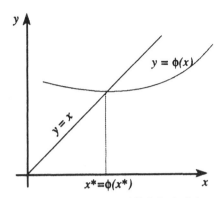

FIGURE 8.1 — Fixed Point of $\phi(x)$

METHOD 8.1 Iteration Method for Finding the Fixed Point

For some *Initial Iterate* x_0 we use the *Iteration Scheme*

$$x_1 = \phi(x_0), x_2 = \phi(x_1), \ldots, x_{n+1} = \phi(x_n) \ldots$$

THEOREM 8.1 Convergence of Iteration Scheme

If ϕ has a continuous derivative and $|\phi'(x)| \le \alpha < 1$, then the iteration scheme $x_{n+1} = \phi(x_n)$ converges.

METHOD 8.2 Bisection Method for Solving $f(x) = 0$

Let a, b be approximations to a root of $f(x) = 0$ such that f differs in sign at these points, $f(a) f(b) < 0$; then we choose the next approximation to be the average of a, b, $c = \frac{1}{2}(a + b)$, examine the sign of $f(c)$ and select that one of the values a, b at which f differs in sign from $f(c)$ to serve with c as the new pair of values to be averaged. (See Figure 8.2.)

METHOD 8.3 Newton-Raphson Method

If f, f' are continuous and f' is not zero at the root of $f(x) = 0$, then for an initial iterate x_0 close to the desired root, we compute the iterations

$$x_{n+1} = x_n - f(x_n)/f'(x_n), n = 0, 1, 2, \ldots.$$

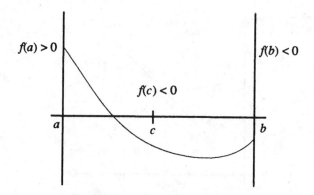

FIGURE 8.2 — Bisection Method

METHOD 8.4 Rule of False Position

If the derivative of f needed in the Newton-Raphson method is difficult to compute, then it can be replaced by a finite difference giving us

$$x_{n+1} = x_n - f(x_n) / \{[f(x_n) - f(x_{n-1})] / [x_n - x_{n-1}]\}$$

8.2 FUNCTION APPROXIMATION

DEFINITION 8.6 Interpolation Polynomial $\phi(x)$

Find a polynomial $\phi(x)$ of degree $\leq N$ assuming given values $f_0, f_1, ..., f_N$ of the function $f(x)$ at $N + 1$ distinct points $x_0, x_1, ..., x_N$:

$$\phi(x_0) = f_0, \phi(x_1) = f_1, ..., \phi(x_n) = f_N.$$

$x_i, i = 0, N$ are the *Interpolation Points*.

THEOREM 8.2 Uniqueness of Interpolation Polynomial

The interpolation polynomial is unique.

THEOREM 8.3 Error of Interpolation Polynomial

For any x there is a point Γ such that

$$f(x) - \phi(x) = (x - x_0)(x - x_1) ... (x - x_N) f^{N+1}(\Gamma) / (N + 1)!$$

DEFINITION 8.7 Lagrange Interpolation Polynomial

For any $k = 0, 1, 2, \ldots, N$, the *Exclusive Product* P_k is the product of all terms $(x - x_i)$ *except for* $(x - x_k)$.

$$P_0(x) = (x - x_1)(x - x_2) \ldots (x - x_N)$$
$$P_1(x) = (x - x_0)(x - x_2) \ldots (x - x_N)$$
$$\ldots$$
$$P_N(x) = (x - x_0)(x - x_1) \ldots (x - x_{N-1}).$$

For

$$Q_k(x) = P_k(x) / P_k(x_k),$$
$$Q_k(x_j) = 0 \text{ for } j = k, = 1 \text{ for } j = k.$$

The *Lagrange Interpolation Polynomial* is

$$\phi(x) = \sum_{j=0}^{N} f_j Q_j(x).$$

REMARK 8.1 Other Approaches

As the order of the interpolation polynomial grows larger, the approximation may become poorer. One alternative is the *Method of Least Squares*, defined as:

DEFINITION 8.8 Least Squares Problem

Let $f(x)$ be continuous on $[a, b]$. Find the "closest" line

$$\phi(x) = Ax + B$$

to f in the sense that it minimizes the integral

$$\delta(f, \phi) = \{\int_a^b [\phi(x) - f(x)]^2 \, dx\}^{1/2}$$

METHOD 8.5 Solution of the Least Squares Problem

Minimizing $\delta(f, \phi)$ over all polynomials ϕ is equivalent to minimizing its square, considered as a function of A, B:

$$F(A, B) = \int_a^b [Ax + B - f(x)]^2 \, dx.$$

Setting the first derivatives equal to zero yields

$$\alpha_{11} A + \alpha_{12} B = \beta_1$$
$$\alpha_{21} A + \alpha_{22} B = \beta_2$$

$$\alpha_{11} = \int_a^b x \, dx = \tfrac{1}{2}\{b^2 - a^2\}$$

$$\alpha_{12} = \int_a^b dx = b - a$$

$$\alpha_{21} = \int_a^b x^2 \, dx = (\tfrac{1}{3})\{b^3 - a^3\}$$

$$\alpha_{22} = \alpha_{11}$$

$$\beta_1 = \int_a^b f(x) \, dx, \quad \beta_2 = \int_a^b x f(x) \, dx.$$

The solution is then

$$A = [\beta_1 \alpha_{22} - \beta_2 \alpha_{12}] / D, \quad B = [\beta_2 \alpha_{11} - \beta_1 \alpha_{21}]/D,$$

where $D = \alpha_{11} \alpha_{22} - \alpha_{12} \alpha_{21}$.

REMARK 8.2 Least Squares for Data Points

To approximate data points (x_j, f_j) by a line in the least squares sense we use the "distance"

$$\delta(f, \phi) = \left\{ \sum_{j=1}^N [Ax_j + B - f_j]^2 \right\}^{1/2}.$$

REMARK 8.3 Regression Line of Statistics

The closest least squares line is the *Regression Line*.

REMARK 8.4 Fourier Series and Least Squares

The Fourier polynomial is the nearest trigonometric polynomial in the least squares sense,

$$\phi(x) = a_0 + \sum_{n=1}^{N} \{a_n \cos(nx) + b_n \sin(nx)\}.$$

8.3 NUMERICAL INTEGRATION

We now turn to the evaluation of the definite integral

$$I = \int_a^b f(x)dx.$$

To do this we divide $[a, b]$ into N equal subintervals of length $h = (b - a)/N$ via the *Mesh Points* $x_0 = a$, $x_1 = a + h$, ..., $x_N = a + Nh = b$. I is the sum of the integrals over each of these subintervals:

$$I = \sum_{n=1}^{N} I_n$$

$$I_n = \int_{x_{n-1}}^{x_n} f(x)dx.$$

The simplest methods are the *Trapezoid* and *Simpson's Rule*. In the first, we replace $f(x)$ on each subinterval by the line joining the endpoints of its curve (see Figure 8.3). Simpson's Rule replaces f by an interpolating parabola at the endpoints and center of each subinterval (see Figure 8.4).

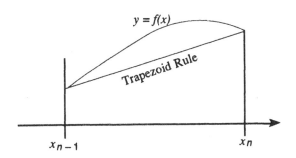

Figure 8.3 — Trapezoid Rule

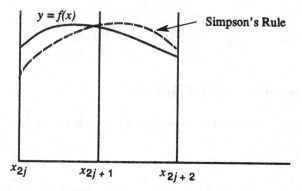

FIGURE 8.4 — Simpson's Rule

DEFINITION 8.9 Trapezoid Approximation for I

For $f_0 = f(x_0), \ldots, f_N = f(x_N)$ the Trapezoid Rule is

$$T_N = \tfrac{1}{2}h\,(f_0 + f_N) + h\sum_{n=1}^{N-1} f_n.$$

THEOREM 8.4 Error of Trapezoid Approximation

If f, f', f'' are continuous on $[a, b]$ with $|f''(x)| \le M_2$,

$$|I - T_N| \le (1/12)h^2 M_2(b - a).$$

REMARK 8.5 Trapezoid Rule Has Second Order Accuracy

The error is proportional to h_2.

DEFINITION 8.10 Simpson's Rule Approximation for I

Suppose the number N of subdivisions of $[a, b]$ is even: $N = 2M$.

Simpson's rule S_N is

$$S_N = (h/3)(f_0 + f_{2m}) + (4h/3)(f_1 + f_3 + \ldots + f_{2m-1})$$
$$+ (2h/3)(f_2 + f_4 + \ldots + f_{2m-2}).$$

THEOREM 8.5 Error of Simpson Approximation

For $f, \ldots, f^4$ continuous with $|f^{(4)}(x)| \leq M_4$,

$$|I - S_N| \leq (1/180)h^4 M_4 (b-a).$$

REMARK 8.6 Simpson's Rule Has 4-th Order Accuracy

REMARK 8.7 Monte-Carlo Integration

A different numerical integration approach is based on a "shooting" exercise using random numbers. For simplicity assume that the integral I is on the unit interval $[0, 1]$, with $0 \leq f(x) \leq 1$. Suppose that we randomly "shoot" at the square $0 \leq x, y \leq 1$. The chance (probability) of hitting a point below the graph of $y = f(x)$ is equal to the area under this graph, or the integral of $f(x)$ (see Figure 8.5). But in M random selections of points in the square, if N is the number of hits below the graph then $N/M \to I$ as $M, N \to \infty$. Thus, for M sufficiently large N/M is an approximation to I. The implementation of such a "Monte-Carlo method" on a computer (or even by hand) is straightforward.

8.4 NUMERICAL LINEAR ALGEBRA

DEFINITION 8.11 System of Linear Equations

$$a_{11}x_1 + \ldots + a_{1N}x_N = b_1$$
$$a_{21}x_1 + \ldots + a_{2N}x_N = b_2$$
$$\ldots$$
$$a_{N1}x_1 + \ldots + a_{NN}x_N = b_N \tag{8.1}$$

The *Coefficients* $\{a_{ij}\}$ and *Right Hand Side* $\{b_j\}$ are given. The system is $N \times N$ and the $\{x_i\}$ form its *solution*.

REMARK 8.8 Cramer's Rule

Using Cramer's rule for large N is impractical because of the large number of multiplications and roundoff errors.

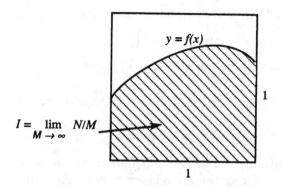

FIGURE 8.5 — Monte-Carlo Method

METHOD 8.6 Gauss Elimination for Solving a System

We successively eliminate unknown variables by an ordered subtraction of equations from each other. We illustrate this approach for the system of order 3.

i) $x_1 + 4x_2 + 2x_3 = -2$

ii) $5x_1 + x_2 - 2x_3 = 1$

iii) $-x_1 - 2x_2 + 6x_3 = 0$

Multiply i) by 5 and subtract it from ii); similarly add the original i) to iii); we obtain the system

i) $x_1 + 4x_2 + 2x_3 = -2$

ii) $-19x_2 - 12x_3 = 11$

iii) $2x_2 + 8x_3 = -2$

x_1 appears only in i) while ii), iii) is a system of two equations in two unknowns. Now multiply ii) by 2/19 and add the result to iii), yielding

iii) $(-24/19 + 8) x_3 = -2 + 22/19$.

This is now solved for x_3:

$$x_3 = -1/8.$$

From ii)

$$x_2 = -(1/19)(11 + 12x_3)$$
$$= -1/2.$$

while from i),

$$x_1 = -2 - 4x_2 - 2x_3 = 1/4.$$

DEFINITION 8.12 Pivotal Coefficient

The coefficient of the first term reached in each equation. In the above example, these pivots are 1 and -19.

REMARK 8.9 The Case of a Small Pivotal Coefficient

Suppose the coefficient of x_2 in ii) is a small number such as -0.00019. Elimination then requires multiplication of the equation by $2/0.00019$, introducing the possibility of a large roundoff error in the final equation for x_3. For this reason *Pivoting Methods* are introduced. In these we choose the *largest* leading coefficient as the next pivot. Thus, if the coefficient of x_2 is indeed -0.00019, we would consider equations ii), iii) as

ii) $\quad -12x_3 - 0.00019x_2 = 11$

iii) $\quad\quad 8x_3 + \quad\quad 2x_2 = -2$

so that the next pivot is -12 instead.

DEFINITION 8.13 Upper Triangular Matrix

$$A = \begin{pmatrix} a_{11} & a_{12} & \cdots & a_{1N} \\ a_{21} & a_{22} & \cdots & a_{2N} \\ \cdots & & & \\ a_{N1} & a_{N2} & \cdots & a_{NN} \end{pmatrix}$$

is upper triangular if terms below the diagonal are zero:

$$a_{ij} = 0, \text{ for } i > j.$$

DEFINITION 8.14 Lower Triangular Matrix

All the terms above the diagonal are zero:

$a_{ij} = 0$, for $i < j$.

THEOREM 8.6 Cholesky Decomposition

Under certain conditions A can be written as

$A = LU$

for L, U lower and upper triangular matrices. L and U are uniquely determined if we specify the diagonal elements of one of them. If A is symmetric ($a_{ij} = a_{ji}$) then L is the transpose of U and vice versa.

METHOD 8.7 Cholesky's Method

Solving system (8.1) is equivalent to solving the vector equation

$A\mathbf{X} = \mathbf{B}$.

Using the Cholesky decomposition this becomes

$LU\mathbf{X} = \mathbf{B}$

and premultiplying by the inverse L^{-1} of L,

$U\mathbf{X} = L^{-1}\mathbf{B}$.

But this is the form obtained by Gauss elimination, which can be solved for each variable in turn.

REMARK 8.10 Matrix Inversion

Finding the inverse $C = A^{-1}$ of a square, nonsingular matrix A, is equivalent to finding C such that

$AC = I$

for I the identity matrix. Let $\mathbf{Y}_1$, $\mathbf{Y}_2$, ..., $\mathbf{Y}_N$ be the vectors formed from the first, second, ..., N-th columns of C. Define the vectors $\mathbf{i}_1$, $\mathbf{i}_2$, ..., $\mathbf{i}_N$ as those formed from the corresponding columns of the identity matrix I. Then the matrix equation (8.2) can be regarded as N linear systems of equations with the same coefficient matrix. Each can be solved by Gauss Elimination to yield the columns of C as their solution.

DEFINITION 8.15 Sparse System of Equations

Systems for which the coefficient matrix is of large order but with few non-zero terms. Sparse systems arise frequently in engineering applications.

REMARK 8.11 Elimination Versus Iterative Methods

Elimination methods such as that of Gauss require handling and storage of the system coefficients, whether the system is sparse or not. For this reason, it is often preferable to use iterative methods for solving such systems.

REMARK 8.12 Iterative Methods for a System of Equations

In iterative methods we represent the system in the form of the fixed point equation $x = \phi(x)$ examined earlier.

METHOD 8.8 Jacobi Iteration

The system $A\mathbf{X} = \mathbf{B}$ can be written

$$\mathbf{X} = (I - A)\mathbf{X} + \mathbf{B}.$$

Starting with an initial iterate $\mathbf{X}_0$, we define the iteration sequence $\{\mathbf{X}_n\}$ by

$$\mathbf{X}_{n+1} = (I - A)\mathbf{X}_n + \mathbf{B}.$$

METHOD 8.9 Gauss-Seidel Iteration

In practice we find the coefficients of $\mathbf{X}_{n+1}$ in turn, starting from the first. The latest known approximation to each is not used until the next complete cycle. The Gauss-Seidel method uses the latest known values of each element at each step.

REMARK 8.13 On Iteration Methods

A key extension of the Gauss-Seidel method is the SOR (Successive Over-Relaxation) method, in which convergence is accelerated by using information about the error of the previous iteration.

Recently, with the growing popularity of *Parallel Computers*, it has become clear that the choice of a numerical method must take into account the architecture of the computer being used. Thus, Jacobi's method, which is poor for a serial computer, is readily suitable to a parallel computer.

REMARK 8.14 Ill-Conditioned Systems

Corresponding to the concept of non-well-posedness is that of ill conditioning of systems of equations. A system is ill conditioned if small errors in the coefficients or introduced in the solution process have a large effect on the solution. This corresponds to the case of a determinant whose value is close to zero.

DEFINITION 8.16 Eigenvalue of a Matrix

A number Γ for which $\det(A - \Gamma I) = 0$ or equivalently, there is a vector (*eigenvector*) $\mathbf{v}$ such that $A\mathbf{v} = \Gamma \mathbf{v}$.

THEOREM 8.7 Eigenvalues of a Symmetric Matrix

The eigenvalues of a symmetric matrix are real numbers.

THEOREM 8.8 Gershgorin's Theorem

Let Γ be any eigenvalue of A. Then for some j,

$$|a_{jj} - \Gamma| \le |a_{j1}| + |a_{j2}| + \ldots + |a_{j,\,j-1}| + |a_{j,\,j+1}| + \ldots + |a_{jN}|$$

THEOREM 8.9 Rayleigh Quotient for Symmetric Matrix A

The *largest eigenvalue* is given by the Rayleigh Quotient

$$\Gamma = \max_{X} \{(AX, X) / (X, X)\}.$$

Here (X, Y) is the inner product of X and Y. Furthermore, suppose that X^* is the maximizing vector. The *second largest eig-*

envalue μ is given by

$$\mu = \max_{X \perp X^*} \{(AX,X)/(X,X)\}.$$

8.5 SOLVING ORDINARY DIFFERENTIAL EQUATIONS

REMARK 8.15 Initial Value Problem

Find $y(x)$ for $x > a$ satisfying the initial condition $y(a) = y_0$ and the ODE

$$y'(x) = f(x, y(x))$$

REMARK 8.16 Definition of a Mesh

For $h > 0$ a (small) *Mesh Length* define *Mesh Points*

$$x_0 = a, x_1 = a + h, \ldots, a_j = a + jh.$$

METHOD 8.10 Euler's Method

At any point x_j

$$y'(x_j) \approx [y(x_{j+1}) - y(x_j)] / h.$$

Hence, the differential equation yields

$$[y(x_{j+1}) - y(x_j)] / h \approx f(x_j, y(x_j))$$

or

$$y(x_{j+1}) \approx y(x_j) + hf(x_j, y(x_j)).$$

Let Y_j be the (desired) approximation to $y(x_j)$. In Euler's method Y_j is found by replacing "$\approx$" by =:

$$Y_0 = y_0, Y_1 = Y_0 + hf(x_0, Y_0) \ldots.$$
$$Y_n = Y_{n-1} + hf(x_{n-1}, Y_{n-1}).$$

EXAMPLE 8.1 An Initial Value Problem

$$y' = y, y(0) = 1.$$

Euler's method takes the form

$$Y_n = (1 + h)Y_{n-1} = \ldots = (1 + h)^n.$$

Let x^* be any point, and let h, n be so related that

$$hn = x^* = x_n.$$

Then

$$Y_n = (1 + x^*/n)^n$$

and this expression converges to x^* as $n \to \infty$.

REMARK 8.17 Accuracy of Euler's Method

The error of Euler's method is proportional to h. This can be improved, most notably by the *Romberg Method*, combining approximations obtained for various h.

A collection of methods due to Runge-Kutta is the standard computer approach to initial value problems for ODE's.

METHOD 8.11 Runge-Kutta Method with Error of Order h^4

To go from Y_n to Y_{n+1} we define

$A = hf(x_n, Y_n)$

$B = hf(x_n + \frac{1}{2}h, Y_n + \frac{1}{2}A)$

$C = hf(x_n + \frac{1}{2}h, Y_n + \frac{1}{2}B)$

$D = hf(x_n + h, Y_n + C)$

$Y_{n+1} = Y_n + (1/6)[A + 2B + 2C + D]$.

8.6 SOLVING PARTIAL DIFFERENTIAL EQUATIONS

THEOREM 8.10 Approximations to the First Derivative

$$f'(x) \approx (1/h)\{f(x + h) - f(x)\}$$

and

$$f'(x) \approx (1/h) \{f(x) - f(x - h)\}.$$

These approximations are, respectively, the *Forward* and *Backward* difference approximations to the derivatives. Each is in error proportional to h.

THEOREM 8.11 An Approximation to the Second Derivative

$$f''(x) \approx (1/h^2) \{f(x + h) - 2f(x) + f(x - h)\}$$

This is the *Centered Difference Approximation* to the second derivative. Its error is proportional to h^2.

THEOREM 8.12 An Approximation to the Heat Equation

For h, k positive (small) values, the heat equation

$$T_t(x, t) = \alpha\, T_{xx}(x, t)$$

can be approximated by

$$(1/k)\,[T(x, t + k) - T(x, t)]$$
$$\approx (\alpha/h^2)\,[T(x + h, t) - 2T(x, t) + T(x - h, t)]. \tag{8.3}$$

Here T_t has been replaced by a forward difference in t while for T_{xx} we have written its centered difference approximation.

METHOD 8.12 A Numerical Scheme for the Heat Equation

For given *Mesh Widths* h, k, introduce the two-dimensional mesh (x_i, t_j) (see Figure 8.6). Let T_{ij} denote the approximation to $T(x_i, t_j)$. Then for each i, j the approximation relation (8.3) leads to the equation

$$(1/k)T_{i, j+1} - T_{ij} = (\alpha/h^2)\,[T_{i+1, j} - 2T_{ij} + T_{i-1, j}]$$

and hence to the numerical scheme

$$T_{i, j+1} = T_{ij} + (\alpha k/h^2)\,[T_{i+1, j} - 2T_{ij} + T_{i-1, j}] \tag{8.4}$$

This is a *Marching Scheme* enabling us to move to later j values, knowing the solution for earlier j. It is an *Explicit Scheme* since it represents a single later value of temperature in terms of earlier values. The scheme is suitable for solving initial-boundary value problems over a finite region (see Figure 8.7).

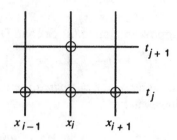

FIGURE 8.6 — Mesh for Heat Equation

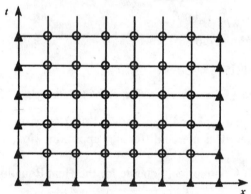

FIGURE 8.7 — Marching Simulation: △ = Data Points, O = Calculated

REMARK 8.18 Instability of the Explicit Scheme

If the ratio

$$= \alpha k / h^2$$

is greater than 1/2 the scheme (equation 8.4) is *Unstable*, meaning that it responds exponentially to small changes in data. To use the scheme one must select h, k such that

> Stability Condition $\alpha k / h^2 \leq 1/2$.

In this case the approximations converge to the solution of the PDE as h, k tend to zero.

REMARK 8.19 Shortcoming of the Explicit Scheme

The stability condition is a strong limitation on the size of the time step k of the explicit scheme. One wants the spatial mesh size h to be small in order that the centered difference scheme accurately approximate the second derivative T_{xx}. Hence, k would have to be yet smaller, since it would be bounded by h^2. For this reason, it is often preferable to seek a scheme which is not limited by a stability condition.

METHOD 8.13 Implicit Difference Scheme for the Heat Equation

This scheme is found by replacing the time derivative T_t by a *Backward Difference*, giving us

$$(1/k)T_{ij} - T_{i,j-1} = (\alpha/h^2)[T_{i+1,j} - 2T_{ij} + T_{i-1,j}] \qquad (8.5)$$

Now assuming that we are moving forward in time, there are three unknowns in the equation: T_{ij}, $T_{i\pm1,j}$. When solving the heat equation with boundary conditions on an interval there is one equation for each mesh point. Hence, the resulting linear system of equations for the unknowns at the time step j can be solved for these unknowns. The methods used include Gauss Elimination and iteration. The system is sparse. In addition, the scheme is *Unconditionally Stable* for all $h, k > 0$.

REMARK 8.20 Other Schemes

Other schemes for the heat equation include the *Crank-Nicholson* scheme, balancing the implicit scheme of equation (8.5) and the explicit scheme of equation (8.4). Another scheme of interest is that of *Dufort-Frankel*, which is explicit, unconditionally stable, but may not correctly model the heat transfer process.

REMARK 8.21 Laplace's Equation

For the Laplace equation $u_{xx} + u_{yy} = 0$ in a domain of the x, y plane, let h be any mesh width and introduce points (x_i, y_j) (see Figure

8.8). Replacing the derivatives by centered differences at any point x, y yields

$$(1/h^2) [u(x + h, y) - 2u(x, y) + u(x - h, y)]$$
$$+ (1/h^2) [u(x, y + h) - 2u(x, y) + u(x, y - h)] \approx 0.$$

For u_{ij} the desired approximation to $u(x_i, y_j)$

$$(1/h^2) [U_{i+1,j} - 2U_{ij} + U_{i-1,j}]$$
$$+ (1/h^2) [U_{i,j+1} - 2U_{ij} + U_{i,j-1}] = 0$$

or

$$U_{ij} = (1/4) [U_{i+1,j} + U_{i-1,j} + U_{i,j+1} + U_{i,j-1}]$$

that is, each value is the average of its neighboring values. Again we have an equation for each point interior to the domain. Assigning boundary values uniquely determines the solution at every mesh point. The system can be solved by iteration.

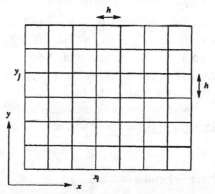

FIGURE 8.8 — Mesh for Laplace Equation

METHOD 8.14 Ritz Method and Finite Elements

An alternate approach to solving the Laplace equation in a domain R is based on the fact that it is the Euler equation for the functional

$$I[u] = \iint_R [u_x^2 + u_y^2] \, dx dy.$$

Suppose that any function on R can be expanded as a linear combination of *Basis Functions* $\{\phi_j\}$. Consider a finite linear combination of these functions

$$\sum_{n=1}^{N} a_j \phi_j(x, y)$$

and seek the coefficients minimizing the functional

$$F(a_1, a_2, ..., a_N) = I[\sum_{n=1}^{N} a_j \phi_j(x, y)].$$

Their determination rests merely on differentiation of F with respect to each one of them and solving the resulting equation system. This approach constitutes the *Rayleigh-Ritz-Galerkin* method and is the foundation of *Finite Element Methods*.

REMARK 8.22 The Wave Equation

For the wave equation $u_{tt} = c^2 u_{xx}$, we replace both second order derivatives by centered differences. As in the case of the heat equation, there is a stability condition

$ck/h \leq 1.$

Unlike the heat equation this condition is not overly restrictive. Moreover, attempts to use implicit approaches must be treated with caution because of the phenomenon of *Overstability* or artificial damping of wave motion that is not actually damped.

CHAPTER 9

STATISTICS AND PROBABILITY

9.1 ON STATISTICS AND PROBABILITY

REMARK 9.1 On Statistics and Probability

The fall of dice is a "random event": the rules governing their flight are so complex that to model them is essentially fruitless, but over many throws the outcomes become statistically predictable. If a die is fair then on the average each of its faces appears once in six throws.

Many phenomena encountered in engineering practice are statistical in the same way as the throw of dice. Significant examples include defective items on a production line, distribution of materials in mixtures, and the anticipated failures in communications networks.

DEFINITION 9.1 Experiment

A process carried out following a well-defined set of rules that can be repeated arbitrarily often and whose results cannot be predicted.

DEFINITION 9.2 Event

A possible outcome of an experiment.

DEFINITION 9.3 Sample Space S

The set of all possible outcomes of the experiment.

DEFINITION 9.4 Complement E^- of an Event E

The event "E does not happen."

DEFINITION 9.5 Venn Diagram

The graphical representation of a sample space and its events.

DEFINITION 9.6 Union $A \cup B$ of Events A, B

The event "either A or B or both occur."

DEFINITION 9.7 Intersection $A \cap B$ of Events A, B

The event "both A and B occur."

DEFINITION 9.8 Impossible Event ϕ

An event that can never occur.

DEFINITION 9.9 Disjoint Events A, B

Events that cannot both be true: $A \cap B = \phi$.

DEFINITION 9.10 Probability Function S

An assignment $P(E)$ of a real number to every element in the sample space S, such that:

i) for all sets E of S, $0 \leq P(E) \leq 1$,

ii) $P(S) = 1$,

iii) If $A \cap B = \phi$, then $P(A \cup B) = P(A) + P(B)$.

THEOREM 9.1 For any events A, B (see Figure 9.1)

$$P(A \cup B) = P(A) + P(B) - P(A \cap B).$$

THEOREM 9.2 For any event A,

$$P(A^-) = 1 - P(A).$$

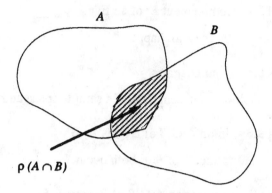

FIGURE 9.1 — Theorem 9.1

DEFINITION 9.11 Conditional Probability $P(A \mid B)$

The probability that A occurs given that B occurs, given by

$P(A \mid B) = P(A \cap B) / P(B)$.

DEFINITION 9.12 Independent Events A, B

Events satisfying $P(A \mid B) = P(A)$ or equivalently

$P(A \cap B) = P(A) P(B)$.

THEOREM 9.3 Bayes Theorem

For events $A, B,$

$P(A \mid B) = P(B \mid A) [P(A) / P(B)]$.

DEFINITION 9.13 Factorial $N!$

$N! = 1 \times 2 \times 3 \times \ldots \times N$,

$0! = 1$.

THEOREM 9.4 Stirling's Formula

For large N,

$N! \sim (2\pi N)^{1/2} (N/e)^N$.

DEFINITION 9.14 Binomial Coefficient

For $N \geq K$,

$$\binom{N}{K} = \frac{N!}{K!(N-K)!}$$

DEFINITION 9.15 Permutation of N Objects

An arrangement of the objects in some order.

THEOREM 9.5 Number of Permutations

The number of permutations of N objects is $N!$.

DEFINITION 9.16 Combination of K Out of N Objects

A selection of K of the N objects without regard to order and with no repetition.

THEOREM 9.6 Number of Combinations

The number of combinations of K objects from a collection of N objects is

$$\binom{N}{K} = \frac{N!}{K!(N-K)!}$$

DEFINITION 9.17 Random Variable X

A real valued variable whose value is determined by the outcome of an experiment.

DEFINITION 9.18 Discrete Random Variable X

A random variable assuming only discrete values (e.g., 1, 2, 3, ...).

DEFINITION 9.19 Continuous Random Variable X

A random variable assuming any values over an interval of the real line.

DEFINITION 9.20 Sample Space for Random Variable X

The collection of events determining the values of X.

DEFINITION 9.21 Probability Distribution Function for X

The function
$$F(x) = P(X \le x).$$

DEFINITION 9.22 Density Function $f(x)$ of a Discrete Random Variable

If X assumes the values $x, y, \ldots$ then
$$f(x) = P(X = x).$$

EXAMPLE 9.1 A Density Function for Dice Throws

Let X assume the values 1, 2, 3, 4, 5, 6 according to the outcome of throwing a fair die. Then
$$f(1) = f(2) = \ldots = f(6) = 1/6$$

DEFINITION 9.23 $f(x)$ for a Continuous Random Variable
$$f(x) = F'(x).$$

THEOREM 9.7 Properties of the Distribution Function

$F(-\infty) = 0, \quad F(\infty) = 1, F(x)$ is increasing

EXAMPLE 9.2 Normal Distribution Function
$$F(x) = (1/\sqrt{2\pi})\int_{-\infty}^{x} \exp(-\tfrac{1}{2}s^2)ds$$

is the distribution function for a *Normally Distributed* random variable (see Figure 9.2). Its density function is

$$f(x) = F'(x) = (1/\sqrt{2\pi})\exp(-x^2/2).$$

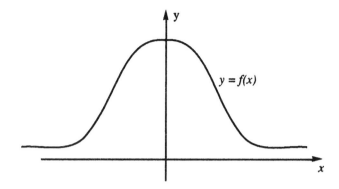

FIGURE 9.2 — Normal Distribution Density

REMARK 9.2 Mean and Standard Deviation

The mean value characterizes the anticipated or expected or average value of a random variable. On the other hand, the standard deviation gives us the spread or deviation of the random variable about its mean value. We define these now for both the discrete and continuous random variable X.

DEFINITION 9.24 Discrete Mean, Variance, and Standard Deviation

Let X assume the values $\{x_j\}$ and have density function $f(x_j)$. Its *Mean* or *Expected Value* $\mu = E[X]$ is

$$\mu = \sum_j x_j f(x_j);$$

its variance is

$$\sigma^2 = \sum_j (x_j - \mu)^2 f(x_j)$$

and its standard deviation is σ.

DEFINITION 9.25 Continuous Mean, Variance, and Standard Deviation

If X is a continuous random variable with density function $f(x)$, then the mean or expected value $\mu = E(X)$ is

$$\mu = \int_{-\infty}^{\infty} x f(x)\, dx$$

while its variance is

$$\sigma^2 = \int_{-\infty}^{\infty} (x - \mu)^2 f(x)\, dx.$$

The standard deviation is again σ.

THEOREM 9.8 Function of a Random Variable

Let $Y = G(X)$ be a function of the random variable X. Then Y is a random variable with the expected value

$$E(G(X)) = \sum_j G(x_j) f(x_j)$$

for the discrete case, while the continuous case

$$E(G(X)) = \int_{-\infty}^{\infty} G(x) f(x)\, dx.$$

DEFINITION 9.26 Binomial Distribution

In a coin toss there is a probability p of observing "heads" and q of observing "tails," $p + q = 1$. Let x be the random variable defined as the number of heads in N coin tosses. Then X is a discrete random variable. Its density function is

$$f(x) = \binom{N}{x} p^x q^{N-x}.$$

The mean value of X is $\mu = Np$, while its variance is $\sigma = (Npq)^{1/2}$.

DEFINITION 9.27 Poisson Distribution

The Poisson random variable is associated with transportation, queues, and service lines. It is a discrete random variable with the density function

$$f(x) = (\mu^x / x!) e^{-\mu}, \quad x = 0, 1, 2, \ldots$$

DEFINITION 9.28 Normal Distribution

The most commonly used distribution is the normal distribution. For mean μ and standard deviation σ, its density function is

$$f(x) = (1/[\sigma\sqrt{2\pi}]) \exp\{-\tfrac{1}{2}[(x-\mu)/\sigma]^2\}.$$

THEOREM 9.9 Law of Large Numbers

As $N \to \infty$ the binomial distributed random variable X tends to a normally distributed random variable.

DEFINITION 9.29 Parameter Estimators

If the values $x_1, x_2, ..., x_N$ of a random variable X are obtained from a series of experiments then estimates of the mean value and variance of X are

$$\mu_{est} = (1/N)\sum_{j=1}^{N} x_j$$

$$\sigma_{est}^2 = (1/[N-1])\sum_{j=1}^{N}(x_j - \mu_{est})^2$$

REMARK 9.3 Applying Statistical Methods

One of the key engineering applications of statistics and probability is to the question of whether a sample taken truly represents the state of the general population of interest. Thus, if we wish to know if a production line is operating properly, we will routinely select a small sample of items and test them. If many are defective does this mean that we must halt production? What if the line is working properly and we just had "bad luck" in our random sample? On the other hand, we might have selected a sample of which very few are defective. Does this mean that the line is working well, or did we just have "bad luck" again, this time on the optimistic side? In any case, we assume that defective items follow a particular probability distribution, and we use the mean and standard deviation as tools for determining the probability that our line is/is not working well as a reflection of our sample.

INDEX

Aberture, D-158
AC,
    analysis A-53
    coupled amplifier, B-103
    load line, B-15—B-18
    servomotor, F-10
Acceptor impurities, B-5
Action plot, F-134
Active filter, B-105
Adder, B-101
    half, B-153
    full, B-154
    look-ahead-carry, B-155
    non-inverting, B-101
Addition,
    matrix, A-34—A-35
    parallel, B-155
Adjustment, gain F-131
Admissible functions, G-115
Admittance, A-50—A-53
    complex frequency, A-67
    parameters, A-100, B-171
Algebra,
    Boolean, B-138—B-144
    matrix, F-44-F-47
    of sets, D-48
    vector, C-1—C-3
Alternating current, A-1—A-6
Alternator, A-61
AM systems, noise, D-116
Ampere, A-1
    circuital law, C-43
Amplification factor, B-57
Amplifier,
    AC coupled, B-103
    audio-frequency linear power, B-61—B-69
    cascade, B-47
    C-E transistor, B-90
    class A common-emitter power, B-61
    class B push-pull power, B-67
    common-drain, B-58-B-87
    common-source, B-57, B-88
    current, B-71
    Darlington, B-46
    difference, B-39—B-45
    emitter follower, B-88
    feedback, B-71—B-78
    frequency response, B-81—B-90
    ideal, B-73
    OP, B-48
    operational, B-92—B-96, B-108-B-114
    small-signal, B-167
    transconductance, B-72
    transconductive, B-102
    transformer-coupled, B-65
    transresistance, B-72
    voltage, B-71
Amplitude, E-56
    modulation, D-63—D-69, D-116
    pulse modulation, D-89
Analog,
    computer, F-163—F-174
    mechanical and electrical, F-5
Analysis,
    AC, A-53
    circuit, A-34—A-41
    feedback amplifier, B-78
    frequency domain, A-66—A-74
    low frequency, B-49—B-60
    mesh, A-41
    of a lag network, F-149
    of a lag-lead network, F-153
    of a lead network, F-151
    of systems, F-63
    Q-point, B-40
    sinusoidal, A-43—A-56
    small-signal, B-49—B-60
    special, D-27—D-34
    state-variable, A-75—A-80
    system, F-89—F-105
    time domain, F-58—F-63
    topological, A-111—A-113
    vector, C-1—C-6
AND function, B-138
Angle,
    curves, F-128
    modulation, D-70
    phase, A-43, A-45
    phasor, A-43
Angular frequency, F-121
Antennas, C-115—C-120, D-155—D-159
    characteristics, D-159
    full-wave, C-118
    half-wave, C-117
    impedance, D-159
    length of, C-120
    properties, C-115
Apparent power, A-55
Applied frequency, E-57
Approximations, G-125
    filter, D-44
    function, G-127
    loss, D-167
Arc length, G-23

I-1

Area,
    under signal, D-18
    under transform, D-18
Armature control, F-11
Armstrong system, D-82
Asymptote, F-73, F-96
Astable multivibrator, B-124—B-127
Attenuation, C-94
Audio-frequency linear power amplifier,
    B-61—B-69
Autocorrelation, D-34, D-58
Average,
    ensemble, D-53
    power, A-53—A-56
    statistical, D-57
    time, D-53

Balanced modulator, D-66
Band,
    frequency, D-38
    limiting of waveform, D-33
Bandwidth, B-82
    and SNR tradeoff, D-150
    full power, B-95
    noise, D-113, D-139
    of an FM signal, D-73
Barkhauserís criterion, B-159
Basic,
    difference amplifier, B-39
    flip-flop, B-173
Basis, G-6
Bayesí theorem, D-51, G-147
Beams, E-63
Beamwidth, D-157
Beat, G-76
Bessel,
    functions, D-74, D-76, E-27,
      G-85—G-86
    modified, E-28
    inequality, G-77, G-81
Beat,
    cutoff frequency, B-26
    function, E-36
B-H curve, A-16
Bias stability, B-28
Binary,
    digits, addition, B-153
    phase-shift keying, D-105
    symmetric channel, D-150
Binomial,
    coefficient, G-148
    distribution, G-151
    probability density function, D-51
Biot-Savart law, C-41

Bipolar,
    junction transistor, B-21—B-28
    transistor, B-169
Bisector method, G-126
Block,
    codes, D-152—D-155
    diagram, F-51—F-55
Bode,
    diagram, B-116
    plots, F-89—F-105
Boolean algebra, B-138—B-144
Boundary conditions, C-34, E-100, G-96
    magnetic field, C-58
Boundary value problem, E-98—E-103,
    G-64—G-72
Brachistrochrone problem, G-119
Break-away points, F-75
Break-point, B-14
Butterworth,
    filter, B-105, D-140
    response, D-44

Calculating, noise, D-133—D-139
Calculus, G-18—G-38
    of one variable, G-18
    of variations, G-110—G-123
Capacitance, C-29—C-36
Capacitor, A-14, D-1
    coupling, B-82
    emitter, B-82
    load, A-55
    parallel plate, A-14
Capture area, D-158
Cascade,
    amplifier, B-47
    compensated systems, F-146—F-153
Carsonís rule, D-75
Catenary problem, G-119
Cauchy,
    integral theorem, G-53
    residue theorem, A-108
    -Riemann equation, G-45
    -Schwartz inequality, G-8, G-80
Cayley-Hamilton theorem, A-80
C-B configuration, B-53, B-24
C-C configuration, B-54
C-E,
    configuration, B-50—B-52
    cutoff currents, B-25
    short circuit, B-85
Central,
    limit theorem, D-61
    moment, D-52
Chain rule, G-21
Changes of scale property, E-8, E-20

Characteristic,
  curve, D-39
  equation, F-67
  function, D-52
  impedance, D-164
  servo, F-58—F-63
Charge, A-3
  density, B-5
  line of, C-9
  point, C-9
  sheet of, C-9
  volume distribution, C-9
Charged particle, B-1
Chart,
  Nichol's, F-133
  Smith, C-85
Chebyshev response, D-45
Choleskyís method, G-135
Circuit,
  analysis, A-34—A-41
  arithmetic, B-153—B-155
  closed, C-51
  diode, B-15—B-18, B-131
  diode logic, B-20
  electric, E-60, F-1
  elements, A-8—A-9
  equivalent, B-56
  linear, A-29
  linear waveshaping, B-185
  loop matrix, A-112
  magnetic, C-55
  multiloop, F-2
  multitransistor, B-39—B-48
  non-linear, B-164
  open, A-13
  parallel, A-24
  phase splitting, B-60
  principles, D-1—D-4
  radio, B-164—B-171
  RC, A-20—A-22, D-2
  regenerative, B-182
  resistive, A-5—A-13
  RL, A-20—A-22, D-4
  RLC, A-24—A-27
  series, A-26, F-2
  short, A-13
  state equation, A-76—A-77
  transient, A-14—A-27
  tuned, B-168
Clapp oscillator, B-159
Class A common-emitter power amplifier, B-61
Class B push-pull power amplifier, B-67
Clear, B-175

Closed circuit,
  force, C-53
  torque, C-53
Closed, form identities, A-104
Code,
  block, D-152
  pulse modulation, D-100—D-107
Codeword, D-153
Coding, D-153
Coefficient, exponential damping, A-25
Cofactor, G-15
Coherent reception, D-143
Collector,
  -coupled flip-flops, B-178
  -coupled multivibrator, B-117, B-129
  dissipation, B-63
Collinearity, G-2
Colpitts oscillator, B-156
Common-drain voltage amplifier, B-58, B-87
Common-mode,
  load line, B-40
  rejection ratio, B-43
Common-source voltage amplifier, B-57, B-88
Communication systems, noise, D-116—D-130
Companding, D-103
Compensated cascade system, F-146—F-153
Compensating,
  network, F-53, F-136—F-145
  network function, F-136
Compensation, F-137
  feedback, F-155—F-160
  lead, B-116, F-141
  lead-lag, F-142
  parallel, F-155
  simple lag, B-115
Compensator,
  comparison of, F-145
  feedback, F-155
  lag, F-139
  lag-lead, F-153
  proportional plus derivative, F-140
  selecting, F-146
Complement set, D-49
Complementary error function, E-30
  property, E-31
Complex,
  form of Fourier series, E-85
  functions, G-40—G-57
  integration theorem, F-32
  inversion formula, A-91
  matrices, G-16

I-3

number, A-47—A-50
plane, F-122
variable, G-43
vector, G-16
Components, electric, F-1
Conditional probability, D-51
Conductance, A-10—A-11
Conductor, A-11, C-29—C-36
   electrostatic field, C-36
Conductivity, B-3
Configuration,
   C-B, B-53
   C-C, B-54
   C-E, B-50—B-52
   transistor, B-23
Conformal mapping, G-60
Conjugate function, D-19
Connected loads, A-64
Connection, A-61—A-63
Constant magnitude circles, F-122
Continuity, G-45
Continuous,
   function, E-4
   random variable, D-53
Control, G-120
   feedback, F-118—F-133
   PI, F-137
   system, F-172
Controller setting, F-174
Convergence, G-18
   of a series, G-55—G-56
   of Fourier series, D-12
   theorem, E-81, E-87, G-77
   theorem, first criterion, E-82
   theorem, second criterion, E-83
Converter,
   current-to-voltage, B-102
   voltage-to-current, B-102
Convolution, D-25
   integral equation, E-92
   in time domain, D-19—D-20
   inverse Laplace transform, E-23
   property, A-94, E-22—E-23
   theorem, A-89, G-83
Coordinate system, G-92—G-93
   Laplace equation, C-37—C-38
   vector, C-6
Correlation,
   cross, D-58
   function, D-58
   power, D-34
Cosine,
   Fourier formula, E-88
   Fourier series, E-79
   integrals, E-31
   integration, A-82

Coulomb, A-1, A-3
   law, A-3, C-7—C-11
Counter, B-148—B-155
   Ripple, B-151
   synchronous, B-152
Coupling capacitor, B-81
Cramerís rule, A-37, G-15, G-132—G-133
Criterion,
   Nyquist stability, F-107—F-117
   Routhís stability, F-39
Critical point, G-70
Cross,
   correlation, D-58
   product, vector, C-3, G-5
Crosstalk, D-90, D-94
   factor, D-96
Crystal oscillator, B-159
Curl, vector field, C-44, G-28
Current, A-5—A-8
   alternating A-6
   amplifier, B-71
   and line voltage, D-164
   capacitor, A-14
   density, B-4, C-29—36
   diffusion, B-2
   direct, A-6
   displacement, C-60
   division, A-12
   drift, B-2
   flow, A-5
   gain, B-85
   independent source, A-9
   Kirchhoffís law, A-13
   offset error, B-94
   sinusoidal, A-43—A-44
   source, difference amplifier, B-43
   to-voltage converter, B-102
   transistor components, B-23
   transmission line, D-162
   voltage-to converter, B-102
Curve,
   angle, F-128
   characteristic, D-39
   gain, F-97
   hysteresis, F-169
   log-magnitude, F-97, F-160
   magnitude, F-128
   saturation, F-168
   system type, F-97
Cycle per second, A-1

DíAlembertís formula, G-108—G-109
Damped vibration, E-53
Damping, A-25—A-26, F-4
   critical, A-25, A-27

over, A-25, A-27
ratio, F-22
under, A-25, A-27
Darlington amplifier, B-46
Data transmission, D-141—D-146
DC,
    load line, B-15
    voltage follower, B-103
    servomotor, F-10
Dead time, F-105
Decade, F-93
Decibel, A-1, F-92
Degree,
    Celcius, A-1
    Fahrenheit, A-1
    Kelvin, A-1
Delay,
    group, D-39
    phase, D-39
Del operator, C-19
Delta,
    alternator, A-61, A-63
    connector, A-61, A-63
    modulation, D-104
    modulation, noise, D-130
Demodulator,
    FM, D-84
    square law, D-63
De Moivreís formula, E-85, G-42
Density,
    charge, in semiconductor, B-5
    current, C-29—C-36
    electric flux, C-13—C-20
    flux, A-16
    flux, magnetic, C-46
    function, probability, D-51, D-53--D-57, G-149
    Gaussian probability, D-61
    power spectral, D-30
Dependent,
    current source, A-9
    voltage source, A-9
Depletion region width, B-9
Depth, skin, C-74
Derivative, G-45
    inverse Laplace transform, E-20
    Laplace transform of E-10
    of Laplace transforms, E-12
Design,
    low frequency, B-49—B-60
    small signal, B-49—B-60
Determinant, A-36—A-37, G-5
Determination, power spectra, D-59
Deterministic, signal, D-8
Diagonal matrix, A-35, G-9

Diagram,
    block, F-51—F-55
    log-magnitude, F-94
    phase, F-94
    pole-zero, F-38
Dielectric, C-29—C-36
    constant, A-4, A-14
    wave motion in, C-66—C-68
Difference,
    differential equation, E-96
    equation, E-95
    equations, Laplace transforms, E-91—E-96
    of sets, D-49
Difference amplifier, B-39—B-45
    basic, B-39
    constant current source, B-43
    emittor resistor, B-45
Differential,
    current, element, C-49—C-50
    -difference equation, E-96
    equation, A-22
        Laplace transform, F-32
        ordinary, G-138
        partial, E-98—E-99, G-139
        system's response, F-16—F-28
    integro- equation, E-96
    methods, E-66
    ordinary linear, E-42—E-63
    phase shift keying, D-106
    volume, C-6
Differentiation,
    Fourier transform, D-20
    general criterion, E-83
    in time domain, D-18
    partial, equation, G-90—G-106
    with respect to a parameter, E-70
Differentiator, D-112
Diffusion,
    capacitance, B-13, B-84
    current, B-2
    in a semiconductor, B-6
Digital filter, D-45
Diode,
    AND gate, B-20
    circuits, B-15
    junction, B-11-B-20
    logic circuits, B-20, B-131
    OR gate, B-20
    rectifier, B-31—B-34
    resistance, B-13
    resistor logic, B-133
    Schnottky, B-18
    Zener, B-18
Dipole, C-27

I-5

Dirac delta function, E-35
Direct,
    current, A-1, A-6
    method, E-66
    polar plot, F-99, F-118, F-131
Directional, D-156
Directivity, D-157
Dirichlet,
    conditions, A-81, E-83, G-77
    problem, E-100
Displacement current, C-60
Discrete random variable, D-47
Discrete system, A-102—A-110
Discriminator, D-123
Distortion frequency, B-81
Distortionless line, D-165
Divergence, C-13—C-20, G-27
    theorem, C-20, G-39
Division,
    by T property, E-71
    current, A-12
    voltage, A-11
Domain, G-43
    frequency, A-46—A-47, A-66—A-74
    time, A-46—A-47
Dominant,
    complex poles, F-80
    poles, F-80—F-87
Donor impurities, B-5
Dot,
    notation, A-18
    product, vector, C-3
Double integral, G-28
Double-sideband
    modulator, D-66
    suppressed carrier modulation, D-119
Drain-source resistance, B-57
Drift,
    current, B-2
    speed, B-4
D-type flip-flop, B-175
Duality, D-17, D-20
Dump PCM receiver, D-141
Dynamic
    programming, G-122
    resistance, B-15

Effective,
    aperature, D-161
    length, C-120
    noise temperature, D-136
    temperature, D-138
    value, A-56
Efficiency, A-7, B-64
    rectifier, B-33

Eigenvalue, G-16, G-72—G-73, G-137
Eigenvector, G-16—G-17, G-72
Einstein relationship, B-6
Elastance, F-3
Electric,
    circuits, E-60, F-1
    transverse waves, C-88
Electric field, C-7—C-11
    intensity, B-3, C-8, C-117
    peak, C-103
Electric flux density, C-13—C-20
Electrical analogs, F-5
Electromagnetic,
    power, C-70
    transverse waves, C-90
    waves, C-73—C-76
Electron, B-4
    volt, A-1
Elementary,
    complex function, G-47
    Laplace transform, E-3
    mappings, G-58
EMF, C-59—C-61
Emitter,
    bypass capacitor, B-82
    coupled flip-flop, B-179
    coupled logic, B-137
    coupled multivibrator, B-120, B-125
    follower configuration, B-54, B-88
    resistor, B-45
Energy, A-2, A-5—A-8, C-22—C-27
    gap, B-1
    potential, C-56
    transmission of, D-46
Ensemble averaging, D-53
Entropy, D-148
Equation,
    characteristic, F-67
    difference, E-91—E-96
    differential, A-22—A-23, F-16—F-28
    differential-difference, E-96
    Faradayís, C-59
    first order wave, G-103
    for line voltage and current, D-162
    Fredholm integral, E-91
    hybrid, B-49
    integral, E-91—E-96
        of convolution type, E-92
    integro-differential, E-93
    KCL, A-24
    KVL, A-27
    Lagrange, F-14
    Laguerre, E-46
    Maxwell's first, C-20, C-59—C-61,
        C-63

normal-form, A-78
ordinary differential, E-42—E-63, G-64—G-72, G-138
partial differential, E-99, G-90—G-106, G-139
Poisson's, C-37—C-38
radar, D-160
second order partial differential, E-99
solutions of, G-125
state, A-76—A-77
systems of, G-9
transmission-line, C-78
Volterra integral, E-91
wave, C-63, G-106
Equivalence, A-40
  Norton, A-31
  Thevenin, A-30
Equivalent, circuit, B-56
Ergodicity, D-53
Error, G-126
  complementary function, E-30
  detection, D-150
  equation, F-27
  evaluation of, F-60
  function, E-29
  function, properties, E-30
  mean square, E-84
  probability, D-143
  series, F-61
Euler's
  equation, G-116
  formula, G-76
  method, G-138—G-139
  theorem, A-49
Even,
  extension, E-80
  function, D-13, E-77
Event, G-145
Excitation table, B-176
Existence, Laplace transform, E-6
Explicit scheme, G-142—G-143
Exponential,
  damping coefficient, A-25
  Fourier series, A-83
  function, B-184
  integral, E-31
  order, E-4
Extension,
  even, E-80
  odd, E-80
  periodic, E-81

Factor, amplification, B-57
Factorial, G-147
Factoring, polynomial, F-87

Families, logic, B-128—B-137
Fan-out, B-130
Farad, A-1
Faraday's law, A-18, C-59
Far zone radiation, C-107
Feedback,
  compensation, F-155—F-160
  compensator, F-155
  control, F-118—F-133
  frequency response, B-11
  of OP amps, B-108—B-114
  systems, F-59
  unity, F-63
Feedback amplifier, B-71—B-78
  analogies, B-78
  negative, B-75
  single-loop, B-74
  topologies, B-74
FET, B-56
Field,
  control, F-13
  electric, C-7—C-11
  electric intensity, C-8
  intensity, B-2
  magnetic, C-41—C-47
  radiation, C-118
  streamline of, C-11
  time-varying, C-59—C-61
  vector, C-3, G-27
  vector, curl, C-44
Figure, noise, D-137
Filter, B-34—B-38
  active, B-105
  approximation, D-44
  Butterworth, B-105
  digital, D-45
  ideal, B-105, D-38
  ideal low-pass, D-111
  L-section, B-37
  matched, D-142
  narrow band, D-110
  Pi, B-36
  RC, B-36, D-110
  rectangular bandpass, D-112
  shunt-capacity, B-34
Filtering, D-37—D-46, D-110
Final value theorem, A-96, D-25, E-65
First,
  fundamental form of S, G-36
  order system, F-24
  order wave equation, G-103
  shift property E-8, E-18
  translation property, E-18
Flat-top,
  antenna, C-115
  sampling, D-90

Flexural rigidity, E-63
Flip-flop, B-173—B-182
  basic, B-173
  collector-coupled, B-178
  D-type, B-175
  emitter-coupled, B-179
  J-K, B-176
  master-slave, B-177
  R-S, B-173
  switching speed, B-180
  synchronous, B-174
  timing, B-177
  T-type, B-176
Flow, graphs, F-51—F-55
Flux,
  density, A-16, C-13
  electric, C-13—C-20
  intensity, magnetic, C-46
  magnetic, C-46
FM
  demodulation, D-84
  discriminator, D-85
  generation, D-81
  receiver, D-121
  signal, D-73, D-80
  system, noise, D-121
Follower, DC voltage, B-103
Foot, A-1
Force, A-2, C-49—C-58
  damping, E-53
  external, E-33
  magnetic, C-49—C-50
  restoring, E-52
Forced response, A-22
Formula,
  DeMoivre, E-85
  Fourier cosine, E-88
  Fourier sine, E-88
  Heaviside expansion, E-76
Forward bias, B-12
  junction, B-9
Fourier analysis, A-81—A-89, G-74—G-85
Fourier series, D-8—D-21, E-77, G-76—G-79
  complex form, E-85
  convergence, D-12
  cosine, E-79
  existence, A-81
  expansion, D-29
  exponential, A-83
  integral, F-89
  sine, E-79
  trigonometric, A-81
Fourier transform, A-86—A-89, D-8—D-21, E-77—E-89, G-82

integral, E-86
inversion theorem, E-87
pairs, D-21
properties, D-16
relationship to Laplace transforms, E-89
theorems, D-20
Fredholm integral equation, E-91
Free,
  nodal analysis, A-41
  vibration, E-54
Frenet-Serret formula, G-24
Frequency,
  applied, E-57
  bandlimiting, D-33
  bands, D-38
  carrier, D-66
  compensation of OP amps, B-108—B-114
  complex, A-66—A-67
  distortion, B-76, B-81
  domain, A-46—A-47, F-89—F-105, F-118-F-133
  domain analysis, A-66—A-67
  high, B-85—B-88
  low, design and analysis, B-49—B-60
  modulation, D-20, D-70—D-84
  natural, E-57
  noise, D-108
  radio circuits, B-164—B-171
  resonant, A-25
  response, F-38, F-78, F-89
  response, low, B-82
  response of amplifier, B-81—B-90
  response, plots, F-90. F-146—F-153
  shifting, D-17
  shift keying, D-107, D-145
  translation, D-20, D-84
Frobenius method, G-86
Full,
  adder, B-154
  power bandwidth, B-95
  wave antenna, C-118
Full-wave,
  rectifier, B-31
  symmetry, D-15
Function,
  admissible, G-115
  analytic, G-62
  AND, B-138
  approximation, G-125
  Bessel, D-76, E-27—E-28, G-85
  beta, E-36
  characteristic, D-52
  complementary error, E-30—E-311
  complex, G-40—G-57

I-8

conjugate, D-19
correlation, D-58
dirac delta, E-35
error, E-29—E-30
even, D-13, E-77
exponential, B-184, E-4
gamma, E-25—E-26, G-85
generating, E-29
Hankel, G-87
harmonic, G-46, G-60
Heaviside, E-31
impulse, F-41
logic, B-138
member, D-53
NAND, B-142
Neumann, G-86
NOR, B-142
NOT, B-138
null, E-16
odd, D-13, E-78
of a complex variable, G-43
of compensating networks, F-136
of two or more variables, G-25
of two variable, E-98
OR, B-139
periodic, E-13
piecewise smooth, E-81
probability density, D-51
ramp, B-184
rational, E-72
sequences, G-56
series, G-56
sinusoidal forcing, A-43
special, E-25—E-38, G-85
stationary, G-116
step, B-184, E-31
transfer, D-37—D-46, F-51—F-55
unit impulse, E-35
vector, G-23
Fundamental existence and uniqueness theorem, G-66

Gain, B-74, D-157
   adjustment, F-131
   crossover, F-113
   curves, F-97
   margin, F-113
   transfer, B-74
Gamma function, E-25, G-85
   properties, E-26
Gates,
   advanced active, B-134
   logic, B-128—B-137
Gauss,
   elimination, G-133
   law, C-13—C-20
   -Seidel iteration, G-136
Gaussian,
   channel, D-149
   probability density, D-61
General nodal analysis, A-41
Generating function, E-29
Generator,
   sweep, B-191
   waveform, B-184—B-191
Geometric series, G-55
Gershgoinís theorem, G-137
Gibbs phenomenon, G-77
Graded semiconductor, B-6
Gradient, C-25, G-26, G-114—G-115
Gram, A-1
Graph, signal flow, F-51—F-55
Graphical method, C-85, F 35
Greenís theorem, G-34
Group delay, D-39
Guided waves, C-87—C-91
Guides, waves, C-99—C-102

Half, adder, B-153
   rectified sine wave, D-41
Half-wave,
   antenna, C-117
   rectifier, B-31
   symmetry, D-14
Hamilton's principle, G-120
Hamming,
   distance, D-152
   weight, D-152
Hankel function, G-87
Harmonic,
   functions, G-46, G-60
   oscillators, B-156—B-160
   waves, G-75
Hartley oscillator, B-158
Hazony rule, F-35
Heat equation, G-100
Heaviside,
   expansion formula, E-76
   expansion theorem, A-95
   function, E-31
Helmholt wave equation, G-109
Henry, A-1
Hermitian matrix, G-16
Hertz, A-1
Holding, D-94
Holes, in semiconductor, B-1, B-4
Homogeniety, G-65
Homogeneous, E-99
Hour, A-1
H-parameters, A-98, B-50

Hybrid,
   equations, B-49
   parameters, A-98, B-27, B-49—B-50
   -Pi model, B-25, B-84
Hydraulic transmission, F-9
Hyperbolic function, G-49—G-50
Hysteresis curve, F-169

Ideal,
   amplifier, B-73
   antenna, D-157
   filter, B-105, D-38—D-39
   low pass filter, D-11
Identity,
   closed form, A-104
   matrix, A-35
   Parseval's A-88, E-85
Imaginary number, A-47
Impedance A-50—A-53, A-97
   antenna, D-159
   characteristic, D-164
   complex frequency, A-67
   input, D-164, D-166
   intrinsic, C-68
Imperfect,
   bit synchronization, D-145
   phase synchronization, D-144
Impulse function, F-41
Impurities,
   acceptor, B-5
   donor, B-5
Incident matrix, A-111
Independence,
   of path, G-53
   statistical, D-51
Independent,
   current source, A-9
   voltage source, A-8
Index, E-27
Inductance, A-17, C-57
   external, C-81
   internal, C-81
   magnetic fields, C-49—D-58
   mutual, A-17, A-19, C-57
   self, A-16
Inductor, A-15—A-20, D-3
   iron-core, A-15
   load, A-55
Infinity, Laplace transform, E-6
Information rate, D-148
Information theory, D-148—D-152
   concepts, D-148
Initial value problem, E-42, G-69, G-138
Initial value theorem, A-96, D-25, E-65, F-32
Inner product, G-3, G-80

Input,
   impedance, D-164, D-166
   resistance, B-76
   standardized, F-16
Instantaneous power, A-53
Integral,
   control, F-137
   cosine, E-31
   definite, G-21—G-23
   double, G-28
   equations, Laplace transforms, E-91—D-96
   exponential, E-31
   Fourier, F-89
   Fourier transform, E-86
   improper, G-22
   inverse Laplace transform, E-21
   Laplace transform, E-1
   line, G-32
   of complex function, G-51
   sine, E-31
   transform, G-74—G-85
   volume, G-38
Integrate receiver, D-141
Integration,
   cosine, A-82
   numerical, G-130
   sine, A-82
   time domain, D-18, D-20
Integrator, D-113
Integro-differential equation, E-93
Intensity, field, B-2
Interpolation, G-127
Intersection, D-49
Inverse, G-14
   matrix, A-36
   polar plane, F-128
   polar plots, F-101, F-111
   transform, F-33
   z-transform, A-108
Inverse Laplace transform, A-91, D-24, E-15—E-23
   definition, E-15
   methods of solutions, E-65—E-76
   of convolution, E-23
   of derivatives, E-20
   of integrals, E-21
   properties, E-18
   uniqueness, E-16
Inversion,
   formula, F-49
   theorem for Fourier transforms, E-87
Inverter, B-100
Iterated integral, G-30
Iteration method, G-126

Jacobian, G-31
Jacobi iteration, G-136
J-K flip-flop, B-176
Joint probability, D-50
Jordan curve theorem, G-52
Joule, A-1
Junction diode, B-11-B-20
    bipolar transistor, B-21—B-28
    P-N, B-11

Karnaugh map, B-144
Kernel, E-1, E-91
Keying,
    binary phase shift, D-105
    differential phase-shift, D-106
    frequency shift, D-107, D-145
    phase shift, D-144
kilogram, A-1
Kirchkoff's law, A-12—A-13, E-62
    current, A-13, D-6
    voltage, A-13, D-3
Kronecker delta sequence, A-102

Lag,
    compensator, F-53, F-139
    network, F-149
    phasor angle, A-43
Lag-lead compensator, F-153
Lagrange,
    equation, F-14
    multiplier, G-113
Laguerre equation, E-46
Laplace, equations, C-37—C-38, G-96
    different coordinate systems, C-58
    solution methods, E-65—E-76
Laplace transform, A-90—A-96, D-22—
    D-25, E-1—E-13, F-29—F-41,
    G-71—G-72
    applications, E-91—E-96
    behavior at infinity, E-6
    definition, F-29
    differential equation, F-32
    existence, E-6
    functions, D-22
    impulse functions, F-41
    inverse, A-91, D-24, E-15—E-23, F-33
    linearity, E-7
    of derivative, E-10
    operators, A-93
    pairs, A-92
    properties, E-7
    relationship to Fourier transform, E-89
    solutions of boundary value problems,
    E-101
    tables of, E-104—E-119

Laplacian, G-28
Large-signal current gain, B-23
Laurent expansion, G-57
Law,
    Ampere's circuital, C-43
    Biot-Savart, C-41
    Coulomb's, C-7—C-11
    Faraday's, A-18, C-59
    Gauss', C13—C-20
    Kirchkoff's, A-13
    of units, A-3
    Ohm's, A-12
    parallelogram, G-3
    square demodulator, D-63
Lead,
    compensation, F-141
    compensator, F-54
    -lag compensation, F-142
    network, F-151
    phasor angle, A-43
Least square,
    distance, G-80
    problem, G-128—G-129
Legendre polynomial, G-89
Length, A-2
    effective, C-120
L'Hopital's rule, G-19
Limits, D-122, G-45
Line,
    current, D-162
    distortionless, D-165
    integral, G-32
    lossless, D-165
    of charge, C-9
    transmission, C-78—C-85, D-162—D-
    167
    voltage, D-162, D-164
Linear,
    algebra, numerical, G-132
    dependence, G-68
    differential equations, ordinary, E-42—
    E-63
    programming, G-123
    system, D-28
    time-invariant network, A-77
Linear-power amplifier, B-61—B-69
    waveshaping circuit, B-185
Linearity, A-29, D-17
    property, E-7, E-18
Link, A-42
Load,
    capacitive, A-55
    connective, A-64
    floating, B-102
    ground, B-102
    inductive, A-55

Load-line,
    AC, B-15—B-18
    common-mode, B-40
    DC, B-15
Locus, root, F-67—F-78, F-136—F-145
Logarithm, G-48—G-49
Logarithmic plots, F-92
Logic,
    basic active, B-132
    basic passive, B-130
    diode circuits, B-20, B-131
    families, B-128—B-137
    functions, B-138
    gates, B-128—B-137
Log-magnitude,
    curves, F-97, F-160
    diagram, F-94
Look-ahead-carry adder, B-155
loop,
    method, F-2
    single, A-38
    single, analysis, A-38
Lorentz force equation, C-49
Loss approximation, D-167
Lossless lines, D-165
Louville theorem, G-55
Low frequency, analysis and design, B-49—B-60
L-section filter, B-37

MacLaurin series, G-56
Magnetic,
    circuit, C-55—C-56
    field, C-49—C-58
    field density, A-16
    field, steady, C-41—C-47
    flux, A-16, C-46
    flux density, C-46
    force, C-49—C-50, C-56
    materials C-53
    potential, vector, C-47
    transverse waves, C-89
Magnitude,
    curves, F-128
    maximum, F-120—F-122
Mapping,
    Conformal, G-60
    elementary, G-58
    of linear transformation, G-12
    Riemann theorem, G-63
Margin,
    gain, F-113
    phase, F-113
Marginal probability, D-50
Masonís rule, F-56

Mass, A-2
    -action law, B-7
Master-slave flip-flop, B-177
Matched filter, D-142
Materials,
    forces, C-56
    magnetic, C-53
Matrix, A-34—A-35, G-1—G-16
    algebra, F-44-F-47
    circuit loop, A-112
    diagonal, A-35
    fundamental loop, A-113
    identity, A-35
    incident, A-111
    inverse of, A-36
    inversion, G-135
    power of, A-36
    rectangular, A-34
    state transition, A-79
    transpose, A-35
    unit, A-35
Maxim,
    theory of, G-11
    principle, G-97
Maximum,
    dissipation hyperbola, B-64
    magnitude, F-120, F-122
    modulus theorem, G-54
    power transfer theorem, A-31
Maxwell,
    curl equations, C-79
    equations, C-59—C-61, C-63
    first equation, C-20
Mean, D-61, G-150
    square error, E-84
Mean value theorem, G-19
Mechanical,
    and electrical analogs, F-5
    rotation systems, F-6
    translation systems, F-3
Medium, conductive, C-73
Member junction, D-53
Mercury thermometer, F-8
Mesh, G-138
    analysis, A-41
Meter, A-1
Method,
    bisection, G-126
    direct, E-66
    Frobenius, G-86
    gradient, G-115
    inverse Laplace transform, E-65—E-76
    iteration, G-126
    Laplace transform, E-65—D-76
    loop, F-2

Newton-Raphson, G-126
Newtonís, A-114
node, F-2
numerical, A-114—A-115, G-125—G-139
of differential equations, E-68—E-70
partial fractions, E-72
partial, F-80—F-87
power series, E-66
simplex, G-124
ultimate-cycle, F-175
Ziegler-Nichols, F-174
Metric system, A-2
Mho, A-1
Miller,
    integration, B-194
    sweep circuit, B-193
Millmanís theorem, A-31
Minima, theory of, G-111
Minimum problem, G-118
Minute, A-1
Mixing, B-79, B-166
Mobility, B-3
Mobius transformation, G-58
Mode, common, load line, B-40
Modeling, system, F-1—F-14
Modulation,
    amplitude, D-63—D-69
    angle, D-70
    delta, D-104
    double sideband suppressed carrier, D-119-D-120
    frequency, D-70—D-84
    index, D-72
    phase, D-70
    pulse code, D-100—D-107
    pulse systems, D-86—D-97
    single sideband D-69, D-117
    sinusoidal, D-81
Modulator,
    balanced, D-66
    double-sideband, D-66
Moments,
    one-dimensional random, D-56
    second, D-52
    second central, D-52
Monopole, C-116
Monostable multivibrator, B-120—B-123
Monte Carlo integration, G-132
Motion,
    simple harmonic, E-55
    wave, in lossy dielectric, C-68
    wave, imperfect dielectric, C-66
Moving particles, magnetic field, C-49
Multiloop electric circuit, F-2

Multiplication,
    by T property, E-71
    frequency, D-81
    in time domain, D-19—D-20
    matrix, A-34—A-35
Multiplier, voltage, B-37
Multitransistor circuits, B-39—B-48
Multivibrators, B-117-B-125
    collector-coupled, B-117—B-119, B-124
    emittor-coupled, B-120—B-123, B-125
Mutual inductance, C-57

NAND functions, B-142
Narrow band filter, D-110
Natural,
    frequency, E-57
    response, A-22
    sampling, D-90
near zone radiation, C-107
negative feedback amplifier, B-75
Neper, A-1
    frequency, A-66
Network,
    compensating, F-53, F-136—F-145
    elements, F-7
    lag, F-149
    lead, F-151
    linear time-invariant, A-77
    noise, D-134
    parameters, A-97—A-101
    sampling, B-73
    stabilizing, B-114
    theorems, A-29—A-33
    two-port, B-49
Neumann function, G-86
Newton, A-1
    method, A-114
    Raphson method, G-126
Nicholís chart, F-133
Node, F-55
    analysis, A-40—A-41
    method, F-2
    pair analysis, A-39
    single, pair analysis, A-38
Noise,
    bandwidth, D-113, D-139
    calculating, D-133—D-139
    figure, D-137
    in AM systems, D-116
    in communication system, D-116—D-130
    in delta modulator, D-130
    in FM systems, D-121
    in Pcm systems, D-125

I-13

margin, B-131
quantization, D-126
representation of, D-108—D-114
resistor, D-133
shot, D-114
temperature, D-136
thermal, D-127
Non-inverting adder, B-101
Nonperiodic,
   signal, D-8
   waveform, D-30, D-34
Non-linear circuits, B-164
NOR function, B-142
Norm, G-2, G-80
Normal distribution function, G-149
Normal-form equation, A-78
Normalized power, D-29, D-109
Norton′s theorem, A-30
NOT function, B-138
Null,
   function, E-16
   set, D-48
Number,
   complex, A-47—A-50
   imaginary, A-47
   sequence, G-55
   series, G-55
Numerical,
   integration, G-130
   linear algebra, G-132
Numerical methods, A-114—A-115,
   G-125—G-139
Nyquist stability criterion, F-107—F-117

Octave, F-93
Odd,
   extension, E-80
   function, D-13, E-78
Offset error voltage, B-92
Ohm, A-1
Omnidirectional, D-156
One-dimensional,
   probability density function, D-53
   random variables, D-56—D-57
OP amplifier, B-48
Open circuit, A-13
   P-N junction, B-11
   transistor, B-21
Operations, vector, C-2
Operational amplifiers, B-92—B-96, F-163
   feedback, B-108—B-114
   frequency response, B-108—B-114
   systems, B-100—B-105
Operators,
   del, C-19
   Laplace transform, A-93

Order, E-27—E-28
Ordinary differential equations, G-64—G-72, G-138
   first order, G-69
   linear, G-66
   of first order, G-64
   solving, G-71
Ordinary linear differential equations, E-42—E-63
   with constant coefficients, E-42
   with variable coefficients, E-46
OR function, B-139
Orthogonality, G-5, G-80
Orthonormal vectors, G-5, G-81
Oscillation, E-56
Oscillators, B-156—B-162
   Clapp, B-159
   Colpitts, B-157
   crystal, B-159
   harmonic, B-156—B-161
   Hartley, B-158
   RC phase shift, B-156
   relaxation, B-162
   tunnel diode, B-160
Output,
   resistance, B-77
   signal power, D-128

Pairs, Fourier transform, D-21
Parabolic function, F-59
Parallel,
   addition, B-155
   compensation, F-155-F-160
   -in shift register, B-149
   places, waves between, C-87
   -plate capacitor, C-30
   resonance, A-70
Parallelogram law, G-3
Parameter,
   admittance, A-100, B-171
   H, A-100—A-101, B-27, B-50
   hybrid, A-98, B-49
   method of differentiation, E-70
   operational amplifier, B-96
   transmission-line, C-81
   two-port network, A-97—A-101
   Y, A-100—A-101
   z, A-97, A-101
Parseval's,
   identity, A-88, E-85, G-78
   theorem, D-31, E-84, E-86
Partial differential equation, E-98, G-90—G-106, G-139
Partial fraction method, E-72
Partial sum of order N, E-78

Particle,
 charged, B-1
 moving, C-49
Partition, method, F-80—F-87
Passband, D-38
Passive logic, B-130—B-132
PCM system, noise, D-125
Peak electric field, C-103
Pendulum, spring, E-58
Performance, evaluation, F-118—F-133
Periodic,
 extension, E-81
 function, E-13
 signal, D-8
 waveform, D-30, D-34
Permeability, C-53
 tensor, C-55
Permittivity,
 of free space, A-4, C-7
 of material, A-4
Permutation, G-148
Phase
 angle, A-43, A-45, G-75
 crossover, F-113
 diagram, F-94
 margin, F-113
 modulation, D-70, D-105
 PSK, D-146
 shifter, B-101
 shift keying, D-105—D-107, D-144
 system, A-58—A-62
Phase, splitting circuit, B-60
 delay, D-39
Phasor,
 angle, A-43
 representation, F-129
 voltage, A-52
Picardís theorem, G-58
PI control, F-137
Piecewise smooth function, E-81, G-77
Pi-filter, B-36
Pivotal coefficient, G-134
PIV rating, B-32
Plane, G-25
 complex, G-42
 inverse polar, F-128
 parallel, C-87
 S, A-67—A-68
 waves, C-75
Plateau problem, G-119
Plots,
 action, F-134
 Bode, F-89—F-105
 direct polar, F-99, F-118
 frequency response, F-90, F-146—F-153

 inverse polar, F-102, F-111
 logarithm, F-92
 polar, F-89—F-105
P-N junction, B-11—B-12
Point, charges, C-9
Poisson
 distribution, G-151
 equation, C-37—C-38
 integral formula, G-62
 probability density function, D-52
Polar,
 form, A-53
 plane, inverse, F-128
 plots, F-89—F-105
 plots, direct, F-99, F-118
 plots, inverse, F-101, F-111
Poles, A-68, G-57—G-58
 adding a, F-117
 complex, F-80
 dominant, F-80—F-87
 zero, F-80—F-87
 zero diagram, F-38, F-85
Polynomial, F-87, G-43, G-127—G-129
 Fourier, G-80
 Legendre, G-89
Polyphase system, A-58—A-64
Positive-displacement rotational
 hydraulic transmission, F-9
Potential, B-2, C-22—C-27
 energy, C-56
 magnetic, C-47
 variation, B-6
Potentiometer, F-165
Power, A-2, A-5—A-8
 apparent, A-55
 available, D-135
 average, A-53—A-56
 complex, A-56
 complex number, A-48
 connected loads, A-64
 correlations, D-34
 density, D-155
 electromagnetic, C-70
 factor, A-1, A-54
 flow, D-167
 instantaneous, A-53
 matrix, A-36
 normalized, D-29
 output, D-135
 reactive, A-55
 relations, D-159
 series input, F-18
 series method, E-66, F-49
 spectral density, D-30, D-35

spectral determination, D-59
spectrum, D-46
supplies, B-31—B-38
triangle, A-55
total radiated, C-108
waves, C-74
Poynting,
   theorem, C-63—C-70
   vector, C-74
p-region, B-10
Preset, B-175
Principles, basic circuit, D-1—D-4
Probability, G-145
   conditional, D-51
   density function, D-51
   error, D-143
   Gaussion density, D-61
   joint, D-50
   marginal, D-50
   one-dimensional density function, D-53
   theory, D-50
Probability density function,
   binomial, D-51
   Poisson, D-52
Problem,
   boundary-value, E-98—E-103
   brachistrochrone, G-119
   catenary, G-119
   Cauchy, E-101
   Dirichlet, E-100
   initial value, E-42
   one-dimensional boundary value, E-101
   plateau, G-119
   solutions of boundary-value, E-101,
   Steiner, G-119
   two-dimensional boundary value, E-103
Process, random, D-47—D-61
Product,
   inner, G-3
   of sets, D-49
   vector, cross, C-3
   vector, dot, C-3
Programming,
   dynamic, G-122
   linear, G-123
Propagation,
   in conductible medium, C-73
   of electromagnetic waves, C-73—C-76
Proper rational function, E-72
Property,
   Bessel function, E-27
   changes of scale, E-8, E-20
   complementary error function, E-30
   convolution, E-22—E-23
   error function, E-29
   first shift, E-8, E-18

gamma function, E-26
inverse Laplace transform, E-18, E-37
linearity, E-7, E-18
matrix, A-35—A-36
   of antennas, C-115
   of Fourier transform, D-16
   of root loci, F-69
   of waveguides, C-99
   second shift, E-9, E-19
   waveform symmetry, A-85
   unit impulse function, E-35
   unit step function, E-32
Proportional control, F-137
   plus derivative compensation, F-140
PSK, D-146
Pulse,
   amplitude modulation, D-89
   code modulation, D-100—D-107
   modulation systems, D-86—D-97
   position modulation, D-97
   wave, B-185

Q-point,
   analysis, B-40
   placement, B-62
   variation, B-28
Quality factor, A-73—A-74
Quantization, D-126
   error, D-101
   noise, D-126
   of signal, D-100

Radar equation, D-160
Radian, A-1
Radiation, C-107—C-109
   resistance, D-159
   pattern, D-156
Radio-frequency circuits, B-164—B-171
Ramp function, B-184, F-58
   input, F-19
Random,
   processes, D-47—D-61
   signal, D-8
   variables, D-47—D-61
   variables, sum, D-61
Rate, information, D-148
Rating, PIV, B-32
Ratio,
   SNR, D-150
   transfer, B-74
Rational function, E-74
Rayleigh quotient, G-137—G-138
RC circuit, D-2
   simple, A-21—A-22
RC filter, B-36, D-110
   phase shift oscillator, B-156

Reactive power, A-55
Receiver,
　amplifier modulation, D-116
　dump PCM, D-141
　FM, D-121
　integrate, D-141
Reception, D-143
Reciprocity theorem, A-33
Recovery,
　signal, D-94
　time, B-119
Rectangular bandpass filter, D-112
Rectifier, B-12
　diode, B-31—B-33
　efficiency, B-33
　full-wave, B-31
　half-wave, B-31
Reduction of noise, B-76
Reflection,
　coefficient, C-76
　uniform plane waves, C-75
Regenerative circuit, B-182
Register, B-148—B-155
　parallel-in shift, B-148
　serial-in shift, B-148
　shift, B-148—B-150
　universal shift, B-150
Relaxation oscillator, B-162
Residue, G-57
Resistance, A-10—A-11
　diode, B-13
　drain source, B-57
　dynamic, B-15
　-inductance-capacitance, A-1
　input, B-76
　Norton, A-30
　output, B-77
　radiation, C-109, D-159
　Thevenin, A-30
Resisitivity, A-10
Resistor, A-11
　emitter, B-45
　logic, B-130
　noise, D-133
　-transistor logic, B-132
Resonance, A-70—A-72, E-56, E-58
　parallel, A-70
　series, A-72
Resonant frequency, A-25, F-120, F-122
Response,
　complete, A-23
　forced, A-22
　frequency, F-38, F-78, F-85, F-89
　frequency of amplifiers, B-81—B-90, B-111
　frequency plots, F-146—F-153

ideal filter, D-39
low frequency, B-82
natural, A-22
of linear system, D-28
parallel RLC circuit, A-25
steady state, F-16
time, F-85, F-89
time specification, F-28
transient, F-20, F-80
Restoring force, E-52
Reverse bias, B-10, B-12
Revolution, F-112
　per second, A-1
Riemann
　-Lebesque theorem, G-77
　mapping theorem, G-63
Right hand rule, G-5
Rigidity, flexural, E-63
Riley rule, F-35
Ripple,
　counter, B-151
　factor, B-33
RL circuit, D-4
　forced response, A-22
　simple, A-20—A-21
RLC circuit, A-24—A-27
　complete response, A-27
　parallel, A-24—A-25
　series, A-26
Root, locus, F-67—F-78, F-136—F-145
　loci, properties, F-78
　of characteristic equation, F-67
Root-mean square, A-1
　value, A-56
Rotation, G-12—G-13
Rotational,
　hydraulic transmission, F-9
　systems, mechanical, F-6
Rouche's theorem, G-63
Roundoff error, G-125
Routh's stability criterion, F-39
R-S flip-flop, B-173
　clocked, B-174

Sampling,
　flat top, D-90
　function, D-27
　natural, D-90
　network, B-73
　theorem, D-86
Saturation curve, F-168
Sawtooth pulse, D-43
Scalar,
　magnetic potentials, C-47
　product, G-3

I-17

Scale changes, B-100, D-20
   property, E-8, E-20
Scaling, A-74
   time, D-17
Schmitt trigger, B-182
Schnottky diode, B-18
Second, A-1
   -order, partial differential equation, E-99
   -order system, F-24, F-120
   shift property, E-9, E-19
   translation property, E-19
Self-inductance, A-16
Semiconductor devices B-1—B-6
Sensitivity, B-75
Sequences,
   function, G-56
   series, G-55
Serial-in shift register, B-148
Series,
   cosine, E-79
   exponential Fourier, A-83
   Fourier, E-77
   function, G-56
   geometric, G-55
   number, G-55
   resonance, A-72
   R-L circuit, F-2
   sine, E-79
   trigonometric, E-77
Servo characteristics, F-58—F-63
Servomotor, AC and DC, F-10
Set,
   algebra of, D-48
   bounded point, G-28
   connected point, G-28—G-29
   in a complex plane, G-42
   null, D-48
   point, G-28—G-29
   theory, D-47
   universal, D-47
Setting, controller, F-174
Shannon's theorem, D-149
Sheet of charge, C-9
Shift register, B-148—B-150
Shifting,
   frequency, D-17
   property, E-18—E19
   time, D-17
Short circuit, A-13
Shot noise, D-114
Shunt,
   capacity filter, B-34
   conductance, C-81
Sideband,
   double, modulator, D-66
   single, modulator, D-69

Signal,
   deterministic, D-8
   flow graphs, F-51—F-55
   nonperiodic, D-8
   periodic, D-8
   quantization, D-100
   terminology, D-8
   test, F-58
Sign changes, B-100
Simple,
   harmonic motion, E-55
   pendulum, E-58
Simplex method, G-124
Simpson rule, A-115, G-131
Simulation system, F-163—F-174
Sine,
   Fourier formula, E-88
   Fourier series, E-79
   integral, E-31
   integration, A-82
Single,
   diode compensation, B-29
   loop circuit analysis, A-38
   loop feedback amplifier, B-74
   node pair analysis, A-39
   phase system, A-58
   sideband modulator, D-69, D-117
   stage CE transistor amplifier, B-90
Singular points, G-62
Singularity, G-58
Sinusoidal,
   analysis, A-43—A-56
   current, A-43—A-44
   forcing function, A-43
   input, F-16
   modulation, D-80
   voltages, A-43—A-44
Skew matrix, G-11
Skin depth, C-74, D-167
Slew rate, B-95
Slope error, B-191
Small-signal,
   design, B-16, B-49—B-60
   model, transistor, B-25
   RF amplifier, B-167
Smith chart, C-85
SNR ratio, D-123, D-150
Solutions,
   of differential equations, F-16—F-28
   of equations, G-125
Source transformation, A-39
Specifications, time-response, F-28
Spectra determination, D-59
Spectral
   components of noise, D-108
   power density, D-30

Spectrum, G-16
 of an FM signal, D-73, D-80
 power, D-46
S-Plane, A-67—A-68
Spring,
 constant, E-52
 vibration of, E-51
Square,
 law demodulator, D-65
 wave, D-41
 wave function, E-32
Stability, F-115
 bias, B-28
 condition, G-141
 factor, B-29
 Nyquist criterion, F-107—F-117
 Routh's criterion, F 39
 system, F-107, F-114
Stabilization system, F-136—F-145
Stabilizing network, B-114
Standard deviation, G-150
Standardized inputs, F-16
Standing waves, C-76, D-166
State,
 equations, A-76—A-77
 transition matrix, A-79—A-80
 -variable, analysis, A-75—A-80
 -variable, method, A-75
Static loop sensitivity, F-67
Stationary point, G-112
Statistical,
 average, D-57
 independence, D-51
Statistics, G-145
Steady magnetic field, C-41—C-47
Steady state response, F-16
Steineris problem, G-119
Step function, B-184, F-58
 input, F-19
 unit, E-31
Stirling's formula, G-147
Stokes' theorem, C-46, G-38
Stopband, D-39
Streamlines of fields, C-11
Sturm-Liouville problem, G-72—G-73
Subset, D-47
Substitution theorem, A-32
Subtraction matrix, A-34—A-35
Sum,
 of random variables, D-61
 of sets, D-48
Superposition, A-30, D-17, D-20, F-31
 of harmonic waves, G-75
 principle, G-66, G-95
 theorem, A-30

Surface,
 area integral, G-36
 level, G-26
 in three dimensions, G-35
Sweep generators, B-191
Switching speed, flip-flop, B-180
SWR, C-76
 standing wave ratio, C-77
Symmetric matrix, G-11
Symmetry,
 conditions, D-13
 full-wave, D-15
 half-wave, D-14
Synchronization,
 imperfect bit, D-145
 imperfect phase, D-144
Synchronous counter, B-152
System,
 analysis of, F-63, F-89—F-105
 cascade compensated, F-146—F-153
 communication, D-116—D-130
 control, F-174
 discrete, A-102—A-110
 feedback, F-59
 first order, F-24
 linear, D-28
 linear differential equation, E-48
 mechanical rotational, F-6
 mechanical translation, F-3
 modeling, F-1—F-14
 polyphase, A-58—A-64
 pulse modulation, D-86—D-97
 response, differential equations, F-16—F-28
 second order, F-24, F-120
 simulation, F-163—F-174
 stability, F-107, F-114
 stabilization, F-136—F-145
 thermal, F-7
 time response, F-157
 tuning, F-172
 type 0, F-63, F-97
 type 1, F-64, F-97
 type 2, F-65, F-98
 type curve, F-97

Tables, E-104—E-119
 of general properties, E-104
 of Laplace transform, E-107
 of special, function, E-119
Tangent, unit, G-24
Taylor series, G-20
Temperature, A-3
 control, G-120—G-121
 noise, D-136, D-138
 V-I characteristics, B-13

I-19

Tensor permeability, C-55
Terminology, signal, D-8
Test signals, F-58
Theorem,
  Bayes', D-51
  Cauchy's residue, A-108
  Cayley-Hamilton, A-80
  central limit, D-61
  complex integration, F-32
  convergence, E-81
  convolution, A-89, E-89
  Divergence, C-20
  Euler's, A-49
  Final value, A-96, D-25, E-65
  Fourier transform, D-20
  fundamental existence and uniqueness, G-66
  Green's G-34
  Heaviside expansion, A-95
  Initial value, A-96, D-25, E-65
  inversion, E-87
  Louville's, G-55
  maximum modulus, G-54
  maximum power transfer, A-31
  Millman's, A-31
  Morera, G-54
  network, A-29—A-33
  Norton, A-30
  Parseval's, D-31, E-84, E-86
  Picard's, G-58
  Poynting, C-63—C-70
  reciprocity, A-33
  Riemann-Lebesque, G-77
  sampling, D-86
  Shannon's, D-149
  Stokes', C-46
  substitution, A-32
  Thevenin, A-30
  uniqueness, C-38
  Weierstrass, G-111
  Weiner-Khintchine, D-59
Theory,
  information, D-148—D-152
  junction transistor, B-21
  of maximum and minima, G-111
  probability, D-50
  set, D-47
Thermal
  noise, D-127
  systems, F-7—F-8
Thermometer, mercury, F-8
Thevenin theorem, A-30
Three phase system, A-59
  source, A-62
  voltage, A-60
  Y-connection, A-61

Time, A-2
  average, D-53
  constant, A-22, F-23
  dead, F-105
  delay, D-20
  division multiplexing, D-89
  invariant network, A-77
  pulse modulation, D-97
  recovery, B-119
  response, F-85, F-89, F-157
  reversal, D-20
  scaling, A-94, D-17
  shifting, A-93, D-17
  slot, D-85
  transition, B-180
  varying field, C-59—C-61
Time domain, A-46—A-47, D-31
  analysis, F-58—F-63
  convolution, D-19
  differentiation, D-18
  integration, D-18
  multiplication, D-19
Timing flip-flops, B-177
Topological analysis, A-111—A-113
Topology, amplifier, B-74
Torque, magnetic field, C-49—C-58
Total, radiated power, C-108
Trajectory, G-70
Transconductance, B-56
  amplifier, B-72
Transconductive amplifier, B-102
Transfer,
  curve, B-182
  function, B-90, D-37—D-46, F-51—F-55, F-142
  gain, B-74
  ratio, B-73
Transform,
  Fourier, D-8—D-21, E-77—E-89, G-74—G-85
  Fourier integral, E-86
  integral, E-1, G-74—G-85
  inverse, F-33
  Laplace, A-90—A-96, D-22-D-25, E-1—E13, F-29—F-41
  z, A-102—A-110, F-44-F-47
Transformation,
  Mobius, G-38
  source, A-39
Transformer, A-15—A-20
  -coupled amplifier, B-65
  iron-core, A-20
Transient,
  circuit, A-14—A-27
  response, F-20, F-80

Transition,
 band, D-39
 diode, B-13
 time, B-180
Transistor,
 biased, B-22
 bipolar, B-169
 bipolar junction, B-21—B-28
 -diode logic, B-132
 junction, theory, B-21
 open-circuited, B-21
 -resistor logic, B-132
 single-stage, CE, B-90
 -transistor logic, B-134
Translation,
 property, first, E-18
 property, second, E-19
 systems, mechanical, F-3
Transmission,
 coefficient, C-76
 data, D-141—D-146
 of energy, D-46
 of power spectra, D-60
 path, D-160
 positive-displacement rotational hydraulic, F-9
Transmission lines, C-78—C-85, D-162—D-167
 equations, C-78
 examples, C-83
 parameters, C-83
Transpose matrix, A-35, G-10—G-11
Transresistance amplifier, B-72
Transverse,
 electric waves, C-88
 electromagnetic waves, C-90
 magnetic waves, C-89
Trapezoid approximation, G-131
Triangle inequality, G-4
Trigonometric,
 Fourier series, A-81—A-82
 function, G-48—G-50
 polynomials, G-75
 series, E-77
Truth table, B-143
T-type flip-flop, B-176
Tuned circuits, B-168
Tuning, system, F-172
Tunnel diode oscillator, B-160
Turns ratio, A-20
Two-diode compensation, B-30
Two-phase system, A-58
Two-port network, B-49
 gain, D-137
 parameter, A-97—A-101

Type 0 system, F-63, F-97
Type 1 system, F-64, F-97
Type 2 system, F-65, F-98

Ultimate cycle method, F-174
Undamped vibration, E-54
Uniform,
 converge, E-82—E-83
 plane wave, C-63—C-70, C-75
Unijunction transistor oscillator, B-162
Union, D-48
Uniqueness,
 of inverse Laplace transform, E-16
 theorem, C-38
Unit, A-1—A-4
 alternating sequence, A-103
 impulse function, E-35
 matrix, A-35
 ramp sequence, A-104
 SI, A-1
 step function, E-31
 step sequence, A-102
 vector, C-1, G-2
Unity feedback, F-63
Universal,
 set, D-47
 shift register, B-150

Variable,
 complex, G-43
 -discrete, D-47
 random, D-47—D-61
 spatial, G-91
Variance, G-150
Variation, calculus of, G-110—G-123
Vector, G-1—G-6
 algebra, C-1
 analysis, C-1—C-6
 complex, G-16
 cross product, C-3
 dot product, C-3
 field, C-2, G-27
 field, curl, C-44
 function, G-23
 in N-dimension, G-7
 in three dimensions, G-1
 magnetic potential, C-47
 space, G-79
 unit, C-1, G-79
Velocity,
 group, C-91
 phase, C-91
Vibration,
 damped, E-53
 free, E-54

I-21

of spring, E-51
undamped, E-54
V-I characteristics, B-12
Volt, A-1
Voltage, A-5—A-8
   amplifier, B-71
   amplifier, common-drain, B-58
   amplifier, common-source, B-57
   capacitor, A-14
   current-to converter, B-102
   DC follower, B-103
   dependent source, A-9
   division, A-11—A-12
   independent source, A-8
   Kirchkoff's law, A-13
   line, D-162
   multiplier, B-37
   offset error, B-94
   sinusoidal, A-43, A-44
   three phase, A-60
   -to-current converter, B-102
   transformer, A-20
Voltampere, A-1
Volterra integral equation, E-91
Volume,
   charge, C-9
   differential, C-6
   integrals, G-38

Watthour, A-1
Wave,
   between parallel planes, C-87
   electromagnetic, C-73—C-76
   equation, C-63, G-103—G-106, G-144
   guided, C-87—C-91
   guides, C-99—C-102
   motion, C-63, C-68
   sine, D-41
   square, D-41
   standing, C-76, D-166
   transverse electric, C-88
   transverse electromagnetic, C-90
   transverse magnetic, C-89
   uniform plane, C-63—C-70
Waveform, B-67, B-120—B-123
   bandlimiting, D-32
   correlation between D-33
   generator, B-184—B-191
   nonperiodic, D-30, D-34
   periodic, D-11, D-30, D-34
   symmetry, A-85
Waveshaping,
   circuits, B-185
   generators, B-184—B-191
Weierstrass theorem, G-111

White noise, D-110
Wiener-Khintchine theorem, D-59
Work, A-2, C-22
Wronskian, G-67
WYE,
   alternators, A-61
   connections, A-61

Y connected load, A-64
Y parameter, A-100-A-101, B-171
Y-Y connection, A-61

Zener diode, B-18
Zero, A-68
   adding a zero, F-117
   of analytic function, G-62
Ziegler-Nichols method, F-174
Z-parameter, A-97
Z-transform, A-102—A-110, F-44-F-47
   pairs, A-110
   properties, A-107